International Workshop on Quantum Chromodynamics Theory and Experiment (QCD@Work 2016)

EPJ Web of Conferences Volume 129 (2016)

Martina Franca, Italy
27 - 30 June 2016

Editors:

G.E. Bruno
G. Chiodini
P. Colangelo
C. Coriano
D.M. Creanza
F. De Fazio
E. Nappi
S. Spagnolo

ISBN: 978-1-5108-3366-1

Printed from e-media with permission by:

Curran Associates, Inc.
57 Morehouse Lane
Red Hook, NY 12571

Some format issues inherent in the e-media version may also appear in this print version.

Printed by Curran Associates, Inc. (2017)

For additional information, please contact EDP Sciences – Web of Conferences at the address below.

EDP Sciences – Web of Conferences
17, Avenue du Hoggar
Parc d'Activité de Courtabœuf
BP 112
F-91944 Les Ulis Cedex A
France

Phone: +33 (0) 1 69 18 75 75
Fax: +33 (0) 1 69 28 84 91

contact-edps@webofconferences.org

Additional copies of this publication are available from:

Curran Associates, Inc.
57 Morehouse Lane
Red Hook, NY 12571 USA
Phone: 845-758-0400
Fax: 845-758-2633
Email: curran@proceedings.com
Web: www.proceedings.com

TABLE OF CONTENTS

Preface

The 2016 edition of **QCD@Work - International Workshop on Quantum Chromodynamics - Theory and Experiment** was held in Martina Franca, Italy, from 27 to 30 June 2016. The series started in 2001 in Martina Franca, and continued in 2003 and 2005 in Conversano, 2007 and 2010 in Martina Franca, 2012 in Lecce, and 2014 in Giovinazzo, Italy. The aim of all the eight meetings was to bring together senior and young researchers, theorists and experimentalists, putting QCD to work in different frameworks, for a common discussion.

The topics debated during the workshop were:

- low energy and non perturbative aspects of QCD
- progress in perturbative QCD
- heavy quark physics
- QCD at finite temperature and baryon density
- new holographic methods for strong interactions.

About 80 researchers (with a large component of young scientists) lively participated to the workshop presentations and to the poster session. This volume registers the outcome of the meeting.

As organizers of the event, we would like to thank the Institutions and individuals who have contributed to its successful realization. We acknowledge the financial support of Istituto Nazionale di Fisica Nucleare (INFN), Mathematics and Physics Department "Ennio De Giorgi" of Salento University in Lecce, Commissione Scientifica Nazionale Teorica INFN, theory group of INFN Sezione di Bari. We thank the directors of Sezione INFN di Bari and Bari Physics Department, Prof. Mauro de Palma and Prof. Salvatore V. Nuzzo, and of Sezione INFN di Lecce and Lecce Physics Department, Dr. Fabio Bossi and Prof. Giovanni Mancarella. We are grateful to Dr. Mariarosa Buonfiglio, INFN Sezione di Bari, and Dr. Loredana Napolitano, Physics Department "Michelangelo Merlin" of Bari University and Polytechnic, for advice in the administrative management. We thank Drs. Konstantinos Bachas, Floriana Giannuzzi, Stefano Nicotri and Giuseppe Trombetta for support in the scientific organization. Finally, we are grateful to Prof. Giuseppe Sylos Labini, Prof. Mimmo Attademo and Ms Rosa Guarini, of Bari Academy of Fine Arts, for the Martina Franca picture used as the logo of the 2016 edition.

Giuseppe E. Bruno, Gabriele Chiodini, Pietro Colangelo, Claudio Corianò,
Donato M. Creanza, Fulvia De Fazio, Eugenio Nappi and Stefania Spagnolo

Editors

Giuseppe E. Bruno
Dipartimento Interateneo di Fisica "M. Merlin", Università e Politecnico di Bari
Via Orabona 4, 70126 Bari, Italy
e.mail: giuseppe.bruno@ba.infn.it

Gabriele Chiodini
Istituto Nazionale di Fisica Nucleare – Sezione di Lecce
Via Arnesano, 73100 Lecce, Italy
e.mail: gabriele.chiodini@le.infn.it

Pietro Colangelo
Istituto Nazionale di Fisica Nucleare – Sezione di Bari
Via Orabona 4, 70126 Bari, Italy
e.mail: pietro.colangelo@ba.infn.it

Claudio Corianò
Dipartimento di Matematica e Fisica "E. De Giorgi", Università del Salento
Via Arnesano, 73100 Lecce, Italy
e.mail: claudio.coriano@le.infn.it

Donato M. Creanza
Dipartimento Interateneo di Fisica "M. Merlin", Università e Politecnico di Bari
Via Orabona 4, 70126 Bari, Italy
e.mail: donato.creanza@ba.infn.it

Fulvia De Fazio
Istituto Nazionale di Fisica Nucleare – Sezione di Bari
Via Orabona 4, 70126 Bari, Italy
e.mail: fulvia.defazio@ba.infn.it

Eugenio Nappi
Istituto Nazionale di Fisica Nucleare – Sezione di Bari
Via Orabona 4, 70126 Bari, Italy
e.mail: eugenio.nappi@ba.infn.it

Stefania Spagnolo
Dipartimento di Matematica e Fisica "E. De Giorgi", Università del Salento
Via Arnesano, 73100 Lecce, Italy
e.mail: stefania.spagnolo@le.infn.it

Scientific Advisory Committee

Roberto Casalbuoni (Firenze University)
Fernando Ferroni (Roma Sapienza University and INFN)
Michelangelo Mangano (CERN Geneva)
Nello Paver (Trieste University)
Mikhail A. Shifman (Minnesota University)
George Sterman (SUNY Stony Brook)
Sheldon Stone (Syracuse University)

Local Organizing Committee

Leonardo Angelini (Bari University)
Giuseppe E. Bruno (Bari Polytechnic)
Gabriele Chiodini (INFN Lecce)
Pietro Colangelo (INFN Bari)
Claudio Corianò (Università del Salento)
Donato M. Creanza (Bari Polytechnic)
Fulvia De Fazio (INFN Bari)
Eugenio Nappi (INFN Bari)
Stefania Spagnolo (Università del Salento)

Scientific organization

Konstantinos Bachas (INFN Lecce)
Floriana Giannuzzi (Bari University)
Stefano Nicotri (INFN Bari)
Giuseppe Trombetta (Bari University)

Participants

ABBAS, Gauhar	IFIC, Valencia, SPAIN
ABUKI, Hiroaki	Aichi University of Education, Kariya, JAPAN
ALFORD, Mark	Washington University, St Louis, USA
ANGELINI, Leonardo	University of Bari and INFN - Sezione di Bari, ITALY
BACHAS, Konstantinos	INFN - Sezione di Lecce, ITALY
BANDYOPADHYAY, Priyotosh	INFN - Sezione di Lecce, ITALY
BARABANOV, Mikhail	JINR, Dubna, RUSSIA
BELLANTUONO, Loredana	University of Bari and INFN - Sezione di Bari, ITALY
BERAUDO, Andrea	INFN - Sezione di Torino, ITALY
BIGAZZI, Francesco	INFN - Sezione di Pisa, ITALY
BIZZETI, Andrea	Univ. of Modena - Reggio Emilia and INFN - Firenze, ITALY
BRUNO, Giuseppe Eugenio	Politecnico di Bari and INFN - Sezione di Bari, ITALY
BURAS, Andrzej	TUM- IAS, Munich, GERMANY
CAFAGNA, Francesco Saverio	INFN - Sezione di Bari, ITALY
CHELABI, Kaddour	Institute of theoretical Physics, Beijing, CHINA
CHIODINI, Gabriele	INFN - Sezione di Lecce, ITALY
CHIRILLI, Giovanni Antonio	University of Regensburg, Regensburg, GERMANY
COLANGELO, Pietro	INFN - Sezione di Bari, ITALY
CORIANÓ, Claudio	University of Salento, Lecce, ITALY
COSTANTINI, Antonio	University of Salento and INFN - Sezione di Lecce, ITALY
CREANZA, Donato	Politecnico di Bari, ITALY
D'AMBROSIO, Giancarlo	INFN - Sezione di Napoli, ITALY
DE FAZIO, Fulvia	INFN - Sezione di Bari, ITALY
DELLE ROSE, Luigi	University of Southampton, Southampton, UK
DESTEFANIS, Marco	University of Torino and INFN - Sezione di Torino, ITALY
DVOEGLAZOV, Valery	Universidad de Zacatecas, Zacatecas, MEXICO
FABBRICHESI, Marco	INFN - Sezione di Trieste, ITALY
FAZIO, Angelo Raffaele	Universidad Nacional de Colombia, Bogotá, COLOMBIA
FINI, Rosa Anna	INFN - Sezione di Bari, ITALY
FRAMPTON, Paul	University of North Carolina, Chapel Hill, USA
GIANNUZZI, Floriana	University of Bari, Bari, ITALY
GREYNAT, David	University of Napoli and INFN - Sezione di Napoli, ITALY
GUZZI, Marco	University of Manchester, Manchester, UK
HUANG, Mei	IHEP, CAS, Beijing, CHINA
IVANOV, Mikhail	JINR, Dubna, RUSSIA
KALAYDZHYAN, Tigran	University of Illinois at Chicago, Chicago, USA
KEYES, Robert	McGill University, Montreal, CANADA
KÖNIG, Matthias	JGU Mainz, Mainz, GERMANY
LUCHA, Wolfgang	Austrian Academy of Sciences, Vienna, AUSTRIA
MAMMADOV, Shahin	Baku State University, Baku, AZERBAIJAN
MANZARI, Vito	INFN - Sezione di Bari, ITALY

MARGETIS, Spyridon	Kent State University, Kent, USA
MARZO, Carlo	University of Salento, Lecce, ITALY
MAZUMDAR, Kajari	Tata Institute of Fundamental Research, Mumbai, INDIA
MAZZILLI, Marianna	University of Bari and INFN - Sezione di Bari, ITALY
MEGIAS, Eugenio	Max Planck Institut fur Physik, Munich, GERMANY
MELIKHOV, Dmitri	Moscow State University, Moscow, RUSSIA
MENINNO, Elisa	University of Salerno, Salerno, ITALY
NAPPI, Eugenio	INFN - Sezione di Bari, ITALY
NICOTRI, Stefano	INFN - Sezione di Bari, ITALY
OCALAN, Kadir	Necmettin Erbakan University, Konya, TURKEY
OZPINECI, Altug	Middle East Technical University, Ankara, TURKEY
PALANO, Antimo	University of Bari and INFN - Sezione di Bari, ITALY
PALESE, Marcella	University of Torino, Torino, ITALY
PAPPAGALLO, Marco	University of Bari and INFN - Sezione di Bari, ITALY
PASQUINI, Barbara	University of Pavia and INFN - Sezione di Pavia, ITALY
PASTORE, Alessandra	University of Bari and INFN - Sezione di Bari, ITALY
PRELOVSEK, Sasa	Institute Jozef Stefan, Ljubljana, SLOVENIA
QAMESH, Ahmed	British University in Egypt, Cairo, EGYPT
REINOSA, Urko	Ecole Polytechnique - CNRS, Palaiseau, FRANCE
ROMANIOUK, Anatoli	MEPhI/CERN, Geneva, SWITZERLAND
RUGGIERI, Marco	University of Chinese Academy of Sciences, Beijing, CHINA
SANTOPINTO, Elena	INFN - Sezione di Genova, ITALY
SANTORELLI, Pietro	University of Napoli and INFN - Sezione di Napoli, ITALY
SCRIMIERI, Egidio	University of Bari and INFN - Sezione di Bari, ITALY
SERINO, Mirko	Institute of Nuclear Physics, Cracow, POLAND
SHIVAJI, Ambresh	INFN - Sezione di Pavia, ITALY
TANHA, Milad	Goethe University, Frankfurt, GERMANY
TASSI, Enrico	Universitá della Calabria and INFN - Cosenza, ITALY
TOIA, Alberica	University of Frankfurt and GSI, Frankfurt, GERMANY
TRELEANI, Daniele	University of Trieste and INFN - Sezione di Trieste, ITALY
TRENTADUE, Luca	University of Parma and INFN - Milano Bicocca, ITALY
TROMBETTA, Giuseppe	University of Bari and INFN - Sezione di Bari, ITALY
WANG, Wei	Shanghai JiaoTong University (SJTU), Shanghai, CHINA
WANG, Yuming	University of Vienna, Vienna, AUSTRIA
WATANABE, Akira	Academia Sinica, Taipei, TAIWAN
WINTERROTH, Ekkehart	University of Torino, Torino, ITALY
YAMAGUCHI, Yasuhiro	INFN - Sezione di Genova, ITALY
ZAMIRALOV, Valeriy	Moscow State University, Moscow, RUSSIA

edp sciences

EPJ Web of Conferences

Statement of Peer review

In submitting conference proceedings to EPJ Web of Conferences, I certify to the Publisher that I adhere to the **Policy on Publishing Integrity** of the journal in order to safeguard good scientific practice in publishing.

1. All articles have been subjected to peer review administered by the proceedings editors.
2. Reviews have been conducted by expert referees, who have been requested to provide unbiased and constructive comments aimed, whenever possible, at improving the work.
3. Proceedings editors have taken all reasonable steps to ensure the quality of the materials they publish and their decision to accept or reject a paper for publication has been based only on the merits of the work and the relevance to the journal.

Title, date and place of the conference

QCD @ Work 2016 - International Workshop on Quantum Chromodynamics - Theory and Experiment
27-30 June 2016 Martina Franca (Italy)

Proceedings editor(s):

G.E. Bruno, G. Chiodini, P. Colangelo, C. Coriano', D.M.Creanza, F. De Fazio, E. Nappi, S. Spagnolo

Date and editor's signature

Bari, 20 October 2016

EPJ Web of Conferences **129**, 00001 (2016) DOI: 10.1051/epjconf/201612900001
QCD@Work 2016

The LHC di-photon excess at 750 GeV

Marco Fabbrichesi[1,⋆]

[1] *INFN, Sezione di Trieste, Italy*

Abstract. I review the main features and some of the theoretical attempts to explain the excess around an invariant mass of 750 GeV seen in 2015 at the LHC in the di-photon channel. As this hint to new physics has now all but disappeared from the higher-luminosity 2016 data, the statistical analysis nicely illustrates why only a high level of significance can be trusted in a discovery. The various explanations that has been suggested remain interesting examples of our current understanding of physics beyond the standard model and also of the challenging task of discriminating among them.

1 Introduction

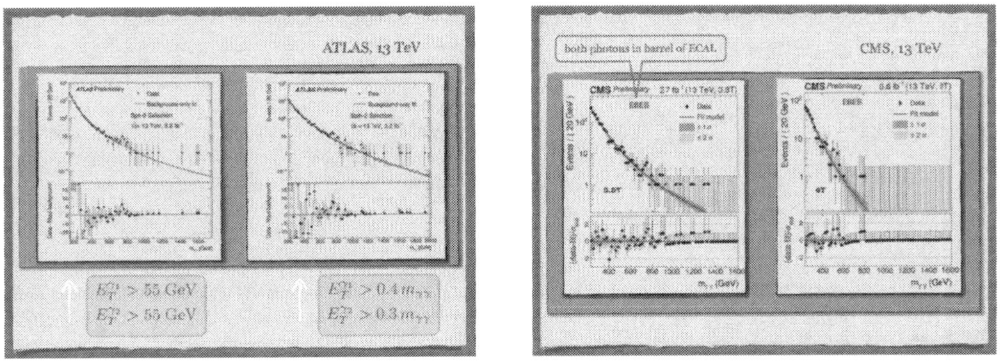

Figure 1. The data: di-photon invariant mass distribution as presented by the ATLAS (left panel) and CMS (right panel) collaborations.

The extraordinary interest aroused by the excess in the di-photon invariant mass distribution revealed by the LHC data collected in 2015 at the center-of-mass energy $\sqrt{s}$ = 13 TeV [1] is easily explained by the desire of finally seeing something beyond the standard model. Never mind that this something was coming out from left field and instead of some missing transverse energy event was just what appeared to be a resonance decaying into two photons.

The analysis of the data—the fitting of the resonance shape to the di-photon invariant mass distribution and the subtraction of the continuum background—shows the many statistical subtleties

⋆e-mail: marco@sissa.it

involved and why an exceptionally high level of significance—like the one currently agreed on of 5σ—is necessary before claiming any discovery.

A review of the models put forward to explain and understand the resonance take us across most of the possible scenarios for physics beyond the standard model; these models taken together make clear both the abundance of ideas and the difficulty of discriminating among the various possibilities. This latter difficulty is a problem that was present in this instance and will be there for any, if any, future discovery.

I have included in this write up most of the talk slides as figures and some of the spoken remarks as text. As a listing of all the theoretical papers devoted to the di-photon excess would fill more than the allotted pages, I have included only those to the papers I used in this review by referring to their arXiv number in the slides where they were mentioned.

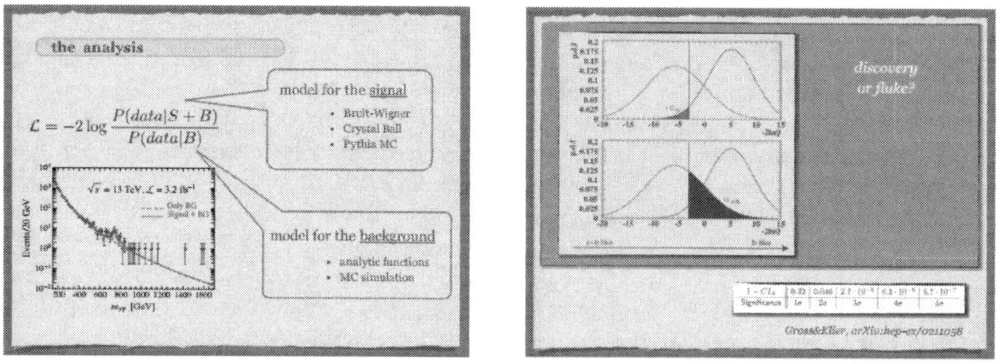

Figure 2. The fit: a likelihood function is defined by the ratio of the probabilities of the model of the signal plus the background over that of the background alone (left panel). Probability distribution functions (p.d.f.) for the hypothesis testing, confidence levels (CL_b) and the corresponding significances (right panel).

2 The facts

As many newspapers like to claim, let us try and keep the facts separated from the opinions—bearing in mind that "the fewer the facts, the stronger the opinions." [2] The facts are contained in the experimental observation of the distribution of the invariant mass of the two final photons. It is shown in Fig. 1 for, respectively, the ATLAS (left) and CMS (right) collaborations.

The plots are not homogenous. In particular the ATLAS collaboration presented two independent analyses and, accordingly, two different cuts on the photon energies. The one with looser cuts (shown on the right of the left panel of Fig. 1) was performed within the exotic group and aimed to possible spin 2 particles, the one with sharper cuts (shown on the left of the left panel of Fig. 1) was performed within the Higgs boson group in searching for spin 0 particles. The spin of the particle is here just a name holder with no physical implications. Similarly, the CMS collaboration presented two sets of data—depending of where the two photons ended up within the detector. On the right panel of Fig. 1, I have reproduce their data for the case in which both photons ended in the barrel of the electromagnetic calorimeter, as presented in two batches of, respectively, 2.7 and 0.6 inverse femto-barns of luminosity.

Having the data, we can fit them to our favorite model. The procedure followed by both collaborations is that indicated in the left panel of Fig. 2, namely of the maximization of a likelihood

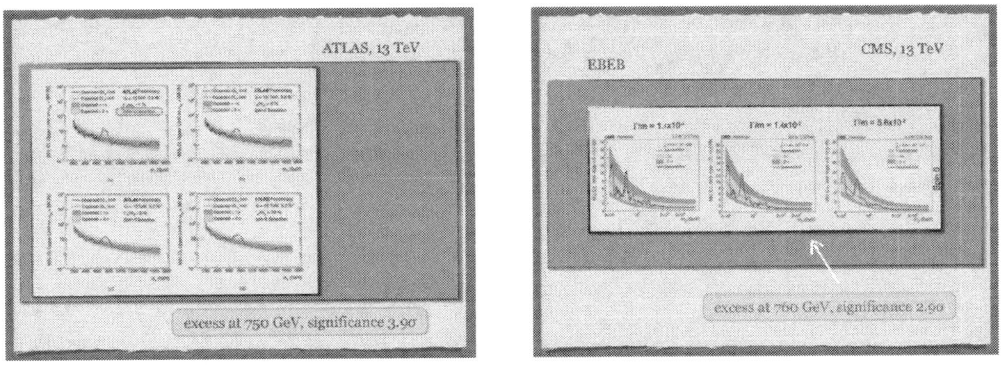

Figure 3. The significance of the bump around 750 GeV in the Brazilian flag plots: 3.9σ for ATLAS (left panel) and 2.9σ for CMS (right panel).

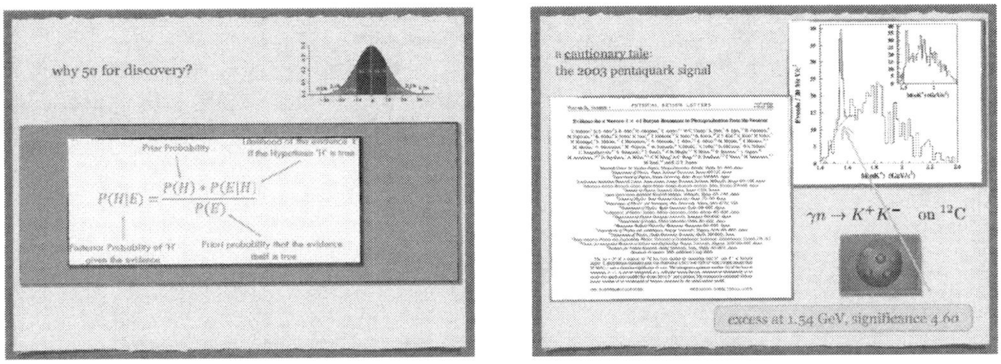

Figure 4. Bayes' theorem linking prior to posterior probabilities (left panel). The 2003 pentaquark signal: an excess at 1.45 GeV with a significance of 4.9σ (right panel).

function in which the model of the signal is included on top of the background (itself described by an appropriated model) against the case of background only. Models for the signal can vary, from the simplest Breit-Wigner resonance to a more sophisticated (read, with more parameters) Crystal Ball model or even a full Monte Carlo simulation. The final significance does not depend critically from this choice. The model for the background can either be an analytical function or, again, a Monte Carlo simulation.

The result of this procedure is shown in Fig. 3 where the significance is counted in terms of sigmas (the so-called Brazilian flag plots made popular by the Higgs boson search). A significance of 3.9σ is found by ATLAS and one of 2.9σ by CMS. This means that the probability for the data to mimic the signal is, respectively, of about 1 out of 15800 and 1 out of 370!

A number of considerations come into play before being carried away by the enthusiasm. The discussion is usually framed in terms of hypothesis testing, like in right panel of Fig. 2 where the probability distribution functions for the competing hypotheses are compared. How strong a significance we need before claiming a discovery? In every day life a significance of 3σ is clearly good enough for all common tasks. After all, the probability of a statistical fluke in this case is only of about 1 out of 370. Nevertheless, there exist examples of very unlikely fluctuations: one of them being the

2003 pentaquark signal, a cautionary tale of a signal with a significance of 4.9σ that has melted away like a snowball in the sun after more data were collected (see right panel of Fig. 4).

The significance shown above is local in the sense that it does not take into in account the entire range of the possible parameters. We were not looking at that particular value of the invariant mass and we would have been equally happy if the excess had shown up at a different value. This is what has been called the look-elsewhere effect. It is an effect that seems difficult to quantify, at least to me, because of the ambiguous definition of what should be considered elsewhere. It is the invariant mass range in the di-photon channel or all searches on all channels performed at that time at the LHC? In any case, the local significance should be tuned down by some factor and it has been by the experimental collaborations.

On the other hand: the probability we estimated by the likelihood procedure is that of the data given the model while what we really care for is the probability of the the model given the data, the posterior probability. Bayes' theorem links these two probabilities (see left panel of Fig. 4), if we know the prior probabilities for the data and the model—that we do not. Again, some factor reducing the final significance is floating around.

These and other considerations behove us to constrain the significance in high-energy physics to higher values than in ordinary life. In particular, the threshold for discovery has been put to 5σ, that is a value for which the probability of a statistical fluke is only about 1 out of 1.7 millions, and just high enough to prevent turning the significance of the 2003 pentaquark excess into a discovery.

3 The opinions

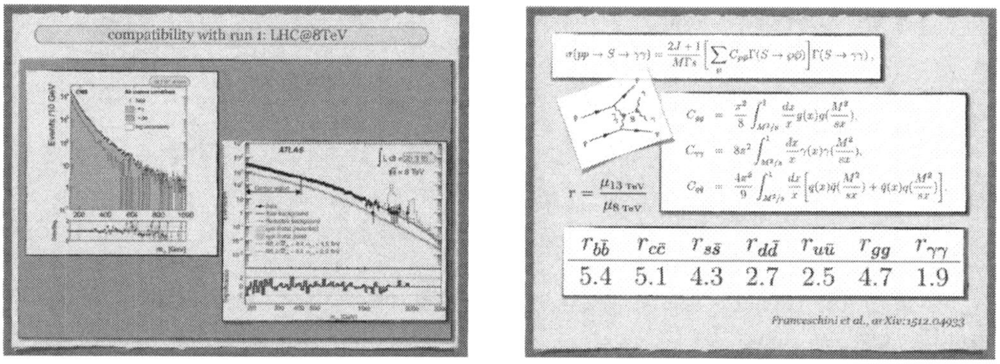

Figure 5. The data at 8 TeV (left panel) and parton distribution functions and the compatibility of the data at 13 and 8 TeV in terms of a rescaling function r (right panel).

Assume, if you will, that we do have a resonance. Let us then come to the opinions—and "that willing suspension of disbelief for the moment, which constitutes poetic faith" [3]—and to the models that have been suggested to make room for the 750 GeV resonance as seen in the 2015 LHC data.

A first problem all models have to face is that no significative excess was found in the previously collected data in the same channel at the LHC with center-of-mass energy of $\sqrt{s} = 8$ TeV. These data are shown in the left panel of Fig. 5.

In order to accommodate this lack of evidence, the rescaling of the cross section in going to the new center-of-mass energy must be taken into account. As it can be seen in the right panel of Fig. 5 different production mechanisms scale by different amounts and, depending on the model you have

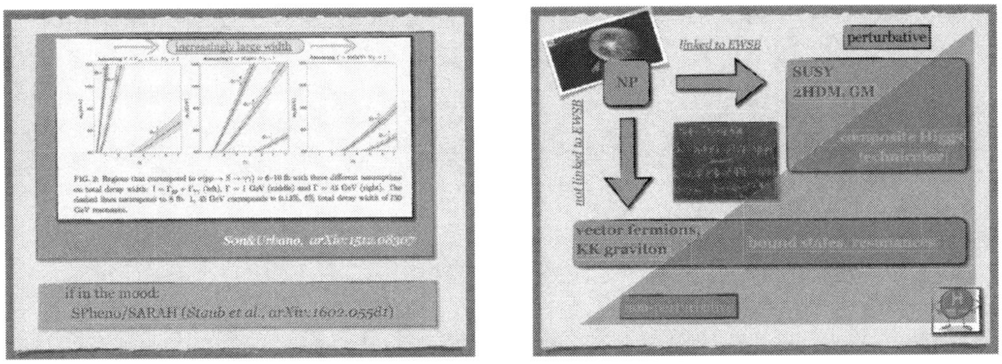

Figure 6. The running of the wilson coefficients (left panel) and classifying the new physics (NP) models: perturbative vs. non-perturbative and linked to electro-weak symmetry breaking (EWSB) vs. unrelated physics (right panel).

in mind, the compatibility between the two sets of data may or may not be an issue. At face value, models in which the production of the resonance is dominated by gluon fusion can have at 13 TeV a much larger cross section than at 8 TeV and are therefore favored.

There are many ways of explaining a resonance with the features of that glimpsed in the di-photon channel at the LHC and we need some sort of classification. Perhaps the simplest one is according to whether the model behind the resonance can be treated by perturbative methods or not; perturbative physics leads to lagrangians with known additional degrees of freedom and calculable amplitudes. Another useful criteria is whether it takes part or not in the breaking of the electro-weak symmetry: if it does not take part, there are less constraints and its signatures are going to be more elusive.

The right panel of Fig. 6 tries to summarize in a single slide these various possibilities naming some of the proposed models.

The discussion can be framed in terms of effective actions. We know that the putative resonance must couple to the photons and the gluons and therefore the effective action must contains the terms shown in Fig. 7. Different terms describe different possible parity and spin symmetries of the resonance: scalar, pseudo-scalar or tensor (spin 2). The case of spin 1 is excluded by the Landau-Yang theorem [4].

Non-perturbative models can be based on such effective actions where the couplings are tuned to a value determined by the experimental fit. Such couplings and the form of the lagrangian are then embedded within the framework of some model of strongly interacting dynamics of some constituents. By construction these models can easily fit the data but remain vague about the ultraviolet completion and the computation details.

If you want to interpret these effective actions in terms of a perturbative lagrangian, you must face the problem that the relatively large cross section implied by the data requires rather large couplings of whatever particles are present in the loop (left panel of Fig. 8) or the presence of many of them. Large couplings can easily run to even larger values thus driving the theory to a non-perturbative regime (see left panel of Fig. 6). Moreover, as explained in the right panel of Fig. 8, the masses of these particles must be large enough to prevent the opening of the threshold above which they can be produce on shell—since they have not been seen.

That said, the possible mechanisms to produce the resonance are many, the main ones shown in the left panel of Fig. 9: from the loop of higher-mass states to the cascade decay of an heavier resonance

 DOI: 10.1051/epjconf/201612900001

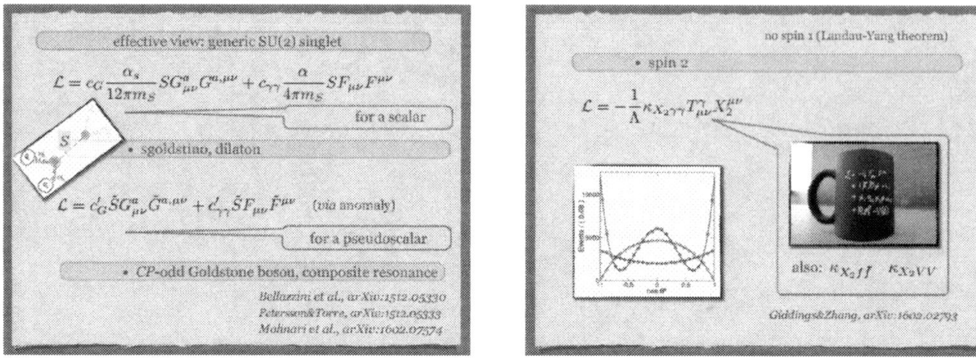

Figure 7. Effective lagrangians for scalar and pseudo-scalar (left panel) and spin 2 (right panel) resonance. The spin 2 case, implies, if the resonance is a heavy graviton, a similar coupling to all others particles.

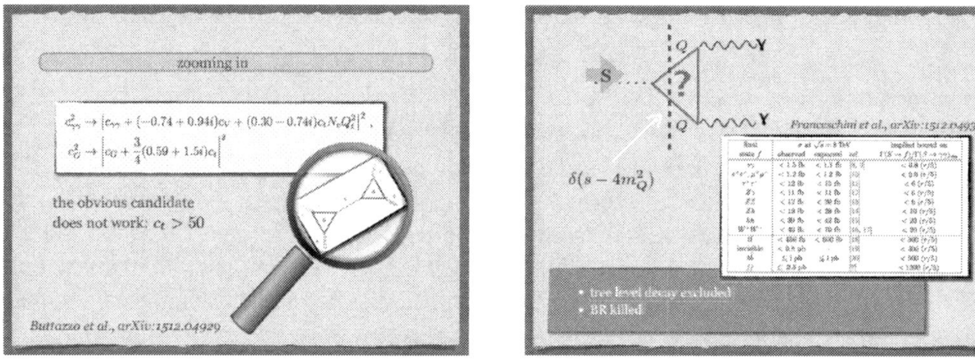

Figure 8. Zooming into the interaction vertex: size of the running coefficients (left panel) and constraints on the masses of the possible constituents in the loop (right panel) from the threshold for their production.

(that would help in explaining the negative result at 8 TeV) to the case in which both or one of the photons seen are actually a collimated pair of photons.

Remaining with the case of a loop of new states providing the effective coupling of the (neutral) resonance to photons and gluons, various models have been explored: from 2 Higgs doublet and Georgi-Machacek models (left panel of Fig. 10) to the always favored supersymmetric model in some of its manifold instances (right panel of Fig. 10). All these examples are linked to the breaking of the electro-weak symmetry and their parameters constrained by precision measurements.

While it is possible to fit the data with the above models, it is true that the fit can happen only in particular corners of the parameter space. For this reason, the most popular class of models has been the one where new (vector-like) fermions are simply added and their couplings and charges tuned to the required cross section (see right panel of Fig. 9). A drawback of this class of models is their *ad hoc* nature and independence of the physics of symmetry breaking at the electroweak scale.

Among the non-perturbative explanations, the resonance as a bound state of some standard, or non-standard constituents has also been advocated. Some of these are listed on the left panel of Fig. 11.

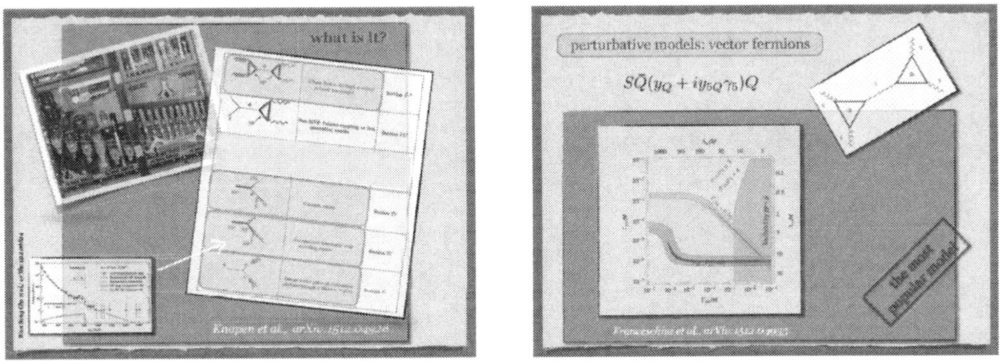

Figure 9. Possible mechanisms (left panel) and the most popular model (right panel).

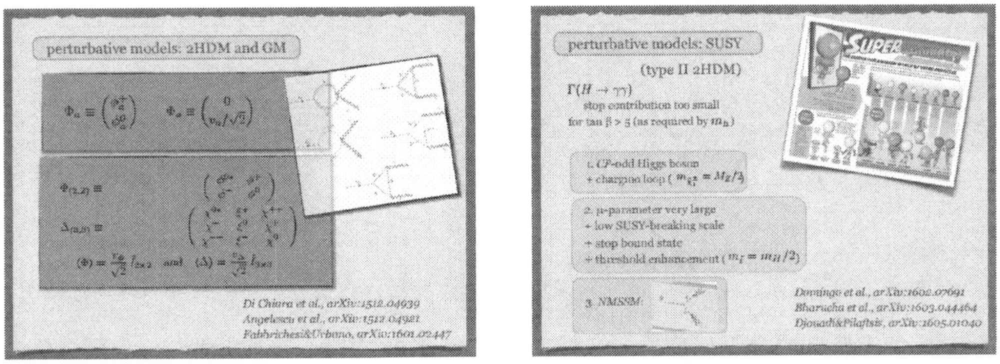

Figure 10. Other perturbative models: 2 Higgs doublets, Georgi-Machacek (left panel) and supersymmetry (SUSY) (right panel).

Finally, the possible role played by the interference between the resonance and the continuum in shaping the signal has been analysed (see right panel of Fig. 11).

4 Conclusions

To the the simple question; "What is it?" (or perhaps better: "What could it have been?"), we must answer that it can be (it could have been) almost anything (see left panel of Fig. 12). Such a resonance, with such a large cross section, was unexpected and it could be fitted only with some tugging among the known perturbative models of physics beyond the standard model. Its, by now, infamous "discovery" has offered us the chance of measuring the many models concocted over the years against a concrete example, made us more aware of the many subtleties (statistical and not) to be faced by going from the data to our favorite lagrangian, and to the importance of many analyses (see right panel of Fig. 12) in addition to the simple fit of the invariant mass distribution—the place where bumps are first seen—which are necessary to claim and understand any new discovery.

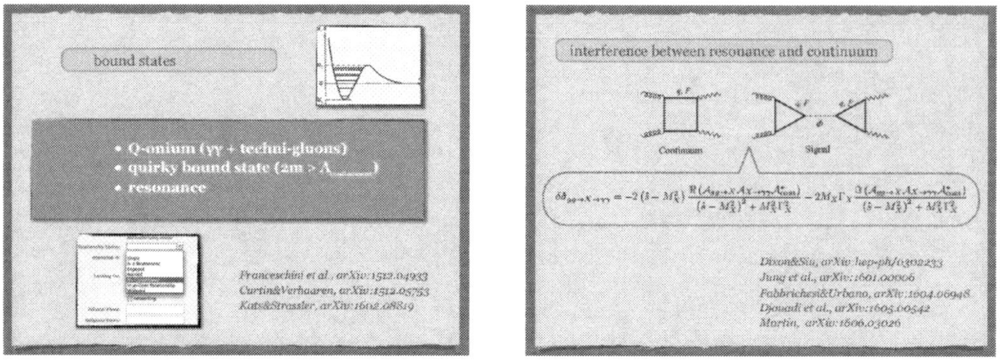

Figure 11. Non-perturbative bound states (left panel) and the interference between resonance and continuum (right panel).

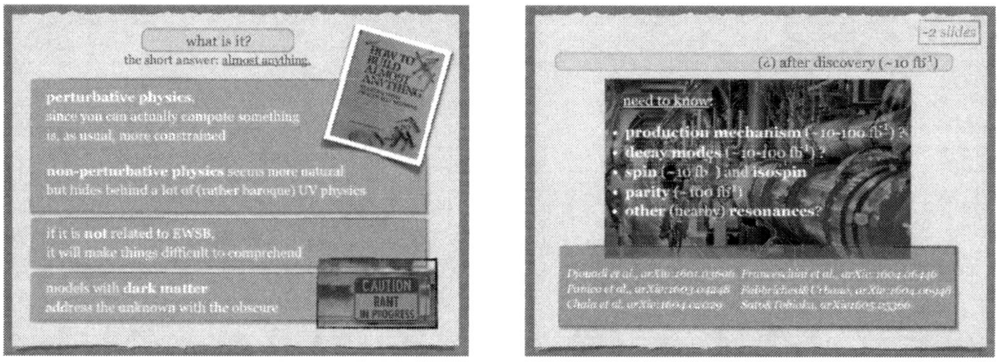

Figure 12. A kind of summary (left panel) and the need-to-know data after a discovery (right panel).

References

[1] ATLAS collaboration, ATLAS-CONF-2015-081; CMS collaboration, CMS-PAS-EXO-15-004.
[2] Arnold H. Glasow.
[3] Samuel T. Coleridge.
[4] L. D. Landau, Dokl. Akad. Nauk Ser. Fiz. **60**, no. 2, 207 (1948); C. N. Yang, Phys. Rev. **77**, 242 (1950).

EPJ Web of Conferences **129**, 00002 (2016) DOI: 10.1051/epjconf/201612900002
QCD@Work 2016

Z/W + jets production at 8/13 TeV in CMS

Kadir Ocalan[1,★] for the CMS Collaboration

[1] *Necmettin Erbakan University, Faculty of Aviation and Space Sciences, Koycegiz Campus, Meram, 42090 Konya, Turkey*

Abstract. The production of vector bosons (Z and W) in association with jets provides a stringent test of perturbative QCD and is a background process in searches for new physics. Differential cross section measurements of vector bosons produced in association with jets and heavy flavour quarks in proton-proton collisions at center of mass energies of 8 TeV and 13 TeV in the CMS experiment at the LHC are presented. The measurements are compared to next-to-leading order calculations and event simulations that devise matrix element calculations interfaced with parton showers.

1 Introduction

The production of vector bosons (Z and W) in association with jets is an important Standard Model (SM) benchmark providing precision tests for perturbative Quantum Chromodynamics (pQCD) calculations and inputs for improving parton distribution functions (PDFs). Vector boson and associated jets (V+jets) data can be used for testing Monte Carlo (MC) based event generators. Productions of V+jets constitute prominent background for several SM processes such as Higgs boson production in association with a vector boson, single top production, top pair production, and vector boson fusion. Moreover, studies of V+jets are important for modelling backgrounds for beyond the SM (BSM) searches such as supersymmetry (SUSY) and dark matter.

In V+jets measurements, Z and W bosons are reconstructed through their leptonic decays ($Z/\gamma^* \rightarrow l^+l^-$, $W \rightarrow l\nu$) and their productions in association with jets are characterised by measured differential cross sections as a function of various observables including jet multiplicity and kinematics, angular correlations, and invariant masses. Measured V+jets distributions are unfolded to particle level for detector effects and compared to several theoretical predictions. The theoretical predictions include leading order (LO) and next-to-leading order (NLO) matrix element (ME) calculations interfaced with parton showers (PS) and NLO and next-to-next-to-leading order (NNLO) fixed order calculations. In this report, the latest results on the V+jets production are reviewed based on mostly 8 TeV and 13 TeV proton-proton (pp) collisions data recorded by the Compact Muon Solenoid (CMS) detector [1] at the CERN Large Hadron Collider (LHC).

2 Z + jets differential cross sections

The differential cross sections of Z boson production in association with jets, in the dimuon and dielectron decay channels [2] and in the dimuon decay channel for an extended jet rapidity region [3],

★e-mail: kadir.ocalan@cern.ch

are measured in pp collisions at a center-of-mass energy of $\sqrt{s} = 8$ TeV. The data sample used corresponds to an integrated luminosity of 19.6 fb^{-1} collected by the CMS detector. The differential cross sections are measured in a phase space where transverse momenta and pseudorapidities of leptons and jets are required as $p_T(l) > 20$ GeV and $p_T(\text{jets}) > 30$ GeV and $|\eta(l, \text{jets})| < 2.4$ ($|y(\text{jets})| < 4.7$ in the extended measurement). The measured differential cross sections are compared to predictions by MADGRAPH5 [4] (+ PYTHIA 6 [5]) tree level ME event generator at LO accuracy and by SHERPA 2 [6] ME event generator at NLO accuracy up to 2 partons, where both generators are interfaced for PS. The total cross section for MADGRAPH5 is normalised to the NNLO cross section computed with FEWZ [7]. The predictions are generally able to describe data within uncertainties, however as shown in figure 1 (left), MADGRAPH5 shows deviation from data for a range of 150–450 GeV and SHERPA 2 slightly underestimates data for the first leading jet p_T.

The first differential cross section measurement of Z+jets at 13 TeV is performed in the dimuon decay channel using pp collisions data corresponding to an integrated luminosity of 2.5 fb^{-1} [8]. The analysed Z+jets events are required to have isolated muon pairs with $p_T(\mu) > 20$ GeV and $|\eta(\mu)| < 2.4$ and jets with $p_T(\text{jets}) > 30$ GeV and $|y(\text{jets})| < 2.4$. The measured differential distributions are compared to predictions by MADGRAPH5_aMC@NLO [9], interfaced with PYTHIA 8 [10] for PS and hadronization, at NLO accuracy up to 2 additional partons merged using FxFx scheme [11] and by an NNLO fixed order calculation [12, 13] for Z + 1 jet production. The MADGRAPH5_aMC@NLO and NNLO predictions show remarkable agreement with data on the first leading jet p_T as shown in figure 1 (right).

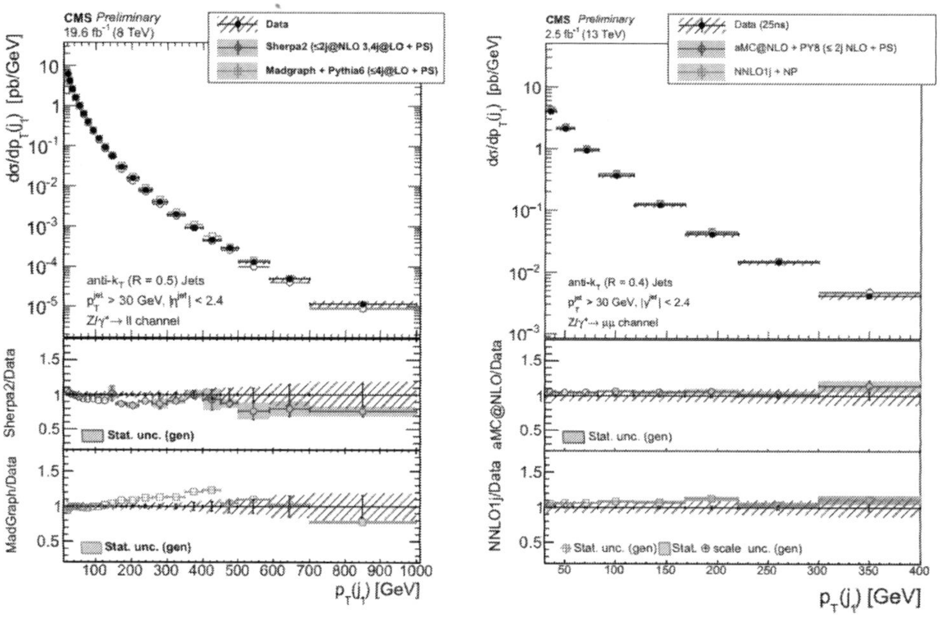

Figure 1. Left: 8 TeV Z+jets differential cross section measurement as a function of the first jet p_T compared to the SHERPA 2 and MADGRAPH5 predictions. Right: 13 TeV Z+jets differential cross section measurement as a function of the first jet p_T compared to the MADGRAPH5_aMC@NLO and Z + 1 jet NNLO predictions. Error bars around the experimental points show the statistical uncertainty, while the crosshatched bands indicate the statistical and systematic uncertainties added in quadrature. The color filled band around the NNLO prediction shows the statistical and systematic uncertainties added in quadrature, while representing only the statistical uncertainty of the generated sample for the other predictions.

DOI: 10.1051/epjconf/201612900002

3 W + jets differential cross sections

The measurement of the differential cross sections for W boson production in the muon decay channel and jets is presented with data corresponding to an integrated luminosity of 19.6 fb^{-1} recorded by the CMS detector at $\sqrt{s} = 8$ TeV [14]. W+jets events are selected by the requirements for single muons with $p_T(\mu) > 25$ GeV and $|\eta(\mu)| < 2.1$ and jets with $p_T(\text{jets}) > 30$ GeV and $|\eta(\text{jets})| < 2.4$, in addition to W boson transverse mass requirement of $M_T(W) > 50$ GeV. Differential distributions are measured as a function of several observables including the jet kinematics and angular correlations among the jets and the muon. Such angular distributions are sensitive to the modelling of higher order corrections and parton emissions. The measured W+jets cross sections are compared to the predictions of MADGRAPH5 (+ PYTHIA 6) tree level LO, MADGRAPH5_aMC@NLO (+ PYTHIA 8) merged NLO up to 2 partons, and SHERPA 2 multileg NLO (+ BLACKHAT [15, 16]) MC generators and the fixed order theoretical predictions provided by BLACKHAT + SHERPA [17] at NLO accuracy. The NNLO calculation [18, 19] in pQCD for W + 1 jet production is also provided. The predictions from MADGRAPH5, MADGRAPH5_aMC@NLO, SHERPA 2, and NNLO calculation show reasonable agreement with data within uncertainties for the H_T variable (defined as scalar sum of jets p_T) for one inclusive jet production, while BLACKHAT + SHERPA systematically underestimates data as shown in figure 2 (left). The measured differential cross section as a function of azimuthal separation between the first leading jet and the muon is well described by the multileg LO and NLO predictions, however fixed order NLO shows lower trend in describing data as shown in figure 2 (right).

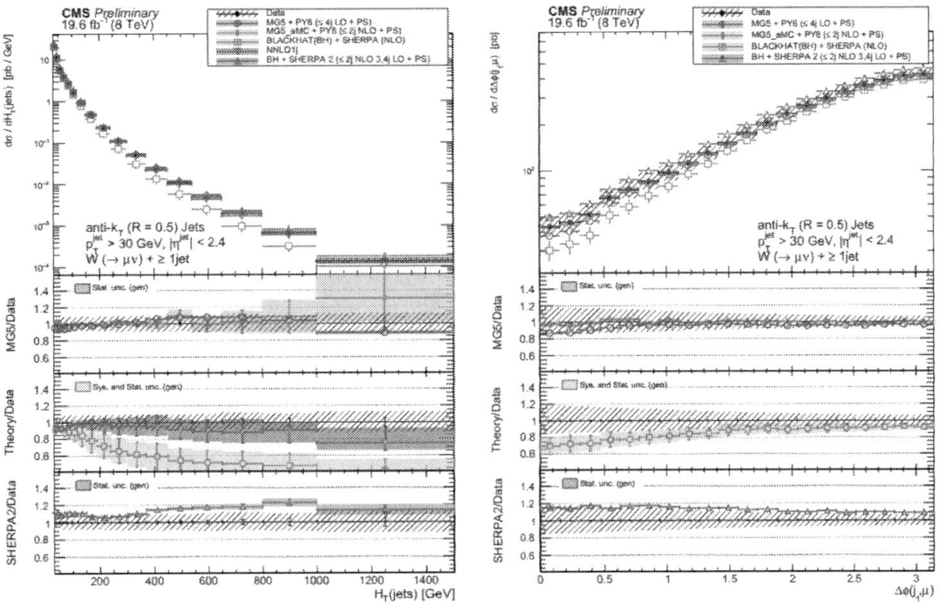

Figure 2. W+jets differential cross section measurement as a function of H_T for one inclusive jet production (left) and azimuthal correlation between the first jet and the muon (right). The measured distributions are compared to several predictions including W + 1 jet NNLO for H_T variable. Black circular markers with the grey hatched band represent the data measurement and its total experimental uncertainty. The color filled bands around the BLACKHAT + SHERPA and NNLO predictions show their statistical and systematic uncertainties added in quadrature, while representing only the statistical uncertainty of the generated sample for the other predictions. BLACKHAT + SHERPA and NNLO are corrected for non-perturbative effects.

4 Z + b jets cross sections

Associated production of a vector boson and heavy flavour jets is sensitive for probing heavy quark content in the proton and constitutes important background processes to Higgs studies and BSM searches. The production of a Z boson in association with at least one jet originating from a b quark is measured in pp collisions at $\sqrt{s} = 8$ TeV [20]. Cross sections are measured from data collected by the CMS, for a total integrated luminosity of 19.6 fb^{-1}, using Z boson decays into electron or muon pairs, and identified b quark jets. Z+b jets events are required to have leptons with $p_T(l) > 20$ GeV and $|\eta(l)| < 2.4$ and b-tagged jets with $p_T(\text{b}) > 30$ GeV and $|\eta(\text{b})| < 2.4$. The fiducial cross sections for Z boson in the combined lepton channel with at least one and two b jets are measured. The differential cross sections for the cases of at least one and two b jets are measured as a function of observables characterising the b jet and Z boson kinematics. Moreover, the ratios of differential cross sections for the associated production with at least one b jet and with any jet inclusive in flavour are also measured. The measurement results are compared to the theoretical predictions including MADGRAPH5 (+ PYTHIA 6) tree level LO generator and POWHEG [21–23] (+ PYTHIA 6) NLO, testing two different flavour schemes for the choice of initial state partons as the four-flavour scheme (4FS) [24] and the five-flavour scheme (5FS) [25]. The differential cross sections for Z+b jets and the cross section ratios for the associated one inclusive b jet and any jets are better described by the predictions using 5FS approach than 4FS as shown in figure 3.

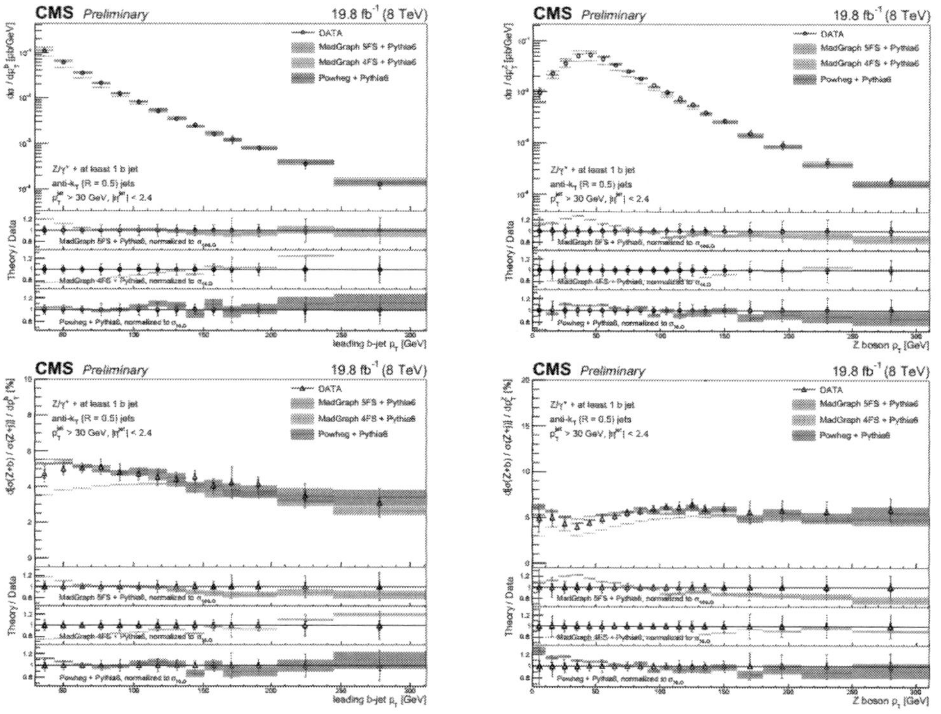

Figure 3. Differential Z+1 b jet cross sections as a function of p_T for the leading b jet and Z boson (top) and the cross section ratios for p_T of the leading b jet and Z boson between Z+1 b jet and Z+jets (bottom). The measured distributions are compared with the MADGRAPH5 (4FS and 5FS) and POWHEG (5FS) theoretical predictions. For each data point the statistical and the total (statistical plus systematic) uncertainty are represented by the double error bar. The width of shaded bands represents the statistical error on the theoretical predictions.

5 W + 2 b jets cross section

The production cross section of the W boson in association with two b jets (W+b$\bar{\text{b}}$) is measured using a data sample of pp collisions collected by the CMS experiment at $\sqrt{s}$ = 8 TeV [26]. The data sample corresponds to an integrated luminosity of 19.8 fb^{-1}. The W bosons are reconstructed using their leptonic decays to muons (W $\to \mu\nu$) and electrons (W $\to e\nu$). The W+b$\bar{\text{b}}$ events are required to have leptons with $p_{\text{T}}(l) > 30$ GeV and $|\eta(l)| < 2.1$ and exactly two b-tagged jets with $p_{\text{T}}(\text{b}) > 25$ GeV and $|\eta(\text{b})| < 2.4$ in the signal region and extracted using a likelihood fit to $M_{\text{T}}(\text{W})$. The measured cross section in the combined lepton channel, $\sigma(\text{pp} \to \text{W}(\text{l}\nu) + \text{b}\bar{\text{b}}) = 0.69 \pm 0.02(\text{stat.}) \pm 0.11(\text{syst.}) \pm 0.07(\text{theo.}) \pm 0.02(\text{lumi.})$ pb, is compared to theoretical predictions of MADGRAPH5 (+ PYTHIA 6) using both 4FS and 5FS, MADGRAPH5 (+ PYTHIA 8) using 4FS, and MCFM [27, 28] NLO. The predicted cross sections agree each other and are consistent with the measured cross section within one standard deviation as shown in figure 4.

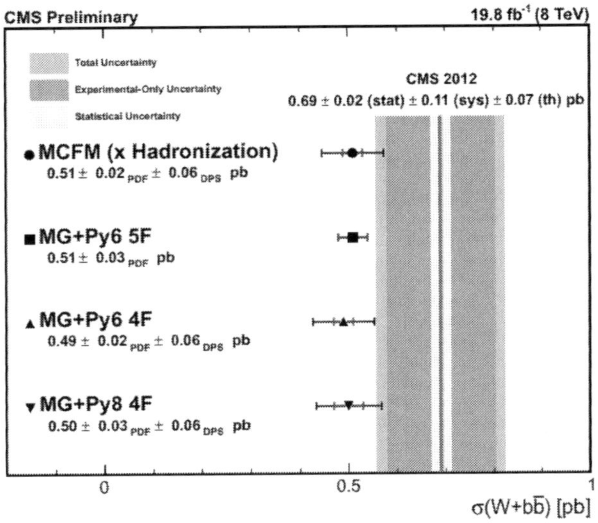

Figure 4. Comparison between the measured W+b$\bar{\text{b}}$ cross section and various theoretical predictions. The blue error bars on the predictions represent the uncertainty in the given sample associated with PDF choice and the black bars represent the total uncertainty. The effects of double parton scattering (DPS) are not included in the generated samples of the predictions except MADGRAPH5 + PYTHIA6 (5F), so the extra DPS factor is included to account for the DPS effect in the theoretical cross section predictions.

6 Conclusions

The CMS collaboration has explored V+jets productions including Z and W bosons in the most relevant areas using 8 TeV and recently 13 TeV pp collisions data at the LHC. The measured cross sections differential in various observables are confronted by several predictions based on multileg LO and NLO ME + PS generators and NLO and NNLO fixed order calculations. The V+jets measurements including associated heavy flavour jets provide unique tests for pQCD calculations to high precision and inputs for tuning MC based event generators with constant developments. Overall, the differential cross section distributions measured as a function of the observables are well modelled by the predictions. NLO ME + PS and fixed order NNLO predictions show good agreement with data distributions, while LO (ME + PS) predictions show some discrepancies. In the associated heavy flavour jets measurements, Z+b jets data spectra are better described by the predictions using 5FS approach. The measured W+$b\bar{b}$ fiducial cross section is in agreement with the SM predictions. The CMS physics program will continue delivery of V+jets results with higher precision using 13 TeV pp collisions data.

The author acknowledges the financial support for the conference by Necmettin Erbakan University Scientific Research Projects Unit (BAP) under the AGED1 project with number 162518001-211.

References

[1] The CMS Collaboration, JINST **3**, S08004 (2008).
[2] The CMS Collaboration, CMS-PAS-SMP-13-007, (2014).
[3] The CMS Collaboration, CMS-PAS-SMP-14-009, (2014).
[4] J. Alwall et al., JHEP **1106**, 128 (2011).
[5] T. Sjostrand et al., JHEP **0605**, 026 (2006).
[6] T. Gleisberg et al., JHEP **0902**, 007 (2009).
[7] R. Gavin et al., Comput. Phys. Commun. **182**, 2388-2403 (2011).
[8] The CMS Collaboration, CMS-PAS-SMP-15-010, (2015).
[9] J. Alwall et al., JHEP **07**, 079 (2014).
[10] T. Sjostrand et al., Comput. Phys. Commun. **178**, 852-867 (2008).
[11] R. Frederix and S. Frixione, JHEP **12**, 061 (2012).
[12] R. Boughezal et al., Phys. Rev. Lett. **116**, 152001 (2016).
[13] R. Boughezal, X. Liu, and F. Petriello, arXiv:1602.08140 (2016).
[14] The CMS Collaboration, CMS-PAS-SMP-14-023, (2016).
[15] C. Berger et al., Nucl.Phys.Proc.Suppl. **183**, 313-319 (2008).
[16] C. Berger et al., Nucl.Phys.Proc.Suppl. **205-206**, 92-97 (2010).
[17] C. F. Berger et al., Phys. Rev. D. **80**, 074036 (2009).
[18] R. Boughezal et al., Phys. Rev. Lett. **115**, 062002 (2015).
[19] R. Boughezal, X. Liu, and F. Petriello, arXiv:1602.06965 (2016).
[20] The CMS Collaboration, CMS-PAS-SMP-14-010, (2015).
[21] P. Nason, JHEP **0411**, 040 (2004).
[22] S. Frixione, P. Nason and C. Oleari, JHEP **0711**, 070 (2007).
[23] S. Alioli et al., JHEP **1006**, 043 (2010).
[24] F. F. Cordero et al., Phys. Rev. D. **80**, 034015 (2009).
[25] J. M. Campbell et al., Phys. Rev. D. **69**, 074021 (2004).
[26] The CMS Collaboration, CMS-PAS-SMP-14-020, (2016).
[27] J. M. Campbell and R. Ellis, Nucl. Phys. Proc. Suppl. **205**, 10 (2010).
[28] S. Badger, J. M. Campbell, and R. Ellis, JHEP **03**, 027 (2011).

EPJ Web of Conferences **129**, 00003 (2016) DOI: 10.1051/epjconf/201612900003
QCD@Work 2016

QCD with jets and heavy flavour in pp and PbPb collisions in ATLAS

Robert Keyes[1,*] on behalf of the ATLAS collaboration

[1] *McGill University*

Abstract. ATLAS has studied different aspects of QCD in pp and PbPb collisions. A summary of interesting recent results at centre-of-mass energies of 7 TeV, 8 TeV and 13 TeV (pp), and 2.76 TeV (PbPb) per nucleon pair is presented. Using pp collisions, measurements of Z, W, photon, quarkonia and open charm production differential cross sections in a variety of variables are presented. Jet and heavy flavour muon measurements in PbPb collisions, aimed to test the properties of the Quark Gluon Plasma, with the view of better understanding of jet quenching, are also presented.

1 Introduction

The Large Hadron Collider (LHC) is the energy frontier of high energy particle physics, providing the highest energy and highest luminosity environment for proton-proton (pp) and heavy ion (HI) collisions to date. ATLAS is a multipurpose particle detector with a broad physics program that makes use of precision tracking, calorimetry and muon detectors [1]. Signals arising from collisions of coloured constituents of the matter are dominated by Quantum Chromodynamic (QCD) processes. Measuring broad aspects of QCD production provides constraints on QCD theoretical predictions as well as requisites for the empirical description of the proton and its constituents. Finally, having an accurate description of the QCD background of non-QCD measurements, such as Higgs measurements or searches beyond the Standard Model, is critical to their success.

2 Standard Model Boson measurements

2.1 Inclusive W and Z production at 13 TeV

W and Z boson production cross sections are measured at 13 TeV, providing an important benchmark of QCD and electroweak processes that are sensitive to the choice of PDF [2]. The measurements consider leptonic decays in the electron and muon channels: single-lepton with missing transverse energy for W boson candidate events and dilepton that fall within the Z mass window for Z boson candidate events. The primary backgrounds for the W boson measurements are multijets, which are estimated using a data driven approach. The primary backgrounds for the Z boson measurements are diboson signatures (WW), Z to di-tau and top quark decays, all of which are derived from simulations. The

*e-mail: rkeyes@cern.ch

DOI: 10.1051/epjconf/201612900003

measurements display a good agreement with predictions from NNLO QCD and NLO+EW corrections. The electron and muon channels measured separately in each case are found to be consistent. Figure 1 (left), displaying the ratio of predicted to measured cross sections shows tension between different PDFs. In Figure 1 (right), the W^+ to W^- cross section ratio enhances this tension due to cancellations of correlated measurement uncertainties.

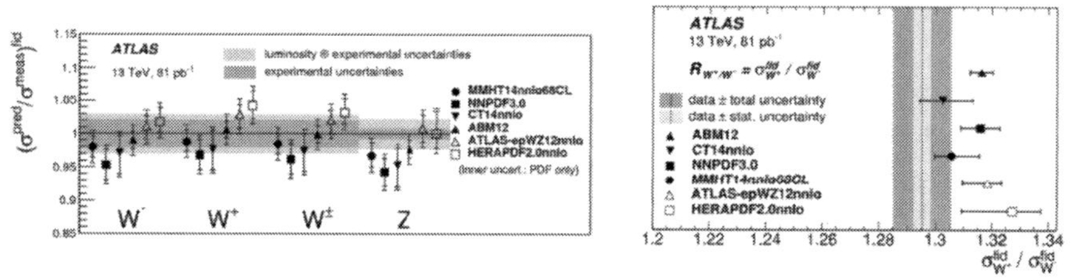

Figure 1. Comparison of the measured and predicted W and Z cross sections (left) and the W^+/W^- ratio of the measured cross-section (right) using various PDF descriptions [2].

2.2 Inclusive Z p_T at 8 TeV

The measurement of the inclusive Z boson p_T spectrum at 8 TeV provides a comprehensive test of QCD predictions spanning the non-perturbative regime at low p_T to the perturbative regime at high p_T [3]. W and Z bosons constitute a significant background in many analyses and previous simulations of these backgrounds have required data-driven corrections [4, 5]. The measurement is conducted by selecting dilepton final states, a clean signature that benefits from a precise p_T determination and low background. Measurements were compared to the predictions from Pythia and Sherpa, as shown in Figure 2 (left), both displaying a poor agreement in the high p_T tail. Comparisons to a higher order pQCD computation DYNNLO yielded a better description of the shape but with a constant offset, as seen in Figure 2 (right).

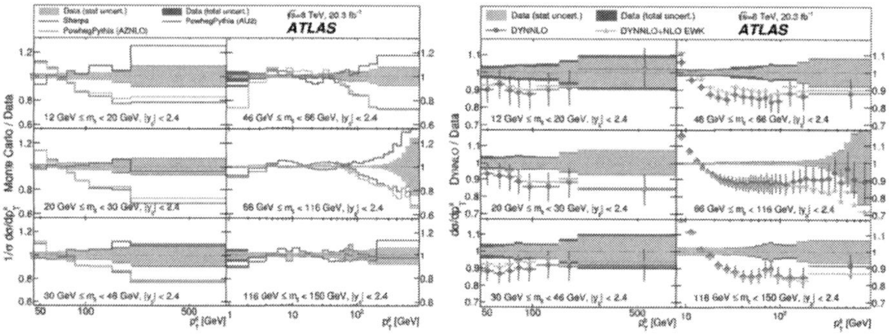

Figure 2. Comparison of the measured cross sections using various generator level descriptions (left) and DYNNLO with and without electroweak corrections (right) [3].

2.3 Inclusive photon E_T at 8 TeV

Prompt photons originating from hard scattering are a clean probe of the parton level dynamics and of the gluon PDF. Arising predominantly through Compton process like $qg \to q\gamma$ scattering, prompt photon signatures are targeted experimentally by requiring that the photon is isolated, i.e. well separated from other calorimetric signatures. Events simulated using Pythia and CTEQ6L1 PDF are used to correct for detector effects and signal leakage in order to extract the central value of the measurements [6]. A significant improvement in experimental uncertainties, compared to previous results, is obtained thanks to improved photon calibrations [7]. The result is compared to theory predictions at NLO from JetPhox using various PDF sets, and at NLO+NNNLL (which is comparable to NNLO) from PeTeR, as seen in Figure 3. Comparisons yield a good shape agreement and a better normalization with PeTeR. This measurement provides an important constraint on the global fit of the gluon PDF.

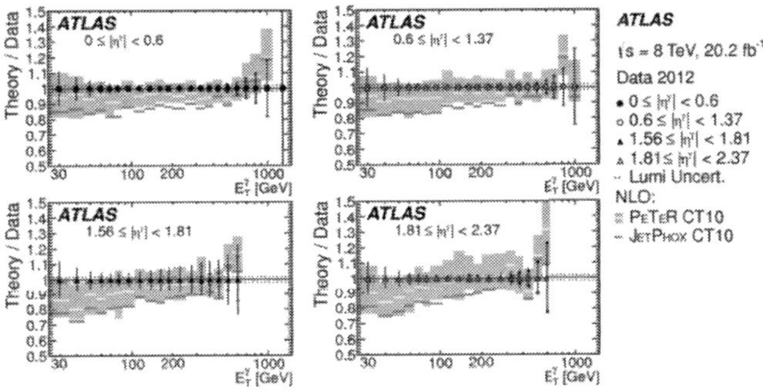

Figure 3. The measured photon E_T differential cross section compared to JetPhox and PeTeR [6].

3 Heavy flavour production measurements

3.1 J/ψ and $\psi(2S)$ at 7 and 8 TeV

Heavy flavour QCD bound states (quarkonia) probe the boundary of perturbative and non-perturbative QCD. Charmonium production can arise through "prompt" short lived sources or through "non-prompt" long lived sources such as the decay of a beauty hadron. The $\psi(2S)$ state is of interest because it is the only vector charmonium state that has mostly direct production. The measurements are performed by selecting di-muon decay events, a final state that exhibits efficient triggering and clean reconstruction [8]. The pseudo-proper time τ observable measures the displacement of the hadron before its di-muon decay, as seen in Figure 4 (left). A two dimensional maximum likelihood fit is performed over the invariant mass and τ, the former discriminating signal from background by isolating the bound state resonance, the latter separating the prompt from the non-prompt components. The results of these cross section measurements agree well with results from other LHC experiment measurements. The comparison of the data to NLO NRQCD predictions for prompt production agree well as seen in Figure 4 (middle), the comparison to FONLL predictions for non-prompt production shows that the predictions slightly overestimate the data at high p_T, as seen in Figure 4 (right).

DOI: 10.1051/epjconf/201612900003

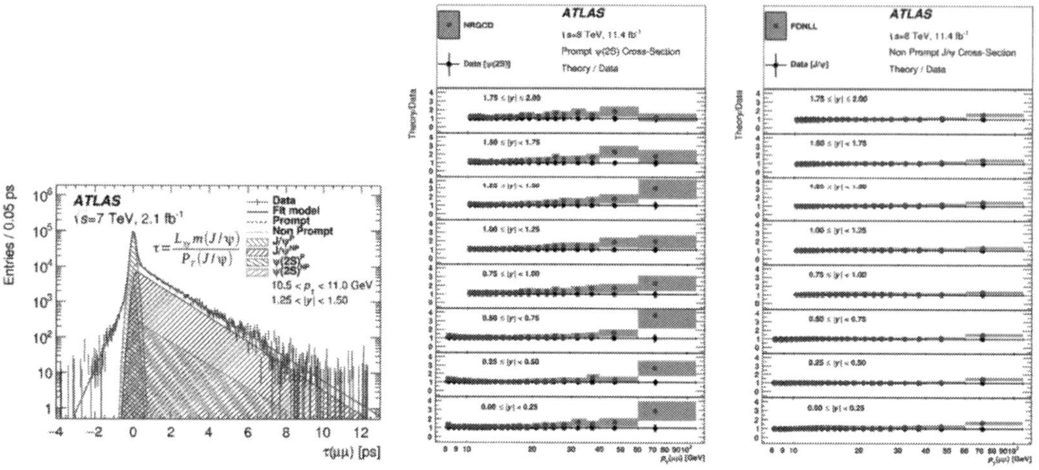

Figure 4. The proper time distribution (left) and comparisons of the measured $\psi(2S)$ (middle) and J/ψ (right) cross section ratios to NRQCD and FONLL predictions [8].

3.2 Prompt and non-prompt J/ψ at 13 TeV

During the long shut down between Run 1 and Run 2 ATLAS installed additional layers of silicon pixel tracking by virtue of shrinking the beam pipe. This new "Inner B Layer" (IBL) provides enhanced vertexing and heavy flavour tagging capabilities in anticipation of more challenging pile-up conditions for Run 2. The ratio of prompt to non-prompt J/ψ production was measured at 13 TeV using a small preliminary dataset [9] and found to be consistent with the Run 1 results in all pseudorapidity regions, providing a promising first look of ATLAS heavy flavour performance in Run 2.

3.3 Open charm production at 7 TeV

D mesons arising in the fragmentation of c and b quarks provide a test of pQCD heavy flavour production, of heavy flavour content of the proton, and serve as an important constraint for heavy flavour backgrounds for Higgs measurements. At LHC the charm cross section is large cross and a clean signatures is provided by the precise tracking and vertexing capability of ATLAS [10]. Comparing the results to various predictions, we see that they typically underestimate the data, with FONLL and POWHEG showing the best shape agreement, as shown in Figure 5. The total measured cross section is consistent with those reported by ALICE [11]. The strangeness suppression factor and the charged non-strange vector D-meson fraction are consistent with previous results from ALICE [12].

4 Heavy ion measurements

Heavy Ion (HI) collisions create a hot dense medium of strong nuclear matter with deconfined color charges referred to as the Quark Gluon Plamsa (QGP). This new testing ground for QCD bears an important correspondence to the state and evolution of the early universe. Jet quenching, a recently discovered QGP phenomena, refers to jets interacting with the QGP and becoming more diffuse. Measurements are compared to analogous measurements derived from pp collisions in place of HI simulations. HI collisions are parametrized in terms of the Centrality observable. Centrality is measured in ATLAS using the total energy collected in the forward calorimeters: the more central or

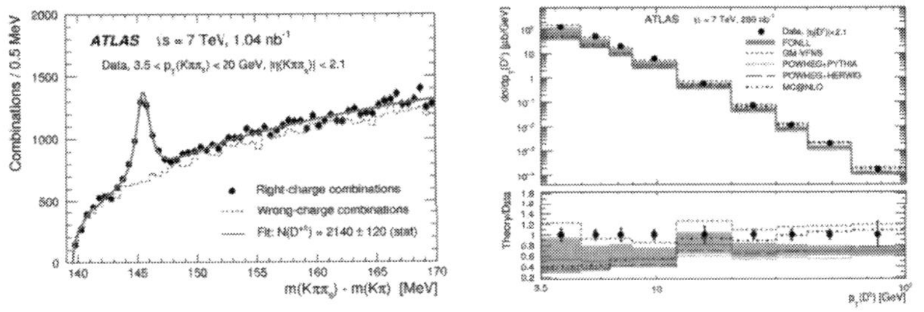

Figure 5. Reconstructed mass peak for the $D^{*\pm}$ meson (left) and the differential cross section in p_T for the $D^{\pm}$ meson (right) [10].

head-on the collision, the more interacting nucleons there are, and in turn, the more outgoing particles and energy deposited [13].

4.1 Dijet p_T correlations in PbPb at 2.76 TeV

Dijets provide a convenient probe of the QGP since in pp collisions they are typically well balanced. The dijet asymmetry, $x_J = \frac{p_T^2}{p_T^1}$, was measured to quantify the effect of jet quenching as a function of jet p_T and centrality [14]. The x_J distribution was found to strongly deviate from the nominal distribution observed in pp collisions with decreasing jet p_T and decreasing collision centrality, as shown in Figure 6 (left). The low p_T and high centrality region show the strongest asymmetry with a peak at $x_J = 0.5$.

4.2 Internal jet structure in PbPb at 2.76 TeV

The internal structure of jets in PbPb collisions is measured to provide constraints on jet quenching models and in-medium modifications for parton shower models [15]. Charged particles with p_T down to 1 GeV falling inside a jet are measured, correcting for tracking inefficiencies. The resulting jet fragmentation functions are reported as a function of longitudinal and transverse charged particle p_T, with respect to the jet axis, in different regions of centrality, jet η and jet p_T. In comparison to pp collision measurements, as shown in Figure 6 (right), an enhancement is observed at low and high charged particle p_T and a suppression in the $4 < p_T < 25$ GeV range. The enhancement at high p_T diminishes with increasing jet p_T and with increasing jet η, and, as would be expected, all the effects diminish with decreasing centrality.

4.3 Heavy flavour muons in PbPb

Heavy flavour signals are produced perturbatively in the hard scatter (as their masses are much greater than the temperature of the medium) and subsequently interact with the medium [16]. Heavy flavour muons are identified using an energy loss variable that takes the difference between the p_T measured by the inner detector and by the muon system. The yields show a centrality dependent suppression that is slightly more significant than that measured by the inclusive jet HI result [17]. Following its creation the QGP expands and cools. By measuring the distribution of the heavy flavour muons in azimuthal angle ϕ, the azimuthal anisotropy, relevant to elliptical flow, is extracted by fitting a 2nd order Fourier coefficient, v_2. v_2 is measured to be up to 8% at low p_T and the overall variation is consistent with previous measurements [18]. Overall the findings are consistent with, and much more precise than, measurements performed by ALICE in the forward region (2.5 < y < 4.0) [19].

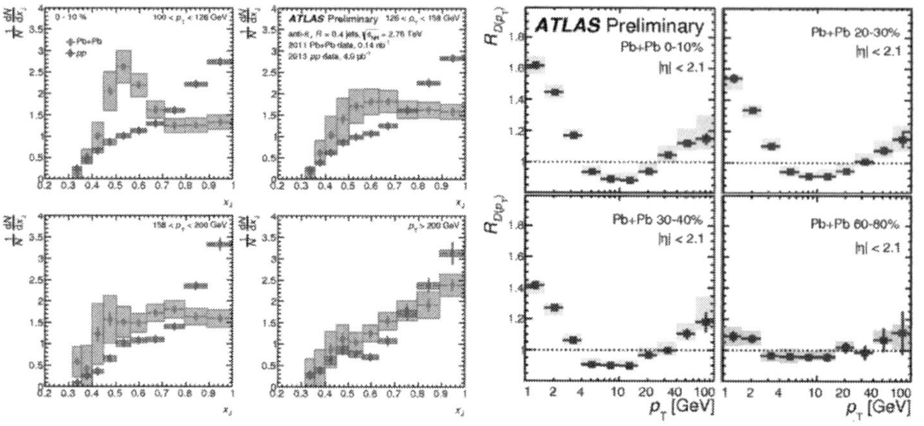

Figure 6. Comparison of the measured dijet asymmetry $x_J = \frac{p_T^2}{p_T^1}$ in PbPb to pp (left) [14], and the ratio of the jet fragmentation functions measured in PbPb to pp (right) [15], in different centrality regions. Statistical uncertainties are displayed using error bars while systematic uncertainties use shaded boxes.

5 Conclusions

The ATLAS experiment has a comprehensive program of QCD studies in the contexts of Standard Model, Heavy Flavour and HI. These results emphasise that ATLAS is performing well, providing new and extended measurements, which are cross checks with previous results. Simulations have also progressed considerably but improvements are still needed, motivating these and future studies.

References

[1] ATLAS Collaboration, JINST **3**, S08003 (2008).
[2] ATLAS Collaboration, Phys. Lett. B **759**, 601 (2016).
[3] ATLAS Collaboration, Eur. Phys. J. C **76**, 291 (2016).
[4] ATLAS Collaboration, J. High Energy Phys. **1**, 69 (2015).
[5] ATLAS Collaboration, ATLAS-CONF-2014-038, https://cds.cern.ch/record/1735215 (2014).
[6] ATLAS Collaboration, J. High Energy Phys. **8**, 5 (2016).
[7] ATLAS Collaboration, Phys. Rev. D **89**, (2014).
[8] ATLAS Collaboration, Eur. Phys. J. C **76**, 283 (2016).
[9] ATLAS Collaboration, ATLAS-CONF-2015-030, https://cds.cern.ch/record/2037967 (2015).
[10] ATLAS Collaboration, Nucl. Phys. B **907**, 717 (2016).
[11] ALICE Collaboration, J. High Energy Phys. **7**, 191 (2012).
[12] ALICE Collaboration, J. Phys. Lett. B **718**, 279 (2012).
[13] ATLAS Collaboration, Phys. Rev. Lett. **105**, 252303 (2010).
[14] ATLAS Collaboration, ATLAS-CONF-2015-052, https://cds.cern.ch/record/2055673 (2015).
[15] ATLAS Collaboration, ATLAS-CONF-2015-055, https://cds.cern.ch/record/2055676 (2015).
[16] ATLAS Collaboration, ATLAS-CONF-2015-053, https://cds.cern.ch/record/2055674 (2015).
[17] ATLAS Collaboration, Phys. Rev. Lett. **114**, (2015).
[18] ATLAS COllaboration, Phys. Rev. C **86**, (2012).
[19] ALICE Collaboration, Phys. Rev. Lett. **109**, (2012).

EPJ Web of Conferences **129**, 00004 (2016) DOI: 10.1051/epjconf/201612900004
QCD@Work 2016

Off-shell amplitudes and four-jet production in k_T-factorization

Mirko Serino[1],*

[1] *The Henryk Niewodniczański Institute of Nuclear Physics, Polish Academy of Sciences, ul. Radzikowskiego 152, 31-342 Kraków, Poland*

Abstract. We perform a study of both Single and Double Parton Scattering contributions to the inclusive 4-jet production in the k_T-factorization framework at Leading Order and $\sqrt{s} = 7$ TeV and 13 TeV. We discuss the importance of double parton scattering for relatively soft cuts on the jet transverse momenta and find out that symmetric cuts do not suit well k_T-factorization predictions, because of a kinematic effect suppressing the double parton scattering contribution.

1 Introduction

Using a proper generalisation of the Britto-Cachazo-Feng-Witten recursion relation [1, 2] for on-shell QCD amplitudes, we can recursively compute fully gauge-invariant amplitudes with initial state off-shell particle both analytically [3, 4] and numerically [5] and use them to assess both the Single Parton Scattering and Double Parton Scattering contributions to four-jet production. This allows us to expand the analysis of Ref. [6] and assess the differences between the collinear approach and the high-energy factorization (HEF) (or k_T-factorization) framework.

2 Calculations and comparison to experimental data

2.1 Single-parton scattering production of four jets

The collinear factorization formula for the calculation of the inclusive partonic 4-jet cross section reads

$$\begin{aligned} \sigma^{B}_{4-jets} &= \sum_{i,j} \int \frac{dx_1}{x_1}\frac{dx_2}{x_2} x_1 f_i(x_1,\mu_F) x_2 f_j(x_2,\mu_F) \\ &\times \frac{1}{2\hat{s}} \prod_{l=1}^{4} \frac{d^3 k_l}{(2\pi)^3 2E_l} \Theta_{4-jet} (2\pi)^4 \,\delta\left(x_1 P_1 + x_2 P_2 - \sum_{l=1}^{4} k_l \right) \overline{|\mathcal{M}(i,j \to 4\,\text{part.})|^2}\,. \end{aligned} \tag{1}$$

Here $f_i(x_{1,2},\mu_F)$ are the collinear PDFs for the $i-th$ parton, carrying $x_{1,2}$ momentum fractions of the proton and evaluated at the factorization scale μ_F; the index l runs over the four partons in the

*e-mail: mirko.serino@ifj.edu.pl

final state, P is the total initial state partonic momentum, associated to the center of mass energy $\hat{s} = P^2 = (P_i + P_j)^2 = 2P_i \cdot P_j$; the Θ function takes into account the kinematic cuts applied and $\mathcal{M}$ is the partonic on-shell matrix element, which includes symmetrization effects due to the possible identity of the final state particles.

The analogous formula to (1) for HEF is

$$\sigma^B_{4-jets} = \sum_{i,j} \int \frac{dx_1}{x_1} \frac{dx_2}{x_2} d^2k_{T1} d^2k_{T2} \mathscr{F}_i(x_1, k_{T1}, \mu_F) \mathscr{F}_j(x_2, k_{T2}, \mu_F)$$

$$\times \frac{1}{2\hat{s}} \prod_{l=1}^{4} \frac{d^3k_l}{(2\pi)^3 2E_l} \Theta_{4-jet} (2\pi)^4 \delta \left(x_1 P_1 + x_2 P_2 + \vec{k}_{T1} + \vec{k}_{T2} - \sum_{l=1}^{4} k_l \right) \overline{|\mathcal{M}(i^*, j^* \to 4\,\text{part.})|^2}. \tag{2}$$

Here $\mathscr{F}_i(x_k, k_{Tk}, \mu_F)$ is a transverse momentum dependent (TMD) parton density for a given type of parton. Similarly as in the collinear factorization case, x_k is the longitudinal momentum fraction and μ_F is a factorization scale. The new degree of freedom is introduced via the transverse k_{Tk}, which is perpendicular to the collision axis. The formula is valid when the x's are not too large and not too small and, in order to construct a full set of TMD parton densities, we apply the Kimber-Martin-Ryskin (KMR) prescription [7, 8], which, at the end of the day, amounts to applying the Sudakov form factor onto the PDFs.

$\mathcal{M}(i^*, j^* \to 4\,\text{part.})$ is the gauge invariant matrix element for $2 \to 4$ particle scattering with two initial off-shell legs. We rely on the numerical Dyson-Schwinger recursion in the AVHLIB[1] for its computation. If a complete calculation of 4-jet production in k_T factorization was still missing in the literature, it was mainly because computing gauge-invariant amplitudes with off-shell legs is definitely non trivial. Techniques to compute such amplitudes in gauge invariant ways are by now analytically and numerically well established [3–5].

We use a running α_s provided with the MSTW2008 PDF sets and set both the renormalization and factorization scales equal to half the transverse energy, which is the sum of the final state transverse momenta, $\mu_F = \mu_R = \frac{\hat{H}_T}{2} = \frac{1}{2}\sum_{l=1}^{4} k_T^l$, working in the $n_F = 5$ flavour scheme.

There are 19 different channels contributing to the cross section at the parton-level, of which the dominant ones, contributing together to $\sim 93\%$ of the total cross section, are

$$gg \to gggg\,, gg \to q\bar{q}gg\,, qg \to qggg\,, q\bar{q} \to q\bar{q}gg\,, qq \to qqgg\,, qq' \to qq'gg\,. \tag{3}$$

2.2 Double-parton scattering production of four jets

The SPS contribution is expected to dominate for high momentum transfer because, as it is intuitively clear, it is highly unlikely that two partons from one proton and two from the other one are energetic enough for two very hard scatterings to take place, as the well-known behaviour of the PDFs for large momentum fractions suggests. However, if the cuts on the transverse momenta of the final state are lowered, a window opens to observe significant double parton scattering effects, as often stated in the literature on the subject and recently analysed for 4-jet production in the framework of collinear factorization [6]. Here we perform the same analysis in HEF, with the goal to assess the differences in the predictions.

[1] available for download at https://bitbucket.org/hameren/avhlib

First of all, let us present the standard, phenomenologically motivated formula for the computation of differential DPS cross sections, tailored to a four-parton final state,

$$\frac{d\sigma^B_{4-jet,DPS}}{d\xi_1 d\xi_2} = \frac{m}{\sigma_{eff}} \sum_{i_1,j_1,k_1,l_1;i_2,j_2,k_2,l_2} \frac{d\sigma^B(i_1 j_1 \to k_1 l_1)}{d\xi_1} \frac{d\sigma^B(i_2 j_2 \to k_2 l_2)}{d\xi_2}, \tag{4}$$

where the $\sigma(ab \to cd)$ cross sections are obtained by restricting formulas (1) and (2) to a single channel and the symmetry factor m is $1/2$ if the two hard scatterings are identical, in order to avoid double counting, and is otherwise 1, whereas ξ_1 and ξ_2 denote generic kinematical variables for the first and second scattering, respectively.

The effective cross section σ_{eff} can be loosely interpreted as a measure of the transverse correlation of the two partons inside the hadrons. In this paper we stick to the widely used value σ_{eff} = 15 mb.

We also have to use an ansatz for Double Parton Distribution Functions; for collinear-factorization this is

$$D_{1,2}(x_1,x_2,\mu) = f_1(x_1,\mu)\, f_2(x_2,\mu)\, \theta(1-x_1-x_2), \tag{5}$$

where $D_{1,2}(x_1,x_2,\mu)$ is the Double Parton Distribution Function and $f_i(x_i,\mu)$ are the ordinary PDFs. The subscripts 1 and 2 distinguish the two generic partons in the same proton. Of course this ansatz can be automatically generalised to the case when parton transverse momenta are included by simply including the dependence on the transverse momentum.

Coming to DPS contributions, in principle we must include all the possible 45 channels which can be obtained by coupling in all possible distinct ways the 8 channels for the $2 \to 2$ SPS process, i.e.

$$\#1 = gg \to gg\,, \quad \#2 = gg \to q\bar{q}\,, \quad \#3 = qg \to qg\,, \#4 = q\bar{q} \to q\bar{q}$$
$$\#5 = q\bar{q} \to q'\bar{q}'\,, \quad \#6 = q\bar{q} \to gg \quad \#7 = qq \to qq\,, \quad \#8 = qq' \to qq'$$

We find that the pairs $(1,1)$, $(1,2)$, $(1,3)$, $(1,7)$, $(1,8)$, $(3,3)$ $(3,7)$, $(3,8)$ together account for more than 95 % of the total cross section. This was tested for all the sets of cuts considered in this paper.

2.3 Comparison to the collinear approach and to ATLAS data with hard central kinematic cuts

Our HEF calculation was first tested against the 8 TeV data recently reported by the ATLAS collaboration [9]. The kinematic cuts are $p_T > 100$ GeV for the leading jet and $p_T > 64$ GeV for the first three subleading jets; in addition $|\eta| < 2.8$ is the pseudorapidity cut and $\Delta R > 0.65$ is the constraint on the jet cone radius parameter.

We employ the running NLO α_s coming with the MSTW2008 sets. For such cuts, not much difference is expected between the two approaches and indeed we find none; DPS effects are irrelevant with this kinematics.

2.4 Comparison to CMS data with softer cuts

As discussed in Ref. [6], so far the only experimental analysis of four-jet production relevant for the DPS studies was realized by the CMS collaboration [10]. The cuts used in this analysis are $p_T > 50$ GeV for the first and second jets, $p_T > 20$ GeV for the third and fourth jets, $|\eta| < 4.7$ and the jet cone radius parameter $\Delta R > 0.5$. In the rest of this section, we present our results for such cuts.

As for the total cross section for the four jet production, the experimental and theoretical LO results are:

$$\begin{aligned}
\text{CMS collaboration:} \quad & \sigma_{tot} = 330 \pm 5\,(\text{stat.}) \pm 45\,(\text{syst.})\,nb \\
\text{LO collinear factorization:} \quad & \sigma_{SPS} = 697\,nb\,, \quad \sigma_{DPS} = 125\,nb\,, \quad \sigma_{tot} = 822\,nb \\
\text{LO HEF } k_T\text{-factorization:} \quad & \sigma_{SPS} = 548\,nb\,, \quad \sigma_{DPS} = 33\,nb\,, \quad \sigma_{tot} = 581\,nb
\end{aligned} \tag{6}$$

The LO results need refinements from NLO contributions, much more than it does in the case of the ATLAS hard cuts, as we are of course not that deep into the perturbative region. For this reason, in the following we will always perform comparisons only to data normalised to the total (SPS+DPS) cross sections. We find that this is better than introducing fixed K-factors, whose phase-space dependence is never really under control. What is immediately apparent in the HEF total cross section is the dramatic damping of the DPS contribution with respect to the collinear case. This damping effect is of kinematical nature. The point is that the emission of gluon radiation, which is taken into account via the TMDs in our approach and via the real contribution in a collinear NLO calculation, alters the exact momentum balance of the final state two-jet system, so that a lot of events are not taken into account for the higher transverse momentum just above the cut. In Fig. 1 we compare the predictions

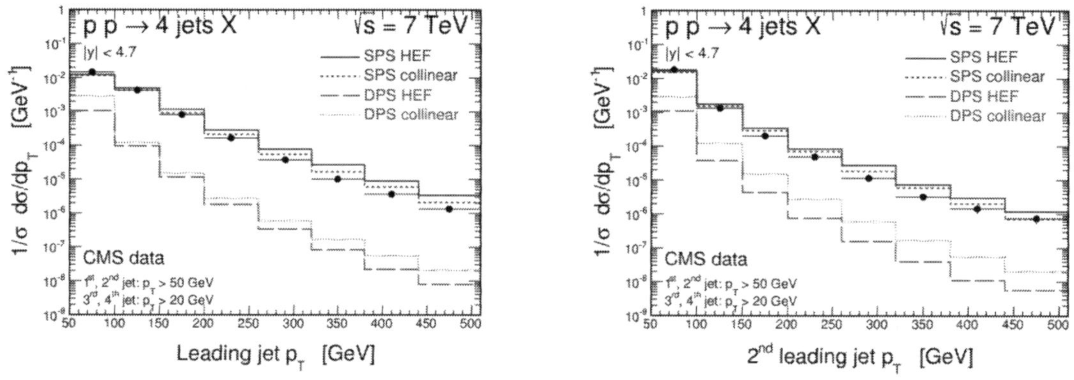

Figure 1. Comparison of the LO collinear and HEF predictions to the CMS data for the 1st and 2nd leading jets.

in HEF to the CMS data for the 1st and 2nd leading jets transverse momenta spectra. Here both the SPS and DPS contributions are normalized to the total cross section, i.e. the sum of the SPS and DPS contributions. In all cases the renormalized transverse momentum distributions agree quite well with the CMS data.

2.5 HEF predictions for a possible set of asymmetric cuts

Now we propose a set of asymmetric cuts. Specifically, we require $p_T > 35$ GeV for the leading jet, $p_T > 20$ GeV for all the other jets and we stick to $|\eta| < 4.7$, $\Delta R > 0.65$ for rapidity and jet size parameter. An experimental analysis with such cuts is not available at the moment. Of course it would be desirable to have such an analysis in the future.

The theoretical total cross sections for these cuts for four-jet production are:

$$\begin{aligned}
\text{LO collinear factorization:} \quad & \sigma_{SPS} = 1969\,nb\,, \quad \sigma_{DPS} = 514\,nb \\
\text{LO HEF } k_T\text{-factorization:} \quad & \sigma_{SPS} = 1506\,nb\,, \quad \sigma_{DPS} = 297\,nb
\end{aligned} \tag{7}$$

When comparing to (6), it is apparent that now the drop in the total cross section for DPS when moving from LO collinear to HEF approach is considerably smaller, as argued.

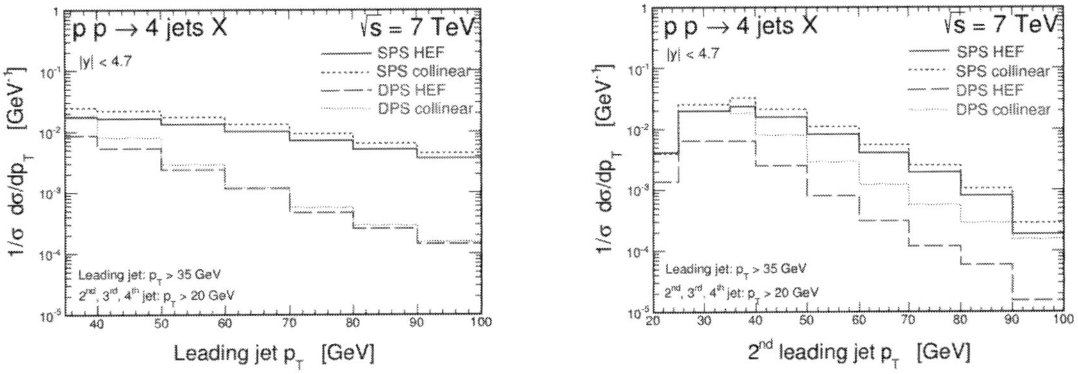

Figure 2. LO collinear and HEF predictions for the 1st and 2nd leading jets with the asymmetric cuts.

In Fig. 2 we show our predictions for the normalized transverse momentum distributions with the new set of cuts.

3 Looking for new variables to enhance DPS

We proposed in [12] that the analysis of cross sections which are differential with respect to three other variables might be interesting in view of the goal of clearly identifying DPS.

The first of such variables was found to be the maximum rapidity distance

$$\Delta Y \equiv max_{\substack{i,j\in\{1,2,3,4\} \\ i\neq j}} |\eta_i - \eta_j| \,. \tag{8}$$

A second candidate is the azimuthal correlations between the jets which are most remote in rapidity

$$\varphi_{jj} \equiv |\varphi_i - \varphi_j| \,, \quad \text{for} \quad |\eta_i - \eta_j| = \Delta Y \,. \tag{9}$$

Finally, we propose that the minimal sum of two azimuthal distances, defined as

$$\Delta\varphi_{3j}^{min} \equiv min_{\substack{i,j,k\in\{1,2,3,4\} \\ i\neq j\neq k}} \left(|\varphi_i - \varphi_j| + |\varphi_j - \varphi_k|\right) \,, \tag{10}$$

can be specially interesting. Lack of space prevents us from presenting a more thorough analysis here, but details can be found in the paper [12].

4 Conclusions

We have compared the perturbative predictions for four-jet production at the LHC in leading-order collinear and high-energy (k_T-)factorization. We find that there is no significant difference between the collinear and HEF approach for hard central cuts, but significant differences show up, especially for DPS, when the cuts on the transverse momenta are lowered. Our approach is able to describe existing CMS data on jet rapidity distributions and we have presented our predictions for differential distributions with respect to other variables as well. We observed that HEF severely tames DPS for symmetric cuts, due to gluon-emission effects encoded in the PDFs which alter the transverse-momentum balance between final state partons. We have found that the damping is sensibly reduces when cuts are not identical.

We find that, for sufficiently small cuts on the transverse momenta, DPS effects are enhanced relative to the SPS contribution when rapidities of jets are large, for large rapidity distances between the most remote jets, for small azimuthal angles between the two jets most remote in rapidity and, finally, for large values of $\Delta\varphi_{3j}^{min}$. In general, the relative effects of DPS in the k_T-factorization approach are somewhat smaller than those found previously in the LO collinear approach. A complete treatment of the subject of this proceeding is extensively given in [11, 12].

Acknowledgments

This work has been supported by Narodowe Centrum Nauki with Sonata Bis grant DEC-2013/10/E/ST2/00656 and by the "Angelo della Riccia" foundation.

References

[1] R. Britto, F. Cachazo, and B. Feng, *Nucl.Phys.* **B715** (2005) 499–522, [arXiv:hep-th/0412308].

[2] R. Britto, F. Cachazo, B. Feng, and E. Witten, *Phys.Rev.Lett.* **94** (2005) 181602, [arXiv:hep-th/0501052].

[3] A. van Hameren, *JHEP*, **07**(2014), 138, arXiv:1404.7818.

[4] A. van Hameren and M. Serino, *JHEP*, **07**(2015), 010, arXiv:1504.00315.

[5] M. Bury and A. van Hameren, *Comput. Phys. Commun.*, **196**(2015), 592, arXiv:1503.08612.

[6] Rafal Maciula and Antoni Szczurek, *Phys. Lett.*, **B749**(2015), 57, arXiv:1503.08022.

[7] M. A. Kimber, Alan D. Martin, and M. G. Ryskin, *Phys. Rev.*, **D63**(2001), 114027, hep-ph/0101348.

[8] M. A. Kimber, Alan D. Martin, and M. G. Ryskin, *Eur. Phys. J.*, **C12**(2000), 655, hep-ph/9911379.

[9] Georges Aad *et al.* (ATLAS), *JHEP*, **12**(2015), 105, arXiv:1509.07335.

[10] Serguei Chatrchyan *et al.* (CMS), *Phys. Rev.*, **D89**(2014) (9), 092010, arXiv:1312.6440.

[11] Krzysztof Kutak, Rafal Maciula, Mirko Serino, Antoni Szczurek, and Andreas van Hameren, *JHEP*, **04**(2016), 175, arXiv:1602.06814.

[12] Krzysztof Kutak, Rafal Maciula, Mirko Serino, Antoni Szczurek and Andreas van Hameren, *Phys. Rev.*, **D94**(2016) (1), 014019, arXiv:1605.08240.

EPJ Web of Conferences **129**, 00005 (2016) DOI: 10.1051/epjconf/201612900005
QCD@Work 2016

Gluon fusion contribution to HBB ($B = H, \gamma, Z$) at the LHC

Ambresh Shivaji[1,★], *Pankaj* Agrawal[2], and *Debashis* Saha[2]

[1] *INFN, sezione di Pavia, Via Agostino Bassi, 6-27100 Pavia (PV) Italy*
[2] *Institute of Physics, P.O.: Sainik School, Bhubaneswar - 751005, India*

Abstract. We have calculated one-loop amplitudes for the production of Higgs boson in association with two electroweak bosons (H, γ, Z) via gluon-gluon fusion. We present preliminary results for the total cross section at 8, 13 and 100 TeV center-of-mass energies at pp colliders. We study the interference effect and, also comment on the effect of new physics in terms of anomalous couplings of the Higgs boson in these processes.

1 Introduction

There are many standard model (SM) decay or scattering processes which begin at loop-level at the leading order itself. Such loop-induced SM processes are expected to be sensitive to new physics scales. For example, new heavy particles which cannot be produced directly can contribute in the loop in these processes leading to a deviation from the SM predictions. We are particularly interested in loop-induced gluon fusion processes, which can be important at high energy hadron colliders such as the LHC and its future upgrades due to the availability of a large gluon flux.

In the past we have studied $gg \to VVj,\ VHj\ (V = \gamma, Z, W)$ processes at the LHC [1–3]. In this talk we will focus on $gg \to HBB\ (B = H, \gamma, Z)$ processes. Some results on these processes are reported in [4–9]. Observing HHH process would provide us direct information on quartic self-Higgs coupling, while HHZ is a background to HHH in $HHb\bar{b}$ channel. On the other hand, $HVV\ (V = \gamma, Z, W)$ processes are backgrounds to $gg \to HH$ when one of the two Higgs bosons decays into a pair of vector bosons ($\gamma\gamma$, γZ, ZZ^*, WW^*).

The processes under consideration are one-loop at the leading order and proceed via quark loop diagrams. We have triangle, box and pentagon one-loop amplitudes which contribute to them. The one-loop topologies involved are shown in figure 1. In most cases, we can identify prototype amplitudes and generate all other amplitudes by permuting the external momenta and polarizations. Various symmetries can also be utilized to simplify such complex calculations. For example, due to charge conjugation, $\mathcal{M}(gg \to HH\gamma) = 0$. For the same reason in $gg \to HHZ$ case only the axial-vector part of qqZ coupling contributes, while in $gg \to HZ\gamma$ and $gg \to HZZ$ cases only the vector type of amplitude gives non-zero contribution.

2 Calculation and checks

We have calculated the quark loop traces in `FORM` [10] in 4−dimensions. Except top quark, all other quarks are considered massless. The HBB amplitude at this stage is expressed in terms of various

★e-mail: ambresh.shivaji@pv.infn.it

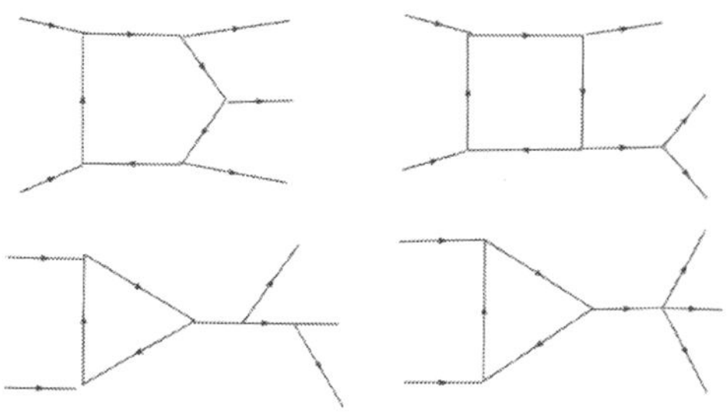

Figure 1. Pentagon, box and triangle topologies contributing to $gg \to HBB$ amplitude. In case of $H\gamma\gamma$ only pentagon contributes, while $HZ\gamma$ does not receive any triangle contribution.

tensor integrals. One of the most difficult parts of one-loop amplitude calculations is the reduction of tensor integrals into a suitable set of scalar integrals. In our processes, we have one-loop five point tensor integral of rank four as the most complicated tensor integral,

$$\mathcal{E}_{\mu\nu\rho\sigma} = \int \frac{d^n l}{(2\pi)^n} \frac{l_\mu l_\nu l_\rho l_\sigma}{D_0 D_1 D_2 D_3 D_4}. \tag{1}$$

Using 4-dimensional *Schouten Identity*, we reduce pentagon tensor and scalar integrals into lower rank box tensor and scalar integrals. For example, the pentagon scalar integral can be written as a linear combination of five box scalar integrals [11, 12]. Reduction of box and lower tensor integrals into appropriate set of scalar integrals is done numerically using the methods of Oldenborgh and Vermaseren (OV) [12, 13] in n $(= 4 - 2\epsilon)$-dimensions. Finally, all the required scalar integrals are calculated using the `OneLOop` package [14].

As we expect, the one-loop HBB amplitude is both ultraviolet (UV) and infrared (IR) finite. This is an important check on our calculation. As an ultimate check, we also check the gauge invariance of the amplitude with respect to the gauge currents. This check is done numerically by replacing the polarizations with their respective 4-momenta, $\epsilon_\mu(k) \to k_\mu$ for a given phase space point. We calculate the amplitude numerically before squaring it to get the cross section.

3 Numerical results

We now discuss some preliminary results for our processes at pp colliders. In our calculations we use following SM input parameters,

$$\begin{aligned} M_H &= 125\ \text{GeV},\ M_Z = 91.188\ \text{GeV},\ M_W = 80.419\ \text{GeV}, \\ m_t &= 173\ \text{GeV},\ G_\mu = 1.166389 \times 10^{-5}\ \text{GeV}^{-2}. \end{aligned} \tag{2}$$

We also use following basic kinematic cuts to produce results,

$$\mathrm{p_T^{H/Z}} > 1\ \text{GeV},\ \mathrm{p_T^{\gamma}} > 20\ \text{GeV},\ |y_{\mathrm{H/Z}}| < 5.0,\ |y_\gamma| < 2.5,\ \Delta\mathrm{R}_{\gamma\gamma} > 0.4. \tag{3}$$

Note that the 1 GeV cut on p_T of H and Z is applied mainly to improve the numerical stability of the code. Further, we have used `cteq6l1` pdf set [15], and have set the partonic center-of-mass energy (cme) as the common scale for renormalization and factorization, $\mu_\mathrm{F} = \mu_\mathrm{R} = \sqrt{\hat{s}}$.

Table 1. SM cross sections at various collider center-of-mass energies with scale uncertainties. All cross sections are in *ab*.

$\sqrt{s}$ (TeV)	8	13	100
$\sigma\,(HHH)$	$7.048^{+34\%}_{-24\%}$	$31.87^{+30\%}_{-22\%}$	$3093^{+17\%}_{-14\%}$
$\sigma\,(HHZ)$	$10.11^{+34\%}_{-24\%}$	$42.76^{+30\%}_{-22\%}$	$3468^{+17\%}_{-14\%}$
$\sigma\,(H\gamma\gamma)$	$1.240^{+37\%}_{-23\%}$	$4.852^{+29\%}_{-22\%}$	$265.8^{+16\%}_{-13\%}$
$\sigma\,(HZ\gamma)$	$1.401^{+32\%}_{-22\%}$	$4.931^{+28\%}_{-21\%}$	$241.3^{+15\%}_{-13\%}$
$\sigma\,(HZZ)$	$83.7000^{+36\%}_{-21\%}$	$471.636^{+36\%}_{-24\%}$	$102573^{+20\%}_{-15\%}$

Table 2. Contributions from pentagon, box and triangle amplitudes at $\sqrt{s}$ = 100 TeV. All cross sections are in *ab*.

	PEN	BX	TR	FULL
$\sigma\,(HHH)$	8110	4319	274.2	3039
$\sigma\,(HHZ)$	17214.5	116996	125552	3468.37
$\sigma\,(H\gamma\gamma)$	265.8	–	–	265.8
$\sigma\,(HZ\gamma)$	78.04	216.2	–	241.3
$\sigma\,(HZZ)$	18677.3	23684.9	31998.7	102573

3.1 SM prediction

In table 1, we report hadronic cross sections for $gg \to HBB$ processes at various collider center-of-mass energies. We also mention the percentage scale uncertainties when the scale is changed by a factor of two around its central value. Due to heavy particles in the final state and presence of many electroweak couplings, these processes have very small cross sections even at 100 TeV. It should be noted that the triangle, box and pentagon amplitudes are separately gauge invariant *with respect to the gluons* in all the processes. To understand the interference effect among these amplitudes, we have computed their individual contributions at the cross section level in table 2. It should be kept in mind that only the full contribution is meaningful and consistent with the complete SM symmetry. We note that except in HZZ case, in all other cases there is a destructive interference between amplitudes. This destructive interference is weakest in $HZ\gamma$ while strongest in HHZ. The HZZ amplitude displays a very strong constructive interference. Similar feature we observe in HWW, however, our calculation is not yet complete [16]. In figure 2 we have selected some kinematic distributions to highlight the variation of the interference effect between amplitudes with respect to a scale like p_T.

3.2 Higgs Anomalous couplings

New physics beyond the SM can induce modifications to couplings among SM particles. The couplings of the Higgs boson with heavy fermions, massive gauge bosons and the Higgs self couplings are of particular interest in this regard. To demonstrate the effect of modifying Higgs couplings in our processes, we have scaled the SM Higgs couplings by factor C_i ($i = ttH,\ ZZH,\ ZZHH,\ 3H,\ 4H$). In table 3 we report the percentage deviation in total cross section when C_i is changed by $\pm 10\%$ from

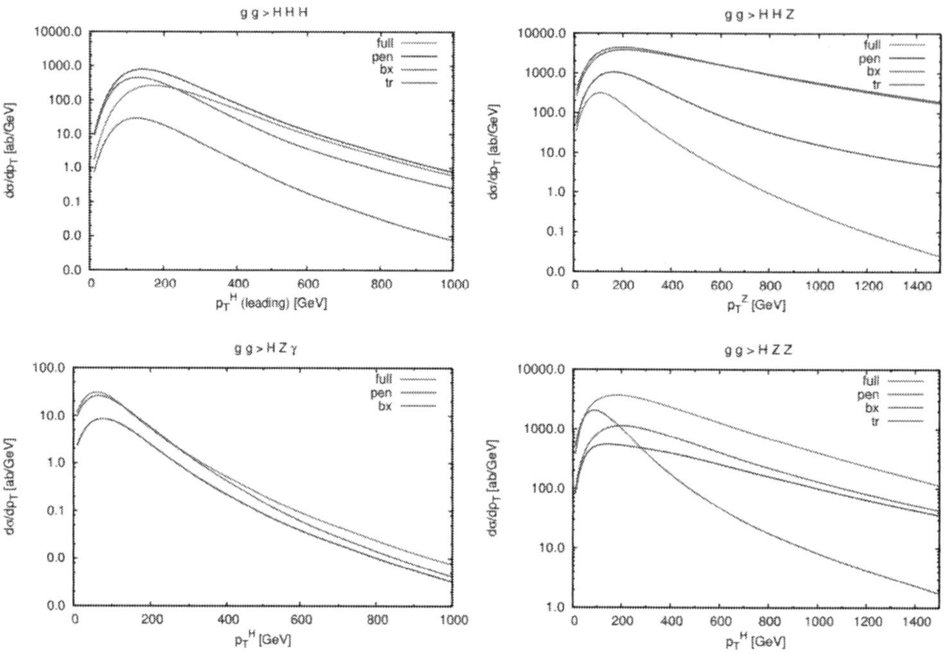

Figure 2. A comparison among various pieces of the amplitude at the level of kinematic distributions in $gg \to HBB$ processes.

its SM value ($C_i^{\rm SM} = 1$), which is consistent with current LHC data on Higgs. We modify one coupling at a time which indeed provides valuable information on its role in changing the interference pattern among various diagrams. The table can be more or less understood with the help of table 2 which has the information on triangle, box and pentagon contributions which in presence of anomalous couplings depend on C_i in a specific manner. We can see that some of our processes are quite sensitive to modifications in ttH and ZZH couplings. Note that we have ignored correlations among the couplings which may arise in a given model of new physics or in presence of higher dimensional operators introduced to capture new physics [16].

4 Conclusion

We have computed loop-induced gluon fusion contributions to HBB ($B = H, \gamma, Z$) processes at pp colliders. One would like to observe HHH to probe the quartic self-coupling of the Higgs boson. Others are also backgrounds to double Higgs production which carries the direct information on the trilinear self-coupling of the Higgs boson. We find that due to small rates, their observation would require a very large luminosity. Some of these processes display a strong interference between different classes of diagrams. We have seen that any modification to the SM Higgs couplings due to new physics effects can spoil the interference and lead to a very different prediction. The effect of anomalous couplings can be studied more systematically using higher dimension operators which would inherently take care of possible correlations among various Higgs couplings.

Table 3. The effect of changing various couplings by 10% of their SM values at $\sqrt{s}$ = 100 TeV. The first and second entries in each parentheses correspond to C_i = 0.9 and 1.1 respectively.

ANML (0.9,1.1)	$C_{\rm ttH}$	$C_{\rm 3H}$	$C_{\rm 4H}$	$C_{\rm ZZH}$	$C_{\rm ZZHH}$
HHH	(−52%, +92%)	(+8%, −5%)	(+1%, −1%)	–	–
HHZ	(+22%, +81%)	(−1%, +1%)	–	(+127%, +140%)	(−6.3%, +18%)
$H\gamma\gamma$	(−1%, +1%)	–	–	–	–
$HZ\gamma$	(−4%, +4%)	–	–	(−15%, +15%)	–
HZZ	(−21%, +25%)	(+0.5%, −0.4%)	–	(−26%, +34%)	(+4%, −3%)

References

[1] P. Agrawal *et al.*, Phys. Rev. D **86**, 073013 (2012).
[2] P. Agrawal *et al.*, JHEP **1301**, 071 (2013).
[3] P. Agrawal *et al.*, Phys. Lett. B **741**, 111 (2015).
[4] T. Plehn *et al.*, Phys. Rev. D **72**, 053008 (2005).
[5] T. Binoth *et al.*, Phys. Rev. D **74**, 113008 (2006).
[6] S. Mao *et al.*, Phys. Rev. D **79**, 054016 (2009).
[7] F. Maltoni *et al.*, JHEP **1411**, 079 (2014).
[8] V. Hirschi *et al.*, JHEP **1510**, 146 (2015).
[9] A. Papaefstathiou *et al.*, JHEP **1602**, 006 (2016).
[10] J. A. M. Vermaseren, math-ph/0010025.
[11] W. L. van Neerven *et al.*, Phys. Lett. B **137**, 241 (1984).
[12] A. K. Shivaji, arXiv:1305.4926 [hep-ph].
[13] G. J. van Oldenborgh *et al.*, Z. Phys. C **46**, 425 (1990).
[14] A. van Hameren, Comput. Phys. Commun. **182**, 2427 (2011).
[15] P. M. Nadolsky *et al.* Phys. Rev. D **78**, 013004 (2008).
[16] P. Agrawal *et al.*, *in preparation*.

Phenomenology of minimal Z' models: from the LHC to the GUT scale

Elena Accomando[1,⋆], *Claudio* Corianò[2,⋆⋆], *Luigi* Delle Rose[1,3,⋆⋆⋆], *Juri* Fiaschi[1,⋆⋆⋆⋆], *Carlo* Marzo[2,†], and *Stefano* Moretti[1,‡]

[1] *School of Physics and Astronomy, University of Southampton, Highfield, Southampton SO17 1BJ, UK*
[2] *Dipartimento di Matematica e Fisica Ennio De Giorgi, Università del Salento and INFN-Lecce, Via Arnesano, 73100 Lecce, IT*
[3] *Dept. of Particle Physics, Rutherford Appleton Laboratory, Chilton, Didcot, OX11 0QX, UK*

Abstract. We consider a class of minimal abelian extensions of the Standard Model with an extra neutral gauge boson Z' at the TeV scale. In these scenarios an extended scalar sector and heavy right-handed neutrinos are naturally envisaged. We present some of their striking signatures at the Large Hadron Collider, the most interesting arising from a Z' decaying to heavy neutrino pairs as well as a heavy scalar decaying to two Standard Model Higgses. Using renormalisation group methods, we characterise the high energy behaviours of these extensions and exploit the constraints imposed by the embedding into a wider GUT scenario.

1 Introduction

In this work we address a class of minimal models which are the most economical Abelian and renormalisable extensions of the Standard Model (SM) with only few additional free parameters. If the extra $U(1)'$ gauge symmetry is broken at energies accessible at the Large Hadron Collider (LHC), the corresponding Z' gauge boson could provide a variety of new signatures. For instance, in addition to the di-lepton channel, the augmented flavour sector allows for the possibility of a Z' decaying into long-lived heavy neutrinos with very clear multi-leptonic final states. The model under study is determined by the (SM) gauge group augmented by a single Abelian gauge factor $U(1)'$ which can always be described by a linear combination of the hypercharge and of the $B-L$ quantum number, with coefficients given by the new Abelian gauge coupling $\tilde{g}$ and g'_1. We choose the $U(1)_{B-L}$ as a reference gauge group and we explore an entire class of minimal Abelian models through the ratio of the extra Abelian gauge couplings. We investigate an Abelian extension of the SM with the only minimal matter content, namely a Right-Handed (RH) neutrino, one for each generation, with $B-L=-1$ and singlet under the SM group. This is necessary to ensure the cancellation of the gauge and gravitational

⋆e-mail: e.accomando@soton.ac.uk
⋆⋆e-mail: claudio.coriano@le.infn.it
⋆⋆⋆Speaker. e-mail: l.delle-rose@soton.ac.uk
⋆⋆⋆⋆e-mail: juri.fiaschi@soton.ac.uk
†e-mail: carlo.marzo@le.infn.it
‡e-mail: s.moretti@soton.ac.uk

anomalies. In the scalar sector we introduce a complex scalar field χ, with $B-L=2$ and singlet under the SM group, to achieve the spontaneous breaking of the extra Abelian symmetry. Its vacuum expectation value x, which we choose in the TeV range, provides the mass to the Z' boson and to the RH neutrinos. The latter acquire a Majorana mass through the Yukawa interactions, dynamically realising the type-I seesaw mechanism.
The interactions of the charged fields with the neutral gauge bosons are described by the covariant derivative $\mathcal{D}_\mu = \partial_\mu + i g_1 Y B_\mu + i(\tilde{g} Y + g_1' Y_{B-L}) B'_\mu + \ldots$, where B_μ and B'_μ are the gauge fields of the $U(1)_Y$ and $U(1)_{B-L}$ gauge groups, while g_1, Y and g_1', Y_{B-L} are the corresponding couplings and charges. The parameter $\tilde{g}$ represents the mixing between the two Abelian groups. The scalar potential is given by $V(H,\chi) = m_1^2 H^\dagger H + m_2^2 \chi^\dagger \chi + \lambda_1 (H^\dagger H)^2 + \lambda_2 (\chi^\dagger \chi)^2 + \lambda_3 (H^\dagger H)(\chi^\dagger \chi)$, where H is the SM Higgs doublet. After spontaneous symmetry breaking the masses of the physical scalars are denoted by $m_{H_{1,2}}$, where m_{H_1} is identified with the SM-like Higgs boson, while the mixing angle between the two degrees of freedom is denoted by α. The Yukawa Lagrangian is $\mathcal{L}_Y = \mathcal{L}_Y^{SM} - Y_\nu^{ij} \overline{L^i} \tilde{H} \nu_R^j - Y_N^{ij} \overline{(\nu_R^i)^c} \nu_R^j \chi + h.c.$ where $\mathcal{L}_Y^{SM}$ is the SM contribution. The light physical neutrinos are given by a combination of the left-handed SM neutrinos and a highly suppressed sterile RH component, while the heavier ones, with mass $m_{\nu_h} \simeq \sqrt{2}\, x Y_N$, are mostly RH.

2 Constraints from EWPTs and LHC searches

Electroweak precision tests (EWPTs) from LEP2 data provide constraints on the $(\tilde{g}, g_1', M_{Z'})$ parameter space [1]. More stringent bounds can be extracted from the recent data of the Run-I of LHC at 8 TeV and $\mathcal{L} = 20$ fb^{-1} using a signal-to-background analysis for the Drell-Yan (DY) channel (both electrons and muons are considered). We have included the Next-to-Next-to-Leading-Order QCD corrections to the DY computations through a k-factor and we have explicitly verified that the impact of the heavy neutrino mass and the parameters of the enlarged scalar potential have a marginal effect on the exclusion bounds. We show in fig. 1(a) the 95% Confidence Level (CL) exclusion limits from EWPTs and di-lepton analysis for $M_{Z'} = 2.5$ TeV. As mentioned above, the LHC analysis significantly improves the previous studies while comparable results are obtained only for high values of the Z' mass. A similar study can be performed in the scalar sector. The present exclusion limits (combining LEP, Tevatron and LHC analyses) and the compatibility of the Z' extension with the signal measurements of the discovered 125.09 GeV Higgs boson are taken into account, respectively, using `HiggsBounds` [2] and `HiggsSignals` [3]. The `HiggsBounds` results are shown in the (m_{H_2}, α) parameter space in fig. 1(b). The most sensitive channels, covering almost all the considered mass interval, are $H \to W^+W^-$ and $H \to ZZ$ [4] (blue region). The requirement of the Higgs signal agreement at 2σ is found to provide a weaker bound with respect to the exclusion limits and therefore is not shown here.

3 The high energy behaviour

The tools of the Renormalisation Group (RG) theory can be exploited to establish a direct connection between accessible EW scale spectra and a potential underlying GUT structure. Therefore these methods can delineate the viable parameter space from both a phenomenological perspective and its theoretical consistency and eventually guide the experimental investigations towards key analyses enabling one to make an assessment of the high energy structure of the model. In this respect we explore the parameter space available at the EW scale under the requirement of perturbativity of the couplings and stability of the vacuum up to the GUT scale and beyond through the RG running. We define $Q_{\max}$ as the maximum scale up to which the model possesses a stable scalar potential and

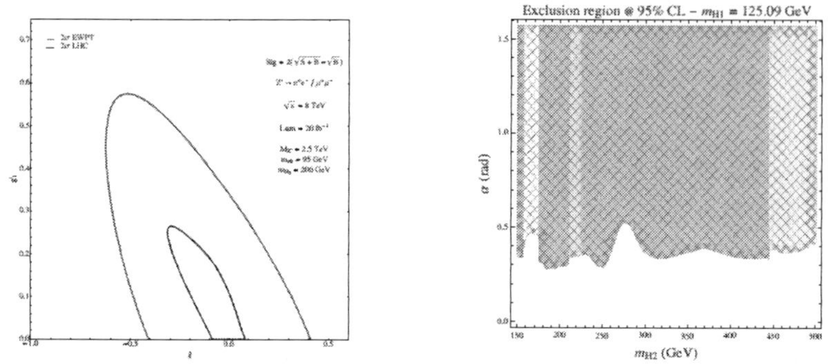

Figure 1. (a) EWPTs and LHC bounds for $M_{Z'} = 2.5$ TeV. (b) Excluded region in the (m_{H2}, α) parameter space for $m_{\nu_h} = 95$ GeV and $M_{Z'} = 2$ TeV from `HiggsBounds`. The most sensitive exclusion channels are the four leptonic decay of two Z bosons [5] (red region), the full leptonic decay of two $W^{\pm}$ bosons [6] (yellow region), the heavy Higgs decays into two Zs or $W^{\pm}$ s[4] (blue region) and a combined search in five decay modes: $\gamma\gamma$, ZZ, W^+W^-, $\tau\tau$ and bb [7] (green region).

perturbative couplings. The corresponding stability and perturbativity regions used in our analysis are depicted in fig. 3(b). For an easy comparison with the SM case we have introduced the region (denoted with the cyan colour) in which the stability and/or the perturbativity of these Abelian extensions is lost at a scale $Q_{\rm max}$ lower than the instability scale of the SM. The parameter space lying in this region compulsorily requires an ultraviolet completion, such a GUT scenario, with effects already visible at the 10^8 GeV scale or even below. On the other hand, the green and yellow regions (where stability and perturbativity are maintained up to the GUT scale and above) define the parameter space in which a more stable configuration is realised with respect to the SM, thus identifying them as viable extensions of the low energy theory.

We show in fig. 2 the stability and perturbatibity regions in the parameter space defined by the two Abelian gauge couplings $\tilde{g}$, g'_1 and for three different values of the mixing angle α. The results have been obtained for $M_{Z'} = 3$ TeV and $m_{\nu_h} = 95$ GeV. We refer to [8–10] for the analysis of the impact of the Z' and heavy neutrino masses. The dashed lines represent three particular $U(1)'$ charge assignments, namely the pure $U(1)_{B-L}$, the $U(1)_\chi$ and the $U(1)_R$ extensions, while the sequential SM coincides with the g'_1 axis. The three dots represent the corresponding benchmark models usually addressed in the literature. From the three plots in fig. 2 one can infer the impact of the scalar mixing angle α in improving the stability of the vacuum. This is, indeed, a common behaviour of scalar degrees of freedom. For $\alpha = 0$ the extra scalar χ is completely decoupled from the SM Higgs and the RG running of these Z' extensions is similar to the SM if the additional $U(1)'$ gauge couplings are sufficiently small. For $\alpha > 0$ the two scalars interact with each other, thus preventing the instability of the vacuum up to GUT scale or above.

To highlight even more the role of the extended scalar sector we present in fig. 2(a) the same RG study in the (m_{H_2}, α) plane. Notice that the cyan region, in which high energy behaviour is worse than the SM case, is excluded for $m_{H_2} \lesssim 500$ GeV. Moreover, we remark that the impact of the one-loop matchings on the initial conditions and of the two-loop corrections on the β functions is critical in the understanding of the high energy behaviour of the theory due to the sizeable corrections they provide on the scale at which the vacuum decay occurs [9].

DOI: 10.1051/epjconf/201612900006

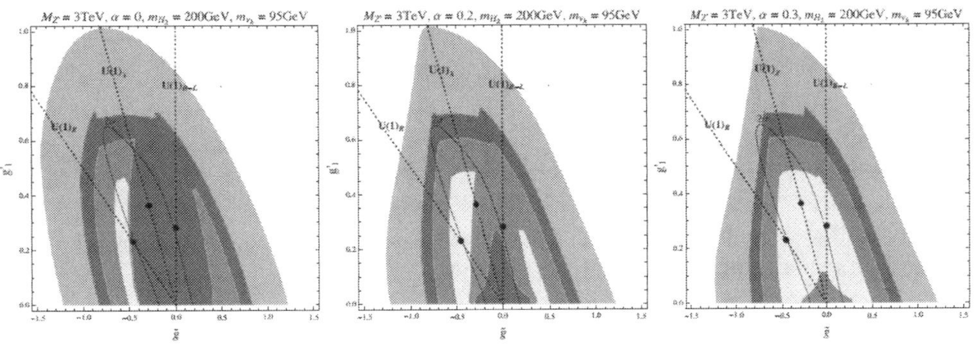

Figure 2. Stability and perturbativity regions in the $(\tilde{g}, g'_1)$ parameter space for three different values of the scalar mixing angle α. The colour legend is depicted in fig. 3(b)

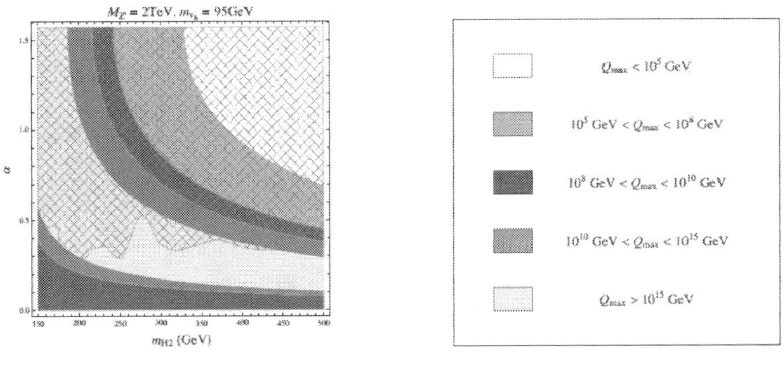

Figure 3. (a) Stability and perturbativity regions in the (m_{H_2}, α) plane. The hatched area is excluded by `HiggsBounds`. (b) Legend of the different regions defined by the maximum scale $Q_{\max}$ up to which the model is stable and perturbative.

4 The LHC phenomenology

The decay mode hierarchy of the Z' is strongly controlled by the gauge mixing $\tilde{g}$. For $\tilde{g} = 0$ the pure $B - L$ configuration is recovered. In this case the decay channel into charged leptons is the favourite one providing nearly the 40% of the Branching Ratio (BR). For $\tilde{g} \gg g'_1$ we obtain the sequential Z' in which the new gauge boson prefers the light quark decay mode with the corresponding BR reaching 60%. The BR into two heavy neutrinos shows an interesting variability with g'_1. Indeed, the partial width is found to be independent of $\tilde{g}$ and it is solely controlled by the Abelian coupling g'_1. Therefore, the corresponding BR ranges from zero, in the sequential limit, to $\sim$ 30%, its highest contribution, in the pure $B - L$ case. The production of heavy neutrinos from Z' decay, which we recall are required to cancel the anomalies associated to the new gauge group, is a hallmark of $U(1)'$ scenarios. The successive decays of the heavy neutrino may result in striking multi-lepton signature (see, for instance, [11] for the 2-lepton, [12] for the 3-lepton and [13, 14] for the 4-lepton channel). In this respect we show in fig. 4(a) the contour plots of the $\sigma \times$ BR for the production of heavy neutrinos from a decaying Z' in the plane defined by the new Abelian gauge couplings. The cross section is computed for a Centre-of-Mass (CM) energy of 13 TeV for different values of $M_{Z'}$. Notice that the Z' on-shell production is characterised by a cross section up to $\sigma = 5$ fb for $M_{Z'} = 2$ TeV and up to

$\sigma = 10$ fb for $M_{Z'} = 3$ TeV in the pure $B-L$ case, where $\tilde{g} = 0$. Changing $\tilde{g}$ modifies the Z' coupling to the light quarks and may also allow more sizeable cross sections. For instance, we approach $\sigma = 100$ fb for $M_{Z'} = 3$ TeV and $\tilde{g} = -0.6$.

The possibility to explore different Z' model scenarios, provided by a non vanishing gauge mixing, opens new decay channels of the Z' into SM bosons, which are absent in the pure $B-L$ case, namely, $Z' \to W^+W^-$, ZH_1 and ZH_2. Among the three, the decays into W^+W^- and ZH_1 represent the main channels, with BR $\sim 2\%$, regardless of the value of the scalar mixing angle α (at least in its allowed interval) with kinematics accounting for the main differences. On the other hand, the BR of the ZH_2 channel does not exceed the 0.1% value.

The existence of an augmented scalar sector of a $U(1)'$ origin offers the chance to explore completely novel signatures with respect to many beyond SM extensions. Among the others, the heavy Higgs decay into heavy neutrinos, $H_2 \to \nu_h\nu_h$, represents a hallmark of $U(1)'$ scenarios. Moreover, a non-zero scalar mixing angle α provides, if kinematically allowed $m_{H_2} > 2m_{H_1}$, the decay of the heavier scalar into two light ones, $H_2 \to H_1H_1$, which is a unique opportunity to investigate the mechanism of spontaneous symmetry breaking. As already stated above, the α parameter plays a key role in the phenomenology of scalars, scaling all H_1 (H_2) couplings with SM-like particles by $\cos(\alpha)$ ($\sin(\alpha)$) and by $\sin(\alpha)$ ($\cos(\alpha)$) when involving particles peculiar of the $U(1)'$ extension, namely the Z' and heavy neutrinos. When $m_{H_1} > 2m_{\nu_h}$, a new interesting channel become accessible to the SM-like Higgs boson, $H_1 \to \nu_h\nu_h$, otherwise it behaves exactly as the SM Higgs, with the same BRs and a total width rescaled by a factor of $\cos^2\alpha$. The BR into two heavy neutrinos ranges from 0.1% to 1% becoming comparable to, or even exceeding, the $\gamma\gamma$ mode of the SM Higgs. We show in fig. 4(b) the $\sigma \times$ BR as a function of the mixing angle for the process $pp \to H_1 \to \nu_h\nu_h$ at the LHC with 13 TeV, for $m_{\nu_h} = 50$ GeV and for three different benchmark points defined by $M_{Z'}, \tilde{g}$ and g'_1. We have only considered the gluon fusion channel which can provide $\sigma \times$ BR ~ 100 fb. The H_1 production cross section scales with a factor of $\cos^2\alpha$ with respect to the SM case, reproduced by $\alpha = 0$. In such case, $\sigma(gg \to H_1) = 44.08$ pb [15]. The behaviour of the three benchmark points is related to the structure of the $H_1\nu_h\nu_h$ vertex which is proportional to $\sin\alpha\,(m_{\nu_h}/x) \sim \sin\alpha\, g'_1(m_{\nu_h}/M_{Z'})$. For a given m_{ν_h}, the interaction of the SM-like Higgs with the heavy neutrinos gets stronger by growing the ratio $g'_1/M_{Z'}$. Taking into account the LHC limits on the Abelian gauge couplings, which are less constraining for higher Z' masses, we find a bigger ratio for $M_{Z'} = 3$ TeV, in which case g'_1 is allowed to vary up to 0.6 for $\tilde{g} = -0.7$.

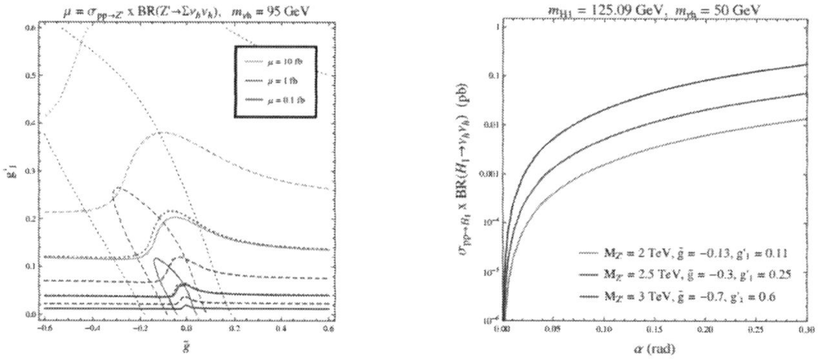

Figure 4. (a) Cross section times BR for the process $pp \to Z' \to \nu_h\nu_h$ at the LHC with $\sqrt{s} = 13$ TeV in the $(\tilde{g}, g'_1)$ plane. Solid, dashed and dotted lines refer, respectively, to $M_{Z'} = 2, 2.5, 3$ TeV. (b) Cross section times BR for $pp \to H_1 \to \nu_h\nu_h$ at the LHC with $\sqrt{s} = 13$ TeV.

Acknowledgments

E.A., L.D.R., J.F. and S.M. are supported in part through the NExT Institute. C.C. thanks The Leverhulme Trust for a Visiting Professorship to the University of Southampton, where part of his work was carried out. The work of L.D.R. has also been supported by the "Angelo Della Riccia" foundation and a COFUND/STFC Rutherford International Fellowship.

References

[1] E. Salvioni, G. Villadoro, and F. Zwirner, *Minimal Z-prime models: Present bounds and early LHC reach*, *JHEP* **11** (2009) 068, arXiv:0909.1320.

[2] P. Bechtle, O. Brein, S. Heinemeyer, G. Weiglein, and K. E. Williams, *HiggsBounds: Confronting Arbitrary Higgs Sectors with Exclusion Bounds from LEP and the Tevatron*, *Comput. Phys. Commun.* **181** (2010) 138–167, arXiv:0811.4169.

[3] P. Bechtle, S. Heinemeyer, O. Stal, T. Stefaniak, and G. Weiglein, *HiggsSignals: Confronting arbitrary Higgs sectors with measurements at the Tevatron and the LHC*, *Eur. Phys. J.* **C74** (2014), no. 2 2711, arXiv:1305.1933.

[4] **CMS** Collaboration, V. Khachatryan *et. al.*, *Search for a Higgs Boson in the Mass Range from 145 to 1000 GeV Decaying to a Pair of W or Z Bosons*, *JHEP* **10** (2015) 144, arXiv:1504.00936.

[5] **CMS** Collaboration, *Properties of the Higgs-like boson in the decay H to ZZ to 4l in pp collisions at sqrt s =7 and 8 TeV*, CMS-PAS-HIG-13-002.

[6] **CMS** Collaboration, *Update on the search for the standard model Higgs boson in pp collisions at the LHC decaying to W + W in the fully leptonic final state*, CMS-PAS-HIG-13-003.

[7] **CMS** Collaboration, *Combination of standard model Higgs boson searches and measurements of the properties of the new boson with a mass near 125 GeV*, CMS-PAS-HIG-12-045.

[8] C. Coriano, L. Delle Rose, and C. Marzo, *Vacuum Stability in U(1)-Prime Extensions of the Standard Model with TeV Scale Right Handed Neutrinos*, *Phys. Lett.* **B738** (2014) 13–19, arXiv:1407.8539.

[9] C. Coriano, L. Delle Rose, and C. Marzo, *Constraints on abelian extensions of the Standard Model from two-loop vacuum stability and* $U(1)_{B-L}$, *JHEP* **02** (2016) 135, arXiv:1510.02379.

[10] E. Accomando, C. Coriano, L. Delle Rose, J. Fiaschi, C. Marzo and S. Moretti, Z', *Higgses and heavy neutrinos in* $U(1)'$ *models: from the LHC to the GUT scale*, *JHEP* **1607** (2016) 086, arXiv:1605.02910.

[11] W. Emam and S. Khalil, *Higgs and Z-prime phenomenology in B-L extension of the standard model at LHC*, *Eur. Phys. J.* **C52** (2007) 625–633, arXiv:0704.1395.

[12] L. Basso, A. Belyaev, S. Moretti, and C. H. Shepherd-Themistocleous, *Phenomenology of the minimal B-L extension of the Standard model: Z' and neutrinos*, *Phys. Rev.* **D80** (2009) 055030, arXiv:0812.4313.

[13] K. Huitu, S. Khalil, H. Okada, and S. K. Rai, *Signatures for right-handed neutrinos at the Large Hadron Collider*, *Phys. Rev. Lett.* **101** (2008) 181802, arXiv:0803.2799.

[14] S. Khalil and S. Moretti, *The* $B-L$ *Supersymmetric Standard Model with Inverse Seesaw at the Large Hadron Collider*, arXiv:1503.08162.

[15] LHC Higgs Cross Section Working Group, *https://twiki.cern.ch/twiki/bin/view/LHCPhysics/CERNYellowReportPageAt13TeV*.

EPJ Web of Conferences **129**, 00007 (2016) DOI: 10.1051/epjconf/201612900007
QCD@Work 2016

Search for Z', vacuum (in)stability and hints of high-energy structures

Elena Accomando[1,⋆], *Claudio* Corianò[2,⋆⋆], *Luigi* Delle Rose[1,3,⋆⋆⋆], *Juri* Fiaschi[1,⋆⋆⋆⋆], *Carlo* Marzo[2,†], and *Stefano* Moretti[1,‡]

[1] *School of Physics and Astronomy, University of Southampton, Highfield, Southampton SO17 1BJ, UK*
[2] *Dipartimento di Matematica e Fisica Ennio De Giorgi, Università del Salento and INFN-Lecce, Via Arnesano, 73100 Lecce, IT*
[3] *Dept. of Particle Physics, Rutherford Appleton Laboratory, Chilton, Didcot, OX11 0QX, UK*

Abstract. We study the high-energy behaviour of a class of anomaly-free abelian extensions of the Standard Model. We focus on the interplay among the phenomenological characterisation of the model and the use of precise renormalisation group methods. Using as boundary conditions regions of the parameter space at the verge of current LHC probe, interesting unification patterns emerge linked to thresholds belonging to a $SO(10)$ grand unification theory (GUT). We stress how the evolution of the mixing between the two abelian factors may provide a valuable tool to address the candidate high-energy embedding. The emerging unification scenarios are then challenged to be perturbative and to allow for a stable vacuum.

1 Introduction

An extra U(1)' gauge symmetry is a common presence in many attempts to go beyond the Standard Model (SM). It represents, from a low-energy perspective, the simplest extension that can be attached to the SM gauge group, at the same time, from the opposite high-energy point of view, an extra abelian factor is almost an unavoidable leftover from the breaking of many GUT scenarios [1].
If we adopt a (grand) unification paradigm, it is therefore feasible that a regime ruled by the gauge structure $SU(3)\otimes SU(2)\otimes U(1)_Y\otimes U(1)'$ could populate the sequence of effective descriptions scaling from the GUT energy, before breaking into the SM one. The last step may be triggered by the non-trivial vacuum expectation of a scalar field χ, that is, consequently, required to be SM-singlet. If such U(1)' breaking is realised at the TeV scale, then there are realistic prospects of an interesting interplay with the current LHC probe, the precise traits of such phenomenological characterisation being dictated by the extended matter content ([2–6]). Beyond the scalar sector, where a SM-singlet accounts for the extra U(1)' breaking, and the neutral vector Z', to accomplish gauge invariance, one

⋆e-mail: e.accomando@soton.ac.uk
⋆⋆e-mail: claudio.coriano@le.infn.it
⋆⋆⋆e-mail: l.delle-rose@soton.ac.uk
⋆⋆⋆⋆e-mail: juri.fiaschi@soton.ac.uk
†Speaker. e-mail: carlo.marzo@le.infn.it
‡e-mail: s.moretti@soton.ac.uk

extra fermion per generation is needed to cancel gauge and gravitational anomalies in a minimal way. This scenario has been the subject of a recent up-to-date investigation [7] where we exploited the bounds and the discovery potential of current and forthcoming collider searches. The more promising regions of the allowed parameter space have supplied the boundary conditions for a Next-to-Leading-Order (NLO) vacuum stability analysis, that we performed extrapolating the model to higher energies with two-loop β functions. As a result, the explored regions have been labeled with the maximal energy scale up to which they would provide a coherent (stable and perturbative) extrapolation of the model.
The role of the Renormalisation Group (RG) extrapolation does not exhaust its insight power with the stability analysis. As we will illustrate, for the particular case of our minimal $SM \otimes U(1)'$ regime, the RG may draw clear indications also about the high-energy regime that is expected to take place. In a combined effort, all the phenomenological and formal aspects of this analysis will contribute to unveil a consistent link between the low-energy model characterisation, and a stable, perturbative, ultraviolet (UV) completion.

2 Structure of the model and constraints from current collider probe

The class of models encompassed in [7] adds, to the SM field content, a massive neutral gauge boson plus a scalar and three extra fermions, all transforming trivially under the SM gauge group. This extended spectrum is naturally introduced to minimally account for gauge invariance, anomaly freedom and the requirement of a massive Z'. We notice, as a valuable consequence of the previous setting, that the presence of a new abelian factor has forced the introduction of states that complete the 16-dimensional representation of SO(10) which is a further motivation to explore the possible UV fate of such model. Anomaly cancellation also rules the possible U(1)' charges, leaving to the ratio of just two parameters the definition of the allowed charge assignments. In a low-energy investigation the common choice is to highlight the Hypercharge operator Y, so that the generator of the extra U(1)' is constrained to the form $Y' = (B-L) + \left(\tilde{g}/g_1'\right) Y$, where B and L are, respectively, the Baryon and the Lepton number of the fields. The overall gauge strength g_1' and the *mixing* $\tilde{g}$ rule, therefore, the content of the $B-L$ and the Hypercharge operators in Y'.
The renormalisable interactions that arise from the extended field content can promptly realise a Type-I seesaw mechanism to account for neutrino masses. The SM Yukawa sector is in fact supplemented by the new contribution

$$\mathcal{L}' = -Y_\nu \overline{L}\, \tilde{H}\, \nu_R - Y_N \overline{(\nu_R)^c}\, \nu_R \chi \tag{1}$$

that, after the sequential breaking of the $U(1)'$ and of the Electro-Weak (EW) symmetry itself, provides the wanted massive state with the typical seesaw hierarchy. The scalar sector, instead, is extended to the form

$$V(H,\chi) = m_1^2 H^\dagger H + m_2^2 \chi^\dagger \chi + \lambda_1 (H^\dagger H)^2 + \lambda_2 (\chi^\dagger \chi)^2 + \lambda_3 (H^\dagger H)(\chi^\dagger \chi) \tag{2}$$

where H is the Higgs doublet state.
It comes therefore with no surprise that, when the vacuum expectation value of χ is at the TeV scale, the model has to deal with the exclusion limits from the more recent collider quest. In particular, the past investigations of the Z' have been strongly affected by bounds coming from EW Precision Tests (EWPTs) from LEP2. The first Run of LHC at 8 TeV and $\mathcal{L} = 20$ fb^{-1} has generated even more stringent bounds at the TeV scale. These can be extracted using a signal-to-background analysis for the Drell-Yan channel. When translated in constraints in the extended $(g_1', \tilde{g})$ plane, the range of the allowed values is visible, for the exemplar case of $M_{Z'} = 3$ TeV, from fig. 1(a). Also the extended

 DOI: 10.1051/epjconf/201612900007

scalar sector creates numerous chances to reveal and characterise the class of models under study. We have limited the related new parameter space, that we parameterised with the new scalar mass m_{H_2} and the mixing angle α, considering the bounds from the direct detection probes, and comparing the signal produced with the one measured of the discovered Higgs at 125.09 GeV. This analysis has been performed using, respectively, `HiggsBounds` [8] and `HiggsSignals` [9] tools, and has constrained the (m_{H_2}, α) space to the regions of fig. 1(b).

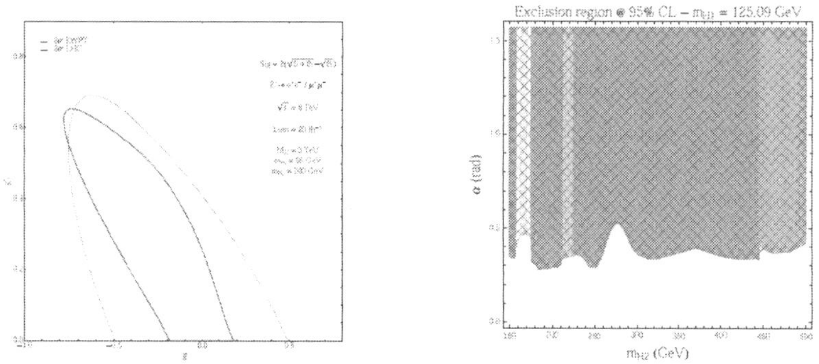

Figure 1. (a) EWPTs and LHC bounds for the case $M_{Z'} = 3$ TeV (b) Excluded region in the (m_{H2}, α) parameter space for $m_{\nu_h} = 95$ GeV from `HiggsBounds`.

3 The kinetic mixing and the breaking of SO(10) thresholds

The presence of multiple abelian factors is a peculiar trait of this class of models. The induced kinetic mixing, absorbed in the contribution of the coupling $\tilde{g}$ to the covariant derivative

$$\mathcal{D}_\mu = \partial_\mu + i g_1 \, Y B_\mu + i(\tilde{g}\, Y + g_1' \, Y_{B-L}) B'_\mu + \ldots, \tag{3}$$

may shed light, supported by a precise RG inspection, on the UV embedding that precedes the U(1)' regime [10–12].

The key for this analysis is in the matching of the low and the high-energy generators basis, used to describe the abelian sector. In our phenomenological survey, adherence with the SM regime suggested the use of the Hypercharge for one of the two $U(1)$. In turn, within the constraints of anomaly cancellation, we chose B-L for the other. The mixing would then provide the component of the Y' generator in the Hypercharge direction. From a high-energy perspective is more appropriate to work with the basis that is naturally provided by the embedding of the abelian factors in the unifying group. For example, a Left-Right (LR) symmetric regime $SU(2)_R \otimes U(1)_{B-L}$, which includes $U(1)_R \otimes U(1)_{B-L}$, would select the corresponding Y_R and Y_{B-L} set of generators. Close to the energy scale of the Left-Right symmetry breaking, the mixing between the Y_R and Y_{B-L} is zero, being protected by the overall *non-abelian* gauge symmetry of $SU(2)_R$. It is possible, with the appropriate normalisation, to match our SM-oriented parameters $(g_1, g_1', \tilde{g})$ with the ones corresponding to the (candidate) high-energy basis $(g_R, g_{B-L}, \tilde{g}_{R/B-L})$. Therefore, following the RG evolution of $\tilde{g}_{R/B-L}$ in terms of g_1, g_1' and $\tilde{g}$, we can recognise, by its zeroing at a given energy, the restoration of a Left-Right symmetry.

This analysis can be promptly extended to include thresholds of SO(10) that represent a realistic UV embedding. These involve, in addition to the Left-Right case discussed, a direct breaking of a Pati-Salam (PS) group into our model, and the flipped SU(5) case (fig 2). Choosing as boundary

DOI: 10.1051/epjconf/201612900007

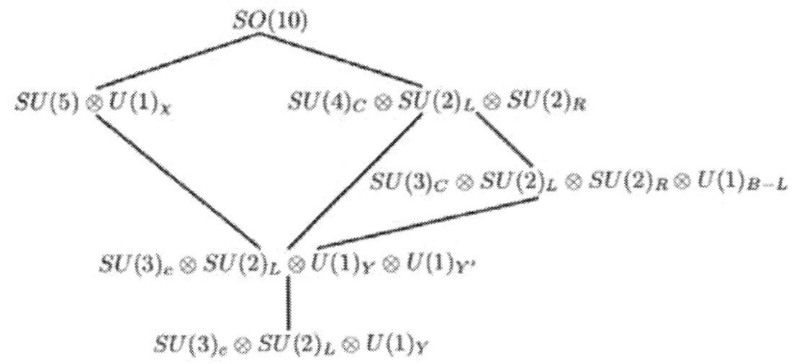

Figure 2. The $SO(10)$ breaking chart illustrating the chains investigated in this work. The cases for the Flipped $SU(5)$ and Pati-Salam required also additional unification conditions involving the non-abelian gauge sector.

conditions, for the RG extrapolation, benchmarks points inspired by our phenomenological analysis, the parameter space offers regions that, in case of discovery, would clearly reveal the presence of one of the previous embeddings (fig. 3).

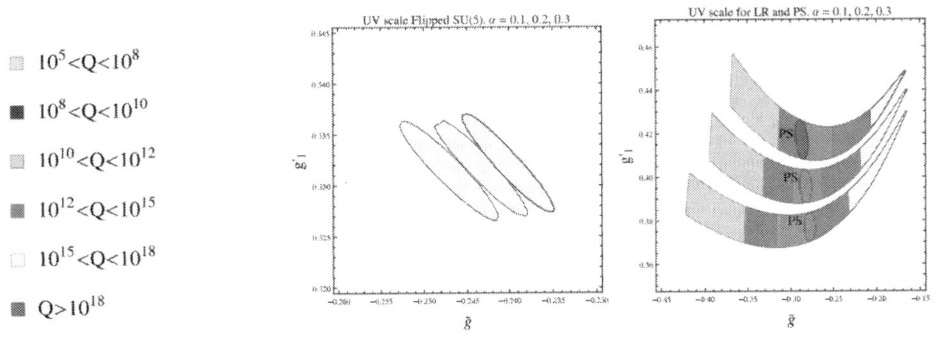

Figure 3. (a) When we consider the stability and perturbativity analysis, the colours refer to the different regions defined by the maximum energy up to which the model is stable and perturbative. The same energy/colour relation is also used for the unification study, referring to regions that fulfil the given unification requirement. (b) Regions with Flipped $SU(5)$ restoration. (c) Regions with LR and PS restoration.

4 Vacuum stability and Unification patterns

The minimal character of this $U(1)'$ extension of the SM makes particularly efficient the use of vacuum stability and perturbativity as constraining requirements to shape the viable parameter space [13–15]. The vacuum stability is addressed asking for the extended scalar potential to be bounded from below,

$$\lambda_1 > 0\,, \quad \lambda_2 > 0\,, \quad 4\lambda_1\lambda_2 - \lambda_3^2 > 0\,. \tag{4}$$

Together with the perturbativity requirement it is also challenges the viability of a given unification scenario. If we accept the minimal content of the model, then a coherent extrapolation asks for the maximum scale of stability and perturbativity to be greater than the one realising a succesful embedding. By relying on the analysis presented in [7], we may exploit this further constrains. Our final results (fig. 4) give an illustration of how the interplay of the tools presented may enrich the forthcoming collider profiling of specific regions of the parameter space. Moreover, in a possible post-discovery phase, frictions with the measured scenarios would help in outlining the degrees of freedom necessary to recover stability, when a promising unification of the gauge sector is at hand.

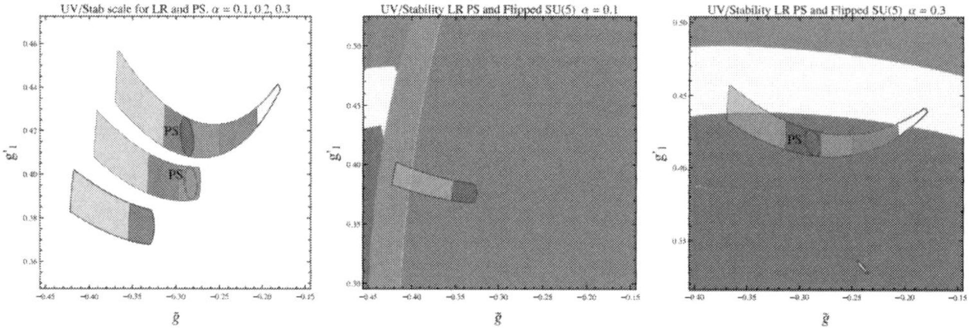

Figure 4. (a) The effect of the stability requirement to the LR and PS restoration. The similar analysis for the Flipped $SU(5)$ case would result trivially in the surviving only of the case with $\alpha = 0.3$. Explicit matching of the stability and perturbativity map with the unification regions. Case $\alpha = 0.1$ (b) and $\alpha = 0.3$ (c).

Acknowledgments

E.A., L.D.R., J.F. and S.M. are supported in part through the NExT Institute. C.C. thanks The Leverhulme Trust for a Visiting Professorship to the University of Southampton, where part of his work was carried out. The work of L.D.R. has also been supported by the "Angelo Della Riccia" foundation and a COFUND/STFC Rutherford International Fellowship.

References

[1] P. Langacker, *The Physics of Heavy Z′ Gauge Bosons*, Rev. Mod. Phys. **81** (2009) 1199, arXiv:0801.1345.

[2] W. Emam and S. Khalil, *Higgs and Z-prime phenomenology in B-L extension of the standard model at LHC, Eur. Phys. J.* **C52** (2007) 625–633, arXiv:0704.1395.

[3] L. Basso, A. Belyaev, S. Moretti and C. H. Shepherd-Themistocleous, *Phenomenology of the minimal B-L extension of the Standard model: Z′ and neutrinos*, Phys. Rev. D **80**, (2009) 055030, arXiv:0812.4313.

[4] L. Basso, S. Moretti and G. M. Pruna, *Phenomenology of the minimal $B - L$ extension of the Standard Model: the Higgs sector*, Phys. Rev. D **83**, (2011) 055014, arXiv:1011.2612.

[5] E. Accomando, D. Becciolini, A. Belyaev, S. Moretti and C. Shepherd-Themistocleous, *Z′ at the LHC: Interference and Finite Width Effects in Drell-Yan, JHEP* **1310** (2013) 153, arXiv:1304.6700.

[6] E. Accomando, A. Belyaev, J. Fiaschi, K. Mimasu, S. Moretti and C. Shepherd-Themistocleous, *Forward-backward asymmetry as a discovery tool for Z′ bosons at the LHC, JHEP* **1601** (2016) 127, arXiv:1503.02672.

[7] E. Accomando, C. Coriano, L. Delle Rose, J. Fiaschi, C. Marzo and S. Moretti, *Z′, Higgses and heavy neutrinos in $U(1)'$ models: from the LHC to the GUT scale, JHEP* **1607** (2016) 086, arXiv:1605.02910.

[8] P. Bechtle, O. Brein, S. Heinemeyer, G. Weiglein, and K. E. Williams, *HiggsBounds: Confronting Arbitrary Higgs Sectors with Exclusion Bounds from LEP and the Tevatron*, *Comput. Phys. Commun.* **181** (2010) 138–167, arXiv:0811.4169.

[9] P. Bechtle, S. Heinemeyer, O. Stal, T. Stefaniak, and G. Weiglein, *HiggsSignals: Confronting arbitrary Higgs sectors with measurements at the Tevatron and the LHC*, *Eur. Phys. J.* **C74** (2014), no. 2 2711, arXiv:1305.1933.

[10] F. del Aguila, M. Masip and M. Perez-Victoria, *Physical parameters and renormalization of* $U(1)^a \times U(1)^b$ *models*, Nucl. Phys. B **456** (1995) 531 hep-ph/9507455.

[11] E. Salvioni, G. Villadoro, and F. Zwirner, *Minimal Z-prime models: Present bounds and early LHC reach*, *JHEP* **11** (2009) 068, arXiv:0909.1320.

[12] P. Pacholek, *Embedding Z-prime models in SO(10) GUTs*, Eur. Phys. J. C **73** (2013) no.6, 2436 arXiv:1306.4517.

[13] L. Basso, S. Moretti and G. M. Pruna, *A Renormalisation Group Equation Study of the Scalar Sector of the Minimal B-L Extension of the Standard Model*, Phys. Rev. D **82**, (2010) 055018, arXiv:1004.3039.

[14] C. Coriano, L. Delle Rose, and C. Marzo, *Vacuum Stability in U(1)-Prime Extensions of the Standard Model with TeV Scale Right Handed Neutrinos*, *Phys. Lett.* **B738** (2014) 13–19, arXiv:1407.8539.

[15] C. Coriano, L. Delle Rose, and C. Marzo, *Constraints on abelian extensions of the Standard Model from two-loop vacuum stability and* $U(1)_{B-L}$, *JHEP* **02** (2016) 135, arXiv:1510.02379.

Search for heavy resonances in the $\ell\ell qq$ final state in pp collisions at $\sqrt{s} = 13$ TeV with ATLAS

Konstantinos Bachas[1,★] on behalf of the ATLAS Collaboration

[1] *INFN Lecce*

Abstract. A search for heavy resonances decaying to a pair of Z bosons is performed using proton–proton collision data produced at $\sqrt{s} = 13$ TeV and recorded by the ATLAS detector at the LHC. The data correspond to an integrated luminosity of 3.2 fb^{-1}. Diboson resonant production is expected in several Standard Model extension scenarios. The ZZ decay mode considered corresponds to one Z boson decaying to a pair of charged leptons and the other decaying to a pair of quarks. No evidence for the production of resonances decaying to a ZZ pair is observed. Upper bounds on the production cross sections at $\sqrt{s} = 13$ TeV times their decay branching ratios to ZZ pairs are derived for a heavy Higgs boson and a spin–2 Randall-Sundrum graviton scenario.

1 Introduction

This note presents a search for heavy resonances decaying to a pair of Z bosons [1]. The ZZ decay mode considered corresponds to one Z boson decaying to a pair of electrons or muons and the other decaying to a pair of quarks. The data used were collected with the ATLAS detector [2] during 2015 at the LHC in proton–proton collisions at a centre-of-mass energy of 13 TeV and correspond to an integrated luminosity of 3.2 fb^{-1}. Heavy ZZ resonances are predicted in several extensions of the Standard Model (SM) such as models with extra Higgs bosons [3, 4], the extended gauge model [5] and warped extra-dimension models [6]. The heavy resonances are modeled using a neutral CP–even spin–0 Higgs boson with SM–like couplings, as well as a spin–2 graviton G*. The gluon-gluon fusion process, gg→H or G*, is used to model the production of both ZZ resonances.

2 Event classification

The online selection of $ZZ \rightarrow \ell\ell qq$ events was based on a selection of various single-electron or single-muon triggers with quality and isolation requirements dependent on the E_T (electron) and p_T (muon) thresholds. The E_T thresholds for electrons were 24, 60 and 120 GeV, while the p_T thresholds for muons were 20 and 50 GeV.

The two leptons from one leg of the ZZ decay are required to be of same flavour, isolated and pass 'loose' identification criteria while at least one of the two is additionally required to fullfill 'medium' selection criteria. The dilepton invariant mass is required to be in the range of 83–99 GeV

★e-mail: konstantinos.bachas@cern.ch

for electrons and wider for muons, 76–106 GeV, in order to accomodate effects of degraded muon momentum resolution at high-p_T. Events with more than two 'loose' leptons are vetoed.

The reconstruction of the Z$\rightarrow qq$ is performed via two exclusive and complementary approaches due to the potentially highly boosted system of the decay. Therefore the object and event selections in the two cases differ. The two strategies are outlined in following sections.

2.1 Resolved selection

Two separate jets (jj) are reconstructed. Candidate events are further categorized based on the number of b-tagged jets; events with two b-tagged jets (tagged category) and events with fewer than two b-tagged jets (untagged category). The selection highlights of the resolved channel are listed below:

- At least 2 jets with $|\eta| < 2.5$.
- Leading jet $p_T \geq 60$ GeV.
- Subleading jet $p_T \geq 25$ GeV.
- Dijet invariant mass in the range $70 \leq m_{jj} \leq 105$ GeV.
- A p_T-balance cut defined as $\sqrt{p_T^2(\ell\ell) + p_T^2(jj)}/m_{\ell\ell jj} > 0.5$, applied only to the untagged category.

2.2 Merged selection

This case comprises of heavy resonances decaying to Z bosons producing a highly boosted system where the two fermions are emitted within a small opening angle in the laboratory frame. The hadronization of the quarks coming from the Z boson in this case produces two partially overlapping jets reconstructed as a single jet. These large-R jets are reconstructed with the anti-k_T algorithm and radius parameter $R = 1.0$, in the rapidity range $|\eta| < 2.0$.

The leading p_T large-R jet in the event, denoted with J, is assumed to originate from the hadronic Z decay and the event selection requires that the following criteria are satisfied:

- Leading jet $p_T \geq 200$ GeV.
- $p_T(\ell\ell)/m_{\ell\ell J} > 0.3$.
- J is tagged as a Z boson in a 2-step process:

  1. A p_T-dependent requirement on the jet substructure variable $D_2^{(\beta=1)}$, which is the ratio of the energy correlation functions of subjets [7].
  2. The large-R jet mass m_J is required to be in a window of ± 15 GeV around 93.4 GeV, the expected value for the Z boson mass based on MC simulations. This requirement is $\sim 68\%$ efficient for the signals studied [1].

The Merged channel has priority over the Resolved channel in the event selection flow. In Table 1 the data and MC yields for both signal and backgrounds after full selection are given for the three non overlappig signal regions (SR).

Table 1. Data and MC yields after full selection.

Process	Merged analysis	Resolved analysis	
		untagged category	tagged category
Z+jets	185 ± 9	4130 ± 50	204 ± 16
Diboson	17.3 ± 3.5	143 ± 14	23.2 ± 3.3
Top quark	1.5 ± 0.6	113 ± 10	151 ± 6
Total background	203 ± 11	4380 ± 50	378 ± 15
Data	201	4484	373
H (750 GeV)	12.9 ± 3.2	7.3 ± 2.1	2.03 ± 0.33
G^* (750 GeV)	15 ± 4	7.9 ± 2.3	2.0 ± 0.4

3 Large-R jets and substructure

In the merged channel, jet-substructure techniques are used to identify the qq pair, reconstructed as a single large-radius jet. The jets are groomed to reduce the effect of pileup and other sources of noise on the resolution. The chosen grooming technique is the trimming algorithm. Trimming takes the original constituents of the jet, reclusters them using the k_t algorithm with a smaller distance parameter and produces a collection of subjets used to recompute the large-R jet energy and momentum. The decision to reject or keep the subjet is based on its p_T. If the ratio of the jet p_T over the original jet p_T is above a certain threshold (f_{cut}) the subjet is kept. The trimming parameters optimised for this search are $R_{subjet} = 0.2$ and $f_{cut} = 5\%$.

Figure 1 illustrates the steps needed to produce a trimmed jet out of the initial large-R jet.

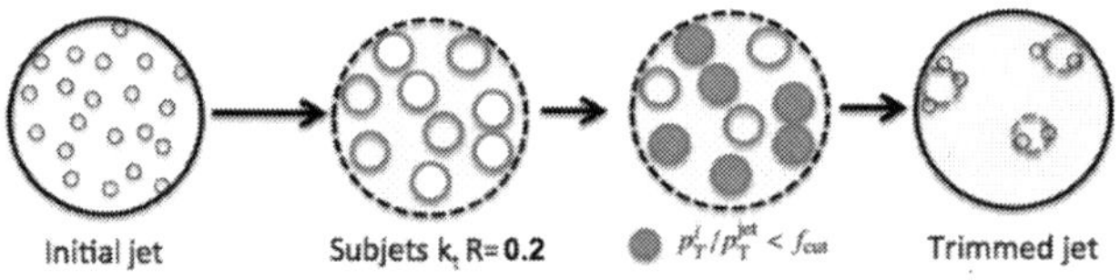

Figure 1. Schematic view of jet grooming technique using the trimming algorithm.

4 Background estimation

A summary of the sources of background and their contribution to each of the three above defined event categories is shown in Figure 2. Background events from Z+jets, diboson and top-quark processes are expected with the Z+jets being the dominant one. The multijet background is estimated to be negligible. The distribution of the $m_{\ell\ell jj}$ and $m_{\ell\ell J}$ variables, for events passing the resolved or merged selections respectively, are used as final discriminants for the signal against the background, which is expected to exhibit a smooth trend for all sources.

The estimation of the background is based both on Monte Carlo (MC) simulation and on data-driven techniques. In all cases, the shapes of kinematic variables are taken from MC simulations. Background from Z+jets and top events are estimated from four control regions (CR) enriched each in one or two of the main backgrounds.

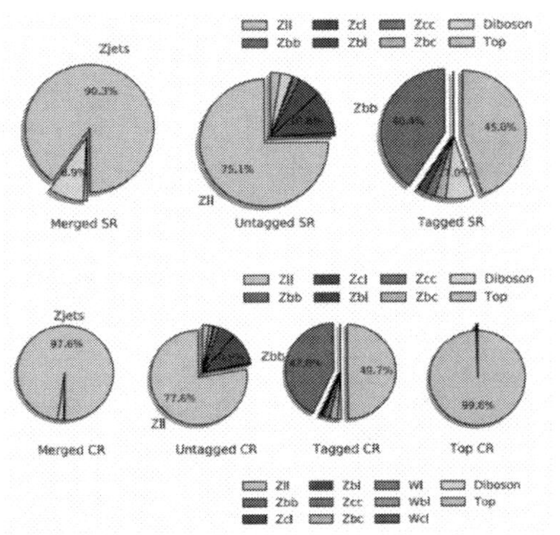

Figure 2. Breakdown of contributions to SR and CR from background sources.

Z+jets CR are defined using the side-bands of the m_J distribution for the merged analysis and the side-bands of the m_{jj} distribution for the resolved analysis. All other selections are kept the same as for the SR definition.

For the merged analysis, the leading large-R jet is required to have its mass value outside the 15 GeV Z boson mass window. For the resolved analysis, a requirement of $50 < m_{jj} < 70$ GeV or $105 < m_{jj} < 150$ GeV is applied.

Top-quark production is a significant background in the tagged category of the resolved analysis. Its contribution is normalised using a top-quark control region where events are selected requiring two different-flavour leptons with invariant mass between 76 and 106 GeV, plus two b-tagged jets with their invariant mass in the range [50,150] GeV.

Diboson production, mainly ZZ and WZ processes, are estimated completely from MC simulations. The compatibility of the data with the hypothesis of a heavy resonance is tested with a simultaneous maximum likelihood fit to the binned distributions of $m_{\ell\ell jj}$ or $m_{\ell\ell J}$ in the three signals and the four control regions devised to contrain the normalization of the main background processes. Systematic uncertainties and their correlations are taken into account in the fit through nuisance parameters.

Figure 3 shows the invariant mass distributions of the ZZ decay products in the three Z+jets and one top-quark control regions in the data. Fit results in scale factors, with respect to the MC predictions, of 1.03 ± 0.15 for Z+jets of the merged analysis, 0.92 ± 0.07 for Z + jj of the resolved analysis, 1.40 ± 0.14 for Z + bb of the resolved analysis, and 1.03 ± 0.07 for top-quark in the tagged category of the resolved analysis.

5 Results

The data are found to be consistent with the background expectations and no evidence for heavy resonance production is observed. Figure 4 shows upper limits at 95% CL on the production cross section of the heavy resonance times its branching ratio to ZZ derived for H→ZZ and $G^\star \to$ ZZ as a function of the resonance mass. The observed (expected) limits range from 5.6 (3.3) pb at 300 GeV to 0.21 (0.11) pb at 1000 GeV for the heavy Higgs boson and from 0.62 (0.52) pb at 500 GeV to 0.20 (0.11) pb at 1000 GeV for the graviton. Limits are derived also for gg→H→ZZ production of Higgs

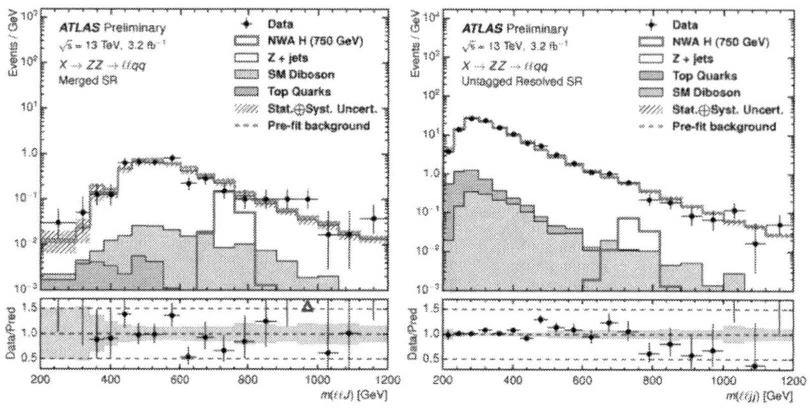

Figure 3. Post-fit of $m_{\ell\ell J}$ and $m_{\ell\ell jj}$ distributions [1].

bosons with a narrow width and three different values of the width: 5%, 10%, and 15% of the Higgs boson mass.

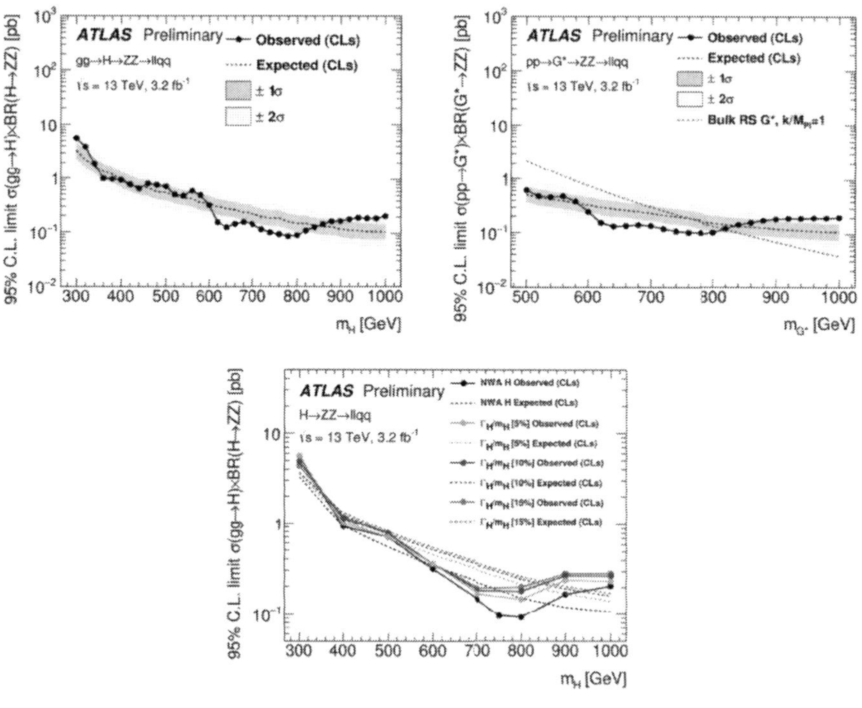

Figure 4. Observed and expected 95% CL upper limits on the production cross section of a heavy resonance at $\sqrt{s}$ = 13 TeV times its decay branching ratio to a Z boson pair for H→ZZ (left) and $G^{\star} \rightarrow$ ZZ (middle) as functions of the resonance mass, combining the merged and resolved analyses. The observed and expected upper limits for gg→H→ZZ production of Higgs bosons with a narrow width (black) and three different values of the width: 5% (green), 10% (blue), and 15% (red) of the Higgs boson mass [1].

References

[1] ATLAS Collabolation, **ATLAS-CONF-2016-016**.
[2] ATLAS Collaboration, JINST **3**, S08003 (2008).
[3] G. C. Branco et al., Phys. Rept. **516**, 1 (2012), arXiv: 1106.0034 [hep-ph].
[4] Hill, A. and van der Bij, J. J., Phys. Rev. **D36**, 3463 (1987).
[5] G. Altarelli, B. Mele, M. Ruiz-Altaba, Z. Phys. **C45**, 109 (1989).
[6] L. Randall, R. Sundrum, Phys. Rev. Lett.**83**, 3370 (1999).
[7] I. Moult A. J. Larkoski, D. Neill, JHEP **12**, 009 (2014).

Extended Higgs sectors in the context of supersymmetry at the LHC

Priyotosh Bandyopadhyay[1,*]

[1] *Dipartimento di Matematica e Fisica "Ennio De Giorgi",*
Università del Salento and INFN-Lecce, Via Arnesano, 73100 Lecce, Italy
and
Indian Institute of Technology Hyderabad, Kandi, Sangareddy-502287, Telengana, India

Abstract. We study an extension of the Higgs sector of the Standard Model (SM) with $SU(2)$ triplet and a SM gauge singlet in the context of supersymmetry. For a Z_3 symmetric superpotential where the singlet field generates the bilinear Higgs mixing term between the two Higgs doublets dynamically. Models like this, often have a very light pseudoscalar as a pseudo-Nambu Goldstone model of an extra $U(1)$ symmetry. We investigate the possibility of such light hidden Higgs boson with the data of the discovered Higgs boson around 125 GeV. Along with the neutral sector, the charged Higgs bosons in the triplet extended case are really interesting. For instance, the triplet charged Higgs bosons give rise to new decay mode to $ZW^{\pm}$ along with an $a_1 W^{\pm}$ final state, due to the existence of a light pseudoscalar. These new modes can be looked for at the LHC in the search for Higgs bosons in other representations of $SU(2)$.

1 Introduction

The discovery of the Higgs boson around 125 GeV has certainly proved the existence of at least one scalar taking part in the process of electroweak symmetry breaking (EWSB) [1]. The presence of such a scalar has been experimentally verified in its decay modes into WW^*, ZZ^* and $\gamma\gamma$ [1] at the Large Hadron Collider (LHC). In the midst of searching for SM decay modes, it is also important to investigate other non standard searches which are often motivated by various theoretical constructs.

In this article we are going to illustrate the theoretical possibilities which remain, at this time, wide open from a phenomenological perspective and which define the possible scenarios for an extended Higgs sector, in the context of supersymmetry. These constructions also address other theoretical issues, with exciting phenomenological implications. The extension of the Higgs sector with a singlet superfield can solve the μ problem in the minimal supersymmetric SM (MSSM) dynamically. Similarly one can introduce an $SU(2)$ triplet superfield, but in this case such dynamical solution is not possible, due to the constraints imposed by the ρ parameter on the vacuum expectation value (v.e.v.) of the triplet [2]. Extending the Higgs sector by a singlet/triplet also gives the possibility of a spontaneous violation of the CP symmetry.

We consider a supersymmetric extension of the Higgs sector by a $SU(2) \times U(1)_Y$ singlet and a $Y = 0$ hypercharge $SU(2)$ triplet. Additional Z_3 symmetry of the Lagrangian predicts the existence

*e-mail: priyotosh.bandyopadhyay@le.infn.it,bpriyo@iith.ac.in

 DOI: 10.1051/epjconf/201612900009

of a light pseudoscalar in some limit of the theory. Such possibility leads to additional decay modes of the Higgs boson around 125 GeV into a pair of light pseudoscalars and/or into a light pseudoscalar accompanied by a Z boson, if kinematically allowed.

The charged Higgs sector is extended due to the triplet but they are fermiophobic leading to distinguishing features. A light charged Higgs boson may now decay into a pseudoscalar and one $W^\pm$ boson. This particular decay is allowed for both doublet and triplet-like charged Higgs bosons. However, the $Y = 0$ triplet brings in an additional coupling of the charged Higgs boson to the Z and $W^\pm$ at tree-level, which breaks the custodial symmetry. This can generate a totally new decay mode for the charged Higgs boson into $Z W^\pm$, which is not present in the 2-Higgs doublet model (2HDM) or in the MSSM. Here we are going to discuss such possibilities and elaborate on how to explore them at current and future experiments.

In section 2 we are going to briefly illustrate the model. In section 3 we address the possibility of having a light pseudoscalar and its phenomenology. The charged Higgs boson phenomenology has been discussed in section 4 with conclusions in section 5.

2 Model

As explained in [3] the model contains a SU(2) triplet $\hat{T}$ of zero hypercharge ($Y = 0$) together with a SM gauge singlet $\hat{S}$, added to the superfield content of the MSSM,

$$W_{TNMSSM} = W_{MSSM} + W_{TS}. \tag{1}$$

The structure of its triplet and singlet extended superpotential can be written as

$$W_{TS} = \lambda_T \hat{H}_d \cdot \hat{T}\hat{H}_u + \lambda_S \hat{S}\, \hat{H}_d \cdot \hat{H}_u + \frac{\kappa}{3}\hat{S}^3 + \lambda_{TS}\hat{S}\,\mathrm{Tr}[\hat{T}^2]. \tag{2}$$

The triplet and doublet superfields are given by

$$\hat{T} = \begin{pmatrix} \sqrt{\frac{1}{2}}\hat{T}^0 & \hat{T}_2^+ \\ \hat{T}_1^- & -\sqrt{\frac{1}{2}}\hat{T}^0 \end{pmatrix}, \qquad \hat{H}_u = \begin{pmatrix} \hat{H}_u^+ \\ \hat{H}_u^0 \end{pmatrix}, \qquad \hat{H}_d = \begin{pmatrix} \hat{H}_d^0 \\ \hat{H}_d^- \end{pmatrix}. \tag{3}$$

Here $\hat{T}^0$ denotes a complex neutral superfield, while $\hat{T}_1^-$ and $\hat{T}_2^+$ are the charged Higgs superfields. The MSSM Higgs doublets are the only superfields which couple to the fermion multiplet via Yukawa coupling. All the coefficients involved in the Higgs sector are chosen to be real in order to preserve CP invariance. The breaking of the $SU(2)_L \times U(1)_Y$ electroweak symmetry is then obtained by giving real vevs to the neutral components of the Higgs field

$$< H_u^0 >= \frac{v_u}{\sqrt{2}}, \qquad < H_d^0 >= \frac{v_d}{\sqrt{2}}, \qquad < S >= \frac{v_S}{\sqrt{2}}, \qquad < T^0 >= \frac{v_T}{\sqrt{2}}, \tag{4}$$

which give mass to the $W^\pm$ and Z bosons

$$m_W^2 = \frac{1}{4}g_L^2(v^2 + 4v_T^2), \qquad m_Z^2 = \frac{1}{4}(g_L^2 + g_Y^2)v^2, \qquad v^2 = (v_u^2 + v_d^2), \qquad \tan\beta = \frac{v_u}{v_d} \tag{5}$$

and also induce, as mentioned above, a μ-term of the form $\mu_D = \frac{\lambda_S}{\sqrt{2}}v_S + \frac{\lambda_T}{2}v_T$. The triplet vev v_T is strongly constrained by the global fit to the measured value of the ρ parameter [2] which restricts its value to $v_T \leq 5$ GeV and in the numerical analysis we have chosen $v_T = 3$ GeV.

DOI: 10.1051/epjconf/201612900009

In the TNMSSM, we are left with four CP-even, three CP-odd and three charged Higgs bosons as shown below

$$\begin{array}{ccc} \text{CP}-\text{even} & \text{CP}-\text{odd} & \text{charged} \\ h_1, h_2, h_3, h_4 & a_1, a_2, a_3 & h_1^\pm, h_2^\pm, h_3^\pm. \end{array}$$

The neutral Higgs bosons are linear combinations of doublets, triplet and singlets, whereas the charged Higgses are combinations of doublets and of a triplet only. At tree-level the maximum value of the lightest neutral Higgs has additional contributions from the triplet and the singlet sectors respectively. The upper bound on the lightest CP-even Higgs can be extracted from the relation

$$m_{h_1}^2 \leq m_Z^2(\cos^2 2\beta + \frac{\lambda_T^2}{g_L^2 + g_Y^2}\sin^2 2\beta + \frac{2\lambda_S^2}{g_L^2 + g_Y^2}\sin^2 2\beta), \tag{6}$$

which is affected on its right-hand-side by two additional contributions from the triplet and the singlet. These can raise the allowed tree-level Higgs mass and additional contributions coming from the radiative corrections of the triplet and the singlet reduce the required the supersymmetric (SUSY) mass scale and thus the fine-tuning problem. We have calcualted the one-loop Higgs mass for the neutral Higgs bosons in [3] following the Coleman-Weinberg effective potential [4]. The existence of the triplet and singlet fields at quantum level are non-negligible which reduce the SUSY fine-tuning for all $\tan\beta$ values [3, 5, 6].

3 Light pseudoscalar and phenomenology

One of the most interesting feature can be seen in the limit where the trilinear soft parameters A_i in go to zero, the discrete Z_3 symmetry of the Lagrangian is promoted to a continuos $U(1)$ symmetry given by Eq. (7)

$$(\hat{H}_u, \hat{H}_d, \hat{T}, \hat{S}) \to e^{i\phi}(\hat{H}_u, \hat{H}_d, \hat{T}, \hat{S}). \tag{7}$$

An explicit breaking of a continuous global symmetry is expected to be accompanied by pseudo-Nambu Goldstone bosons (pNGB) as in the case of chiral symmetry breaking in QCD where the pions take the role of the corresponding pNGB. In this case, in the TNMSSM we should expect a light pseudoscalar in the spectrum, whose mass is of the same order of the A_i parameters. A similar behaviour is allowed in the NMSSM and such light pseudoscalar is known as the R-axion [7, 8]. However, such light pseudoscalar receives bounds if it decays into fermions. Previous analysis at LEP have searched for such light scalar bosons in $e^+e^- \to Zh$ and $e^+e^- \to Ah$, where h, A are CP-even and odd neutral Higgs bosons respectively [9]. Similar bounds also come from the bottomonium decay [10] for such light pseudoscalar in the mass range of 5.5 -14 GeV. Recently published data from CMS also provide strong bounds for such a light pseudoscalar when it couples to fermions [11]. Being mostly singlet-like, it is easy to evade such bounds but difficult to produce it directly at a hadron collider.

In Figure 1 we show a Higgs $\sim$ 125 GeV which at the same time satisfy the ZZ^*, WW^* bounds at 1σ level and the $\gamma\gamma$ bound at 2σ level from both CMS and ATLAS along with a light pseudoscalar with or without a light scalar. Such singlet type light pseudoscalar couples to the doublet-like Higgs bosons via λ_S (see Eq. 2). This makes it easy to produce by an intermediate Higgs. If such a light pseudoscalar ($m_{a_1} \leq 125/2$) exists, then it can be produced in $gg \to h_{125} \to a_1a_1$ as shown in Figure 2. Such pair production can give rise to final states rich in τ, b and even in muons. In [12]

DOI: 10.1051/epjconf/201612900009

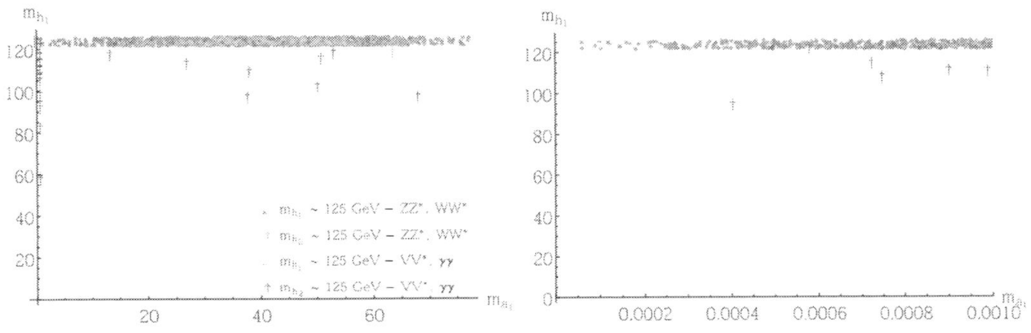

Figure 1. The lightest CP-even Higgs boson mass m_{h_1} vs the lightest pseudoscalar mass m_{a_1} at one-loop consistent with the Higgs data from CMS, ATLAS and LEP. Very light pseudoscalar masses $m_{a_1} \leq 1$ MeV are shown in panel (b)[3].

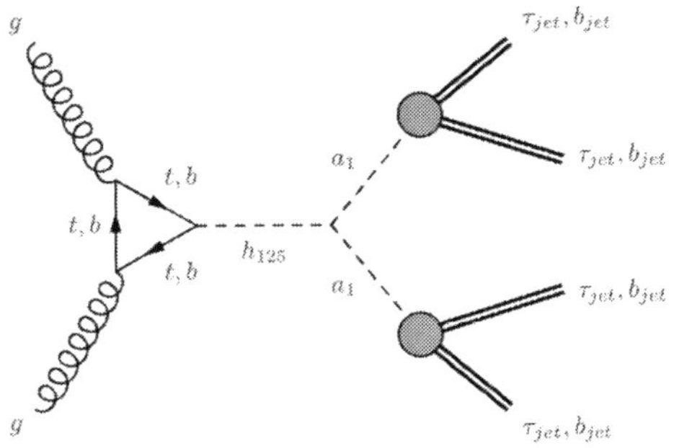

Figure 2. Pseudoscalar (triplet/singlet) pair production from Higgs boson, generated via gluon-gluon fusion, and their decays via their mixing with the doublets [12].

we have investigated such final states by considering all the dominant SM backgrounds at the LHC with a center of mass energy of 13 and 14 TeV. A detailed signal to background analysis of the final states $2\tau + 2b$, $2b + 2\mu$ and $\geq 3\tau$ reveals that some of the benchmark points with the light pseudoscalar ($m_{a_1} \sim 10 - 20$ GeV) can be probed with early data of 25 fb^{-1} at the LHC.

4 Charged Higgs boson phenomenology

As mentioned earlier that unlike in the NMSSM and MSSM, the TNMSSM has three charged Higgs bosons out of which two of them are triplets. Henceforth, they do not couple to fermions and are hard to produce directly. In the mass basis they are generally in mixed states. The presence of a light pseudoscalar affects the charged Higgs decay phenomenology and such light state can be probed via $h_1^\pm \to a_1 W^\pm$. Similarly in the case of the NMSSM with a Z_3 symmetry [7], the non-standard decay mode of the light charged Higgs boson $h^\pm \to a\, W^\pm$ can evade the recent mass bounds from the LHC [13] and can be probed with the τ and b rich final states coming from the light pseudoscalar [14]. The charged Higgs boson in this case (NMSSM) is still doublet-like so its production from the decay of a

 DOI: 10.1051/epjconf/201612900009

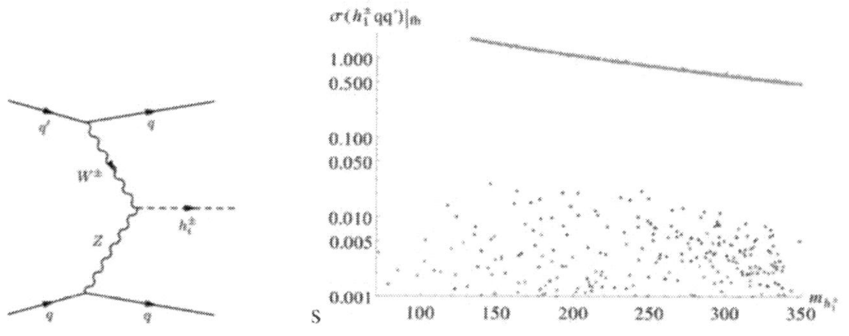

(a) (b)

Figure 3. The Feynman diagram for the charged Higgs production via vector boson fusion at the LHC (a) and the production cross-section of a light charged Higgs boson via vector boson fusion versus the light charged Higgs boson mass $m_{h_1^\pm}$ (b) [16].

top or via $bg \to th^\pm$ is still possible. The possibility of a light scalar is also present in the MSSM with CP-violating interactions, where the light Higgs is mostly CP-odd and can evade the existing LEP bounds. Such Higgs boson with a mass $\lesssim$ 30 GeV can open the decay mode $h^\pm \to h_1 W^\pm$, which can be explored at the LHC [15].

However things can change a lot in a triplet extended supersymmetric model (TESSM), where we have a light charged Higgs triplet. It does not couple to fermions, which makes its production via conventional methods a lot harder. However, its production channels in association with $W^\pm$ bosons or with neutral Higgs bosons are still considerable. The interesting feature of having a charged Higgs triplet is the non-zero $h_i^\pm - Z - W^\mp$ coupling, which not only gives rise to a new decay mode of $ZW^\pm$ but also makes the charged Higgs production possible via vector boson fusion [16] as can be seen from Fig 3. This is not allowed for the doublet-like charged Higgs boson of the MSSM or of the 2HDM. In the TNMSSM we have both the presence of a very light pseudoscalar - as in the NMSSM - and of a charged Higgs triplet, as in the TESSM. If we can pair produce such charged Higgs where one of them decays into $a_1 W^\pm$ and the other into $ZW^\pm$, then we can search for such decay modes. This provides a smoking gun signature for the extended Higgs structure of the TNMSSM. We are currently investigating in detail ways to probe both aspects at the LHC [17].

5 Conclusions

The discovery of a Higgs boson around 125 GeV in mass cannot explain all the shortcomings of the SM, and the quest for extra Higgs bosons are still there. Supersymmetric theories allow additional Higgs bosons which may belong to the usual or to higher representations of $SU(2)$. Phenomenologically, the existence of a very light scalar below 100 GeV in mass is still open, together with the theoretical possibility of various representations. Given that all the SM decay modes of the 125 GeV Higgs boson are yet to be discovered, the question of the existence of non-standard decays in this sector still remains elusive. The other Higgs bosons could have higher mass values (> 125 GeV) but the possibilities of lower mass values (< 100 GeV) also remain quite open. Certainly in such cases, if the $h_{125} \to h_1 h_1 / a_1 a_1$ decay modes are kinematically allowed, then they can be probed at the LHC. Models with Z_3 symmetry, such as the NMSSM or the TNMSSM, where such scalars are naturally light, can be tested once additional data will be made available from the LHC. If an extended Higgs sector exists, then finding a charged Higgs is a direct proof of it. So far, the search for a charged Higgs boson is performed in its decay to $\tau\nu$, but the existence of a light scalar/pseudoscalar gives us the new decay mode $h^\pm \to a_1/h_1 W^\pm$, which should be looked for at the LHC. Triplet charged and neutral Higgs bosons do not couple to fermions, and a charged Higgs triplets can decay to $ZW^\pm$ which can be explored at the LHC. To complete our understanding of electroweak phase transition we need to explore all the theoretical possibilities experimentally.

Acknowledgement

PB wants to thank University of Salento for the invitation and support to attend the workshop.

References

[1] G. Aad *et al.* [ATLAS and CMS Collaborations], arXiv:1503.07589 [hep-ex].
[2] J. Beringer *et al.* [Particle Data Group Collaboration], Phys. Rev. D **86** (2012) 010001.
[3] P. Bandyopadhyay, C. Corianò and A. Costantini, JHEP **1509** (2015) 045 [arXiv:1506.03634 [hep-ph]].
[4] S. R. Coleman and E. J. Weinberg, Phys. Rev. D **7** (1973) 1888.
[5] P. Bandyopadhyay, K. Huitu and A. Sabanci, JHEP **1310** (2013) 091 [arXiv:1306.4530 [hep-ph]].
[6] P. Bandyopadhyay, S. Di Chiara, K. Huitu and A. S. Keceli, JHEP **1411** (2014) 062 [arXiv:1407.4836 [hep-ph]]. S. Di Chiara and K. Hsieh, Phys. Rev. D **78** (2008) 055016 [arXiv:0805.2623 [hep-ph]].
[7] U. Ellwanger, C. Hugonie and A. M. Teixeira, Phys. Rept. **496** (2010) 1 [arXiv:0910.1785 [hep-ph]].
[8] K. Agashe, A. Azatov, A. Katz and D. Kim, Phys. Rev. D **84** (2011) 115024 [arXiv:1109.2842 [hep-ph]].
[9] R. Barate *et al.* [LEP Working Group for Higgs boson searches and ALEPH and DELPHI and L3 and OPAL Collaborations], Phys. Lett. B **565** (2003) 61 [hep-ex/0306033]. S. Schael *et al.* [ALEPH and DELPHI and L3 and OPAL and LEP Working Group for Higgs Boson Searches Collaborations], Eur. Phys. J. C **47** (2006) 547 [hep-ex/0602042].
[10] P. Franzini, D. Son, P. M. Tuts, S. Youssef, T. Zhao, J. Lee-Franzini, J. Horstkotte and C. Klopfenstein *et al.*, Phys. Rev. D **35** (1987) 2883.
[11] S. Chatrchyan *et al.* [CMS Collaboration], Phys. Rev. Lett. **109** (2012) 121801 [arXiv:1206.6326 [hep-ex]].
[12] P. Bandyopadhyay, C. Corianò and A. Costantini, JHEP **1512** (2015) 127 [arXiv:1510.06309 [hep-ph]].
[13] CMS Collaboration [CMS Collaboration], CMS-PAS-HIG-14-020. G. Aad *et al.* [ATLAS Collaboration], JHEP **1503** (2015) 088 [arXiv:1412.6663 [hep-ex]].
[14] P. Bandyopadhyay, K. Huitu and S. Niyogi, JHEP **1607** (2016) 015 [arXiv:1512.09241 [hep-ph]].
[15] P. Bandyopadhyay, JHEP **1108** (2011) 016 [arXiv:1008.3339 [hep-ph]].
P. Bandyopadhyay and K. Huitu, JHEP **1311** (2013) 058 [arXiv:1106.5108 [hep-ph]]. P. Bandyopadhyay, A. Datta, A. Datta and B. Mukhopadhyaya, Phys. Rev. D **78** (2008) 015017 [arXiv:0710.3016 [hep-ph]].
[16] P. Bandyopadhyay, C. Corianò and A. Costantini, arXiv:1512.08651 [hep-ph]. P. Bandyopadhyay, K. Huitu and A. S. Keceli, JHEP **1505** (2015) 026 [arXiv:1412.7359 [hep-ph]].
[17] P. Bandyopadhyay, and A. Costantini, in preparation.

EPJ Web of Conferences **129**, 00010 (2016) DOI: 10.1051/epjconf/201612900010
QCD@Work 2016

Strong Sector in non-minimal SUSY model

Antonio Costantini[1,⋆]

[1] *Dipartimento di Matematica e Fisica "Ennio De Giorgi",*
Università del Salento and INFN-Lecce,
Via Arnesano, 73100 Lecce, Italy

Abstract. We investigate the squark sector of a supersymmetric theory with an extended Higgs sector. We give the mass matrices of stop and sbottom, comparing the Minimal Supersymmetric Standard Model (MSSM) case and the non-minimal case. We discuss the impact of the extra superfields on the decay channels of the stop searched at the LHC.

1 Introduction

The search of supersymmetric partners of the known particles, expecially the quarks, is one of the main purpose of the LHC. They are supposed to be heavy because supersymmerty (SUSY) is broken, otherwise they would be observed at the electroweak symmetry breaking (EWSB) scale. Since the start of LHC with Run 1, the lower bound on the mass of the lightest squark, which is usually taken to be the lightest stop, has increased [1]. In this paper we are going to show how much these bounds are related to the assumptions that the supersymmetric model is minimal in its superfield content. In section 2 we briefly review the squark sector in the MSSM, giving the expression of the mass matrices for stop and sbottom. In section 3 we show how the squark sector changes in the case of non-minimal SUSY and in section 4 we present the results for the branching ratios of the lightest stop decay channels tested at the LHC. Section 5 is left to the conclusions.

2 Squark Mass Matrices in MSSM

As it is well known, in the minimal version of SUSY each particle of the Standard Model (SM) has a superpartner and they are related each other by a SUSY transformation. The superpotential of the MSSM is

$$\mathcal{W} = \mathcal{W}_{MSSM} + \mathcal{W}_H \tag{1}$$

with

$$\mathcal{W}_{MSSM} = y_t \hat{U} \hat{H}_u \cdot \hat{Q} - y_b \hat{D} \hat{H}_d \cdot \hat{Q} - y_\tau \hat{E} \hat{H}_d \cdot \hat{L}\,, \tag{2}$$

and

$$\mathcal{W}_H = \mu\, \hat{H}_u \cdot \hat{H}_d. \tag{3}$$

⋆e-mail: antonio.costantini@le.infn.it

 DOI: 10.1051/epjconf/201612900010

Higgs superfields are given by

$$\hat{H}_u = \begin{pmatrix} \hat{H}_u^+ \\ \hat{H}_u^0 \end{pmatrix}, \qquad \hat{H}_d = \begin{pmatrix} \hat{H}_d^0 \\ \hat{H}_d^- \end{pmatrix}. \tag{4}$$

The extraction of the potential from the superpotential is straightforward and we have $V = V_F + V_D + V_{soft}$ where

$$V_F = \sum_k F_k F_k^*, \quad F_k = \left.\frac{\partial \mathcal{W}^\dagger}{\partial \hat{\Phi}_k^\dagger}\right|_{\theta=\bar{\theta}=0} \tag{5}$$

$$V_D = \frac{1}{2}\sum_k g_k^2 (\phi_i^\dagger t_{ij}^a \phi_j)^2. \tag{6}$$

and V_{soft} is responsible for the soft breaking of supersymmetry. The mass matrices are defined as the second derivative of the potential evaluated on the vacuum. Stop and sbottom mass matrices are given respectively by

$$\mathcal{M}_{\tilde{t}} = \begin{pmatrix} m_t^2 + m_{Q_3}^2 + m_Z^2 \cos(2\beta)(\frac{1}{2} - \frac{2}{3}\sin^2\theta_w) & m_t(A_t - \mu\cot\beta) \\ m_t(A_t - \mu\cot\beta) & m_t^2 + m_{\bar{u}_3}^2 - \frac{1}{3}m_Z^2\cos(2\beta)\sin^2\theta_w \end{pmatrix} \tag{7}$$

$$\mathcal{M}_{\tilde{b}} = \begin{pmatrix} m_b^2 + m_{Q_3}^2 - m_Z^2 \cos(2\beta)(-\frac{1}{2} + \frac{2}{3}\sin^2\theta_w) & m_b(A_b - \mu\tan\beta) \\ m_b(A_b - \mu\tan\beta) & m_b^2 + m_{\bar{b}_3}^2 + \frac{2}{3}m_Z^2\cos(2\beta)\sin^2\theta_w \end{pmatrix} \tag{8}$$

A close inspection of Eq. (7) and (8) shows that the off-diagonal terms are proportional to the quark mass. The mixing $\tilde{q}_R - \tilde{q}_L$ is enhanced in the case of stop and hence $\tilde{t}_1$ is the lightest squark. It's easy to extract the eigenvalues of the matrices in Eq. (7) and (8). In the case of stop we have

$$\begin{aligned} m_{\tilde{t}_{1,2}}^2 &= \frac{1}{2}\Big(m_t^2 + m_{Q_3}^2 + 2m_t^2 + \frac{1}{2}m_Z^2\cos(2\beta) \\ &\pm \sqrt{(m_{Q_3}^2 - m_{\bar{u}_3}^2 + (\frac{1}{2} - \frac{4}{3}\sin^2\theta_w)\cos(2\beta)m_Z^2)^2 + 4m_t^2(A_t - \mu\cot\beta)^2}\Big) \end{aligned}$$

3 Squark Mass Matrices in TNMSSM

In this section we give the expression for the mass matrices of stop and sbottom in the model considered in the analysis which is an extension of the MSSM. We have enlarged the Higgs sector of MSSM including a singlet superfield and an SU(2) triplet superfield. The gauge group is the usual $SU(3)_c\times SU(2)_L\times U(1)_Y$. A detailed analysis of this model, named TNMSSM, can be found in [2–4]. The superpotential of the model is $\mathcal{W}_{TNMSSM} = \mathcal{W}_{MSSM} + \mathcal{W}_{TS}$ with

$$\mathcal{W}_{MSSM} = y_t \hat{U}\hat{H}_u\cdot\hat{Q} - y_b\hat{D}\hat{H}_d\cdot\hat{Q} - y_\tau\hat{E}\hat{H}_d\cdot\hat{L}\,, \tag{9}$$

and

$$\mathcal{W}_{TS} = \lambda_T \hat{H}_d\cdot\hat{T}\hat{H}_u + \lambda_S\hat{S}\hat{H}_d\cdot\hat{H}_u + \frac{\kappa}{3}\hat{S}^3 + \lambda_{TS}\hat{S}\,\mathrm{tr}[\hat{T}^2]. \tag{10}$$

Triplet and doublets superfields are given by

$$\hat{T} = \begin{pmatrix} \sqrt{\frac{1}{2}}\hat{T}^0 & \hat{T}_2^+ \\ \hat{T}_1^- & -\sqrt{\frac{1}{2}}\hat{T}^0 \end{pmatrix}, \qquad \hat{H}_u = \begin{pmatrix} \hat{H}_u^+ \\ \hat{H}_u^0 \end{pmatrix}, \qquad \hat{H}_d = \begin{pmatrix} \hat{H}_d^0 \\ \hat{H}_d^- \end{pmatrix}. \tag{11}$$

In this model the mass matrix for stop is

$$\mathcal{M}_{\tilde{t}} = \begin{pmatrix} m_t^2 + m_{Q_3}^2 + \frac{1}{24}\left(g_Y^2 - 3g_L^2\right)\left(v_u^2 - v_d^2\right) & \frac{1}{\sqrt{2}}A_T v_u + \frac{y_t v_d}{2}\left(\frac{v_T \lambda_T}{\sqrt{2}} - v_S \lambda_S\right) \\ \frac{1}{\sqrt{2}}A_T v_u + \frac{y_t v_d}{2}\left(\frac{v_T \lambda_T}{\sqrt{2}} - v_S \lambda_S\right) & m_t^2 + m_{\tilde{u}_3}^2 + \frac{1}{6}\left(v_d^2 - v_u^2\right)g_Y^2 \end{pmatrix} \tag{12}$$

Comparing Eq. (7) with Eq. (12) we can see that they are similar, the only difference being in the off-diagonal terms. The μ-term of Eq. (7) is now dynamically generated through the terms $\frac{v_T \lambda_T}{\sqrt{2}} - v_S \lambda_S$. This is clear from a comparison of the Higgs part of the superpotential in Eq. (3) and Eq. (10).

4 Branching Ratios of the Stop

In this section we present the main result of our analysis on the decay of the lightest stop. We are going to compare one of the latest result form the ATLAS collaboration on the mass bound for the lightets stop [5] with our phenomenological analysis.

In Fig. 1(a) we can see the correlation plot $m_{\tilde{t}_1} - m_{\chi_1^0}$ of a recent analysis from the ATLAS collaboration [5]. In Fig. 1(b) we present the same correlation plot for a set of $\sim$ 2000 benchmark points used in a previous analysis. The dashed line is the digitized curve delimiting the yellow region of Fig. 1(a). Let us remark that the benchmark points showed in Fig. 1(b) pass a certain number of phenomenological tests. In particular for each point we have a Higgs boson with a mass $\sim$ 125 GeV and the couplings with the Z bosons, the W bosons, the photons and the gluons are within the current experimental limits [4]. Moreover the mass of the lightest neutralino, χ_1^0, which is the lightest supersymmetric particle (LSP) of the model, is compatible with the bounds coming from the relic abundance of dark matter.

The bounds on the mass of the lightest stop are putted under the assumption of $\mathcal{B}r(\tilde{t}_1 \to t\tilde{\chi}_1^0) = 1$ [5]. This is not the case if the model has an extended Higgs sector, as one can see from Fig. 1(b). Another possible decay channel for the stop is $\tilde{t}_1 \to b\tilde{\chi}_1^{\pm}$ and the mass bounds for the stop mass

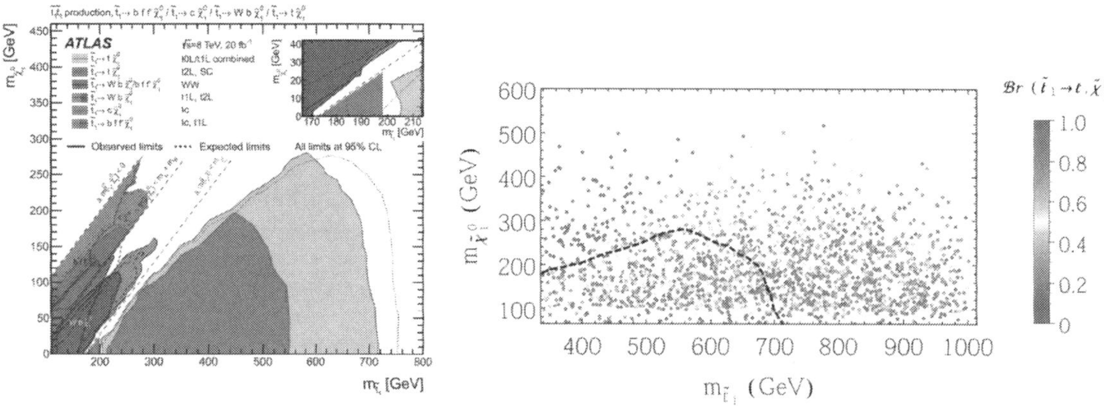

(a) (b)

Figure 1. Plot of the mass correlation $m_{\tilde{t}_1} - m_{\tilde{\chi}_1^0}$ from the ATLAS collaboration [5] (a) and from our phenomenological analysis (b), where we have reported the correspinding branching ratio for $\tilde{t}_1 \to t\tilde{\chi}_1^0$. The dashed line is the digitized curve delimiting the yellow region in (a).

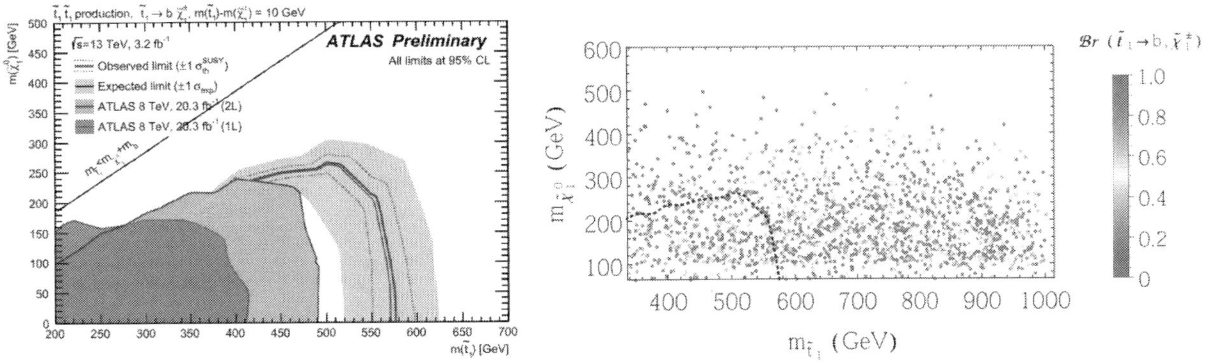

(a)   (b)

Figure 2. Plot of the mass correlation $m_{\tilde{t}_1} - m_{\tilde{\chi}_1^0}$ from the ATLAS collaboration [6] (a) and from our phenomenological analysis (b), where we have reported the correspinding branching ratio for $\tilde{t}_1 \to b\tilde{\chi}_1^{\pm}$. The dashed line is the digitized curve corresponding to the observed limit in (a).

coming from this analysis are presented in [6]. The channel $\tilde{t}_1 \to b\tilde{\chi}_1^{\pm}$ is assumed to have $\mathcal{B}r = 1$. In Fig. 2(a) is shown the correlation plot $m_{\tilde{t}_1} - m_{\chi_1^0}$ in the case of $\tilde{t}_1 \to b\tilde{\chi}_1^{\pm}$ and in Fig. 2(b) we present the results of our analysis. As is the previous case the dashed line is the digitized curve corresponding to the observed limit in Fig. 2(a). In this case there are some points with $\mathcal{B}r_{b\tilde{\chi}_1^{\pm}} \simeq 1$ but they are outside the excluded region. The points below the dashed curve have $\mathcal{B}r_{b\tilde{\chi}_1^{\pm}} < 1$.

We clearly see that the mass bounds for the scalar top considered here are evaded by a large number of benchmark points. In fact in both Fig. 1(b) and Fig. 2(b) there are mostly red points corresponding to branchings lower than 50%.

5 Conclusions

We have presented a phenomenological analysis of the decay channels of the scalar top in the context of a supersymmetric theory with an extended Higgs sector. We have compared our results with the recent analysis done by the ATLAS collaboration [5, 6] for the decay channels $\tilde{t}_1 \to t\tilde{\chi}_1^0$ and $\tilde{t}_1 \to b\tilde{\chi}_1^{\pm}$. In both the cases the branching ratios for these channels is assumed to be 100%. This is not true in the case of a non-minimal SUSY theory, like the TNMSSM [2–4], as shown in Fig. 1(b) and Fig. 2(b). Although the mass matrices of the squarks are almost the same in the MSSM and in the TNMSSM, the branching ratios of the channels studied at the LHC are quite different. This is due to the enlarged Higgs sector of the TNMSSM and to the presence of other representation of $SU(2)_L$ than the doublets, as extensively discussed in [2–4].

References

[1] G. Aad *et al.* [ATLAS Collaboration], Phys. Lett. B **701** (2011) 1, [arXiv:1103.1984 [hep-ex]]. G. Aad *et al.* [ATLAS Collaboration], Phys. Rev. Lett. **109** (2012) 211802, [arXiv:1208.1447 [hep-ex]]. G. Aad *et al.* [ATLAS Collaboration], JHEP **1406** (2014) 124, [arXiv:1403.4853 [hep-ex]]. G. Aad *et al.* [ATLAS Collaboration], JHEP **1409** (2014) 015, [arXiv:1406.1122 [hep-ex]].

[2] P. Bandyopadhyay, C. Corianò and A. Costantini, JHEP **1509** (2015) 045, [arXiv:1506.03634 [hep-ph]].
[3] P. Bandyopadhyay, C. Corianò and A. Costantini, JHEP **1512** (2015) 127, [arXiv:1510.06309 [hep-ph]].
[4] P. Bandyopadhyay, C. Corianò and A. Costantini, arXiv:1512.08651 [hep-ph].
[5] G. Aad *et al.* [ATLAS Collaboration], Eur. Phys. J. C **75** (2015) no.10, 510 Erratum: [Eur. Phys. J. C **76** (2016) no.3, 153], [arXiv:1506.08616 [hep-ex]].
[6] The ATLAS collaboration, ATLAS-CONF-2016-009.

EPJ Web of Conferences **129**, 00011 (2016) DOI: 10.1051/epjconf/201612900011
QCD@Work 2016

CHY Representation of the Massive Six Quark Amplitude in QCD

A.R. Fazio[1,*] and J.P. Vesga Simmons[1,**]

[1] *Departamento de Física*
Universidad Nacional de Colombia
Bogotá, Colombia

Abstract. The Cachazo-He-Yuan (CHY) representation for tree-level scattering amplitudes based on the so called scattering equations is proved to hold for six quarks QCD primitive amplitudes. Since the CHY representation allows to separate information into a factor which depends upon the helicities of the particles and into a factor depending on the ordering of the external particles, we prove that the external orderings of a minimal amplitude basis for three quarks pairs remain linearly independent, when viewed as the external orderings of the pure gluonic amplitudes.

1 Introduction

The CHY representation provides an elegant and compact way of writing tree amplitudes in a variety of quantum field theories, based on the solution of the so-called scattering equations. These algebraic equations introduce a space of auxiliary complex variables, whose solutions are given in terms of the kinematic invariants of the scattering process in question. For a process with n external particles, there are $(n-3)!$ inequivalent solutions to the scattering equations. Thus, one can construct an integral representation of the amplitudes that is completely localized on the solutions to the system. The representation for a variety of theories can be found in [1]-[6].

In this work we introduce the CHY representation for amplitudes in QCD. We review the details of the CHY formalism and its application to QCD and the description of a minimal amplitude basis in the language of words and shuffle algebra. A general primitive amplitude can be expanded in a minimal amplitude basis whose elements have a CHY representation with a consequent factorization of the information on the polarizations and the external ordering. Following [7] the existence of a CHY representation for an arbitrary QCD amplitude is seen to depend on the existence of a pseudoinverse of a matrix F, which is a function of the kinematic invariants of the process. We explicitly calculate the elements of the F matrix, in the case of the six quark amplitude with all different flavours, and show that F has full row rank by proving the existence of a right pseudoinverse in a general kinematic regime.

* e-mail: arfazio@unal.edu.co
** e-mail: jpvesgas@unal.edu.co

DOI: 10.1051/epjconf/201612900011

2 The CHY Formalism

2.1 The Scattering Equations

Let $(p_i)_{1\leq i\leq n}$ denote the momenta of the external particles of an n-particle QCD scattering process. These momenta are massless for gluons, $p_g^2 = 0$, and can be massive for a quark q and its corresponding antiquark $\bar{q}$, $p_q^2 = p_{\bar{q}}^2 = m_q^2$. For each i, a solution to the scattering equations (SE) is a complex n-tuple $z = (z_1, \ldots, z_n)$ satisfying

$$f_i(z,p) = \sum_{j\neq i} \frac{2p_i \cdot p_j + 2\Delta_{ij}}{z_{ij}} = 0 \tag{1}$$

where $z_{ij} = z_i - z_j$ and $\Delta_{q_k\bar{q}_k} = \Delta_{\bar{q}_k q_k} = m_{q_k}^2$, but $\Delta_{ij} = 0$ in other cases. One particular property of the solutions to the SE is that they are invariant under Möbius transformations,

$$z \to z' = \frac{az+b}{cz+d}; \quad ad - bc = 1. \tag{2}$$

There is therefore a redundacy among the solutions of SE which can be fixed [2] such that for a given n there are $(n-3)!$ inequivalent solutions, meaning not related to each other by a Möbius transformation.

2.2 CHY Representation

Let $A_n(p;\epsilon)$ denote an n particle tree amplitude of an arbitrary field theory, written as a function of the external momenta, p, and the wavefunctions of the external particles, generically denoted by ϵ. We say that the theory has a CHY representation if any tree-level amplitude can be written in the form

$$A_n(p;\epsilon) = \int d\Omega_{CHY} \mathcal{I}(z,p,\epsilon) \tag{3}$$

where

$$d\Omega_{CHY} = \frac{d^n z}{\mathrm{vol}\,(SL(2,\mathbb{C}))} z_{ij} z_{jk} z_{ki} \prod_{a\neq i,j,k} \delta(f_a(z,p)) \tag{4}$$

is the integration measure, in which the factor vol $(SL(2,\mathbb{C}))$ cancels the gauge redundancy introduced by the Möbius invariance of the SE. The product of delta functions, independent of the choice of i, j, k, completely localizes the integral on the solutions to the SE. Formally, this is equivalent to a sum over the $(n-3)!$ solutions of the SE. For QCD, it is natural to consider primitive amplitudes, which are obtained from an arbitrary amplitude by stripping its color structure. These amplitudes then have a fixed external ordering. For an arbitrary QCD primitive amplitude, the CHY representation can be written as [8]

$$A_n(p,\epsilon;w) = i \sum_j J(p,z^j)\hat{C}_n(w,z^j)\hat{E}_n(p,\epsilon;z^j) \tag{5}$$

where z^j denotes the j-th solution to the SE, $J(z)$ is the Jacobian arising from the delta functions, and $\hat{C}_n$, $\hat{E}_n$ contain, respectively, all the information on the external cyclic ordering and the information on spin (which we compactly denote by ϵ). The argument $w = \ell_1 \ldots \ell_n$ stands for a word, that is, an ordered string of labels, which we call letters. These provide a compact way to express the dependence of the amplitude on the external ordering and to define a minimal basis of QCD amplitudes, which is fixed by the cyclic invariance, Kleiss-Kuijf (KK) relations and the Bern-Carrasco-Johansson (BCJ) relations satisfied by the amplitudes and, in the multiquark case, the vanishing of primitive amplitudes with crossed fermion lines [8]. In particular, in the gluon case,

$$\hat{C}_n(w, z^j) = C_n(w, z^j) = \frac{1}{z_{\ell_1\ell_2} z_{\ell_2\ell_3} \cdots z_{\ell_n\ell_1}} \tag{6}$$

which is known as the (standard) Parke-Taylor factor. For a fixed n, we can construct the N_b dimensional vectors A_w providing the minimal amplitudes basis for the process under consideration and the $N_s = (n-3)!$ dimensional vector of components $\hat{E}_j$ obtained by evaluating the function $\hat{E}$ at the j-th solution of the SE. Then, we can cast the CHY representation for an arbitrary QCD primitive amplitude as the matrix equation

$$A_w = i\hat{M}_{wj}\hat{E}_j \tag{7}$$

where we have defined the $N_b \times N_s$ dimensional matrix

$$\hat{M}_{wj} = J(p, z^j)\hat{C}(w, z^j). \tag{8}$$

As a consequence the CHY representation for a QCD amplitude exists if (7) can be inverted for $\hat{E}_j$ in terms of A_w. The necessary condition for this to be possible is that $N_b \leq N_s$ (Rouché-Capelli theorem), which can be seen to be satisfied for arbitrary n. Moreover, since in the gluon case $N_b = N_s$, $\hat{M}_{wj}$ is a square matrix and its inverse can be found through a property satisfied by the Parke-Taylor factors, known as Kawai-Lewellen-Tye (KLT) orthogonality [2]. In the general case, the factors $\hat{C}(w, z^j)$ are not equal to the standard Parke-Taylor factors. However, due to the Möbius invariance of the SE, a Faddeev-Popov procedure is needed to fix the redundancy in the integral representation. Therefore, one can work in a "gauge" in which $\hat{C}(w, z^j) = C(w, z^j)$ for $w \in B$, where B is the set of words which form the amplitude basis. Thus, ignoring the information about flavour, we consider a set of letters

$$\mathbb{A} = \{1, 2, \ldots, n\} \tag{9}$$

which we call an alphabet, and define the sets

$$W_0 = \{w = \ell_1 \ldots \ell_n \,|\, \ell_i \in \mathbb{A},\ \ell_i \neq \ell_j, \text{for } i \neq j\}, \tag{10}$$

$$W_2 = \{w \in W_0 |\, \ell_1 = 1,\ \ell_n = n\}, \tag{11}$$

$$B_{n_q \leq 2} = \{w \in W_0 |\, \ell_1 = 1,\ \ell_{n-1} = n-1,\ \ell_n = n\}. \tag{12}$$

The details of the basis for an arbitrary multiquark amplitude can be obtained by using the so-called Melia Basis [9]-[10]. Furthermore, since the Parke-Taylor factors contain all the information on the ordering of the external particles, they satisfy the BCJ relations

$$C(w, z^j) = F_{ww'}C(w', z^j) \tag{13}$$

where $w \in B$, $w' \in W_2$, and the factors $F_{ww'}$ form an $N_b \times N_s$ matrix, whose elements we define later. By the above construction, the function $\hat{E}$ can be found to be [8]

$$\hat{E}(z, p, \epsilon) = -i \sum_{u,v \in B_{n_q \leq 2}} \sum_{w \in B} S[u|\bar{v}]G_{uw}C(\bar{v}, z)A_n(p, w, \epsilon) \tag{14}$$

 DOI: 10.1051/epjconf/201612900011

where $S[u|\bar{v}]$ is the KLT bilinear, which depends only on the kinematic invariants of the process, and the matrix G defines a right pseudoinverse to F, through

$$G = F^T (FF^T)^{-1}. \tag{15}$$

The matrix G always exists, provided that the square matrix FF^T has an inverse. Therefore, to show that there exists a CHY representation for an arbitrary QCD primitive amplitude, it is sufficient to construct the matrix FF^T and show that, for general kinematics, it has a non-vanishing determinant. The existence of the right pseudoinverse G means that the matrix F has full row rank, that is, its rows are linearly independent.

3 Matrix Elements for the Six Quark Amplitude

Now, we consider the explicit calculation of the matrix elements $F_{ww'}$ for the case of the six quark amplitude. If we denote the quarks and their corresponding antiquarks according to

$$q_1 = 1, \quad q_2 = 2, \quad q_3 = 3 \quad \bar{q}_3 = 4, \quad \bar{q}_2 = 5, \quad \bar{q}_1 = 6. \tag{16}$$

We can construct the basis

$$B = \{123456, 125346, 132546, 134256\}. \tag{17}$$

We also define the set

$$B_{n_q \leq 2} = \{123456, 124356, 132456, 134256, 142356, 143256\} \tag{18}$$

which would correspond to the basis for the six gluon amplitude. The matrix $\mathbb{F}_{6q}$, whose elements are $F_{ww'}$, will be of 4×6 size and it can be regarded as the linear operator relating the multiquark amplitude with the purely gluonic one. Moreover, the elements $F_{ww'}$ are the factors appearing in the BCJ relations. Then, in order to determine the non-trivial elements (that is, those different from one or zero), one fixes the word w and using the linear expansion appearing in the BCJ relations, one can find the words w' for which the element $F_{ww'}$ is non-trivial. Since the letters $1, 5, 6$ are fixed in the elements of $B_{n_q \leq 2}$ and the letters $1, 6$ are fixed in the elements of B, it is sufficient to consider the words in B which are not in $B_{n_q \leq 2}$. The amplitudes with those external orderings are the only ones that will have a non-trivial linear expansion in terms of the amplitudes whose external orderings are in $B_{n_q \leq 2}$. As an example of the calculation of the elements, we consider the word $w = 125346$. To determine the words $w' \in B_{n_q \leq 2}$ that contribute to the linear expansion of w, we consider two subwords of w,

$$w_1 = 2, \quad w_2 = 34 \tag{19}$$

these are chosen in such a way that $w = 1w_1 5 w_2 6 \notin B_{n_q \leq 2}$. Then, we calculate

$$s(w_2) = 34 + 43 \tag{20}$$

which, in general, is defined as the sum over all possible permutations of the letters in w_2. Finally, we calculate the shuffle product

$$w_1 \sqcup\!\sqcup\, s(w_2) = 234 + 243 + 324 + 342 + 324 + 432 \tag{21}$$

which, for arbitrary letters $u = \ell_1 \ldots \ell_j$, $v = \ell_{j+1} \ldots \ell_r$, corresponds to the sum over all ordered permutations of the letters in u and v. The shuffle product is distributive. If we let σ denote an

DOI: 10.1051/epjconf/201612900011

arbitrary term appearing on the sum in 21, we construct the word $w' = 1\sigma 56$, and we can calculate the element $F_{ww'}$ for the fixed w and each of those w'. In this particular case, we need to calculate six matrix elements. For example, we can calculate the matrix element for $w = 125346$ and $w' = 123456$. In general, every element will be a product of rational functions of the kinematic invariants. In this case,

$$F_{ww'} = \frac{\mathcal{F}(1\sigma 5|3)}{\hat{s}_{6,\ell_2,\ell_3}} \frac{\mathcal{F}(1\sigma 5|4)}{\hat{s}_{6,\ell_3}} \tag{22}$$

where $\sigma = 234$. Then, we form the string $\rho = 12345$, and we need to calculate the factors $\mathcal{F}(\rho|3)$ and $\mathcal{F}(\rho|4)$, whose structure depends on the position of the letters in w_1 and w_2, relative to the position of the same letters in σ. For a general factor $\mathcal{F}(\rho|\ell_k)$, one needs to determine the position of ℓ_k in the string $\rho = 1\sigma(n-1)$, which we denote by t_{ℓ_k}, and compare it with the positions of ℓ_{k+1} and ℓ_{k-1}; furthermore, if we let ℓ_j be the last letter appearing in w_1, which in our case is the letter 2, the position of said letter is always fixed to $t_{\ell_j} = n$, the number of particles in the scattering process. Finally, for $k = n-4$, we always have the condition $t_{\ell_{n-4}} = t_{\ell_{n-2}}$. In general, $\mathcal{F}(\rho|\ell_k) = \mathcal{F}_1 + \mathcal{F}_2$, where

$$\mathcal{F}_1 = \begin{cases} \sum_{r=1}^{t_{\ell_k}-1} \mathcal{G}(\ell_k, \rho_r) & t_{\ell_k} < t_{\ell_{k+1}} \\ -\sum_{r=t_{\ell_{k+1}}}^{n-1} \mathcal{G}(\ell_k, \rho_r) & t_{\ell_k} > t_{\ell_{k+1}} \end{cases}$$

$$\mathcal{F}_2 = \begin{cases} \hat{s}_{n,\ell_k,\ldots,\ell_{n-3}} & t_{\ell_{k-1}} < t_{\ell_k} < t_{\ell_{k+1}} \\ -\hat{s}_{n,\ell_k,\ldots,\ell_{n-3}} & t_{\ell_{k-1}} > t_{\ell_k} > t_{\ell_{k+1}} \\ 0 & Else \end{cases}$$

being

$$\hat{s}_{i_1,\ldots,i_k} = \sum_{a<b} (2p_{i_a} \cdot p_{i_b} + 2\Delta_{i_a i_b}). \tag{23}$$

On the other hand,

$$\mathcal{G}(\ell_k, \rho_r) = \begin{cases} 2p_{\ell_k} \cdot p_{\rho_r} + 2\Delta_{\ell_k \rho_r} & \rho_r = 1, (n-1) \\ 2p_{\ell_k} \cdot p_{\rho_r} + 2\Delta_{\ell_k \rho_r} & \rho_r = \ell_t, t < k \\ 0 & Else. \end{cases}$$

For the factor $\mathcal{F}(\rho|3)$ in our example, since $\ell_2 = 3$ and $\ell_3 = 4$, $t_{\ell_2} < t_{\ell_3}$. Furthermore, since $\ell_1 = 2$, from the general condition $t_{\ell_j} = n$, we find $t_2 = 6$ so that $\mathcal{F}(\rho|3) = 2(p_1 + p_2) \cdot p_3$. From the general condition $t_{\ell_{n-2}} = t_{\ell_{n-4}}$, we find that $t_{\ell_3} > t_{\ell_4}$ therefore $\mathcal{F}(\rho|4) = -2p_4 \cdot p_5$.
This implies

$$F_{ww'} = -\frac{4(p_1 + p_2) \cdot p_3(p_4 \cdot p_5)}{\hat{s}_{6,3,4}\hat{s}_{6,4}}. \tag{24}$$

The calculation of the rest of matrix elements proceeds similarly. That is, once w is fixed, one constructs the sum in (21) and calculates the element $F_{ww'}$ for each of those two words. In the case $w = w'$, $F_{ww'} = 1$ and the rest of elements are zero. With this in mind, we can arrange the matrix $\mathbb{F}_{6q}$, and find the square matrix

$$\mathbb{F}_{6q}\mathbb{F}_{6q}^T = \begin{bmatrix} 1 & 0 & 0 & D \\ 0 & 1 & A & E \\ 0 & A & A^2 + B^2 + C^2 & AE + BF + CG \\ D & E & AE + BF + CG & D^2 + E^2 + F^2 + G^2 + H^2 + I^2 \end{bmatrix}.$$

The non-zero elements are rational functions of various kinematic invariants and masses, and are given by

$$\begin{aligned}
A &= -\frac{2[(p_1+p_3)\cdot p_4 + m_3^2]}{\hat{s}_{6,4}}, \quad B = \frac{2[p_4\cdot(p_1+p_2+p_3)+m_3^2]}{\hat{s}_{4,6}}, \quad C = \frac{2p_1\cdot p_4}{\hat{s}_{4,6}} \\
E &= -\frac{4\left((p_2+p_5)\cdot p_4\right)(p_1\cdot p_3)}{\hat{s}_{6,3,4}\hat{s}_{6,4}}, \quad F = -\frac{4(p_1\cdot p_3)(p_4\cdot p_5)}{\hat{s}_{6,3,4}\hat{s}_{6,4}} \\
G &= -\frac{2p_1\cdot p_4\left(2p_3\cdot(p_2+p_5)+\hat{s}_{6,4,3}\right)}{\hat{s}_{6,4,3}\hat{s}_{6,4}} \\
H &= -\frac{2[(p_1+p_2)\cdot p_4][2p_3\cdot p_5+\hat{s}_{6,4,3}]}{\hat{s}_{6,4,3}\hat{s}_{6,4}}, \quad I = -\frac{-2p_1\cdot p_4\left(2p_3\cdot p_5+\hat{s}_{6,4,3}\right)}{\hat{s}_{6,4,3}\hat{s}_{6,4}}
\end{aligned} \tag{25}$$

and D is given by the expression on (24). For general kinematics, this matrix has a non-vanishing determinant. Thus, by explicitly computing the matrix elements and the determinant, we have confirmed the claim [8] that the CHY representation exists for the six quark primitive amplitude in QCD.

4 Conclusions

In most phenomenological applications one usually wants the amplitude computed from a given set of external four-momenta. In numerical applications, this is done through the Berends-Giele [11], or Britto-Cachazo-Feng-Witten recursion relations [12]. It is useful to have compact analytical formulae for the scattering amplitudes. Using spinor techniques [13] this can be done for every specific helicity configuration and every specific external ordering. This approach, however, does not give a way to identify how the result for a given amplitude changes when the external ordering or helicity configuration is changed. The CHY representation provides compact expressions for the amplitudes and allows to identify their dependence on the external ordering and the helicities by separating the information of these two set of parameters into pairs of independent factors, thus giving an alternate way to study the properties of tree-level scattering amplitudes.

References

[1] F. Cachazo, S. He and E. Y. Yuan, JHEP **1310**, 141 (2013) [arXiv:1306.2962 [hep-th]].
[2] F. Cachazo, S. He and E. Y. Yuan, Phys. Rev. D **90**, no. 6, 065001 (2014) [arXiv:1306.6575 [hep-th]].
[3] F. Cachazo, S. He and E. Y. Yuan, Phys. Rev. Lett. **113**, no. 17, 171601 (2014) [arXiv:1307.2199 [hep-th]].
[4] F. Cachazo, S. He and E. Y. Yuan, JHEP **1407**, 033 (2014) [arXiv:1309.0885 [hep-th]].
[5] F. Cachazo, S. He and E. Y. Yuan, JHEP **1501**, 121 (2015) [arXiv:1409.8256 [hep-th]].
[6] F. Cachazo, S. He and E. Y. Yuan, JHEP **1507**, 149 (2015) [arXiv:1412.3479 [hep-th]].
[7] S. Weinzierl, JHEP **1503**, 141 (2015) [arXiv:1412.5993 [hep-th]].
[8] L. de la Cruz, A. Kniss and S. Weinzierl, JHEP **1511**, 217 (2015) [arXiv:1508.06557 [hep-th]].
[9] H. Johansson and A. Ochirov, JHEP **1601**, 170 (2016) [arXiv:1507.00332 [hep-ph]].
[10] T. Melia, JHEP **1512**, 107 (2015) [arXiv:1509.03297 [hep-ph]].
[11] F.A. Berends and W.T. Giele, Nucl.Phys.B**306**, 759 (1988).
[12] R. Britto,F. Cachazo,B. Feng and E. Witten, Phys.Rev.Lett. **94**, 181602 (2005).
[13] L.J. Dixon, *Boulder 1995, QCD and beyond* (TASI lectures (1995)) 539-582,[hep-ph/9601359].

 DOI: 10.1051/epjconf/201612900012

Impact of the HERA I+II combined data on the CT14 QCD global analysis

S. Dulat[1,3], T.-J. Hou[2], J. Gao[3], M. Guzzi[4,*] J. Huston[5], P. Nadolsky[2], J. Pumplin[5], C. Schmidt[5], D. Stump[5], and C.-P. Yuan[5]

[1]*School of Physics Science and Technology, Xinjiang University, Urumqi, Xinjiang 830046, China*
[2]*Department of Physics, Southern Methodist University, Dallas, TX 75275-0181, U.S.A.*
[3]*High Energy Physics Division, Argonne National Laboratory, Argonne, Illinois 60439, U.S.A.*
[4]*The University of Manchester, Oxford Road, M13 9PL Manchester, United Kingdom*
[5]*Department of Physics and Astronomy, Michigan State University, East Lansing, MI 48824, U.S.A.*

Abstract. A brief description of the impact of the recent HERA run I+II combination of inclusive deep inelastic scattering cross-section data on the CT14 global analysis of PDFs is given. The new $CT14_{HERA2}$ PDFs at NLO and NNLO are illustrated. They employ the same parametrization used in the CT14 analysis, but with an additional shape parameter for describing the strange quark PDF. The HERA I+II data are reasonably well described by both CT14 and $CT14_{HERA2}$ PDFs, and differences are smaller than the PDF uncertainties of the standard CT14 analysis. Both sets are acceptable when the error estimates are calculated in the CTEQ-TEA (CT) methodology and the standard CT14 PDFs are recommended to be continuously used for the analysis of LHC measurements.

1 Introduction

The CT14 parton distribution functions (PDFs) [1] are obtained in a global analysis of a variety of hadronic scattering experimental data and are suitable for general-purpose QCD calculations at the Large Hadron Collider (LHC) and in other experiments.

In global analyses to determine the PDFs of the proton, inclusive DIS measurements impose the most important constraints. In 2015, the H1 and ZEUS collaborations released a novel combination, "HERA2", of measurements of inclusive deep-inelastic scattering cross sections [2] at the HERA $e^{\pm}p$ collider. The two collaborations employed different experimental techniques and used different detectors and methods for kinematic reconstruction. As a result, the new HERA2 combined measurements exhibit reduced systematic uncertainties with respect to the previous combination of DIS data sets, "HERA1", published in 2009 [3].

In the study of Ref. [4], the impact of the HERA2 measurements on the CT14 global analysis is investigated by replacing the combined HERA1 data set used in the published CT14 PDFs [1], with the HERA2 set and by examining the resulting changes in PDF central values and uncertainties. Moreover, the dependence of the goodness-of-fit upon kinematic cuts on Q and x is also examined, as in Ref. [2] it was suggested that the low Q^2 HERA2 data are not well fitted by the CT10 and CT14

* Presenter, e-mail: marco.guzzi@manchester.ac.uk

PDFs. Related studies of the impact of HERA2 data in the context of MMHT14 and NNPDF3.0 analyses can be found in Refs. [5–7].

The CTEQ-TEA PDFs have been refitted at NLO and NNLO by using the global CT14 data ensemble, but with the HERA2 measurements in place of HERA1. The new PDFs obtained after the refitting procedure are named CT14$_{\rm HERA2}$, to distinguish from CT14.

2 Features of the CT14$_{\rm HERA2}$ analysis

The HERA2 data set has 1120 data points in the fitted region with $Q > 2$ GeV and $W^2 > 12.5$ GeV^2. There are 162 correlated systematic errors, and 7 procedural uncertainties, in addition to the luminosity uncertainty. When HERA2 is included in the global fit, there are in total 3287 data points in the CT14$_{\rm HERA2}$ data ensembles, compared to 2947 in the original CT14 fits. This is because two other changes have been made in the data analysis: 1) the NMC muon-proton inclusive DIS data on F_2^p [8] have been dropped because this set is influenced by some unknown or underestimated systematic errors [1], in concordance with findings in our earlier studies [9]; 2) the data table for the CMS 7 TeV 5 fb^{-1} inclusive jet experiment [10], which became available after the completion of the CT14 study, has been updated and it produces no appreciable effects on the PDFs.

As in CT14, the theoretical predictions for the majority of processes in the CT14$_{\rm HERA2}$ fit are calculated at the NNLO level of accuracy. In particular, a NNLO treatment [11] of heavy-quark mass effects in neutral-current DIS is realized in the S-ACOT-χ scheme [12–15] and is essential for obtaining correct predictions for LHC electroweak cross sections [16–19]. However, the calculations for charged-current DIS and inclusive jet production are included at NLO only; in both cases, the complete NNLO contributions are not yet available. In Sec. II of Ref. [1], various arguments are presented suggesting that the expected impact of the missing NNLO effects in jet production on the PDFs is small relatively to current experimental errors. Similarly, the NNLO contribution to charged-current DIS, including massive charm scattering contributions [20], is modest compared to the experimental uncertainties.

2.1 Parametrization of the PDFs

The CT14$_{\rm HERA2}$ analysis adopts the CT14 PDF functional forms [1] at initial scale $Q_0 = 1.3$ GeV

$$x\, f_a(x, Q_0) = x^{a_1}\,(1-x)^{a_2}\, P_a(x), \qquad (1)$$

where the $P_a(x)$ functions are linear combinations of Bernstein polynomials. One change is made relative to the standard CT14. In Ref. [1] the strange quark PDF is parametrized according to Eq. (1), with $P_s(x)$ being a constant. The parameter a_1 is tied to the common a_1 of $\bar{u}$ and $\bar{d}$, and $s(x) = \bar{s}(x)$ is assumed in the analysis. Therefore, in the standard CT14 analysis, one has only two parameters for the strange quark PDF: a_2 and normalization. With this limitation on $s(x, Q_0)$, it is necessary to extend the strange quark uncertainty by adding two "extreme strange" PDFs to the set of Hessian error PDFs. In the CT14$_{\rm HERA2}$ PDFs, a different technique is used to avoid underestimating the strangeness uncertainty provided by the Hessian error PDF set: while in the published CT14 PDFs the a_1 parameters are such that $a_1(s) = a_1(\bar{s}) = a_1(\bar{d}) = a_1(\bar{u})$, in CT14$_{\rm HERA2}$ $a_1(s) = a_1(\bar{s})$ is allowed to differ from $a_1(\bar{d}) = a_1(\bar{u})$. By freeing the parameter $a_1(s)$, it is found that it is not necessary to construct additional extreme strange PDFs. So, whereas the CT14 error PDFs include two "extreme strange"

[1] In the CT14$_{\rm HERA2}$ analysis of Ref. [4] it has been observed that even if they were included in the fit, they would induce a negligible change in the total χ^2

and two "extreme gluon" PDFs, the CT14$_{\text{HERA2}}$ error PDFs include only two "extreme gluon" PDFs to model the uncertainty of gluon PDFs in the very large x and small x regions, respectively. Thus the total number of error PDFs is the same for CT14 and CT14$_{\text{HERA2}}$.

3 Impact of data selection cuts

The HERA2 publication [2] found that both HERAPDF2.0 PDFs and χ^2 values depend significantly on the choice of Q_{cut}, the minimum value of the four-momentum-transfer Q in the HERA2 analysis. In order to explore the impact of variations of Q_{cut} on the CT14$_{\text{HERA2}}$ global analysis, multiple fits of CT14$_{\text{HERA2}}$ PDFs have been performed in which Q_{cut} is varied from 2 GeV to 6 GeV and the results are compared to those of the CT14 analysis. For every choice of Q_{cut}, the total χ^2, reduced-χ^2 (*i.e.*, χ^2_{re}), and the systematic shift penalty R^2 (defined respectively in Eq. 2 and 3 of Ref. [4]), are illustrated in Fig. 1. The χ^2/N_{pts} accounts for the sum of all four DIS subprocesses and the graphs show the

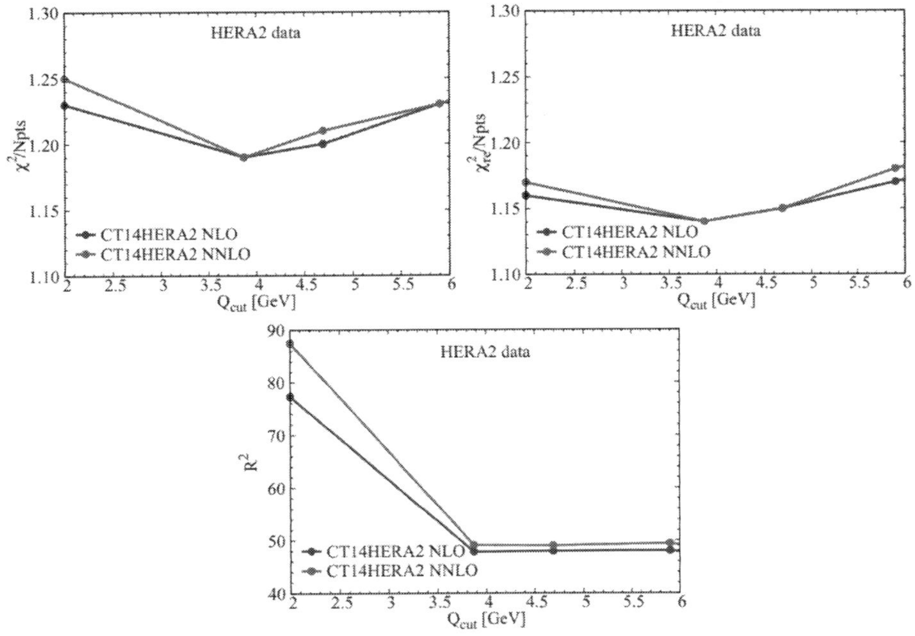

Figure 1. χ^2/N_{pts} (left), reduced-χ^2/N_{pts} (right), and R^2 (bottom) for the HERA2 data and CT14$_{\text{HERA2}}$ PDFs, as a function of Q_{cut}.

dependence on Q_{cut} in the CT14$_{\text{HERA2}}$ analysis at both NLO and NNLO. The values of χ^2/N_{pts} for the HERA2 data exhibit a *shallow minimum* for Q_{cut} in the range $3.5 \lesssim Q_{\text{cut}} \lesssim 4$ GeV. The reduction of χ^2 at $Q_{\text{cut}} \sim 4$ GeV, compared to the CT standard choice of $Q_{\text{cut}} = 2$ GeV, from 1.17 to 1.15, does not seem significant. An interesting feature of the graphs is that near the minimum the NNLO and NLO results are equal, whereas NNLO has larger χ^2 on either side of the minimum. The lower panel in Fig. 1 shows R^2, the total quadratic penalty for the systematic errors, as a function of Q_{cut}. The value of R^2 decreases significantly from $Q_{\text{cut}} = 2$ to 3.87 GeV, from 87 to 49 (for 170 systematic errors). For ideal Gaussian systematic errors one would expect $R^2 \sim 170$. When the low-Q data points are discarded by the cut, the systematic errors become less important. However, this reduction of R^2 is shared by 1120 total data points, so the overall net change in χ^2/N_{pts} is mild.

4 Results for the CT14$_{\mathrm{HERA2}}$ PDFs

The central values and uncertainties of the CT14$_{\mathrm{HERA2}}$ PDFs are compared to those of the CT14 global analysis in Fig. 2, where only the NNLO PDFs at the scale Q equal to the initial scale $Q_0 = 1.3$ GeV are shown. At this low scale, the PDF uncertainties are magnified, and they are reduced at electroweak scales as a consequence of DGLAP evolution. Additional plots can be found on the CTEQ public website [21]. The comparison is summarized below where a few comments are in order.

The central value of the CT14$_{\mathrm{HERA2}}$ gluon in the range $10^{-2} \lesssim x \lesssim 0.2$ is almost unchanged compared to CT14; it is larger by about 30% at $x \approx 10^{-4}$, by a larger factor for $x > 0.5$, and it is smaller by about 10% at $x \approx 0.3$.

The u and d quarks are generally slightly larger than (but close to) CT14 in the range $10^{-2} \lesssim x \lesssim 0.5$ where the CT14$_{\mathrm{HERA2}}$ uncertainty band is comparable to that of CT14; whereas they are both systematically larger by about 5% in the intermediate region of $10^{-4} \lesssim x \lesssim 10^{-2}$. The CT14$_{\mathrm{HERA2}}$/CT14 ratio decreases at $x \lesssim 10^{-4}$ in both cases. The d quark increases at $x > 0.5$, while the u quark decreases slightly at $x \approx 0.5$. The slow oscillations in $d(x, Q_0)$ reflect the behaviour of Bernstein polynomials in Eq. (1).

The strange quark central prediction is reduced over the entire x range, mainly due to the change of adding one additional shape parameter for describing the strange quark PDF; but this reduction is statistically insignificant and completely within the uncertainty of the previous PDF ensemble. In particular a reduction of approximately −50% is observed at both $x \lesssim 10^{-3}$ and $x \gtrsim 0.5$.

The $\bar{u}$ and $\bar{d}$ quarks share similar features. They are almost unchanged for $10^{-2} \lesssim x \lesssim 0.2$ where the uncertainties are comparable to those of CT14. The $\bar{u}$ quark PDF increases by about 10% at x around 0.2, and the $\bar{d}$ quark PDF similarly at x around 0.3. Both of $\bar{u}$ and $\bar{d}$ quarks, similar to s quark, decrease by large factors for $x \gtrsim 0.4$, where both gluon and down quark PDFs increase, as a consequence of momentum sum rule. It is important to keep in mind that at $x > 0.5$ the antiquark PDFs take very small values, their behaviour is very uncertain and strongly depends on the parametrization form.

The individual PDF uncertainties do not change much, except in the unconstrained x regions of the strange quark PDF due to the change made in its parametrization form.

It has been verified that the change seen in gluon, up and down quark PDFs mainly arises from replacing the HERA1 data (in CT14 analysis) by the HERA2 data (in CT14$_{\mathrm{HERA2}}$ analysis). This was explicitly checked by comparing CT14 PDFs to the the result of a new fit which uses exactly the same setup as that in the CT14 global analysis, but with the HERA1 data replaced by the HERA2 data.

Conclusions

A short overview of the CT14 global analysis of proton's PDFs has been given. With rapid advancements in LHC measurements, the focus of the global analysis has shifted toward providing accurate predictions in the wide range of x and Q covered by the LHC data. This development requires long-term multi-prong effort in theoretical, experimental, and statistical areas. On the theory side, a more flexible parametrization has been introduced to better capture variations in the PDF dependence. A series of benchmark tests of NNLO cross sections, carried out in the run-up for the CT14 fit for all key fitted processes, have resulted in better agreement with most experiments and brought accuracy of most predictions to the truly NNLO level. The final CT14 PDFs are presented in the form of 1 central and 56 Hessian eigenvector sets at NLO and NNLO. The 90% C.L. PDF uncertainties for physical observables can be estimated from these sets using the symmetric [9] or asymmetric [22, 23] master formulae. These PDFs are determined for the central QCD coupling of $\alpha_s(M_Z) = 0.118$, consistent with the world-average α_s value.

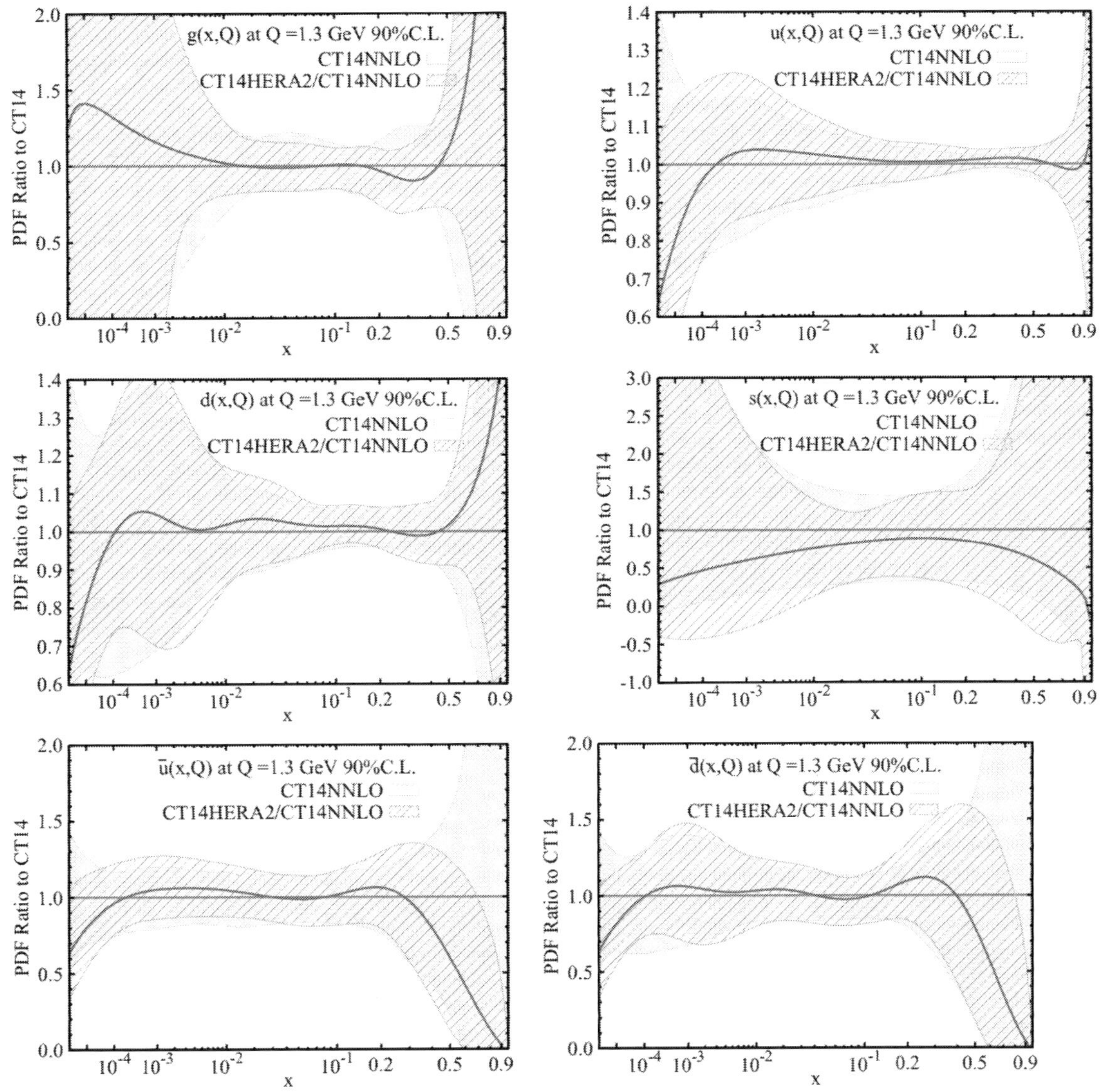

Figure 2. Comparison of $CT14_{HERA2}$ (red) and CT14 (blue) PDFs at $Q = 1.3$ GeV. Flavors $g, u, d, s, \bar{u}, \bar{d}$ are shown. The curves compare the central fits, plotted as ratios to CT14. The uncertainty bands are 90% C.L. uncertainties evaluated from the CT14 (shaded blue) and $CT14_{HERA2}$ (hatched red) error ensembles; both error bands are normalized to the corresponding central CT14 PDFs. All PDFs are from the NNLO QCD analysis.

Acknowledgments

M.G. would like to thank the organizers of the QCD@work2016 conference for the kind invitation. This work was supported by the STFC (ST/L000431/1) and by the Lancaster-Manchester-Sheffield Consortium formFundamental Physics under STFC grant ST/L000520/1; by the U.S. DOE Early Career Research Award DE-SC0003870; by the U.S. Department of Energy under Grant No. DE-FG02-96ER40969, DE-SC0013681, and DE-AC02-06CH11357; by the U.S. National Science Foundation under Grant No. PHY-0855561 and PHY-1417326; by Lightner-Sams Foundation; and by the National Natural Science Foundation of China under Grant No. 11165014 and 11465018.

References

[1] S. Dulat, T.J. Hou, J. Gao, M. Guzzi, J. Huston, P. Nadolsky, J. Pumplin, C. Schmidt, D. Stump, C.P. Yuan, Phys. Rev. **D93**, 033006 (2016), 1506.07443.
[2] H. Abramowicz et al. (ZEUS, H1), Eur. Phys. J. **C75**, 580 (2015), 1506.06042.
[3] F.D. Aaron et al. (ZEUS, H1), JHEP **01**, 109 (2010), 0911.0884.
[4] T.J. Hou, S. Dulat, J. Gao, M. Guzzi, J. Huston, P. Nadolsky, J. Pumplin, C. Schmidt, D. Stump, C.P. Yuan (2016), 1609.07968.
[5] L.A. Harland-Lang, A.D. Martin, P. Motylinski, R.S. Thorne, Eur. Phys. J. **C76**, 186 (2016), 1601.03413.
[6] R.S. Thorne, L.A. Harland-Lang, A.D. Martin, P. Motylinski, PoS **EPS-HEP2015**, 505 (2015), 1508.06621.
[7] J. Rojo (NNPDF), PoS **EPS-HEP2015**, 506 (2015), 1508.07731.
[8] M. Arneodo et al. (New Muon), Nucl. Phys. **B483**, 3 (1997), hep-ph/9610231.
[9] J. Pumplin, D.R. Stump, J. Huston, H.L. Lai, P.M. Nadolsky, W.K. Tung, JHEP **07**, 012 (2002), hep-ph/0201195.
[10] S. Chatrchyan et al. (CMS), Phys. Rev. **D87**, 112002 (2013), [Erratum: Phys. Rev. D87 (2013) 119902], 1212.6660.
[11] M. Guzzi, P.M. Nadolsky, H.L. Lai, C.P. Yuan, Phys. Rev. **D86**, 053005 (2012), 1108.5112.
[12] M.A.G. Aivazis, F.I. Olness, W.K. Tung, Phys. Rev. **D50**, 3085 (1994), hep-ph/9312318.
[13] M.A.G. Aivazis, J.C. Collins, F.I. Olness, W.K. Tung, Phys. Rev. **D50**, 3102 (1994), hep-ph/9312319.
[14] J.C. Collins, Phys. Rev. **D58**, 094002 (1998), hep-ph/9806259.
[15] W.K. Tung, S. Kretzer, C. Schmidt, J. Phys. **G28**, 983 (2002), hep-ph/0110247.
[16] J. Gao, M. Guzzi, P.M. Nadolsky, Eur. Phys. J. **C73**, 2541 (2013), 1304.3494.
[17] H.L. Lai, J. Huston, Z. Li, P. Nadolsky, J. Pumplin, D. Stump, C.P. Yuan, Phys. Rev. **D82**, 054021 (2010), 1004.4624.
[18] P.M. Nadolsky, H.L. Lai, Q.H. Cao, J. Huston, J. Pumplin, D. Stump, W.K. Tung, C.P. Yuan, Phys. Rev. **D78**, 013004 (2008), 0802.0007.
[19] W.K. Tung, H.L. Lai, A. Belyaev, J. Pumplin, D. Stump, C.P. Yuan, JHEP **02**, 053 (2007), hep-ph/0611254.
[20] E.L. Berger, J. Gao, C.S. Li, Z.L. Liu, H.X. Zhu, Phys. Rev. Lett. **116**, 212002 (2016), 1601.05430.
[21] http://hep.pa.msu.edu/cteq/public/index.html, the CTEQ-TEA webpage.
[22] H.L. Lai, M. Guzzi, J. Huston, Z. Li, P.M. Nadolsky, J. Pumplin, C.P Yuan, Phys. Rev. **D82**, 074024 (2010), 1007.2241.
[23] P.M. Nadolsky, Z. Sullivan, eConf **C010630**, P510 (2001), hep-ph/0110378.

EPJ Web of Conferences **129**, 00013 (2016)
QCD@Work 2016
DOI: 10.1051/epjconf/201612900013

Double Parton Interactions in pp and pA Collisions

Daniele Treleani[1,⋆], *Giorgio* Calucci[1,⋆⋆], and *Simona* Salvini[⋆⋆⋆]

[1] *University of Trieste and INFN*

Abstract. As a consequence of the increasingly large flux of partons at small x, Double Parton Interactions (DPI) play an increasingly important role at high energies. A detail understanding of DPI dynamics is therefore mandatory, for a reliable subtraction of the background in the search of new physics. On the other hand, DPI are an interesting topic of research by themselves, as DPI probe the hadron structure in a rather different way, as compared with the large p_t processes usually considered. In this note we will make a short illustration of some of the main features characterizing DPI in pp and in pA collisions.

1 Introduction

The rapid increase of the parton flux at small x causes a dramatic rise of all cross sections with large momentum transfer exchange in high energy hadronic collisions [1,2]. The values of x, which contribute to a hard process, with a fixed lower cutoff in the exchanged momentum, are in fact increasingly smaller at large energies. One thus faces a unitarity problem in large p_t processes at high energy. A good example is the inclusive cross section to produce mini-jets which, as shown in fig.1, may become larger than the total inelastic cross section at high energy.

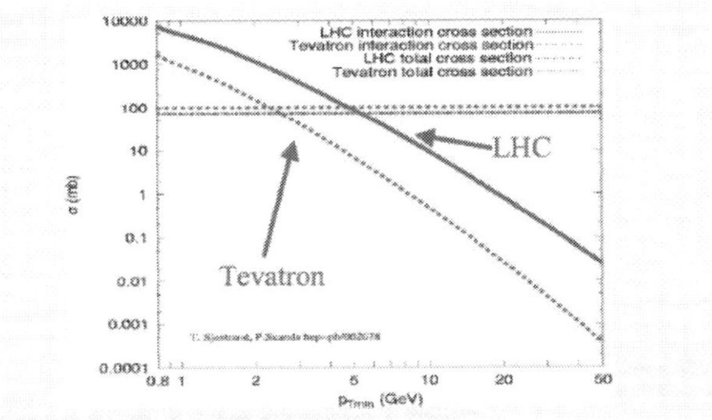

Figure 1. Minijets cross section as a function of the cutoff in p_t at Tevtron and at the LHC.

The problem with unitarity is solved when Multiple Parton Interactions (MPIs) are introduced in the interaction dynamics. MPIs introduce the possibility of different hard partonic interactions in a

⋆e-mail: daniele.treleani@ts.infn.it
⋆⋆e-mail: Giorgio.Calucci@ts.infn.it
⋆⋆⋆e-mail: salmonae@gmail.com

 DOI: 10.1051/epjconf/201612900013

given inelastic event and, while each inelastic event contributes with multiplicity 'one' to the inelastic cross section, each inelastic event contributes with the multiplicity of hard partonic interactions to the inclusive hard cross section. In this way, when the average multiplicity of hard partonic interactions in a inelastic event is large, the inclusive cross section is no more bounded by the value of the total inelastic cross section [3].

The simplest case of MPIs is the Double Parton Interaction (DPI). A possible case to consider is the production of four large p_t jets, where transverse momenta are compensated pairwise. The process can thus be originated by two different pairs of initial state partons, which interact independently with large transverse momentum exchange. The corresponding contribution to the cross section maximizes the incoming parton flux and thus it gives an increasingly important contribution at large energies.

Given the large momentum transfer, hard interactions are localized in a space region much smaller as compared to the hadron size. In a DPI two hard partonic interactions are thus localized in two different points in transverse space, in the overlap region of the matter distribution of the two interacting hadrons. The hard component of the interaction is thus disconnected and the process can be described by the geometrical picture in fig.2.

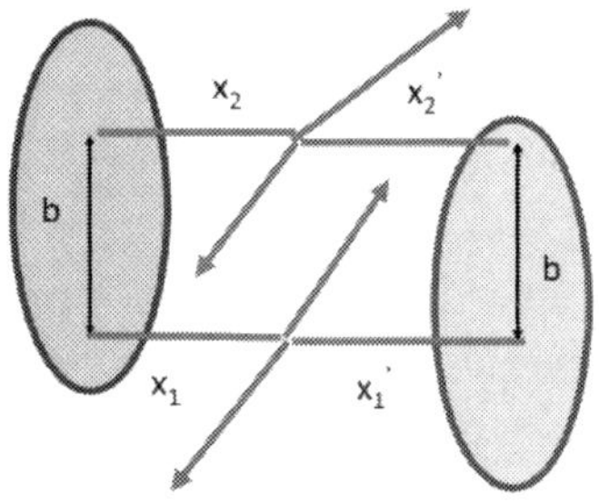

Figure 2. Geometrical picture of a DPI.

As apparent in fig.2, the non-perturbative components of a DPI are in this way factorized into functions which depend on two fractional momenta and on the relative transverse distance b between the two interaction points. The non-perturbative input to DPI, namely the double parton distribution functions, have thus dimensions of an inverse area and contain information on the hadron structure not accessible in a single scattering processes.

When neglecting spin and color, the inclusive double parton-scattering cross-section, for two parton processes A and B in a pp collision, is thus given by [4]

$$\sigma_D^{(A,B)} = \frac{m}{2}\sum_{i,j,k,l}\int \Gamma_{i,j}(x_1, x_2;\, b)\hat{\sigma}^A_{i,k}(x_1, x_1')\hat{\sigma}^B_{j,l}(x_2, x_2')\Gamma_{k,l}(x_1', x_2';\, b)\; dx_1 dx_1' dx_2 dx_2' d^2b \quad (1)$$

where m is a symmetry factor ($m = 1$ when $A = B$ and $m = 2$ when A is different from B), the functions $\Gamma(x, x'; b)$ are the double parton distributions, which depend on two fractional momenta and of b, $\hat{\sigma}^{A,B}$ are the partonic scattering cross sections and the sum is over the different parton species contributing to the process.

Notice that the dependence of σ_D on the transverse momenta of final partons is very well characterised and it is very strong. Actually it is as strong as the square of a single hard scattering cross section. One should stress that this feature represents a rather non trivial experimental test of the DPI interaction dynamics.

By introducing the "effective cross section", σ_{eff}, one may expresses σ_D by the so called "pocket formula", utilised in the experimental analysis:

$$\sigma_D^{pp\,(A,B)} = \frac{m}{2}\frac{\sigma_A\sigma_B}{\sigma_{eff}} \tag{2}$$

where σ_A and σ_B are the single scattering inclusive cross sections. Of course the "pocket formula" makes sense only, if comparing with experiments, σ_{eff} turns out to be weakly dependent on the reaction channel and on kinematics. Which is indeed the case, as it may be seen by looking at fig.3, where the results of different measurements of σ_{eff} are shown [5-9].

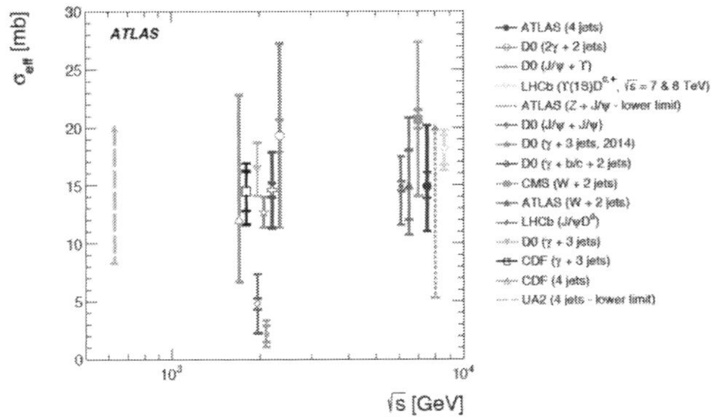

Figure 3. σ_{eff} as measured in different experiments.

When σ_A is small, the ratio σ_A/σ_{inel} represents the probability of having the process A in a inelastic collision. In the pocket formula $\sigma_D^{(AB)} = \sigma_A\sigma_B/\sigma_{eff}$ (here one assumes $A \neq B$) and in the biased case, where the process B takes place in presence of the process A, the effective cross section thus plays the role of the inelastic cross section.

By looking at fig.3, although within large experimental errors, the observed values of the effective cross section are approximately constant ($\sim$ 15 mb) and do not seem to depend on

the C.M. energy (Fermilab, LHC)

the reaction channel ($4j$; $\gamma 3j$; $\gamma\gamma 2j$; γb(c)$2j$; Wjj; $J/\Psi\ J/\Psi$; Z J/Ψ; Υ D^0; J/Ψ D^0)

the values of x and Q^2

There is on the contrary a clear indication that the effective cross section is sizably smaller ($\approx$ 2-4 mb) in the case ($J/\Psi\ J/\Psi$) and ($J/\Psi\ \Upsilon$) production

Notice that σ_{eff} is sizably smaller as compared with σ_{inel}, which is an indication of strong partonic correlations in the hadron structure.

2 σ_{eff} and Partonic Correlations

One may write the double parton distribution functions as

$$\Gamma(x_1, x_2;\ b) = G(x_1, x_2) f_{x_1 x_2}(b), \qquad G(x_1, x_2) = K_{x_1 x_2} G(x_1) G(x_2) \tag{3}$$

where $G(x)$ is a usual "one-body" parton distribution. The function f is normalised to one and the transverse scale, characterising f, may depend on fractional momenta. After integration on the transverse coordinate b, one thus obtains the average multiplicity of pairs with fractional momenta x_1 and x_2:

$$\int f_{x_1x_2}(b)\, d^2b = 1; \quad G(x_1,x_2) = K_{x_1x_2}G(x_1)G(x_2) = \langle n(n-1)\rangle_{x_1x_2}; \quad G(x) = \langle n\rangle_x \tag{4}$$

Parton distributions are in this way understood as average number of partons with a given momentum fraction. In the simplest case one would have $K_{x_1x_2} = 1$ which, after integrating on the transverse coordinate b, would be consistent with a Poissonian multiparton distribution in multiplicity [10]. By using relations in Eqs. (3) and (4), the double parton scattering cross section is given by

$$\begin{aligned}\sigma_D^{pp\,(A,B)}(x_1,x_1',x_2,x_2') &= \frac{m}{2}K_{x_1x_2}K_{x_1'x_2'}G(x_1)\hat{\sigma}_A(x_1,x_1')G(x_1') \\ &\quad \times G(x_2)\hat{\sigma}_B(x_2,x_2')G(x_2')\int f_{x_1x_2}(b)f_{x_1'x_2'}(b)db \\ &= \frac{m}{2}\frac{K_{x_1x_2}K_{x_1'x_2'}}{\pi\Lambda^2(x_1,x_1',x_2,x_2')}\sigma_A(x_1,x_1')\sigma_B(x_2,x_2')\end{aligned} \tag{5}$$

where

$$\int f_{x_1x_2}(b)f_{x_1'x_2'}(b)db = \frac{1}{\pi\Lambda^2(x_1,x_1',x_2,x_2')} \tag{6}$$

and the effective cross section is thus expressed in terms of the typical area of the interaction region Λ, which may depend on the fractional momenta of the interacting partons, and on the multiplicities of the interacting parton pairs:

$$\sigma_{eff}(x_1,x_1',x_2,x_2') = \frac{\pi\Lambda^2(x_1,x_1',x_2,x_2')}{K_{x_1x_2}K_{x_1'x_2'}} \tag{7}$$

Limiting cases are

a) partons are not correlated in multiplicity. In such a case $K_{x_1x_2} = 1$

b) partons are not correlated in the transverse coordinates. In such a case one may write the generalised "one-body" parton distribution $\Gamma(x;b)$ as

$$\Gamma(x;b) = G(x)f_x(b), \quad \int f_x(b)d^2b = 1 \text{ and one has : } f_{x_1x_2} = \int f_{x_1}(b')f_{x_2}(b-b')d^2b' \tag{8}$$

where $f_x(b)$ is the two gluon from factor of the nucleon. If one assumes that the effects of correlations are negligible, there are no unknowns and σ_{eff} can be evaluated. One obtains

$$\sigma_{eff} = \pi\Lambda^2 = 32\text{mb} \tag{9}$$

which is roughly a factor 2 larger as compared with the experimental indications. One may thus conclude that either K is not equal to 1 or partons are correlated in the transverse coordinates or, most likely, that partons are correlated both in multiplicity and in the transverse coordinates.

An additional experimental indication is that, in the kinematical region where DPI are important, the effective cross section depends only weakly on fractional momenta [5]. It makes therefore sense to assume that there is a weak dependence of both Λ and K on fractional momenta.

On the other hand it is clear that, since all new information on the hadron structure is summarized by a single quantity, namely the effective cross section, by studying DPI in pp collisions, one does not obtain enough information to discriminate between Λ and K. In other words one cannot disentangle parton correlations in the transverse coordinates and in multiplicity by studying DPI only in pp.

To obtain additional information on multi-parton correlations one needs to study DPI in pA collisions. In the case of a double parton interaction, in a collision of a proton with a nucleus, the effects of longitudinal and transverse correlations are in fact different when a single nucleon or two different target nucleons participate in the hard process.

3 DPI in pA Collisions

In the case of DPI in proton - nucleus collisions, in the regime where non additive corrections to the nuclear parton distributions are small, one may have a double parton interaction against a single or against two different target nucleons [11,12]. DPI have been measured in the WJJ production channel at the LHC, in pp collisions. It's interesting to compare the results, obtained in pp collisions, with the expectations of WJJ production in $p-Pb$ collisions in the same kinematical regime.

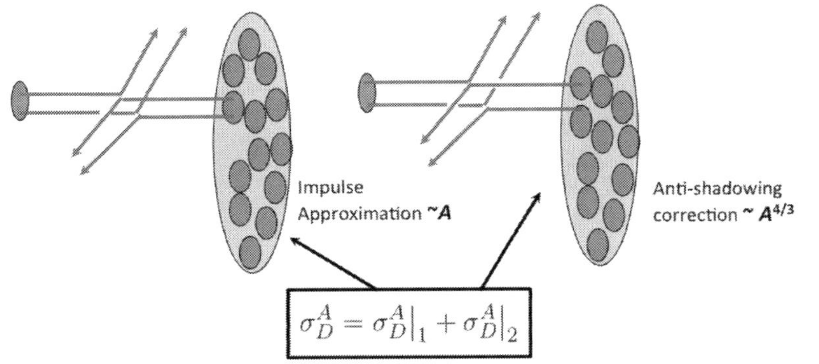

Figure 4. Different contributions to the DPI cross section in pA collision.

When neglecting the effects of the interference terms, which are estimated to produce a correction of the order of 10%, one obtains a simple expression for the DPI cross section in pA collisions:

$$\sigma^{pA}(WJJ) = \sigma_S^{pA}(WJJ) + \sigma_D^{pA}(WJJ)$$
$$\sigma_D^{pA}(WJJ) = \sigma_D^{pA}(WJJ)\big|_1 + \sigma_D^{pA}(WJJ)\big|_2 \tag{10}$$

The labels S and D here above correspond to the single and double parton scattering contributions, while the labels 1 and 2 distinguish the terms where one or two different target nucleons take active part in the process. By neglecting the effects of interference terms, the explicit expressions of $\sigma_D^{pA}(WJJ)\big|_1$ and of $\sigma_D^{pA}(WJJ)\big|_2$ are [13]

$$\sigma_D^{pA}(WJJ)\big|_1 = \frac{1}{\sigma_{eff}}[Z\sigma_S^{pp}(W) + (A-Z)\sigma_S^{pn}(W)]\sigma_S^{pp}(JJ)$$
$$\sigma_D^{pA}(WJJ)\big|_2 = K\Big[\frac{Z}{A}\sigma_S^{pp}(W) + \frac{A-Z}{A}\sigma_S^{pn}(W)\Big]\sigma_S^{pp}(JJ)\int T(b)^2 d^2b \tag{11}$$

where Z is the nuclear charge and $T(b)$ the nuclear thickness, as a function of the impact parameter b. In the kinematical regime of interest the production of jets is dominated by gluons, so in Eq.(11) we have put $\sigma_S^{pp}(JJ) = \sigma_S^{pn}(JJ)$. Notice that, while $\sigma_D^{pA}(WJJ)\big|_1$ grows linearly with A, $\sigma_D^{pA}(WJJ)\big|_2$ represents an additional positive contribution to the cross section, which grows as $A^{4/3}$. Therefore, in the case of DPI, the correction to the "impulse approximation term" is not the typical negative shadowing correction term. On the contrary, in the case of DPI, the correction term is positive and one has a anti-shadowing correction which, with heavy nuclei, may represent the dominant contribution to the cross section.

The first term in the equations above, $\sigma_D^{pA}(WJJ)\big|_1$, is proportional to the effective cross section and does not add much to the information on DPI obtained from pp collisions. The second term, $\sigma_D^{pA}(WJJ)\big|_2$, is on the contrary proportional to K, which measures the multiplicity of pairs of partons in the projectile hadron (one should remind that when $K = 1$ the multiplicity distribution is a Poissonian).

By measuring the amount of anti-shadowing, actually $\sigma_D^{pA}(WJJ)\big|_2$, one can thus obtain information on how the pairs of projectile partons are correlated in multiplicity.

Different contributions to the cross section are therefore possible, depending on the actual value of K. Two extreme cases can thus be considered in proton - lead collisions

a) There are no correlations in multiplicity:

$$K^2 = 1 \text{ and } \pi\Lambda^2 = \sigma_{eff}$$
$$\frac{\sigma_D^{pA}(WJJ)\big|_2}{\sigma_D^{pA}(WJJ)\big|_1} \approx 2 \tag{12}$$

b) There are no correlations in the transverse coordinates:

$$K^2 = 2 \text{ and } \pi\Lambda^2 = K^2\sigma_{eff}$$
$$\frac{\sigma_D^{pA}(WJJ)\big|_2}{\sigma_D^{pA}(WJJ)\big|_1} \approx 3 \tag{13}$$

Notice the huge values of the anti-shadowing contributions to the cross section: 200% and 300% in the two limiting cases.

A more detailed information is obviously obtained when looking at the differential distributions. In fig.5 the transverse distributions in $p-p$ and $p-Pb$ are compared. The leading Standard Model contribution (Single Parton Interaction) is represented by the pinkish histograms. The green histograms are the contributions due to Double Parton Interactions. The black histograms are the sum of Single and Double Parton Interactions.

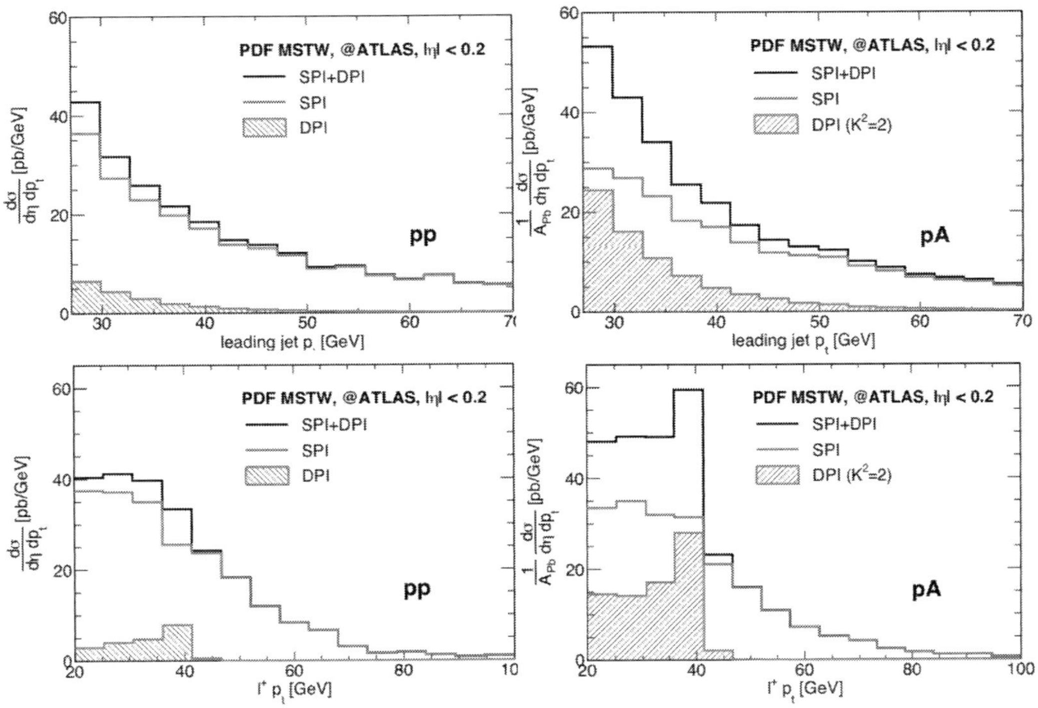

Figure 5. Upper figures: transverse spectra of the leading jet in $p-p$ (left panel) and $p-Pb$ collisions (right panel). Lower figures: transverse spectra of the W decay-lepton in $p-p$ (left panel) and $p-Pb$ collisions (right panel).

In the upper panels in fig.5, one compares the transverse spectra of the leading jet in $p-p$ (left panel) and $p-Pb$ collisions (right panel). One may notice that, consistently with the results of ATLAS and CMS, in $p-p$ DPI can only provide a minor modification to the SPI spectrum. On the contrary, the contribution of DPI is much more important in $p-Pb$, where the slope in p_t of the leading jet is substantially modified at $p_t \leq 40\ GeV$.

In the lower panels in fig.5, one compares the transverse spectra of the W decay-lepton in $p-p$ and $p-Pb$ collisions. Again DPI have a small effect in $p-p$ while, as shown in lower panel on the right side, DPI produce a huge increase of the spectrum at $p_t \lesssim 40\ GeV$ in $p-Pb$. In a DPI the W boson is in fact produced with a rather small transverse momentum, since the observed jets do not recoil against the W. The transverse spectrum of the W decay-lepton produced by DPI (green histogram in the figure) is thus limited at $p_t \lesssim 40\ GeV$ and the huge increase of the spectrum of the decay lepton at 40 GeV can in this way give a direct indication on the contribution of DPI, and thus on the anti-shadowing term, to the cross section.

One may therefore conclude that one expects rather sizable effects from DPI in $p-Pb$ as compared to $p-p$ at the LHC, because of the different interaction mechanism in the two cases. The additional contribution to the cross section, namely $\sigma_D^{pA}(WJJ)\big|_2$, is dominant at large A and it is directly proportional to the multiplicity of interacting pairs of partons in the projectile (the factor K in the present note). In the case of WJJ production in $p-Pb$ one should therefore be able to obtain a rather direct indication on the size of K by simply looking at the inclusive transverse spectrum of the W decay-lepton in $p - Pb$ collisions.

The sizable increase of the fraction of events due to DPI in $p - Pb$ collisions and the significant difference of the resulting value of the cross section, as a function of the correlation parameter K, are rather encouraging indications of the potential of DPI in $p - Pb$ collisions to obtain information on the parton correlation parameter K and, as a consequence, also on Λ. With a joint study of DPI in $p - p$ and $p - Pb$, one might thus be able to obtain information, to a large extent model independent, on the typical values of multiplicity of parton pairs and on the typical transverse distances between partons in the hadron structure; possibly also at different values of the parton's fractional momenta. The option of an experimental study of DPI in $p - Pb$ at the LHC could therefore be highly rewarding, offering a viable possibility of obtaining a remarkable insight into the three dimensional structure of the hadron.

References

[1] A. D. Martin, W. J. Stirling, R. S. Thorne and G. Watt, Eur. Phys. J. C **63**, 189 (2009) doi:10.1140/epjc/s10052-009-1072-5 [arXiv:0901.0002 [hep-ph]].

[2] T. C. Rogers, A. M. Stasto and M. I. Strikman, Phys. Rev. D **77**, 114009 (2008) doi:10.1103/PhysRevD.77.114009 [arXiv:0801.0303 [hep-ph]].

[3] L. Ametller and D. Treleani, Int. J. Mod. Phys. A **3**, 521 (1988). doi:10.1142/S0217751X88000217.

[4] N. Paver and D. Treleani, Nuovo Cim. A **70**, 215 (1982). doi:10.1007/BF02814035.

[5] F. Abe *et al.* [CDF Collaboration], Phys. Rev. D **56**, 3811 (1997). doi:10.1103/PhysRevD.56.3811.

[6] V. M. Abazov *et al.* [D0 Collaboration], Phys. Rev. D **93**, no. 5, 052008 (2016) doi:10.1103/PhysRevD.93.052008 [arXiv:1512.05291 [hep-ex]].

[7] G. Aad *et al.* [ATLAS Collaboration], New J. Phys. **15**, 033038 (2013) doi:10.1088/1367-2630/15/3/033038 [arXiv:1301.6872 [hep-ex]].

[8] S. Chatrchyan *et al.* [CMS Collaboration], JHEP **1403**, 032 (2014) doi:10.1007/JHEP03(2014)032 [arXiv:1312.5729 [hep-ex]].

[9] R. Aaij *et al.* [LHCb Collaboration], JHEP **1607**, 052 (2016) doi:10.1007/JHEP07(2016)052 [arXiv:1510.05949 [hep-ex]].

[10] G. Calucci and D. Treleani, Phys. Rev. D **79**, 074013 (2009) doi:10.1103/PhysRevD.79.074013 [arXiv:0901.3089 [hep-ph]].

[11] M. Strikman and D. Treleani, Phys. Rev. Lett. **88**, 031801 (2002) doi:10.1103/PhysRevLett.88.031801 [hep-ph/0111468].

[12] D. d'Enterria and A. M. Snigirev, Phys. Lett. B **718**, 1395 (2013) doi:10.1016/j.physletb.2012.12.032 [arXiv:1211.0197 [hep-ph]].

[13] S. Salvini, D. Treleani and G. Calucci, Phys. Rev. D **89**, no. 1, 016020 (2014) doi:10.1103/PhysRevD.89.016020 [arXiv:1309.6201 [hep-ph]].

EPJ Web of Conferences **129**, 00014 (2016) DOI: 10.1051/epjconf/201612900014
QCD@Work 2016

Very rare, exclusive, hadronic decays in QCD factorization

Matthias König[1,*]

[1] *Mainz Institute for Theoretical Physics, JGU Mainz*

Abstract. We study exclusive hadronic decays of the electroweak bosons Z, W and h in the framework of QCD factorization. We show that the theory uncertainties in these channels are remarkably small compared to past applications of the QCD factorization framework. While the branching ratios are small, many of the modes are accessible at future colliders. The Higgs decays exhibit interesting dependences on the couplings due to the interferences of different diagram topologies, making the $h \to V\gamma$ decays possible probes of the quark Yukawa couplings and the $h \to VZ$ decays probes of the coupling between the Higgs boson, a photon and a Z-boson.

Talk based on work in collaboration with Stefan Alte, Yuval Grossman and Matthias Neubert [1–4].

1 Introduction

The Standard Model (SM) of particle physics is in remarkable agreement with observation and the discovery of the Higgs boson in 2012 [5, 6] is arguably the biggest triumph of particle physics in recent times. Still many questions remain unanswered and thus the SM has to be extended. When doing so, the Higgs sector is usually modified, making it a crucial experimental task to measure the parameters in this sector as accurately as possible. The exclusive hadronic decays of the form $h \to M\gamma$, $h \to MZ$ and $h \to MW$ can be used to probe several different couplings of the Higgs boson to the SM fermions and gauge bosons [2, 7, 8].

Even within the SM, particle physics still faces the challenge of obtaining rigorous control over non-perturbative effects in QCD at low energy scales. The framework of QCD factorization is a well-established approach to hard exclusive processes with indiviual final state hadrons [9–13]. Within this framework, amplitudes are given as convolutions of hard-scattering functions with non-local hadronic matrix elements, which encode the non-perturbative physics at the low scale. The amplitudes will formally be given as expansions in the ratio of the two scales. Past applications struggled with the fact that this factorization scale was not high enough for power corrections to be neglected and that it was difficult to disentangle these uncertainties from those in the poorly-known hadronic input parameters. In our case, the scale is set by the decaying heavy boson and power corrections are well under control. The decays $Z \to M\gamma$ and $W \to M\gamma$ can be used to test that factorization approach in a theoretically clean environment [1, 3].

*e-mail: m.koenig@uni-mainz.de

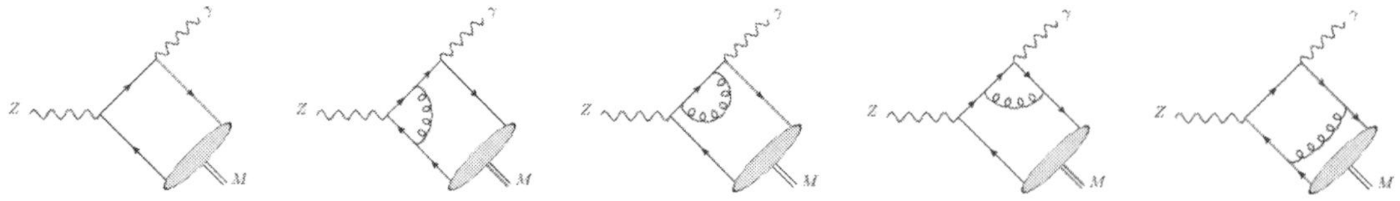

Figure 1. A subset of the diagrams contributing to $Z \to M\gamma$ at $O(\alpha_s)$. Additional mirrored diagrams exist.

2 QCD factorization

In the framework of QCD factorization, the amplitude for a hard exclusive process can be written as a convolution of a hard scattering function with a hadronic function, where the result will formally be an expansion of the scale separation parameter $\lambda \sim \Lambda_{\text{QCD}}/\mu_{\text{hard}}$. While the original derivation of the factorization formula dates back to the early 1980s, it can be rephrased in the language of Soft-Collinear Effective Theory (SCET) [14–17]. The factorization formula for the decays of $h/Z \to MV$ the final state meson M and the gauge boson V can be written as:

$$i\mathcal{A} = \sum_q \int dx\, T_H^q(x,\mu)\, \phi_M^q(x,\mu) + O(\lambda)\,, \tag{1}$$

where $T_H^q(x,\mu)$ is the hard-scattering kernel, $\phi_M^q(x,\mu)$ is the light-cone distribution amplitude (LCDA) and the sum runs over the active quark flavors. Power corrections of higher orders in λ are tiny in our case due to the high scale μ_{hard}.

The LCDAs are expanded in the set of Gegenbauer polynomials

$$\phi_M^q(x,\mu) = 6x\bar{x}\left[1 + \sum_{n=2,4,\ldots} a_n^{M,q}(\mu)\, C_n^{(3/2)}(2x-1)\right], \tag{2}$$

where $a_n^{M,q}(\mu)$ are the scale-dependent Gegenbauer moments [9, 13, 18–21]. By evolving these parameters from the hadronic scale to the high scale using the renormalization group, large logarithms of the form $\alpha_s \log\lambda$ are resummed to all orders. It is noteworthy that the evolution to an arbitrarily high scale lets all moments $a_n^{M,q}$ vanish, decreasing the sensitivity of our prediction to these poorly-known hadronic parameters.

3 Radiative hadronic decays of Z-bosons

We can now write down the amplitude for the decays $Z \to M\gamma$. We will neglect the modes $Z \to \eta^{(\prime)}\gamma$ here since the flavor-singlet components lead to non-trivial matching conditions that are too involved to explain here [3]. The amplitude of a Z-boson decaying into a meson M with momentum k and a photon with momentum q can be written as:

$$i\mathcal{A} = \pm\frac{eg}{2c_W}\left[i\epsilon_{\mu\nu\alpha\beta}\frac{k^\mu q^\nu \varepsilon_Z^\alpha \varepsilon_\gamma^{*\beta}}{k\cdot q}F_1^M - (\epsilon_Z^\perp\cdot\epsilon_\gamma^{\perp *})F_2^M\right]. \tag{3}$$

We can express the form factors as sums over the Gegenbauer moments and the hard scattering coefficients:

$$F_1^M = Q_M \sum_{n=0}^{\infty} C_{2n}^{(+)}(m_Z,\mu)a_{2n}^M(\mu)\,, \quad F_2^M = -Q'_M \sum_{n=0}^{\infty} C_{2n+1}^{(-)}(m_Z,\mu)a_{2n+1}^M(\mu)\,, \tag{4}$$

Table 1. Predicted branching fractions for various $Z \to M\gamma$ decays, including error estimates due to scale dependence (subscript "μ") and the uncertainties in the meson decay constants ("f"), and the shape parameters of the LCDA ("ϕ").

Decay mode	Branching ratio
$Z^0 \to \pi^0\gamma$	$(9.80\,^{+0.09}_{-0.14}{}_{\mu} \pm 0.03_f \pm 1.02_\phi) \cdot 10^{-12}$
$Z^0 \to \eta\gamma$	$(2.36\,^{+0.02}_{-0.04}{}_{\mu} \pm 1.19_f \pm 0.04_\phi) \cdot 10^{-10}$
$Z^0 \to \eta'\gamma$	$(6.68\,^{+0.08}_{-0.11}{}_{\mu} \pm 0.49_f \pm 0.12_\phi) \cdot 10^{-9}$
$Z^0 \to \rho^0\gamma$	$(4.19\,^{+0.04}_{-0.06}{}_{\mu} \pm 0.16_f \pm 0.44_\phi) \cdot 10^{-9}$
$Z^0 \to \omega\gamma$	$(2.89\,^{+0.03}_{-0.05}{}_{\mu} \pm 0.15_f \pm 0.38_\phi) \cdot 10^{-8}$
$Z^0 \to \phi\gamma$	$(8.63\,^{+0.08}_{-0.13}{}_{\mu} \pm 0.41_f \pm 0.92_\phi) \cdot 10^{-9}$
$Z^0 \to J/\psi\,\gamma$	$(8.02\,^{+0.14}_{-0.15}{}_{\mu} \pm 0.20_f\,^{+0.39}_{-0.36}{}_{\phi}) \cdot 10^{-8}$
$Z^0 \to \Upsilon(1S)\,\gamma$	$(5.39\,^{+0.10}_{-0.10}{}_{\mu} \pm 0.08_f\,^{+0.11}_{-0.08}{}_{\phi}) \cdot 10^{-8}$
$Z^0 \to \Upsilon(4S)\,\gamma$	$(1.22\,^{+0.02}_{-0.02}{}_{\mu} \pm 0.13_f\,^{+0.02}_{-0.02}{}_{\phi}) \cdot 10^{-8}$
$Z^0 \to \Upsilon(nS)\,\gamma$	$(9.96\,^{+0.18}_{-0.19}{}_{\mu} \pm 0.09_f\,^{+0.20}_{-0.15}{}_{\phi}) \cdot 10^{-8}$

where $Q^{(\prime)}$ are combinations of the meson decay constants and the quark couplings to the Z-boson and the photon. Evaluating diagrams like the ones shown in figure 1, we find the hard scattering coefficients:

$$\begin{aligned} C_n^{(\pm)}(m_V,\mu) &= 1 + \frac{C_F\alpha_s(\mu)}{4\pi}c_n^{(\pm)}\left(\frac{m_V}{\mu}\right) + O(\alpha_s^2), \\ \text{with}\quad c_n^{(\pm)}\left(\frac{m_V}{\mu}\right) &= \left[\frac{2}{(n+1)(n+2)} - 4H_{n+1} + 3\right]\left(\log\frac{m_V^2}{\mu^2} - i\pi\right) \\ &\quad + 4H_{n+1}^2 - \frac{4(H_{n+1}-1)\pm 1}{(n+1)(n+2)} + \frac{2}{(n+1)^2(n+2)^2} - 9\,. \end{aligned} \tag{5}$$

This is the leading result in our expansion in the scale separation parameter λ, with subleading corrections starting at $O(m_V^2/m_Z^2)$, that we can safely neglect. We present our phenomenological results in table 1.

Similar results are obtained for the $W \to M\gamma$ decays. We omit a discussion of these for briefness and refer the reader to ref. [1] for details.

4 Exclusive hadronic decays of the Higgs boson

In order to investigate the hadronic decays of the Higgs boson and understand to what extend new physics can manifest itself in them, we work in an effective Lagrangian framework, where we allow a rescaling of the relevant parameters in the SM as well as including higher-dimensional operators:

$$\begin{aligned} \mathcal{L}_{\text{eff}}^{\text{Higgs}} &= \kappa_W\,\frac{2m_W^2}{v}\,h\,W_\mu^+W^{-\mu} + \kappa_Z\,\frac{m_Z^2}{v}\,h\,Z_\mu Z^\mu - \sum_f \frac{m_f}{v}\,h\,\bar f\left(\kappa_f + i\tilde\kappa_f\gamma_5\right)f \\ &\quad + \frac{\alpha}{4\pi v}\left(\kappa_{\gamma\gamma}\,h\,F_{\mu\nu}F^{\mu\nu} - \tilde\kappa_{\gamma\gamma}\,h\,F_{\mu\nu}\tilde F^{\mu\nu} + \frac{2\kappa_{\gamma Z}}{s_W c_W}\,h\,F_{\mu\nu}Z^{\mu\nu} - \frac{2\tilde\kappa_{\gamma Z}}{s_W c_W}\,h\,F_{\mu\nu}\tilde Z^{\mu\nu}\right). \end{aligned} \tag{6}$$

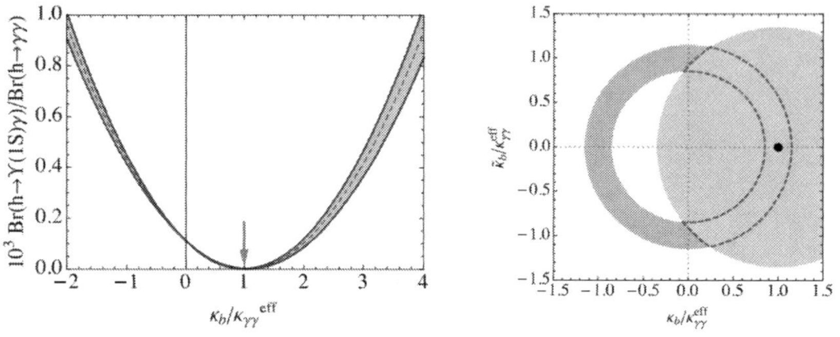

Figure 2. *Left diagram*: The dependence of Br($h \to \Upsilon(1S)\gamma$) on the rescaling of the b-quark Yukawa coupling under the assumption that all other parameters take their SM values. *Right diagram*: An upper bound on Br($h \to \Upsilon(1S)\gamma$) constrains the b-quark Yukawa to the orange region in the κ_b–$\tilde\kappa_b$ plane, whereas the measurement of Br($h \to \bar{b}b$) constrains only the absolute value $\kappa_b^2 + \tilde\kappa_b^2$, giving the blue band. The overlap region yields the allowed range for the b-quark Yukawa coupling in the CP-plane.

In the SM, the parameters κ_W, κ_Z and κ_f are equal to 1 whereas all other κ's vanish.

In the following, we only briefly state the main results of our analysis and direct the interested reader towards refs. [2, 4].

4.1 Radiative hadronic decays of the Higgs

The case of $h \to V\gamma$ is similar to the Z-boson decays, with the difference of an important additional contribution, where the Higgs decays into a pair of photons or into a photon and a Z-boson. Here, either the off-shell photon or Z-boson converts to the final state meson. The $h\gamma\gamma$- and $h\gamma Z$-couplings are generated at 1-loop in the SM or can occur at tree-level in our effective Lagrangian [22]. Generally, these "indirect" contributions are dominant over the ones that directly involve the Yukawa couplings of the valence quarks of the vector meson V. When one wants to use these decays to probe said Yukawa couplings, it is useful to normalize the branching ratios to the branching ratio of $h \to \gamma\gamma$, since all NP effects in the indirect contribution will affect $h \to \gamma\gamma$ in the same way (up to small corrections due to the off-shellness of the photon and the Z-boson diagram). Expanding in small parameters, we find:

$$\frac{\mathrm{Br}(h \to V\gamma)}{\mathrm{Br}(h \to \gamma\gamma)} = \frac{\Gamma(h \to V\gamma)}{\Gamma(h \to \gamma\gamma)} = \frac{8\pi\,\alpha_{\rm EM}^2(m_V)}{\alpha_{\rm EM}(0)} \frac{Q_V^2 f_V^2}{m_V^2} \left(1 - \frac{m_V^2}{m_h^2}\right)^2 \frac{|1-\Delta_V|^2 + \left|r_{\rm CP} - \tilde\Delta_V\right|^2}{1 + |r_{\rm CP}|^2} . \tag{7}$$

The parameters Δ_V and $\tilde\Delta_V$ contain the direct contributions and the corrections to the indirect contributions due to the off-shellness of the photon and the additional diagram with the Z-boson:

$$\overset{(\sim)}{\Delta}_V = \overset{(\sim)}{\kappa}_V F_{\rm direct}^V + \text{indirect corrections} . \tag{8}$$

The direct form factors $F_{\rm direct}^V$ can written in a similar form to the ones in eq. (4). The parameter $r_{\rm CP}$ vanishes in the SM and contains the various CP-odd couplings to the indirect amplitude and the $h \to \gamma\gamma$ rate. The detailed expressions for both F_V and $r_{\rm CP}$ can be found in ref. [2].

Table 2 quotes our SM branching ratios. Note that in the case of $h \to \Upsilon(1S)\gamma$ the direct and indirect contributions cancel almost exactly, suppressing this decay by three orders of magnitude. This cancellation leads to a spectacular sensitivity of this branching ratio on κ_b and $\tilde\kappa_b$, as demonstrated in

Table 2. Predicted branching fractions for various $h \to V\gamma$ decays, including error estimates due to the uncertainties in the meson decay constants ("f"), the direct contribution ("direct") and the branching ratio $\mathrm{Br}(h \to \gamma\gamma)$.

Decay mode	Branching ratio
$h \to \rho^0\gamma$	$(1.68 \pm 0.02_f \pm 0.08_{h\to\gamma\gamma}) \cdot 10^{-5}$
$h \to \omega\gamma$	$(1.48 \pm 0.03_f \pm 0.07_{h\to\gamma\gamma}) \cdot 10^{-6}$
$h \to \phi\gamma$	$(2.31 \pm 0.03_f \pm 0.11_{h\to\gamma\gamma}) \cdot 10^{-6}$
$h \to J/\psi\,\gamma$	$(2.95 \pm 0.07_f \pm 0.06_{\mathrm{direct}} \pm 0.14_{h\to\gamma\gamma}) \cdot 10^{-6}$
$h \to \Upsilon(1S)\,\gamma$	$(4.61 \pm 0.06_f\,{}^{+1.75}_{-1.21}\,{}_{\mathrm{direct}} \pm 0.22_{h\to\gamma\gamma}) \cdot 10^{-9}$
$h \to \Upsilon(2S)\,\gamma$	$(2.34 \pm 0.04_f\,{}^{+0.75}_{-0.99}\,{}_{\mathrm{direct}} \pm 0.11_{h\to\gamma\gamma}) \cdot 10^{-9}$
$h \to \Upsilon(3S)\,\gamma$	$(2.13 \pm 0.04_f\,{}^{+0.75}_{-1.21}\,{}_{\mathrm{direct}} \pm 0.10_{h\to\gamma\gamma}) \cdot 10^{-9}$

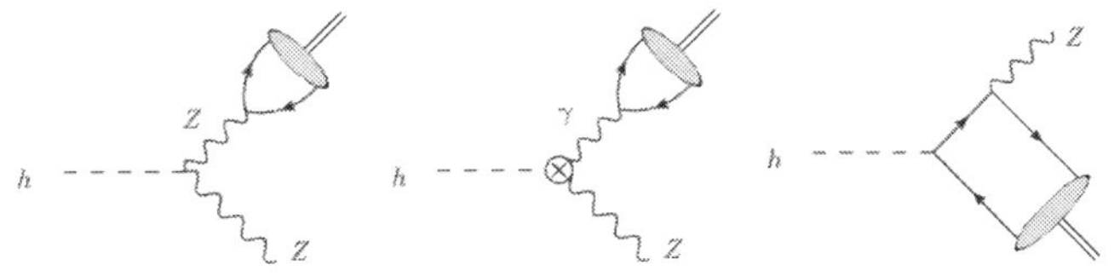

Figure 3. Diagrams contributing to the $h \to ZM$ decays. The suppression from the loops in the second diagram is lifted by the near on-shell photon.

figure 3 for the cases of $\tilde{\kappa}_b$ vanishing (left diagram) and not vanishing (right diagram). When compared to the branching ratio of $h \to \bar{b}b$, even information on the CP-phase of the Yukawa coupling of the b-quark can be obtained.

4.2 Weak radiative Higgs decays

Weak Higgs decays of the form $h \to VZ$ can be used to probe the coupling of the Higgs to a photon and a Z-boson, both loop-induced and from the couplings $\kappa_{\gamma Z}$ and $\tilde{\kappa}_{\gamma Z}$. There are three important contributions, shown in figure 3. While naive dimensional analysis suggests that the tree-level diagram involving the hZZ gauge coupling is the dominant amplitude, the diagram with the loop-induced $h\gamma Z$ vertex is equally important. This is because the loop-suppression is overcome by the photon propagator being almost on its pole. For pseudoscalar mesons, the photon diagram does not exist and thus the decays $h \to PZ$ can only serve as a SM reference values.

The current limits on the parameters $\kappa_{\gamma Z}$ and $\tilde{\kappa}_{\gamma Z}$ from CMS [23] and ATLAS [24] imply upper bounds on the decay rates of 9 and 11 times the SM value, respectively, both at 95% confidence level. The stronger bound from CMS implies the constraint (under the assumption that all other parameters are at their SM value)

$$\sqrt{|\kappa_{\gamma Z} - 2.395|^2 + |\tilde{\kappa}_{\gamma Z}|^2} < 7.2\,. \tag{9}$$

Table 3 quotes our branching ratios both for the SM along with the maximal deviation when the bounds (9) are saturated. Of course it should be noted, that NP effects in $\kappa_{\gamma Z}$ and $\tilde{\kappa}_{\gamma Z}$ will also show in the direct $h \to \gamma Z$ measurement.

Table 3. Predicted branching fractions for various $h \to V\gamma$ decays, including error estimates due to the uncertainties in the meson decay constants ("f"), the direct contribution ("direct") and the branching ratio $\mathrm{Br}(h \to \gamma\gamma)$.

Decay mode	Branching ratio [10^{-6}]	NP effects from $\kappa_{\gamma Z}$
$h \to \pi^0 Z$	$2.30 \pm 0.01_f \pm 0.09_\Gamma$	–
$h \to \eta Z$	$0.83 \pm 0.08_f \pm 0.03_\Gamma$	–
$h \to \eta' Z$	$1.24 \pm 0.12_f \pm 0.05_\Gamma$	–
$h \to \rho^0 Z$	$7.19 \pm 0.09_f \pm 0.28_\Gamma$	1.83 - 53.3
$h \to \omega Z$	$0.56 \pm 0.01_f \pm 0.02_\Gamma$	0.06 - 4.56
$h \to \phi Z$	$2.42 \pm 0.05_f \pm 0.09_\Gamma$	1.77 - 9.12
$h \to J/\Psi Z$	$2.30 \pm 0.06_f \pm 0.09_\Gamma$	1.59 - 13.10
$h \to \Upsilon(1S) Z$	$15.38 \pm 0.21_f \pm 0.60_\Gamma$	13.7 - 20.8

5 Conclusions

Exclusive hadronic decays of heavy electroweak bosons are a theoretically clean playground for the framework of QCD factorization. The hard scale $\mu_{\rm hard}$ is set by the decaying bosons and thus very high, rendering the impact of power corrections of $O(\Lambda_{\rm QCD}/\mu_{\rm hard})$ tiny. Additionally, the hadronic uncertainties are comparably small thanks to the renormalization group evolution of the LCDA. The price to pay is the smallness of the branching ratios. For the Z-decays, the most promising experimental environment would be a future lepton collider running at $\sqrt{s} = m_Z^2$, whereas the Higgs decays are an interesting physics target for the high-luminosity run of the LHC or a future 100 TeV proton collider.

This report skips over a large amount of technical and phenomenological details, like a detailed derivation of the factorization theorem, the renormalization group evolution of the hadronic parameters and the special case of flavor-singlet mesons as well as the $W^\pm \to M^\pm\gamma$ and $h \to M^\pm W^\mp$ decays [1–3]. We nevertheless hope to convince the reader that the physics opportunities in a dedicated program for the decays considered here are compelling, both within and beyond the standard model.

Acknowledgements

M.K is supported by the DFG Graduate School *Symmetry Breaking in Fundamental Interactions* (GRK 1581) and wants to thank the organizers of the workshop for the hospitality and a very interesting conference.

References

[1] Y. Grossman, M. König and M. Neubert, JHEP **1504** (2015) 101.
[2] M. König and M. Neubert, JHEP **1508** (2015) 012.
[3] S. Alte, M. König and M. Neubert, JHEP **1602** (2016) 162.
[4] S. Alte, M. König and M. Neubert, *in preparation*.
[5] G. Aad *et al.* [ATLAS Collaboration], Phys. Lett. B **716** (2012) 1.
[6] S. Chatrchyan *et al.* [CMS Collaboration], Phys. Lett. B **716** (2012) 30.

[7] G. T. Bodwin, F. Petriello, S. Stoynev and M. Velasco, Phys. Rev. D **88** (2013) no.5, 053003.
[8] A. L. Kagan, G. Perez, F. Petriello, Y. Soreq, S. Stoynev and J. Zupan, Phys. Rev. Lett. **114** (2015) 10, 101802.
[9] G. P. Lepage and S. J. Brodsky, Phys. Lett. B **87** (1979) 359.
[10] G. P. Lepage and S. J. Brodsky, Phys. Rev. D **22** (1980) 2157.
[11] A. V. Efremov and A. V. Radyushkin, Theor. Math. Phys. **42** (1980) 97 [Teor. Mat. Fiz. **42**, 147 (1980)].
[12] A. V. Efremov and A. V. Radyushkin, Phys. Lett. B **94** (1980) 245.
[13] V. L. Chernyak and A. R. Zhitnitsky, Phys. Rept. **112** (1984) 173.
[14] C. W. Bauer, S. Fleming, D. Pirjol and I. W. Stewart, Phys. Rev. D **63** (2001) 114020.
[15] C. W. Bauer, D. Pirjol and I. W. Stewart, Phys. Rev. D **65** (2002) 054022.
[16] C. W. Bauer, S. Fleming, D. Pirjol, I. Z. Rothstein and I. W. Stewart, Phys. Rev. D **66** (2002) 014017.
[17] M. Beneke, A. P. Chapovsky, M. Diehl and T. Feldmann, Nucl. Phys. B **643** (2002) 431.
[18] M. V. Terentev, Sov. J. Nucl. Phys. **33** (1981) 911 [Yad. Fiz. **33**, 1692 (1981)].
[19] T. Ohrndorf, Nucl. Phys. B **186** (1981) 153.
[20] M. A. Shifman and M. I. Vysotsky, Nucl. Phys. B **186** (1981) 475.
[21] V. N. Baier and A. G. Grozin, Nucl. Phys. B **192** (1981) 476.
[22] L. Bergstrom and G. Hulth, Nucl. Phys. B **259** (1985) 137 Erratum: [Nucl. Phys. B **276** (1986) 744].
[23] S. Chatrchyan *et al.* [CMS Collaboration], Phys. Lett. B **726** (2013) 587.
[24] G. Aad *et al.* [ATLAS Collaboration], Phys. Lett. B **732** (2014) 8.

On the $h \to V\ell^+\ell^-$ decays

Pietro Santorelli[1,2,⋆]

[1] *Dipartimento di Fisica "Ettore Pancini", Università di Napoli Federico II, Complesso Universitario di Monte Sant'Angelo, Via Cintia, Edificio 6, I-80126 Napoli, Italy*
[2] *INFN, Sezione di Napoli, I-80126 Napoli, Italy*

Abstract. A set of exclusive decay of the Higgs boson into a vector meson and a dilepton pair ($h \to V\ell^+\ell^-$, with $V = \Upsilon, J/\psi, \phi$, and $\ell = \mu, \tau$) are studied in the framework of the Standard Model. We have evaluated the decay rates, the dilepton mass spectra and the V longitudinal helicity fraction distributions. In the same framework, we considered the exclusive modes $h \to V\nu\bar{\nu}$ and the implications of the CMS and ATLAS results for the lepton flavor-changing process $h \to \tau^+\mu^-$ on the $h \to V\tau^+\mu^-$ decay modes.

1 Introduction

The Higgs-like scalar observed at LHC with $m_h = 125.7(4)$ GeV [1–3] seems to fulfilled the prediction of the Standard Model (SM) for the Higgs boson. However, it is important to confirm that the couplings of the observed state to the fermions and gauge bosons are what the SM dictates. The couplings of the observed scalar to top and beauty quarks and to τ leptons are very well studied and are consistent with the SM predictions [4]; less known are the couplings to the light quarks and leptons. Many theoretical papers have been devoted to study how to modify these couplings as a consequences of physic beyond the SM [5–9].

The study of the couplings to the first two generation of fermions is an experimental difficult job. From the theoretical point of view the radiative $h \to f\bar{f}\gamma$ processes have been considered with particular attention. The leptonic modes $h \to \ell^+\ell^-\gamma$ (with $\ell = e, \mu$) have been considered in [10–14]. While the exclusive channels $h \to V\gamma$, with V a vector meson, have been scrutinized in [15–18], and $h \to VZ$ have been studied in [19, 20] as a way to measure the Higgs couplings to the light quarks.

Here we review the results obtained in [21] where we have studied the exclusive Higgs decays $h \to V\ell^+\ell^-$, with $V = \Upsilon, J/\psi, \phi$ and ℓ is a light or a heavy charged lepton. The motivations to study these processes are:

1. there is the possibility of considering, in addition to the decay rates, some distributions encoding important physical information, namely the distributions in the dilepton invariant mass squared;

2. due to the fact that several amplitudes contribute to each process, one can look at kinematical configurations where interferences are enhanced in order to get information on the various Higgs couplings;

⋆e-mail: pietro.santorelli@na.infn.it

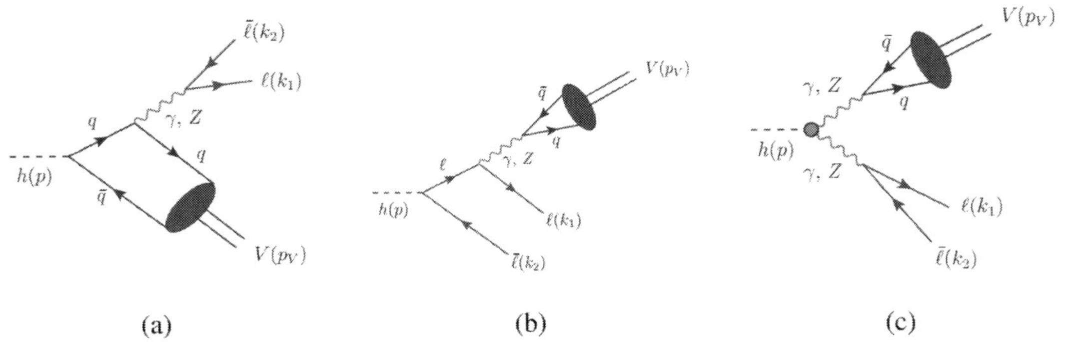

Figure 1. Diagrams contributing to $h \to V\ell^+\ell^-$ decays. In (a) and (b) the diagrams with γ and Z emitted by $\bar{q}$ and $\bar{\ell}$ are considered. In (c) the contributions of the $h \to ZZ$ and the effective $h \to \gamma\gamma$, $h \to \gamma Z$ vertices are considered.

3. they have a clear experimental signature, although the rates are small;
4. deviations from the Standard Model can also be probed through the search of lepton flavor violating signals.

In the next section we will discuss the diagrams contributing to the $h \to V\ell^+\ell^-$ processes, the couplings and the hadronic quantities necessary to evaluate them. In section 3 we show the calculation of the amplitudes, while, in section 4, the numerical results on the branching ratios, decay distributions and the fraction of polarized decay distribution are presented and discussed. Finally, we will give our conclusions.

2 The Relevant Diagrams

The amplitudes contributing to the $h \to V\ell^+\ell^-$ decays contain vertices in which the Higgs couples to quarks, to leptons and to the gauge bosons Z and γ. In SM such couplings are $g_{hf\bar{f}} = i\,m_f/v$ for fermions[1] and $g_{hZZ} = i\,2m_Z^2/v$ for Z ($v = 2m_W/g = (\sqrt{2}G_F)^{-1/2} = 246$ GeV is the Higgs field vacuum expectation value). Fig. 1 displays the three kinds of diagrams that must be taken into account.

In the diagrams of Fig. 1 (a) we have the Higgs coupled to the quark-antiquark pair. The neutral gauge boson, γ or Z, is emitted from the quark or the antiquark line before they hadronize in the vector meson V. When the dilepton invariant mass squared $q^2 = (k_1 + k_2)^2$ is low, the quark hadronization in the vector meson V can be studied by using the formalism of the QCD hard exclusive processes [22–25]. Here we should consider the matrix element of the non-local quark antiquark operator while the vector meson state can be expressed as an expansion in increasing twists, which involves various vector meson distribution amplitudes. In the case of $V = \Upsilon$, J/ψ, ϕ, the leading twist light-cone distribution amplitude (LCDA) $\phi_\perp^V$ is defined from the matrix element of the non-local $\bar{q}(y)\sigma_{\mu\nu}q(x)$ quark current:

$$\langle V(p_V, \epsilon_V)|\bar{q}(y)\sigma_{\mu\nu}q(x)|0\rangle = -f_V^\perp(\epsilon^*_{V\mu}p_{V\nu} - \epsilon^*_{V\nu}p_{V\mu}) \int_0^1 du\, e^{i u\, p_V\cdot x + i\bar{u}\, p_V\cdot y}\phi_\perp^V(u) \quad (1)$$

[1]For the quarks we use the running masses evaluated at the Higgs mass scale $\mu \simeq m_h = 125$ GeV at NNLO in the $\overline{\rm MS}$ scheme.

($\bar{u} = 1 - u$). $u\,p_V$ and $\bar{u}\,p_V$ represent the meson longitudinal momentum fraction carried by the quark and antiquark. $\phi_\perp^V$ is normalized to 1; the hadronic parameter $f_V^\perp$ is discussed below. For more details see [21].
The couplings of the Higgs to leptons are present in the diagrams of Fig. 1 (b), with the $q\bar{q}$ pair emitted by the photon or Z. Such diagrams are important in the case of τ. Here the hadronization of the $q\bar{q}$ pair into the vector meson is represented by the matrix element

$$\langle V(p_V, \epsilon_V)|\bar{q}\,\gamma_\mu\, q|0\rangle = -i f_V m_V \epsilon^*_{V\mu} \,, \tag{2}$$

with p_V and ϵ_V the V meson momentum and polarization vector, respectively. The experimental measurement of the decay width of $V \to e^+e^-$ allow to measure the constant f_V. Less accessible is the hadronic parameter $f_V^\perp$ in (1) and so results from lattice or QCD sum rule computations must be used. In our analysis we use the range for the ratio $R_{f_V} = f_V^\perp/f_V$ quoted in [26], obtained exploiting non-relativistic QCD scaling relations [27, 28]:

$$\begin{aligned} f_\phi &= 0.223 \pm 0.0014 \;\; \text{MeV} \;, & R_{f_\phi} &= 0.76 \pm 0.04 \;, \\ f_{J/\psi} &= 0.4033 \pm 0.0051 \;\; \text{MeV} \;, & R_{f_{J/\psi}} &= 0.91 \pm 0.14 \;, \\ f_\Upsilon &= 0.6844 \pm 0.0046 \;\; \text{MeV} \;, & R_{f_\Upsilon} &= 1.09 \pm 0.04 \;. \end{aligned} \tag{3}$$

The diagrams of Fig. 1 (c) involve the coupling of the Higgs to a pair of gauge bosons, which in turn are coupled to a lepton pair and to a $q\bar{q}$ pair that hadronizes into V. The elementary hZZ coupling can be obtained from the SM Lagrangian. The effective $h\gamma\gamma$ and $hZ\gamma$ vertices can be written as

$$A(H \to G_1 G_2) = i\frac{\alpha}{\pi v} C_{G_1 G_2} \left[g_{\mu\nu}(p_V \cdot q) - p_{V\mu} q_\nu \right] \epsilon_{G_1}^{*\mu} \, \epsilon_{G_2}^{*\nu} \,, \tag{4}$$

with G_1 and G_2 either $\gamma\gamma$ or $Z\gamma$, and ϵ_{G_1}, ϵ_{G_2} polarization vectors. In Eq. (4) p_V is the momentum of the meson V and q the momentum of the dilepton. The effective $h\gamma\gamma$ and $hZ\gamma$ couplings are determined by calculations of loop diagrams: $C_{\gamma\gamma} = -3.266 + i0.021$ and $C_{\gamma Z} = -2.046 + i0.005$ [18]. In the Z propagator we take into account, without considering the uncertainty, the width $\Gamma(Z) = 2.4952$ GeV [3]. It is worth remarking that the possibility to access the hZZ coupling is a feature of the class of modes we are analyzing. Moreover, since a sizeable contribution to $h \to V\ell^+\ell^-$ involves the effective $h\gamma\gamma$ and $hZ\gamma$ couplings from diagrams sensitive to New Physics effects, the exclusive processes also probe deviations from SM.

3 Decay Amplitudes

In this section we give the expressions for the amplitudes corresponding to the diagrams of Fig. 1. At this end, we define

$$C_\gamma = 4\pi\alpha Q_\ell Q_q \,, \qquad C_Z = \frac{4\pi\alpha}{s_W^2 c_W^2} \,, \tag{5}$$

with $s_W = \sin\theta_W$, $c_W = \cos\theta_W$, and θ_W the Weinberg angle, and write the propagators in Fig. 1 in terms of the functions

$$D_1(a, b, \hat{q}^2) = a + b\hat{q}^2 - ab\,\hat{m}_V^2 - \hat{m}_q^2 \,, \qquad D_2(\hat{q}^2) = \hat{q}^2 - \hat{m}_Z^2 + i\,\hat{m}_Z \hat{\Gamma}_Z \,, \qquad D_3(\hat{k}) = 1 - 2n \cdot \hat{k} \,, \tag{6}$$

where $n = (1, \vec{0})$ and we use the notation $\hat{x} = x/m_h$, x being a mass or a momentum. The lepton current, due to the intermediate gauge boson, has various Dirac structures,

$$V_\ell^\mu = \bar{\psi}_\ell(k_1)\gamma^\mu \psi_{\bar{\ell}}(k_2) \,, \tag{7}$$

for the photon; while for the intermediate Z we have

$$A_\ell^\mu = \bar{\psi}_\ell(k_1)\gamma^\mu\gamma_5\psi_{\bar{\ell}}(k_2), \qquad T_\ell^{\mu\nu} = \bar{\psi}_\ell(k_1)\gamma^\mu\gamma^\nu\psi_{\bar{\ell}}(k_2), \qquad \tilde{T}_\ell^{\mu\nu} = \bar{\psi}_\ell(k_1)\gamma^\mu\gamma^\nu\gamma_5\psi_{\bar{\ell}}(k_2). \tag{8}$$

We write the SM neutral current coupled to the Z boson as

$$\mathcal{L}_\mu = \left(-\frac{ie}{s_W c_W}\right)\left(\Delta_V^f \bar{f}\gamma_\mu f + \Delta_A^f \bar{f}\gamma_\mu\gamma_5 f\right), \tag{9}$$

where f denotes a fermion, and

$$\Delta_V^f = \frac{1}{2}\left(T_3^f - 2s_W^2 Q^f\right), \qquad \Delta_A^f = -\frac{1}{2}T_3^f, \tag{10}$$

with T_3^f the third component of the weak isospin and Q^f the electric charge of f. Diagrams in Fig. 1(a) also involve the integrals over the LCDA of the vector meson V:

$$I_1 = I_1(\hat{q}^2) = \int_0^1 du\, \phi_\perp^V(u)\left[\frac{1}{D_1(1-u,u,\hat{q}^2)} + \frac{1}{D_1(u,1-u,\hat{q}^2)}\right], \tag{11}$$

$$I_2 = I_2(\hat{q}^2) = \int_0^1 du\, \phi_\perp^V(u)\left[\frac{u}{D_1(1-u,u,\hat{q}^2)} + \frac{1-u}{D_1(u,1-u,\hat{q}^2)}\right]. \tag{12}$$

We report the various expressions in correspondence with the diagrams in Fig. 1(a), (b) and (c), considering separately the intermediate photon and Z contributions.

- Fig. 1(a), intermediate γ:

$$A_{(a)}^\gamma = C_{(a)}^\gamma m_h \epsilon_V^{*\alpha} V_{\ell\mu}\left\{[n_\alpha \hat{p}_V^\mu - g_\alpha^\mu(n\cdot\hat{p}_V)]I_1 - g_\alpha^\mu \hat{m}_V^2 I_2\right\}, \qquad \text{with} \qquad C_{(a)}^\gamma = \frac{1}{m_h^2}\frac{\hat{m}_q}{v}C_\gamma f_V^\perp \frac{1}{\hat{q}^2}. \tag{13}$$

- Fig. 1(a), intermediate Z:

$$A_{(a)}^Z = C_{(a)}^Z \epsilon_V^{*\alpha}\left[\Delta_V^\ell V_{\ell\mu} + \Delta_A^\ell A_{\ell\mu}\right]\left(g^{\mu\alpha}p_V^\sigma - g^{\alpha\sigma}p_V^\mu\right)[n_\sigma I_1 - \hat{p}_{V\sigma}I_2], \tag{14}$$

with

$$C_{(a)}^Z = -\frac{1}{m_h^2}\frac{\hat{m}_q}{v}C_Z\frac{1}{D_2(\hat{q}^2)}f_V^\perp \Delta_V^q. \tag{15}$$

- Fig. 1(b), intermediate γ:

$$A_{(b)}^\gamma = C_{(b)}^\gamma \epsilon_V^{*\alpha}n^\mu\left[-\frac{1}{D_3(\hat{k}_1)}T_{\ell\mu\alpha} + \frac{1}{D_3(\hat{k}_2)}T_{\ell\alpha\mu}\right], \qquad \text{with} \qquad C_{(b)}^\gamma = \frac{1}{m_h^2}\frac{\hat{m}_\ell}{v}C_\gamma\frac{f_V m_V}{\hat{m}_V^2}. \tag{16}$$

- Fig. 1(b), intermediate Z:

$$A_{(b)}^Z = C_{(b)}^Z \epsilon_{V\alpha}^* n_\mu\left\{-\frac{1}{D_3(\hat{k}_1)}\left[\Delta_V^\ell T_\ell^{\mu\alpha} + \Delta_A^\ell \tilde{T}_\ell^{\mu\alpha}\right] + \frac{1}{D_3(\hat{k}_2)}\left[\Delta_V^\ell T_\ell^{\alpha\mu} - \Delta_A^\ell \tilde{T}_\ell^{\alpha\mu}\right]\right\}, \tag{17}$$

with

$$C_{(b)}^Z = \frac{1}{m_h^2}\frac{\hat{m}_\ell}{v}C_Z\frac{\Delta_V^q}{D_2(\hat{m}_V^2)}f_V m_V. \tag{18}$$

- Fig. 1(c), two intermediate photons:

$$A^{\gamma\gamma}_{(c)} = C^{\gamma\gamma}_{(c)} \epsilon^{*\alpha}_V [g_{\alpha\mu}(q \cdot p_V) - m_h^2 n_\alpha n_\mu] V^\mu_\ell , \qquad \text{with} \qquad C^{\gamma\gamma}_{(c)} = \frac{1}{m_h^4} \frac{\alpha}{\pi v} C_{\gamma\gamma} C_\gamma \frac{f_V m_V}{\hat{m}_V^2} \frac{1}{\hat{q}^2} . \quad (19)$$

- Fig. 1(c), two intermediate Z:

$$A^{ZZ}_{(c)} = C^{ZZ}_{(c)} \epsilon^*_{V\alpha} \left(\Delta^\ell_V V^\alpha_\ell + \Delta^\ell_A A^\alpha_\ell\right) , \qquad \text{with} \qquad C^{ZZ}_{(c)} = \frac{1}{m_h^2} \frac{2\hat{m}_Z^2}{v} C_Z \frac{1}{D_2(\hat{q}^2)} \frac{1}{D_2(\hat{m}_V^2)} \Delta^q_V f_V m_V . \quad (20)$$

- Fig. 1(c), intermediate γZ, with γ converting to leptons:

$$A^{\gamma Z}_{(c)} = C^{\gamma Z}_{(c)} \epsilon^{*\alpha}_V [g_{\alpha\mu}(q \cdot p_V) - m_h^2 n_\alpha n_\mu] V^\mu_\ell , \qquad \text{with} \qquad C^{\gamma Z}_{(c)} = \frac{1}{m_h^4} \frac{\alpha}{\pi v} C_{\gamma Z} \frac{4\pi\alpha Q_\ell}{s_W c_W} \frac{1}{\hat{q}^2} \frac{\Delta^q_V}{D_2(\hat{m}_V^2)} f_V m_V . \quad (21)$$

- Fig. 1(c), intermediate $Z\gamma$, with Z converting to leptons:

$$A^{Z\gamma}_{(c)} = C^{Z\gamma}_{(c)} \epsilon^{*\alpha}_V [g_{\alpha\mu}(q \cdot p_V) - m_h^2 n_\alpha n_\mu] \left(\Delta^\ell_V V^\mu_\ell + \Delta^\ell_A A^\mu_\ell\right) , \quad (22)$$

with

$$C^{Z\gamma}_{(c)} = \frac{1}{m_h^4} \frac{\alpha}{\pi v} C_{\gamma Z} \frac{4\pi\alpha Q_q}{s_W c_W} \frac{1}{\hat{m}_V^2} \frac{1}{D_2(\hat{q}^2)} f_V m_V . \quad (23)$$

The effective couplings $C_{\gamma\gamma}$ and $C_{\gamma Z}$ are defined through Eq. (4).

4 Numerical Analysis

Starting from the expressions obtained in the previous section we are able to compute the widths of the Higgs into the final states. But to calculate the branching fractions is necessary to get rid of the poorly known Higgs full width. One possibility is to use the expression [18]

$$\mathcal{B}(h \to V\ell^+\ell^-) = \frac{\Gamma(h \to V\ell^+\ell^-)}{\Gamma(h \to \gamma\gamma)} \mathcal{B}(h \to \gamma\gamma)_{exp} \quad (24)$$

which employs the computed widths $\Gamma(h \to V\ell^+\ell^-)$ and $\Gamma(h \to \gamma\gamma) = \left(\alpha^2/(64\pi^3 v^2)\right)|C_{\gamma\gamma}|^2 m_h^3$ combined with the measurement $\mathcal{B}(h \to \gamma\gamma)_{exp} = (2.28 \pm 0.11) \times 10^{-3}$ [29]. We obtain

$$\begin{aligned} \mathcal{B}(h \to \phi\mu^+\mu^-) &= (7.93 \pm 0.39) \times 10^{-8} \qquad \mathcal{B}(h \to \phi\tau^+\tau^-) = (2.35 \pm 0.12) \times 10^{-6} \\ \mathcal{B}(h \to J/\psi\mu^+\mu^-) &= (9.10 \pm 0.50) \times 10^{-8} \qquad \mathcal{B}(h \to J/\psi\tau^+\tau^-) = (1.82 \pm 0.10) \times 10^{-6} \quad (25) \\ \mathcal{B}(h \to \Upsilon\mu^+\mu^-) &= (5.60 \pm 0.37) \times 10^{-7} \qquad \mathcal{B}(h \to \Upsilon\tau^+\tau^-) = (5.66 \pm 0.29) \times 10^{-7} . \end{aligned}$$

The errors in the branching ratios take into account the uncertainties on the LCDA parameters (cfr [21]), on the decay constants f_V and on the ratios R_{f_V} in Eqs. (3), and the error on $\mathcal{B}(h \to \gamma\gamma)_{exp}$. The largest contribution to the uncertainties on the branching ratios is due to the uncertainty on $\mathcal{B}(h \to \gamma\gamma)_{exp}$ amounting to 50 – 60% of the total error. The uncertainty on R_{f_V} constitutes 20 – 30% of the total error.

The larger rates in Eq. (25) are predicted for modes with τ pairs, $h \to \phi\tau^+\tau^-$ and $h \to J/\psi\tau^+\tau^-$. Due to the smaller coupling, the processes with muons have rates suppressed by a factor 30 and 20, respectively; however, the higher experimental identification efficiency should cancel this suppression.

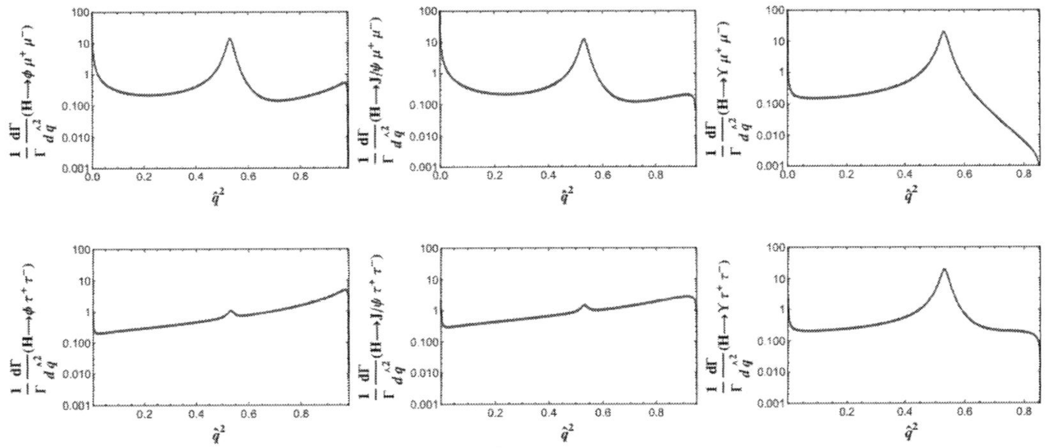

Figure 2. The decay distributions $(1/\Gamma)d\Gamma(h \to V\ell^+\ell^-)/d\hat{q}^2$, with $\hat{q}^2 = q^2/m_h^2$ and q^2 the dilepton mass squared.

In the case of Υ, the branching ratios with $\tau^+\tau^-$ and $\mu^+\mu^-$ are similar. This is an effect of the dominance of the diagram with two intermediate Z, Fig. 1 (c), and with coincident contributions. For the $h \to \Upsilon\mu^+\mu^-$ the next most relevant diagram is the one with the Higgs coupled to quarks, Fig. 1 (a), and it gives contribution very similar to the one coming from the diagram with the Higgs coupled to leptons, Fig. 1 (b), which has the same role for the $h \to \Upsilon\tau^+\tau^-$ mode.

A comparison with the branching ratios of two body modes is in order. In [18] the authors found $\mathcal{B}(h \to \phi\gamma) = (2.31 \pm 0.11) \times 10^{-6}$ and $\mathcal{B}(h \to J/\psi\gamma) = (2.95 \pm 0.17) \times 10^{-6}$, while $\mathcal{B}(h \to \Upsilon\gamma)$ is $O(10^{-9})$. For the $h \to VZ$ modes, $\mathcal{B}(h \to \phi Z) \simeq \mathcal{B}(h \to J/\psi Z) = 2.2 \times 10^{-6}$ are expected in SM [17].

In Fig. 2 the decay distributions are plotted in the normalized dilepton mass squared $\hat{q}^2 = q^2/m_h^2$. The studied modes, with the exception of $h \to \phi\tau^+\tau^-$ and $h \to J/\psi\tau^+\tau^-$, are dominated by the virtual photon and Z contributions in Fig. 1 (c). In the $h \to \phi\tau^+\tau^-$ and $h \to J/\psi\tau^+\tau^-$ the $\hat{q}^2$-distributions show a small Z peak and increase with $\hat{q}^2$: an effect of the diagrams with the Higgs coupled to the leptons.
For all the modes the forward-backward lepton asymmetry is very small in the whole range of $\hat{q}^2$.

In Fig. 3, the $F_L(\hat{q}^2) = \left(d\Gamma_L(h \to V\ell^+\ell^-)/d\hat{q}^2\right) / \left(d\Gamma(h \to V\ell^+\ell^-)/d\hat{q}^2\right)$ distributions of the fraction of longitudinally polarised vector meson are depicted. $F_L \simeq 1$ at the Z mass for the modes with muons in the final state. For the $\phi\tau^+\tau^-$ and $J/\psi\tau^+\tau^-$ one can see a narrow peaks in F_L for $\hat{q}^2 = m_Z^2/m_h^2$, all the other cases present a smooth $\hat{q}^2$ dependence. It is quite simple to modify the diagrams in Fig.1 to be able to study the $h \to V\nu\bar{\nu}$ decay widths. Our predictions are the following

$$\begin{aligned} \mathcal{B}(h \to \phi\nu\bar{\nu}) &= (1.50 \pm 0.075) \times 10^{-7}, \\ \mathcal{B}(h \to J/\psi\nu\bar{\nu}) &= (1.54 \pm 0.085) \times 10^{-7}, \\ \mathcal{B}(h \to \Upsilon\nu\bar{\nu}) &= (1.52 \pm 0.08) \times 10^{-6}, \end{aligned} \tag{26}$$

with a factor 3 included to account for the neutrino species. The measurements of these decay modes are particularly challenging.

Finally, it is interesting to look at the implications on the processes we have studied of possible lepton flavour violating transition $h \to \tau\mu$. The $h \to \tau\mu$ process has been studied at LHC: for such a mode the CMS Collaboration has published $\mathcal{B}(h \to \tau\mu) = \left(0.84^{+0.39}_{-0.37}\right) \times 10^{-2}$ together with the upper bound $\mathcal{B}(h \to \tau\mu) < 1.51 \times 10^{-2}$ at 95% CL [30], while the ATLAS Collaboration quotes the bound

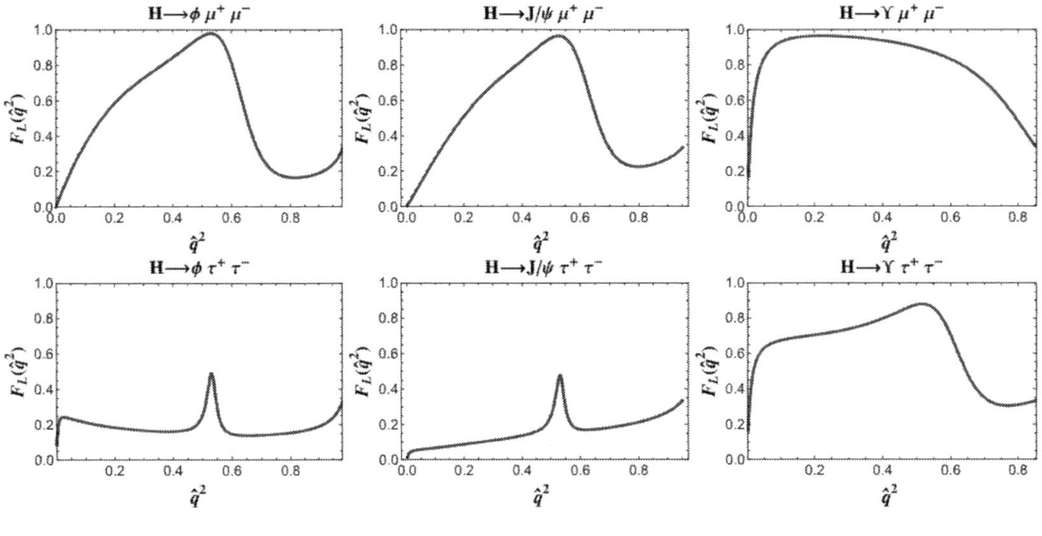

Figure 3. Fraction $F_L(\hat{q}^2)$ of longitudinally polarized meson.

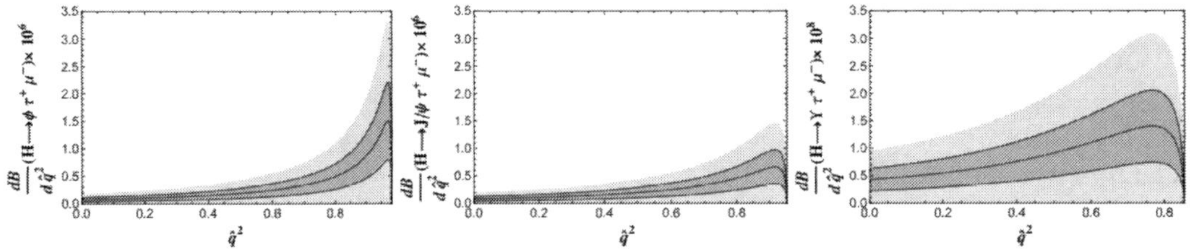

Figure 4. Distributions $d\mathcal{B}(h \to V\tau^+\mu^-)/d\hat{q}^2$ obtained in correspondence to the CMS result for $\mathcal{B}(h \to \tau^+\mu^-)$ [30]. The light shaded area corresponds to the ATLAS bound in [31].

$\mathcal{B}(h \to \tau\mu) < 1.85 \times 10^{-2}$ at 95% CL [31]. By using the CMS results, the effective coupling, $\kappa_{h\tau\mu}$, can be extracted $\kappa_{h\tau\mu} = (2.6 \pm 0.6) \times 10^{-3}$, considering the uncertainties on $\mathcal{B}(h \to \tau^+\mu^-)$ and $\mathcal{B}(h \to \gamma\gamma)$. The ATLAS upper bound, instead, implies $\kappa_{h\tau\mu} < 3.9 \times 10^{-3}$. For these values, the $h \to V\tau^+\mu^-$ branching fractions and their upper bounds can be computed from the diagrams in Fig. 1 (b):

$$\begin{aligned}
\mathcal{B}(h \to \phi\tau^+\mu^-) &= (3.2 \pm 1.5) \times 10^{-7} \;\; (< 6.9 \times 10^{-7}), \\
\mathcal{B}(h \to J/\psi\tau^+\mu^-) &= (2.4 \pm 1.1) \times 10^{-7} \;\; (< 5.2 \times 10^{-7}), \\
\mathcal{B}(h \to \Upsilon\tau^+\mu^-) &= (7.2 \pm 3.4) \times 10^{-9} \;\; (< 1.6 \times 10^{-8}).
\end{aligned} \tag{27}$$

As one can see by looking at Fig. 4, all the decay distributions have an enhancement at large q^2.

5 Conclusions

For the exclusive decay modes $h \to V\ell^+\ell^-$ the obtained branching ratios are in the range $10^{-8} \div 10^{-6}$ in SM, of the same order of magnitude of $h \to \phi\gamma$, $h \to \Upsilon\gamma$, $h \to (\phi, J/\psi)Z$. The largest rate is predicted for $h \to \phi\tau^+\tau^-$. The branching ratios of the neutrino modes have been calculated. We have also studied the lepton flavour-changing process $h \to V\tau\mu$ by using the CMS and ATLAS experimental results on the $h \to \tau\mu$ process.

Acknowledgements

I would like to thank Pietro Colangelo and Fulvia De Fazio for the pleasant collaboration.

References

[1] G. Aad et al. (ATLAS), Phys. Lett. **B716**, 1 (2012), 1207.7214.
[2] S. Chatrchyan et al. (CMS), Phys. Lett. **B716**, 30 (2012), 1207.7235.
[3] K.A. Olive et al. (Particle Data Group), Chin. Phys. **C38**, 090001 (2014).
[4] ATLAS, CMS, ATLAS-CONF-2015-044, CMS-PAS-HIG-15-002 (2015).
[5] R. Contino, M. Ghezzi, C. Grojean, M. Muhlleitner, M. Spira, JHEP **07**, 035 (2013), 1303.3876.
[6] I. Brivio, T. Corbett, O.J.P. Éboli, M.B. Gavela, J. Gonzalez-Fraile, M.C. Gonzalez-Garcia, L. Merlo, S. Rigolin, JHEP **03**, 024 (2014), 1311.1823.
[7] M. Gonzalez-Alonso, A. Greljo, G. Isidori, D. Marzocca, Eur. Phys. J. **C75**, 128 (2015), 1412.6038.
[8] R.S. Gupta, A. Pomarol, F. Riva, Phys. Rev. **D91**, 035001 (2015), 1405.0181.
[9] Y.T. Chien, V. Cirigliano, W. Dekens, J. de Vries, E. Mereghetti (2015), 1510.00725.
[10] A. Abbasabadi, D. Bowser-Chao, D.A. Dicus, W.W. Repko, Phys. Rev. **D55**, 5647 (1997), hep-ph/9611209.
[11] L.B. Chen, C.F. Qiao, R.L. Zhu, Phys. Lett. **B726**, 306 (2013), 1211.6058.
[12] Y. Sun, H.R. Chang, D.N. Gao, JHEP **05**, 061 (2013), 1303.2230.
[13] D.A. Dicus, W.W. Repko, Phys. Rev. **D87**, 077301 (2013), 1302.2159.
[14] G. Passarino, Phys. Lett. **B727**, 424 (2013), 1308.0422.
[15] G.T. Bodwin, F. Petriello, S. Stoynev, M. Velasco, Phys. Rev. **D88**, 053003 (2013), 1306.5770.
[16] A.L. Kagan, G. Perez, F. Petriello, Y. Soreq, S. Stoynev, J. Zupan, Phys. Rev. Lett. **114**, 101802 (2015), 1406.1722.
[17] G. Isidori, A.V. Manohar, M. Trott, Phys. Lett. **B728**, 131 (2014), 1305.0663.
[18] M. Koenig, M. Neubert, JHEP **08**, 012 (2015), 1505.03870.
[19] B. Bhattacharya, A. Datta, D. London, Phys. Lett. **B736**, 421 (2014), 1407.0695.
[20] D.N. Gao, Phys. Lett. **B737**, 366 (2014), 1406.7102.
[21] P. Colangelo, F. De Fazio, P. Santorelli, Phys. Lett. **B760**, 335 (2016), 1602.01372.
[22] G.P. Lepage, S.J. Brodsky, Phys. Lett. **B87**, 359 (1979).
[23] G.P. Lepage, S.J. Brodsky, Phys. Rev. **D22**, 2157 (1980).
[24] A.V. Efremov, A.V. Radyushkin, Phys. Lett. **B94**, 245 (1980).
[25] V.L. Chernyak, A.R. Zhitnitsky, Phys. Rept. **112**, 173 (1984).
[26] Y. Grossman, M. Koenig, M. Neubert, JHEP **04**, 101 (2015), 1501.06569.
[27] W.E. Caswell, G.P. Lepage, Phys. Lett. **B167**, 437 (1986).
[28] G.T. Bodwin, E. Braaten, G.P. Lepage, Phys. Rev. **D51**, 1125 (1995), [Erratum: Phys. Rev.D55,5853(1997)], hep-ph/9407339.
[29] J.R. Andersen et al. (LHC Higgs Cross Section Working Group) (2013), 1307.1347.
[30] V. Khachatryan et al. (CMS), Phys. Lett. **B749**, 337 (2015), 1502.07400.
[31] G. Aad et al. (ATLAS), JHEP **11**, 211 (2015), 1508.03372.

EPJ Web of Conferences **129**, 00016 (2016) DOI: 10.1051/epjconf/201612900016
QCD@Work 2016

Classical Higgs fields on gauge gluon bundles

Marcella Palese[1,★] and *Ekkehart* Winterroth[1,2,★★]

[1] *Department of Mathematics, University of Torino, via C. Alberto 10, 10123 Torino, Italy*
[2] *Lepage Research Institute, University of Prešov, 17 November 1, 081 16 Prešov, Slovak Republic*

Abstract. Classical Higgs fields and related canonical conserved quantities are defined by invariant variational problems on suitably defined gauge gluon bundles. We consider Lagrangian field theories which are assumed to be invariant with respect to the action of a gauge-natural group. As an illustrative example we exploit the 'gluon Lagrangian', *i.e.* a Yang-Mills Lagrangian on the $(1, 1)$-order gauge-natural bundle of $SU(3)$-principal connections. The kernel of the gauge-natural Jacobi morphism for such a Lagrangian, by inducing a reductive split structure, canonically defines a 'gluon classical Higgs field'.

1 Introduction

In this note we shall deal with the definition of a classical Higgs field canonically induced by the invariance of a gluon Yang-Mills Lagrangian with respect to the gauge-natural infinitesimal tranformations of the bundle of $SU(3)$-connections, seen as a $(1, 1)$-order gauge-natural affine bundle.

In particular, the Jacobi equations associated with the gluon Lagrangian define a *canonical classical Higgs field*, that is a reduction of the relevant principal bundle structure. It is noteworthy that the principal bundle structure is not the $SU(3)$-principal bundle, but its $(1, 1)$-order gauge-natural prolongation. For basics on gauge-natural prolongations and applications in Physics, see *e.g.* [1, 2, 4]. Classical physical fields are sections of bundles functorially associated with gauge-natural prolongations of principal bundles, by means of left actions of Lie groups on manifolds.

Let us shortly summarize the geometric frame. Denote by $\boldsymbol{P} \to \boldsymbol{X}$ a principal bundle with structure group $\boldsymbol{G}$, $dim\boldsymbol{X} = n$, by $L_k(\boldsymbol{X})$ the bundle of k–frames in $\boldsymbol{X}$. For $r \leq k$ the *gauge-natural prolongation of* $\boldsymbol{P}$ is $\boldsymbol{W}^{(r,k)}\boldsymbol{P} \doteq J_r\boldsymbol{P} \times_{\boldsymbol{X}} L_k(\boldsymbol{X})$, a principal bundle over $\boldsymbol{X}$ with structure group the semi-direct product $\boldsymbol{W}^{(r,k)}_n\boldsymbol{G} \equiv T^r_n\boldsymbol{G} \rtimes GL_k(n)$, with $GL_k(n)$ group of k–frames in $\mathbb{R}^n$ while $T^r_n\boldsymbol{G}$ is the space of (r, n)-velocities on $\boldsymbol{G}$.

Let $\boldsymbol{F}$ be a manifold and $\zeta : \boldsymbol{W}^{(r,k)}_n\boldsymbol{G} \times \boldsymbol{F} \to \boldsymbol{F}$ be a left action of $\boldsymbol{W}^{(r,k)}_n\boldsymbol{G}$ on $\boldsymbol{F}$. It is associated a *gauge-natural bundle* of order (r, k) defined by $\boldsymbol{Y}_\zeta \doteq \boldsymbol{W}^{(r,k)}\boldsymbol{P} \times_\zeta \boldsymbol{F}$.

Since the group $Diff(\boldsymbol{X})$ is not canonically embedded into the group $Aut(\boldsymbol{P})$ of all automorphisms of the underling principal bundle $\boldsymbol{P}$, there is *a priori* no natural way of relating infinitesimal gauge transformations with infinitesimal base transformations. A *canonical* determination of Noether conserved quantities, without fixing any connection *a priori*, can be performed on a reduced subbundle of

★e-mail: marcella.palese@unito.it
★★corresponding author, e-mail: ekkehart.winterroth@unito.it

$W^{(r,k)}\boldsymbol{P}$ determined by the original $W_n^{(r,k)}\boldsymbol{G}$-invariant variational problem. Connections can be characterized by means of such a canonical reduction and conserved quantities can be characterized in terms of Higgs fields on gauge principal bundles having such a richer structure [6, 8, 9].

2 Higgs field induced by 'gluon Lagrangians'

As well known the Standard Model is a gauge theory with structure group $\boldsymbol{G} = SU(3)\times SU(2)\times U(1)$. One can consider the coupling with gravity by adding the principal spin bundle $\bar{\Sigma}$ with structure group Spin(1, 3); the structure bundle of the whole theory can be then taken to be the fibered product $\boldsymbol{\Sigma} = \bar{\Sigma} \times_X \boldsymbol{P}$. There is an action of Spin(1, 3) on a spinor matter manifold $V = \mathbb{C}^k$ and therefore a representation Spin$(1,3) \times SU(3) \times SU(2) \times U(1) \times V$, given by a choice of Dirac matrices for each component of the spinor field.

A corresponding Lagrangian is therefore given by $\lambda = \bar{\psi}(i\gamma_\mu D^\mu - m)\psi - \frac{1}{4}(\mathcal{F}_{\mu\nu}\mathcal{F}^{\mu\nu} + \mathcal{F}^A_{\mu\nu}\mathcal{F}^{\mu\nu}_A + \mathcal{F}^a_{\mu\nu}\mathcal{F}^{\mu\nu}_a)$. Experimental evidence concerned with symmetry properties of fundamental interactions shows the phenomenon of *spontaneous symmetry breaking* suggesting the presence of a scalar field called the Higgs boson on which the spin group acts trivially. A very clear and precise introduction to those topics can be found in [12].

For an illustrative purpose, let us then restrict to pure gluon fields assumed to be critical sections of the 'gluon Lagrangian' $\lambda_{gluon} = -\frac{1}{4}\mathcal{F}^a_{\mu\nu}\mathcal{F}^{\mu\nu}_a$. In this note, we shall therefore restrict to a principal bundle $\boldsymbol{\Sigma}$ with structure group $\boldsymbol{G} = SU(3)$ such that $\boldsymbol{\Sigma}/SU(3) = \boldsymbol{X}$ and $dim\boldsymbol{X} = 4$.

Recall that $W_4^{(1,1)}\boldsymbol{G}$ is the semi-direct product of $GL(4,\mathbb{R})$ on $T_4^1\boldsymbol{G}$, where $GL(4,\mathbb{R})$ is the structure group of linear frames in $\mathbb{R}^4$.

The set $\{j_0^k\alpha : \alpha : \mathbb{R}^4 \to \mathbb{R}^4\}$, with $\alpha(0) = 0$ locally invertible, equipped with the jet composition $j_0^k\alpha \circ j_0^k\alpha' := j_0^k(\alpha \circ \alpha')$ is a Lie group called the k-th differential group and denoted by G_4^k. For $k = 1$ we have, of course, the identification $G_4^1 \simeq GL(4,\mathbb{R})$. The principal bundle over $\boldsymbol{X}$ with group G_4^k is called the k-th order frame bundle over $\boldsymbol{X}$ and will be denoted by $L_k\boldsymbol{X}$. For $k = 1$ we have the identification $L_1\boldsymbol{X} \simeq L\boldsymbol{X}$, where $L\boldsymbol{X}$ is the usual (principal) bundle of linear frames over $\boldsymbol{X}$.

Unlike $J_1\boldsymbol{\Sigma}$, $W^{(1,1)}\boldsymbol{\Sigma}$ is a principal bundle over $\boldsymbol{X}$ with structure group $W_4^{(1,1)}\boldsymbol{G} \doteq T_4^1SU(3) \rtimes GL(4,\mathbb{R})$, $T_4^1SU(3)$ being the Lie group of (4, 1)-velocities of $SU(3)$ (if $u : \mathbb{R}^4 \to SU(3)$, a generic element of $j_0^1u \in T_4^1SU(3)$ is represented by $g^b = u^b(0)$ and $g^b_\nu = (\partial_\nu(g^{-1}\cdot u(x))|_{x=0})^b$). The group multiplication on $W^{(1,1)}\boldsymbol{G}$ being $(j_0^1\alpha, j_0^1a) \odot (j_0^1\beta, j_0^1b) \doteq (j_0^1(\alpha \circ \beta), j_0^1((a \circ \beta)\cdot b))$ and denoting by $\cdot_r$ the right action of $SU(3)$ on $\boldsymbol{\Sigma}$, the right action of $W_4^{(1,1)}\boldsymbol{G}$ on $W^{(1,1)}\boldsymbol{\Sigma}$ is then defined by $(j_0^1\rho, j_x^1\sigma) \odot (j_0^1\alpha, j_0^1a) \doteq (j_0^1(\rho \circ \alpha), j_x^1(\sigma \cdot_r (a \circ \alpha^{-1} \circ \rho^{-1})))$.

The bundle of principal connections on $\boldsymbol{\Sigma}$ is a gauge-natural bundle associated with the gauge-natural prolongation $W^{(1,1)}\boldsymbol{\Sigma}$. Consider the action ζ induced by the adjoint representation: $\zeta : W_4^{1,1}\boldsymbol{G} \times (\mathbb{R}^4)^* \otimes \mathfrak{su}(3) \to (\mathbb{R}^4)^* \otimes \mathfrak{su}(3) : ((g^b, g^c_\mu, \alpha^\sigma_\rho), f^a_\nu) \mapsto (Ad_g)^a_b(f^b_\sigma - g^b_\sigma)\bar{\alpha}^\sigma_\nu$, where $(Ad_g)^a_b$ are the coordinate expression of the adjoint representation of $\boldsymbol{G} = SU(3)$ and g^b, g^c_μ denote natural coordinates on $T_4^1SU(3)$. The sections of the associated bundle $C(\boldsymbol{\Sigma}) \doteq W^{(1,1)}\boldsymbol{\Sigma} \times_\zeta (\mathbb{R}^4)^* \otimes \mathfrak{su}(3) \to \boldsymbol{X}$ are in 1 to 1 correspondence with the principal connections on $\boldsymbol{\Sigma}$ and are called a $SU(3)$-connections. Clearly, by construction, $C(\boldsymbol{\Sigma})$ is a (1, 1)-order gauge-natural affine bundle [4], see also [3].

Note that the Lie algebra of $W_4^{(1,1)}SU(3)$ is the semi-direct product of $\mathfrak{gl}(4,\mathbb{R})$ with the Lie algebra, $\mathfrak{t}_4^1\mathfrak{su}(3)$, of $T_4^1SU(3)$. It is easy to characterize the semi-direct product of the two Lie algebras, from now on denoted by $\mathcal{S}$, as the direct sum $\mathfrak{t}_4^1\mathfrak{su}(3) \oplus \mathfrak{gl}(4,\mathbb{R})$ with a bracket induced by the right action of $GL(4,\mathbb{R})$ on $T_4^1SU(3)$ given by the jet composition, in particular by the induced Lie algebra homomorphism $\mathfrak{t}_4^1\mathfrak{su}(3) \to hom(\mathfrak{gl}(4,\mathbb{R}))$; given a base of $\mathfrak{t}_4^1\mathfrak{su}(3) \rtimes \mathfrak{gl}(4,\mathbb{R})$; the adjoint representation of the Lie group $W_4^{(1,1)}SU(3)$ is also easily defined (see *e.g.* [15, 16], §1.3). Local coordinates on $W_4^{1,1}SU(3)$ are given by $(g^b, g^b_\sigma; \alpha^\mu_\sigma)$ and induced local coordinates on $\mathcal{S}$ are given by $(Y^a, Y^a_\mu, X^\mu_\sigma)$. Local generators of the tangent space are of course partial derivative with respect to such local coordinates.

Consider the right action $R_{\hat{g}} : W^{(1,1)}\mathbf{\Sigma} \to W^{(1,1)}\mathbf{\Sigma}$, $\hat{g} \in W^{(1,1)}_4 SU(3)$. Let Ξ be a right invariant vector field on $W^{(1,1)}\mathbf{\Sigma}$. In coordinates we have $\Xi = \xi^\lambda(x)\partial_\lambda + \Xi^a(x)\tilde{\mathfrak{b}}_a$ where $(\tilde{\mathfrak{b}}_a)$ is the base of vertical right invariant vector fields on $W^{(1,1)}\mathbf{\Sigma}$ which are induced by the base $(\mathfrak{b}_a)$ of $\mathcal{S}$. They are sections of the bundle $TW^{(1,1)}\mathbf{\Sigma}/W^{(1,1)}_4 SU(3) \to \boldsymbol{X}$. We have $\tilde{\mathfrak{b}}_a = (R_{\hat{g}})^b_a\partial_b$, where the invertible matrix $(R_{\hat{g}})^b_a$ is the matrix representation of $TR_{\hat{g}}$. It is clear that so-called Gell-Mann matrices λ_a are matrix representations of $\mathfrak{b}_a$ and they therefore induce $\tilde{\mathfrak{b}}_a$ in the standard way. Analogously a matrix representation can be obtained for $\mathfrak{b}^\mu_a$, being essentially $T^1_4 SU(3) \rtimes GL(4, \mathbb{R}) \simeq (SU(3) \times (\mathbb{R}^4)^* \otimes \mathfrak{su}(3)) \rtimes GL(4, \mathbb{R})$.

2.1 A variationally featured classical Higgs field associated with the gauge-natural invariant gluon Lagrangian

The functorial linearity properties of the gauge-natural lift $\hat{\Xi}$ of infinitesimal automorphisms of $W^{(1,1)}\mathbf{\Sigma}$ to he bundle $C(\mathbf{\Sigma})$ of $SU(3)$-connections enable to define a *gauge-natural generalized Jacobi morphism* associated with a Lagrangian λ and the *variation vector field* $\hat{\Xi}_V$, the vertical part of $\hat{\Xi}$, *i.e.* the linear morphism $\mathcal{J}(\lambda_{gluon}, \hat{\Xi}_V) \doteq \hat{\Xi}_V \rfloor \mathcal{E}(\hat{\Xi}_V \rfloor \mathcal{E}(\lambda_{gluon}))$, where $\mathcal{E}$ is the Euler-Lagrange operator [5, 7]. Lifted tangent vector fields the vertical part of which are in the kernel $\mathfrak{K} \doteq \ker \mathcal{J}(\lambda_{gluon}, \hat{\Xi}_V)$ are called *generalized Jacobi vector fields* and generate *canonical* covariant conserved quantities [8]. They have the property that the Lie derivative of critical sections of Yang–Mills Euler–Lagrange equations taken with respect to them is zero, *i.e. they leave invariant the equations and their solutions*. It is a subalgebra of the Lie algebra of vertical tangent vector field, which can be interpreted as internal symmetry algebra (see the Remark below). An explicit description of $\mathfrak{K}$ for λ_{gluon} and the corresponding Jacobi quadratic form $\mathcal{J}$ will appear in a separate paper.

A split structure given by $\mathfrak{K} \oplus Im\mathcal{J}$ is defined on $VW^{(1,1)}\mathbf{\Sigma} \,/\, W^{(1,1)}_4 SU(3)$. Furthermore, let $\mathfrak{h}$ be the Lie algebra of right-invariant vector fields on $W^{(1,1)}\mathbf{\Sigma}$ and $\mathfrak{k}$ the Lie subalgebra of generalized Jacobi vector fields; it is easy to see that the split structure is also reductive, being $[\mathfrak{k}, Im\mathcal{J}] = Im\mathcal{J}$ [5, 7, 8].

For each $\boldsymbol{p} \in W^{(1,1)}\mathbf{\Sigma}$, denote $\mathcal{S} \doteq \mathfrak{h}_{\boldsymbol{p}}$, $\mathcal{R} \doteq \mathfrak{k}_{\boldsymbol{p}}$ and $\mathcal{V} \doteq Im\,\mathcal{J}_{\boldsymbol{p}}$; we have a reductive Lie algebra decomposition $\mathcal{S} \doteq \mathfrak{t}^1_4\mathfrak{su}(3) \rtimes \mathfrak{gl}(4, \mathbb{R}) = \mathcal{R} \oplus \mathcal{V}$, with $[\mathcal{R}, \mathcal{V}] = \mathcal{V}$; we recall that $\mathcal{S}$ is the Lie algebra of the Lie group $W^{(1,1)}_4 SU(3)$. $\mathcal{R}$ is a reductive Lie subalgebra of $\mathfrak{t}^1_4\mathfrak{su}(3) \rtimes \mathfrak{gl}(4, \mathbb{R})$. The reductive split structure is responsible of the canonical variational breaking of the symmetry group $W^{(1,1)}_4 SU(3)$ and of the generation of Higgs fields. There exists an isomorphism between $\mathcal{V} \doteq Im\mathcal{J}_{\boldsymbol{p}}$ and $T_{\boldsymbol{x}}\boldsymbol{X}$ so that $\mathcal{V}$ turns out to be the image of an horizontal subspace. We caracterize a principal bundle $\boldsymbol{H} \to \boldsymbol{X}$, with $dim\boldsymbol{H} = dim\,\mathcal{S}$ and such that $\boldsymbol{X} = \boldsymbol{H}/\boldsymbol{R}$, where $\boldsymbol{R}$ is a Lie group of the Lie algebra $\mathcal{R}$ which is a closed subgroup of $W^{(1,1)}_4 SU(3)$. The principal subbundle $\boldsymbol{H} \subset W^{(1,1)}\mathbf{\Sigma}$ is then a *reduced principal bundle* [9].

We have the composite fiber bundle $W^{(1,1)}\mathbf{\Sigma} \to W^{(1,1)}\mathbf{\Sigma}/\boldsymbol{R} \to \boldsymbol{X}$, where $W^{(1,1)}\mathbf{\Sigma}/\boldsymbol{R} = W^{(1,1)}\mathbf{\Sigma} \times_{W^{(1,1)}_4 SU(3)} W^{(1,1)}_4 SU(3)/\boldsymbol{R} \to \boldsymbol{X}$ is a gauge-natural bundle functorially associated with $W^{(1,1)}\mathbf{\Sigma} \times W^{(1,1)}_4 SU(3)/\boldsymbol{R} \to \boldsymbol{X}$ by the right action of $W^{(1,1)}_4 SU(3)$. The left action of $W^{(1,1)}_4 SU(3)$ on $W^{(1,1)}_4 SU(3)/\boldsymbol{R}$ is defined by the reductive Lie algebra decomposition.

According to [13, 14], we call a global section $h : \boldsymbol{X} \to W^{(1,1)}\mathbf{\Sigma}/\boldsymbol{R}$ a *gluon classical Higgs field.*

The following Remark is now noteworthy .

Let ω be a principal connection on $W^{(1,1)}\mathbf{\Sigma}$; $\bar{\omega}$ principal connection on the principal bundle $\boldsymbol{H}$ defines the splitting $T_{\boldsymbol{p}}\boldsymbol{H} \simeq_{\bar{\omega}} \mathcal{R} \oplus \bar{\mathcal{H}}_{\boldsymbol{p}}$, $\boldsymbol{p} \in \boldsymbol{H}$. Note that, for each $\boldsymbol{q} \in W^{(1,1)}\mathbf{\Sigma}$, $T_{\boldsymbol{q}}W^{(1,1)}\mathbf{\Sigma} \simeq_\omega$

$V_q W^{(1,1)}\Sigma \oplus \mathcal{H}_q$. We find that $V_q W^{(1,1)}\Sigma \simeq T_q \boldsymbol{H} \simeq_{\bar{\omega}} \mathcal{R} \oplus \bar{\mathcal{H}}_q$, $\boldsymbol{q} \in \boldsymbol{H}$, *i.e.* it is defined a Cartan connection $\hat{\omega}$ of type $\mathcal{S}/\mathcal{R}$, such that $\hat{\omega}|_{V\boldsymbol{H}} = \bar{\omega}$. It is a connection on $W^{(1,1)}\Sigma = \boldsymbol{H} \times_{\boldsymbol{R}} W_4^{(1,1)} SU(3) \to \boldsymbol{X}$, thus a Cartan connection on $\boldsymbol{H} \to \boldsymbol{X}$ with values in $\mathcal{S}$, the Lie algebra of the gauge-natural structure group of the theory; it splits into the $\mathcal{R}$-component which is a principal connection form on the $\mathcal{R}$-manifold $\boldsymbol{H}$, and the $\mathcal{V}$-component which is a displacement form.

A gluon gauge-natural Higgs field is therefore a global section of the Cartan horizontal bundle $\hat{\mathcal{H}}_p$, with $\boldsymbol{p} \in \boldsymbol{H}$, it is related with the displacement form defined by the $\mathcal{V}$-component of the Cartan connection $\hat{\omega}$ above. A global section h of $W^{(1,1)}\Sigma/\boldsymbol{R} \to \boldsymbol{X}$ defines a vertical covariant differential and therefore the Lie derivative of fields is constrained and it is parametrized by gluon Higgs fields h characterized by $\mathfrak{K}$ [10, 11].

References

[1] D.J. Eck: Gauge-natural bundles and generalized gauge theories, *Mem. Amer. Math. Soc.* **247** (1981) 1–48.

[2] L. Fatibene, M. Francaviglia: *Natural and gauge natural formalism for classical field theories. A geometric perspective including spinors and gauge theories*; Kluwer Academic Publishers, Dordrecht, 2003.

[3] M. Godina, P. Matteucci: Reductive G-structures and Lie derivatives, *J. Geom. Phys.* **47** (1) (2003) 66–86.

[4] I. Kolář, P.W. Michor, J. Slovák: *Natural Operations in Differential Geometry*, (Springer–Verlag, N.Y., 1993).

[5] M. Palese, E. Winterroth: Global Generalized Bianchi Identities for Invariant Variational Problems on Gauge-natural Bundles, *Arch. Math. (Brno)* **41** (3) (2005) 289–310.

[6] M. Palese, E. Winterroth: Covariant gauge-natural conservation laws, *Rep. Math. Phys.* **54** (3) (2004) 349–364.

[7] M. Palese, E. Winterroth: The relation between the Jacobi morphism and the Hessian in gauge-natural field theories, *Theoret. Math. Phys.* **152** (2) (2007) 1191–1200.

[8] M. Palese, E. Winterroth: Lagrangian reductive structures on gauge-natural bundles, *Rep. Math. Phys.* **62** (2) (2008) 229–239.

[9] M. Palese, E. Winterroth: Invariant variational problems and Cartan connections on gauge-natural bundles, *AIP Conf. Proc.* **1191** (2009) 160–165.

[10] M. Palese, E. Winterroth: A variational perspective on classical Higgs fields in gauge-natural theories, *Theor. Math. Phys.* **168** (1) (2011) 1002–1008.

[11] M. Palese, E. Winterroth: Higgs fields on spinor gauge-natural bundles, *Journal of Physics: Conference Series* **411** (2013) 012025.

[12] R.D. Peccei: Exact and Broken Symmetries in Particle Physics, arXiv:hep-ph/0002225.

[13] G. Sardanashvily: Geometry of classical Higgs fields, *Int. J. Geom. Methods Mod. Phys.* **3** (1) (2006) 139–148.

[14] G.A. Sardanashvily: Classical Higgs fields, *Theoret. and Math. Phys.* **181** (3) (2014) 1599–1611.

[15] J. Vondra: *Natural prolongation of principal connections*, PhD thesis, Masaryk University, Brno 2010.

[16] J. Janyška, J. Vondra: Natural principal connections on the principal gauge prolongation of a principal bundle, *Rep. Math. Phys.* **64** (3) (2009) 395–415.

EPJ Web of Conferences **129**, 00017 (2016) DOI: 10.1051/epjconf/201612900017
QCD@Work 2016

Hadron Spectroscopy at BESIII

Marco Destefanis[1,*] for the BESIII Collaboration
[1] *Università degli Studi di Torino and INFN Torino*

Abstract. The BESIII experiment, hosted at the IHEP of Beijing, has collected the world largest data sample in the charmonium energy region. One of the most important physics goals of BESIII is the investigation of the QCD prediction. QCD can be accessed in a unique way by means of hadron spectroscopy, which was extensively studied and many important progresses were achieved in the last years. Charmonium decays provide an excellent scenario for studying nucleons, hyperons and their excited states, XYZ resonances, as well as light hadrons. Some of the most recent results for hadron spectroscopy from BESIII will be reported.

1 Introduction

Particles subject to the strong interaction are defined as hadrons. The hadronic states description is provided by the quark model [1]. The experimental results for the meson sector and for the ground state baryons are in a good agreement with this description, while the interpretation of the excited baryonic spectrum is much less clear. The strong interaction is described by the well estabilished QCD theory. In order to obtain a deep understanding of QCD, its predictions have to be searched for and properly investigated.

Hadron spectroscopy is a poweful tool to understand the quark structure of the particles, and the force which acts between them. This knowledge could be gained by searching for new hadronic states, and by a deep investigation of the light hadron sector. Studies performed in the charmonium energy range offer an appropriate playground to investigate the hadronic interactions, and complementary informations to the existing data.

2 The BESIII Experiment

The BESIII (BEijing Spectrometer III) [2], hosted on the BEPCII $e^+ e^-$ collider at the Institute of High Energy Physics (IHEP) in Beijing, with its excellent performances offers the appropriate scenario to support hadron spectroscopy studies. The Beijing Electron-Positron Collider II (BEPCII), with beam currents up to 0.93 A, has reached in 2016 its design Luminosity of $1{\cdot}10^{33}$cm^{-2} s^{-1}. The beam energy (from 1.0 up to 2.3 GeV) can be tuned according to the required center of mass energy, for example to pin the charmonium resonances down. The wide geometrical acceptance provided by the spectrometer, depicted in Fig.1, of about 93% of the solid angle is a consequence of its shell like structure. Departing from the interaction point, it hosts, inside a 1T solenoidal magnetic field, 43

*e-mail: destefan@to.infn.it

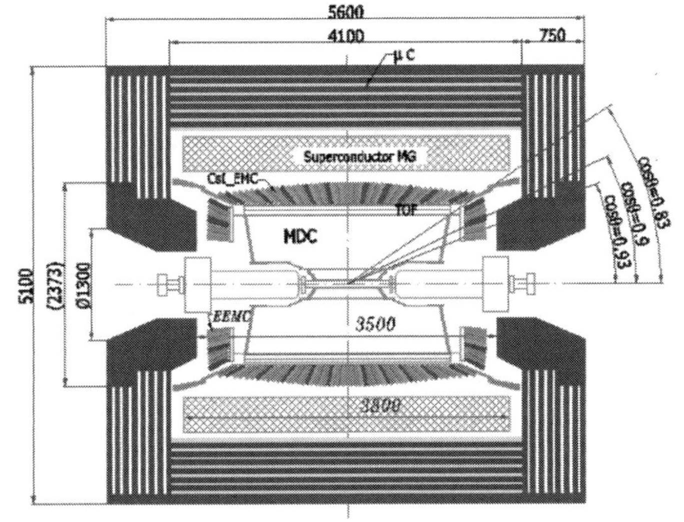

Figure 1. *View of the BESIII spectrometer [2].*

small-celled, helium-based main drift chamber (MDC) chambers for charged tracks reconstruction, a time-of-flight (TOF) system for particle identification, and an electromagnetic calorimeter (EMC) composed of 6240 CsI (Tl) crystals arranged in a cylindrical shape (barrel) plus two end-caps. A muon chamber (MUO) system, composed of resistive plate chambers (for a total of 1000 m^2) arranged in 9 layers in the barrel and 8 layers in the end-caps, is hosted in the semented magnet iron yoke.

The excellent performances provided by the BESIII spectrometer allowed to collect the world largest data sample of J/ψ, ψ(3686), and ψ(3770). Moreover, 1.3fb^{-1} of data were collected for the *R* scan, and 3.9fb^{-1} for the *Y* resonances studies.

3 X(1835) Studies

Since its discovery, the $X(1835)$ was subject of different speculations about its nature, in connection with its possible relationship with the anomaly seen at threshold in J/$\psi \to \gamma p\bar{p}$. The $X(1835)$ resonance was observed for the first time by the BESII experiment [3]. The high statistic data samples collected by the BESIII experiment gave the opportunity to further confirm the resonance and to have a deep insight into QCD dynamics, since possible interpretations include $p\bar{p}$ bound state [4], glueball [5], radial excitation of η' meson [6]. The lowest lying pseudoscalar glueball meson predicted by the Lattice QCD is expected to have a mass around 2.3 GeV/c^2 [7]. Its decay dynamics suggests that it may have properties in common with the η_c, of which one of the strongest decay channel is $\pi^+\pi^-\eta'$. For this reason, the $J/\psi \to \gamma\pi^+\pi^-\eta'$ was subject of studies. As anticipated, those investigations were stimulated by an anomalous $p\bar{p}$ invariance mass enhancement at threshold, reported by BESII [8] and confirmed by BESIII [9, 10] in $J/\psi \to \gamma p\bar{p}$ and $\psi' \to \pi^+\pi^- J/\psi \to \pi^+\pi^-\gamma p\bar{p}$ decays, respectively, and CLEO-c [11].

From the $\pi^+\pi^-\eta'$ invariant mass spectrum, shown in Fig.2, it was possible to notice two already known resonances, the $f_1(1510)$ and the $X(1835)$, and two new peaks defined as $X(2120)$ and $X(2370)$, which need to be included in the Partial Wave Analysis (PWA) performed to describe the experimental

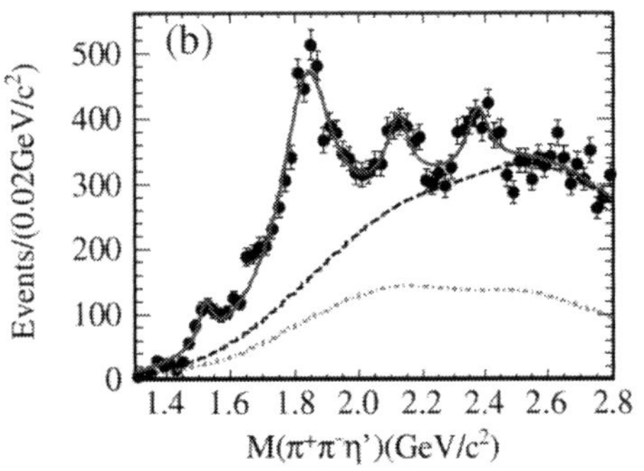

Figure 2. *$\pi^+\pi^-\eta'$ invariant mass spectrum [10].*

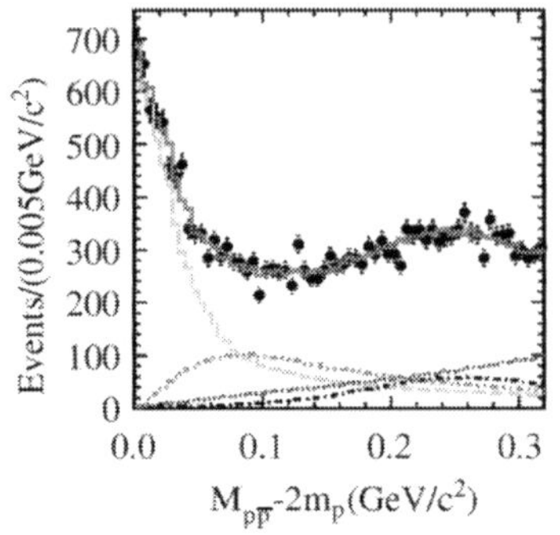

Figure 3. *$p\bar{p}$ threshold enhancement [12].*

data. The angular distribution of the radiative photon suggests the psuedoscalar nature of all those states.

In order to understand the nature of the $p\bar{p}$ threshold enhancement, a dedicated study was performed [12]. The invariant mass spectrum was investigated by means of PWA: the structure was described with a Breit-Wigner (BW) function (green), and $f_2(1910)$ (purple), $f_0(2100)$ (black), 0^{++} phase space (PS) (black), as shown in Fig.3, and a final state interaction (FSI) effect using the Juelich formulation [13] were included as well. Variation of the fit components were studied, and it was found that changes of the log-likelihood values and of the parameters of the $X(p\bar{p})$ were quite small. The mass of the $X(p\bar{p})$ was found to be consistent with the $X(1835)$, but with a significatively narrower width.

A similar enhancement was searched in $J/\psi \to \omega p\bar{p}$ [14] and $J/\psi \to \varphi p\bar{p}$ [15], where $\varphi \to K_S K_L$ or K^+K^-, final states. The decays $J/\psi \to \omega p\bar{p}$, $\varphi p\bar{p}$ restrict the isospin of the $p\bar{p}$ system and could clarify the role of the FSI. The Dalitz plot shows no significant signal close to the $p\bar{p}$ threshold, disfavoring a pure FSI effect giving rise to the $X(p\bar{p})$. The upper limits on the product of branching fractions were $B(J/\psi \to \omega X(p\bar{p}) \to \omega p\bar{p} < 3.7{\cdot}10^{-6}$ and $B(J/\psi \to \varphi X(p\bar{p}) \to \varphi p\bar{p} < 2.1{\cdot}10^{-7}$.

A clear $X(1835)$ signal was seen in $J/\psi \to \gamma K_S K_S \eta$ final state [16]. By means of the PWA method, the best fit to the data was obtained by including the three components: $X(1835)\eta \to f_0(980)\eta$, $X(1560)\eta \to f_0(980)\eta$, and a non-resonant $f_0(980)\eta$ component. A spin parity 0^{-+} was associated to all the components, althought a $J^P = 1^{++}$ for the non-resonant one cannot be excluded. The obtained results were consistent with the $J/\psi \to \gamma\pi^+\pi^-\eta'$ ones.

Further studies on the $J/\psi \to \omega\eta\pi^+\pi^-$ [17], $J/\psi \to \gamma\omega\varphi$ [18], and $J/\psi \to \gamma 3(\pi^+\pi^-)$ [19] final states investigated the $X(1840)$ and $X(1870)$ resonances. The PWA was performed with the tensor covariant amplitudes, and contributions from S-wave and P-wave were considered. The $X(1840)$ properties are in agreement with the $X(1835)$, but their width were found to be significantly different; it is now important to understand if they are the same particle or not. On the other hand, it has been reported that investigations on the $J/\psi \to \eta\varphi\pi^+\pi^-$ final state [20] did not show a remarkable presence of the $X(1835)$ and $X(1870)$ resonances.

In more recent studies on $J/\psi \to \gamma\pi^+\pi^-\eta'$ with the full statistic collected [21] by BESIII it was possible to observe a significant abrupt change in the slope of $\eta\pi^+\pi^-$ invariant mass distribution. The invariant mass spectrum lineshape around 1.85 GeV/c^2 might be characterized by means of two models. The first one takes the advantage of the Flatté formula to describe the opening of an additional decay mode, and a resonance at $X(1920)$ with a significance of 5.7σ has to be included (see Fig.4. The second model is based on the coherent sum of two BW amplitudes, the $X(1835)$ and a narrow

 DOI: 10.1051/epjconf/201612900017

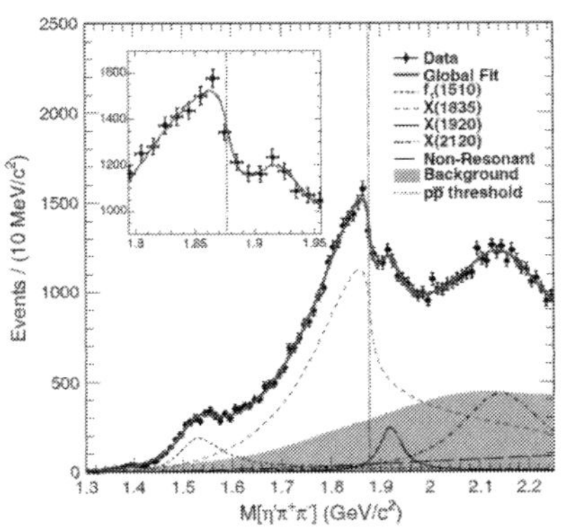

Figure 4. *Fit results from using the Flatté formula and the X(1920) resonance [21].*

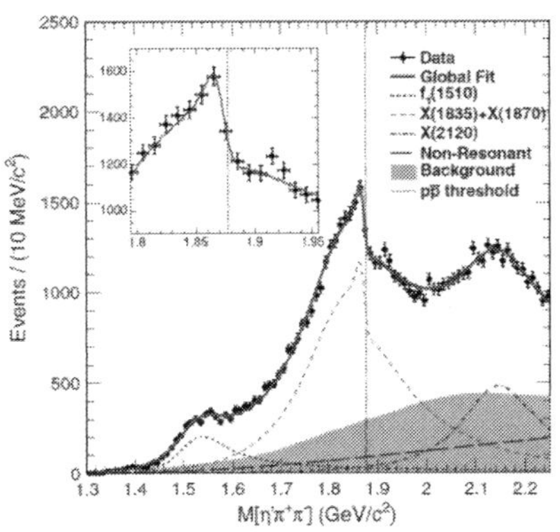

Figure 5. *Fit results from using a coherent sum of two Breit-Wigner amplitudes [21].*

$X(1870)$ (see Fig.5. Both models support the existence of a state, either a narrow state just below the $p\bar{p}$ mass threshold, or a broad state with strong couplings to $p\bar{p}$.

4 XYZ States

Below the $D\bar{D}$ threshold, the charmonium states could be properly described by the $c\bar{c}$ potential model. The scenario becomes more complex above the $D\bar{D}$ threshold: some of the predicted states were not found, while many new states were discovered. Thus the aim of BESIII searches was to estabilish the spectrum of those states and build a connection between them.

4.1 X States

The $\psi(3770)$ is the lightest charmonium state above the $D\bar{D}$ threshold, and it can be identified as the $1\,^3D_1$ state, as member of the D-wave spin-triplet. At the moment, there was no clear observation of the $1\,^3D_2$ and $1\,^3D_3$, the other two spin-triplet states. According to the phenomenological prediction, the $1\,^3D_2$ state should strongly couple to $\gamma\chi_{c1}$ and $\gamma\chi_{c2}$ final states. The narrow $X(3823)$ resonance, observed by BELLE in $X(3823 \to \gamma\chi_{c1})$ decay [22], was investigated as well by the BESIII experiment in $\gamma\chi_{c1,2}$ final states [23]. The analysis, performed at 5 different energy values, agreed with the BELLE discovery indicating the $X(3823)$ as a good candidate to be considered the $1\,^3D_2$ state, but no signal in $\gamma\chi_{c2}$ was found, as shown in Fig.6. The energy dependent cross section was fitted including the $Y(4360)$ or the $\psi(4415)$ line shapes: both hypothesis provide a reasonable description of the data.

4.2 Y States

The discovery of the Y resonances ($Y(4260)$, $Y(4360)$, $Y(4660)$) by BELLE and BABAR poses many questions in understanding their nature. They have relatively narrow widths and strong coupling to hidden-charm final states. The study of hadronic transitions to lower charmonia states together with lineshape measurements should allow us to investigate the different properties of those Y states. In order to deeply understand their nature, a complete PWA has to be performed.

One of the most recent results is the investigation of the $e^+e^- \to \eta(\pi^0)J/\psi$ reaction [24] in the energy range between 3.810 and 4.600 GeV. The measurement was performed at 17 different center

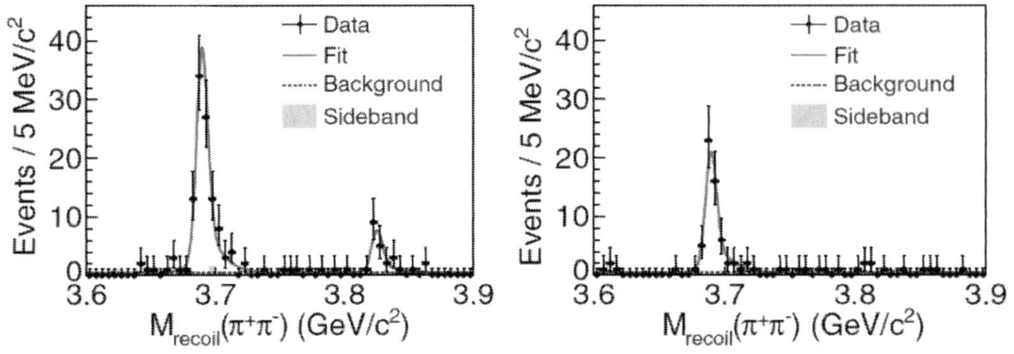

Figure 6. *Fit of the* $\pi^+\pi^-$ *recoil mass distribution of* $\gamma\chi_{c1}$ *(right) and* $\gamma\chi_{c2}$ *(left) events [23].*

of mass energies and a clear enhancement around 4.2 GeV was observed. The obtained cross sections, shown in Fig7, were compared with the previous measurements from BELLE [25] and BESIII [26], and are predicted to be strongly affected by open charm effects. This work provided the most precise measurements available in the litterature so far.

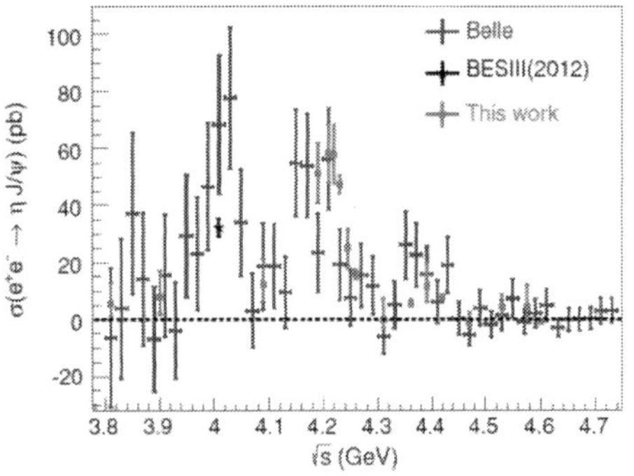

Figure 7. *Measured Born cross sections from the* $e^+e^- \to \eta J/\psi$ *final state [24].*

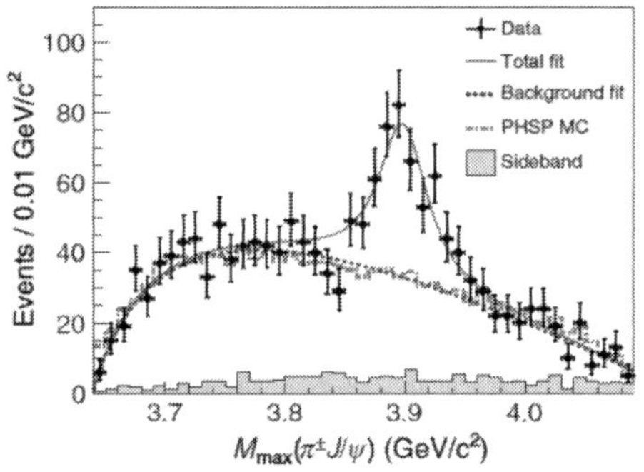

Figure 8. *Fit to the invaiant mass distribution of* $\pi^\pm J/\psi$ *[30].*

4.3 Z States

At the begining of 2013 a new charmonium like structure, the so called $Z_c(3900)$, was announced [27], then confirmed by BELLE [28] and CLEO-c [29]. The analysis was performed on the 515pb^{-1} data collected by BESIII at the center of mass energy of 4.260 GeV. The structure was observed with a significance higher than 8σ through its decay into a charged pion and a J/psi ($Z_c(3900)^\pm \to \pi^\pm J/\psi$), as shown in Fig.8, and its main carachteristic is that it is a charged state. The Z_c could be interpreted as tetraquark or as a hadronic molecule, so its nature is still under discussion.

Similar investigations, aimed to understand the Z_c nature were, performed for different final states. It was possible to discover the neutral partner of Z_c ($Z_c(3900)^0 \to \pi^0 J/\psi$) [30], allowing to complete the $Z_c(3900)$ triplet. Also the decay into open charm final states [31, 32] and its excited state, the $Z_c(4020)^0$ [32–34], were observed for the first time.

5 Baryonic States

Baryons are powerful tools to investigate the hadronic structure. The experimental data on ground baryonic states available in the literature are in good agreement with the quark model. Despite this, our knowledge becomes poor as soon as we investigate their excited states. For example, according to the prediction provided by the relativistic quark model [35] or to lattice QCD calculations [36] some resonances were not experimentally found and some others were unexpectedly observed. There is a huge effort from both theory and experiments devoted to obtain a clear comprehension of the baryonic scenario.

The data collected by BESIII can offer a complementary information by means of charmonium decays studies. In fact, in e^+e^- machines, like BEPCII, the baryon coupling to the conventional production channels (γ-nucleons) can be quite small, but the coupling between 3 gluon and baryons could be larger (ψ decays). The statistics collected by the BESIII experiment could be enough to improve our knowledge.

5.1 Hyperon Studies

The two lowest lying hyperon states were investigated by means of $\psi(3686)$ decays.

The $\psi(3686) \to \bar{p}K^+\Sigma$ decay was observed for the first time [37]. In order to improve our knowledge on hyperon properties, the $\bar{p}K^+\Lambda$ final state was investigated in the decay of χ_{cj}, abundantly produced from $\psi(3686)$ through radiative decays. In the analysis of the Dalitz plot of the χ_{c0}, as shown in Fig.9, a certain discrepancy between the phase space MC and the data due to the presence of an intermediate state, close to the $\bar{p}\Lambda$ threshold was observed [37].

In order to test QCD and its interplay with Perturbative-QCD, the $\Lambda\bar{\Sigma}^{\pm}\pi^{\mp}$ final state was investigated [38]. The experimental data, shown in Fig.10, depict clear clusters, indicating that this process is mediated by excited baryons. In order to determine the correct detection efficiency, a PWA was performed considering 16 possible intermediate states, which were described by a BW function. It was possible to measure for the first time the branching ratios for the $\psi(3686)$ decays into those final states.

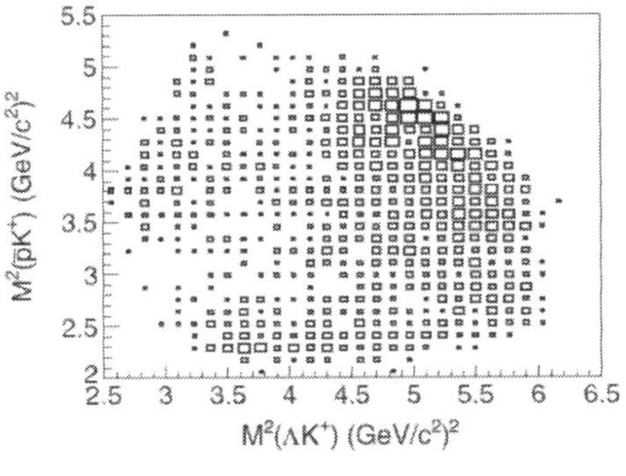

Figure 9. *Dalitz plot of $M^2(\bar{p}K^+)$ versus $M^2(\Lambda K^+)$ for χ_{c0} [37].*

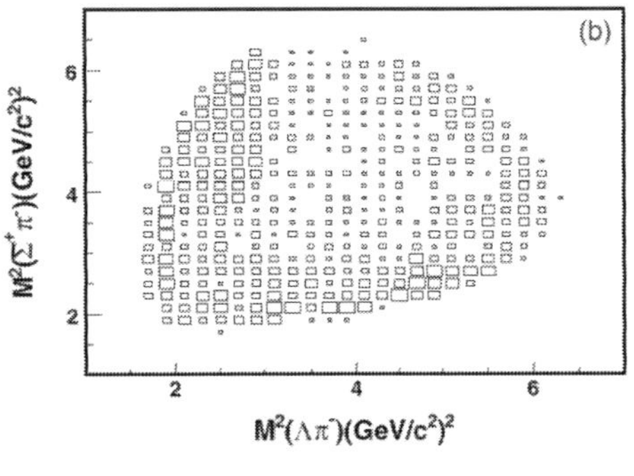

Figure 10. *Dalitz plot of $M^2(\Sigma^+\pi^-)$ versus $M^2(\Lambda\pi^-)$ [38].*

DOI: 10.1051/epjconf/201612900017

In the investigation of the $\psi(3686) \to (\gamma)K^{\mp}\Lambda\bar{\Xi}^{\pm}$ [39], two structures in the invariant mass of the ΛK system at 1690 and 1820 MeV were observed with a significance of 4.9σ and 6.2σ, respectively. The structures, shown in Fig.11, were fitted with a Voigtian function, while the background was estimated with the Λ and Ξ sidebands (red and green), by means of the inclusive MonteCarlo (shaded blue), and the non-resonant contribution are shown by the blue line. The number of events collected was not enough to study the angular distributions and to assign a J^P value to those structures.

J/ψ and $\psi(3686)$ were exploited to reconstruct $\Xi^-\bar{\Xi}^+$ and $\Sigma(1385)^{\mp}\bar{\Sigma}(1385)^{\pm}$ final states by means of the single tag method [40]. The decay of $\psi(3686)$ into $\Sigma(1385)$, shown in Fig.12 was observed for the first time. Their angular distributions, proprtional to $1 + \alpha\cos^2\vartheta$, were investigated as well and it was found that α from J/ψ and $\psi(3686)$ has the same sign for the Ξ production, while it is opposite for the $\Sigma(1385)$ scenario. The obtained results are the most precise measurements available in the litterature.

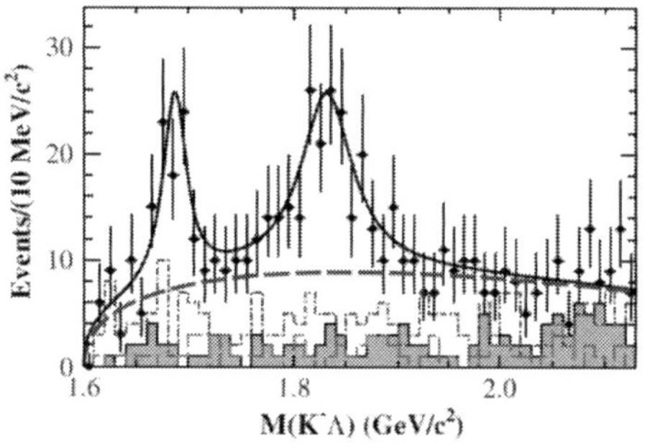

Figure 11. *$K^-\Lambda$ invariant mass distribution, with clear structures at 1690 and 1820 MeV [39].*

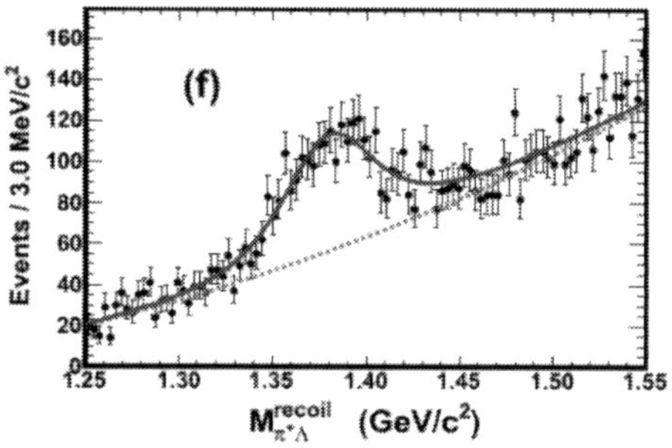

Figure 12. *$\Lambda\pi^+$ recoil mass spectrum for $\psi(3686) \to \Sigma(1385)^+\bar{\Sigma}(1385)^-$ [40].*

6 Conclusions

Charmonium decays at BESIII have proven to be a powerful tool to investigate the hadron spectroscopy, spanning from light hadron to XYZ states and to the baryon resonances sector. The discoveries performed by BESIII are complementary to those coming from the existing data. BESIII is expected to provide a significant contribution exploiting the new and larger data samples which are foreseen to be collected in future.

References

[1] H. Fritzsch, M. Gell-Mann and H. Leutwyler, Phys. Lett. B **47**, 365 (1973).

[2] D.M. Asner et al., *Physics at BES-III*, arXiv:0809.1869v1 [hep-ex] (2008).

[3] M. Ablikim et al., Phys. Rev. Lett. **95**, 262001 (2005).

[4] G.J. Ding and M.L. Yan, Phys. Rev. C **72**, 015208 (2005); G.J. Ding and M.L. Yan, Eur. Phys. J. A **28**, 351 (2006); J.P. Dedonder et al., Phys. Rev. C **80**, 045207 (2009); C. Liu, Eur. Phys. J. C **53**, 413 (2007); Z.G. Wang and S.L. Wan, J. Phys. G **34**, 505 (2007).

[5] G. Hao, C.F. Qiao, and A.L. Zhang, Phys. Lett. B **642**, 53 (2006); B.A. Li, Phys. Rev. D **74**, 034019 (2006); N. Kochelev and D.P. Min, Phys. Lett. B **633**, 283 (2006).

[6] T. Huang and S.L. Zhu, Phys. Rev. D **73**, 014023 (2006).

[7] C. Amsler and N. A. Tornqvist, Phys. Rev. **389**, 61 (2004); E. Klempt and A. Zaitsev, Phys. Rep. **454**, 1 (2007); Y. Chen et al., Phys. Rev. D **73**, 014516 (2006).
[8] J. Z. Bai et al., Phys. Rev. Lett. **91**, 022001 (2003).
[9] M. Ablikim et al., Chinese Phys. C **34**, 421 (2010).
[10] M. Ablikim et al., Phys. Rev. Lett. **106**, 072002 (2011).
[11] J. P. Alexander et al., Phys. Rev. D **82**, 092002 (2010).
[12] M. Ablikim et al., Phys. Rev. Lett. **108**, 112003 (2012).
[13] A. Sirbirtsen et al., Phys. Rev. D **71**, 054010 (2005).
[14] M. Ablikim et al., Phys. Rev. D **87**, 112004 (2013).
[15] M. Ablikim et al., Phys. Rev. D **93**, 052010 (2016).
[16] M. Ablikim et al., Phys. Rev. Lett. **115**, 091803 (2015).
[17] M. Ablikim et al., Phys. Rev. Lett. **107**, 182001 (2011).
[18] M. Ablikim et al., Phys. Rev. D **87**, 032008 (2013).
[19] M. Ablikim et al., Phys. Rev. D **88**, 091502 (2013).
[20] M. Ablikim et al., Phys. Rev. D **91**, 052017 (2015).
[21] M. Ablikim et al., Phys. Rev. Lett. **117**, 042002 (2016).
[22] V. Bhardwaj et al., Phys. Rev. Lett. **111**, 032001 (2016).
[23] M. Ablikim et al., Phys. Rev. Lett. **115**, 011803 (2015).
[24] M. Ablikim et al., Phys. Rev. D **91**, 112005 (2015).
[25] X.L. Wang et al., Phys. Rev. D **87**, 051101(R) (2013).
[26] M. Ablikim et al., Phys. Rev. D **86**, 071101(R) (2012).
[27] M. Ablikim et al., Phys. Rev. Lett. **110**, 252001 (2013).
[28] Z.Q. Liu et al., Phys. Rev. Lett. **110**, 252002 (2013).
[29] T. Xiao et al., Phys Lett B **727**, 336-370 (2013).
[30] M. Ablikim et al., Phys. Rev. Lett. **115**, 112003 (2015).
[31] M. Ablikim et al., Phys. Rev. Lett. **112**, 022001 (2014).
[32] M. Ablikim et al., Phys. Rev. Lett. **111**, 242001 (2013).
[33] M. Ablikim et al., Phys. Rev. Lett. **113**, 212002 (2014).
[34] M. Ablikim et al., Phys. Rev. Lett. **112**, 132001 (2014).
[35] U. Löring et al., Eur. Phys. J. A **10**, 395-446 (2001).
[36] R.G. Edwards et al., Phys. Rev. D **84**, 074508 (2011).
[37] M. Ablikim et al., Phys. Rev. D **87**, 012007 (2013).
[38] M. Ablikim et al., Phys. Rev. D **88**, 112007 (2013).
[39] M. Ablikim et al., Phys. Rev. D **91**, 092006 (2013).
[40] M. Ablikim et al., Phys. Rev. D **93**, 072003 (2016).

EPJ Web of Conferences **129**, 00018 (2016) DOI: 10.1051/epjconf/201612900018
QCD@Work 2016

Hadron spectroscopy and interactions from lattice QCD

Sasa Prelovsek[1,2,⋆]

[1] *Department of Physics, University of Ljubljana, Jadranska 19, Ljubljana, Slovenia*
[2] *Jozef Stefan Institute, Jamova 39, 1000 Ljubljana, Slovenia*

Abstract. Lattice QCD approach to study the hadronic resonances and exotic hadrons is described at an introductory level. The main challenge is that these states decay strongly via one or more decay channels, and they often lie near thersholds. Specific results for conventional and exotic hadrons are shown to illustrate the current status.

1 Introduction

Lattice QCD provides very accurate and reliable results for hadrons that lie well below strong-decay threshold and do not decay strongly. Over past few years, the lattice community provided also rigorous results for the resonances that can decay strongly via one-decay channel $R \to H_1H_2$. Interesting shallow bound states that lie just below H_1H_2 threshold were also investigated. Both require the simulation of H_1H_2 scattering and extraction of the underlying scattering matrix. A resonance that can strongly decay to several channels presents an even more challenging problem, which implies for the majority of the exotic hadrons. First lattice results were obtained very recently on this front as well. Here I illustrate the lattice approach to this problem in the introductory manner. Only few selected results are shown to illustrate various types of cases, while this is not a review of available results.

2 Lattice methodology

The physics information on a hadron (below, near or above threshold) is commonly extracted from the discrete energy spectrum in lattice QCD. The physical system for given quantum numbers is created from the vacuum $|\Omega\rangle$ using interpolator $O_j^\dagger$ at time $t=0$ and the system propagates for time t before being annihilated by O_i. To study a meson or an exotic-meson with given J^P, one can use $O \simeq \bar{q}\Gamma q$, two-meson interpolators $O = (\bar{q}\Gamma_1 q)(\bar{q}\Gamma_2 q)$, $(\bar{q}\Gamma_1 q)(\bar{q}\Gamma_2 q)$ and $O = [\bar{q}\Gamma_1 \bar{q}][q\Gamma_2 q]$ with desired quantum numbers. After the spectral decomposition the correlators are expressed in terms of the energies E_n of eigenstates $|n\rangle$ and their overlaps Z_j^n

$$C_{ij}(t) = \langle\Omega|O_i(t)O_j^\dagger(0)|\Omega\rangle = \sum_n Z_i^n Z_j^{n*}\, e^{-E_n t}\,, \qquad Z_i^n \equiv \langle\Omega|O_i|n\rangle\,. \tag{1}$$

The correlators are evaluated on the lattice and their time-dependence allows to extract E_n and Z_n^i [1, 2].

⋆e-mail: sasa.prelovsek@ijs.si

The energy eigenstates $|n\rangle$ are predominantly "one-meson" states (e.g. $\chi_{c1}(1P)$) or predominantly "two-meson" states (e.g. $D\bar{D}^*$) - in interacting theory they are mixtures of those. Two-meson states have a discrete spectrum due to the periodic boundary condition on the finite lattices. If they do not interact, then the momenta of each meson is $\vec{p} = \frac{2\pi}{L}\vec{N}$ with $\vec{N} \in N^3$, and the non-interacting energies of $M_1(\vec{p})M_2(-\vec{p})$ are $E^{n.i.} = E_1(p) + E_2(p)$ with $E_{1,2}(p) = (m_{1,2}^2 + p^2)^{1/2}$. The energies E_n extracted from the lattice are slightly shifted in presence of the interaction and the shift provides rigorous information on the scattering matrix, as discussed in Section 4. In experiment, two-meson states correspond to the two-meson decay products with a continuous energy spectrum.

3 Hadrons well below strong-decay threshold

Masses are straightforwardly determined for the hadrons that can not decay strongly and lie well below strong decay threshold. These masses $m_i = E_i(P = 0)$ are extracted from the energies obtained with $\bar{q}q$ or qqq interpolating fields, which are extrapolated $a \to 0$, $V \to \infty$ and $m_q \to m_q^{phys}$. The resulting masses for all such hadrons (i.e. π, K, D, B, B^*, B_s, J/ψ, η_b, n, p, Λ ,...) are in very good agreement with the experimental measurements. The main remaining uncertainty is the omission of charm and beauty annihilation for heavy quarkonia.

4 Scattering approach to resonances and bound states

Most of the hadrons are resonances, they are located above strong decay threshold(s) and decay strongly. In the energy region near or above threshold, the masses of bound-states and resonances have to be inferred from the infinite-volume scattering matrix of the one-channel (elastic) or multiple-channel (inelastic) scattering. The simplest example is a one-channel elastic scattering in partial wave l, where the scattering matrix is parametrized in terms of the phase shift $\delta_l(p)$ and satisfies unitarity $SS^\dagger = 1$

$$S(E) = e^{2i\delta_l(E)}\,, \quad S(E) = 1 + 2iT(E)\,, \quad T(E) = \frac{1}{\cot(\delta_l(E)) - i}\,. \tag{2}$$

Lüscher has shown that the energy E of two-meson eigenstate in finite volume L gives the elastic phase shift $\delta(E)$ at that energy in infinite volume [3]. This relation and its generalizations are at the core of extracting rigorous information about the scattering from the recent lattice simulation. It leads $\delta(E)$ only for specific values of E since since spectrum of two-meson eigenstates is discrete. The $\delta(E)$ or $T(E)$ (2) provide the masses of resonances and bound states:

- In the vicinity of a *hadronic resonance* with mass m_R and width Γ, the cross section $\sigma \propto |T(p)|^2$ has a Breit-Wigner-type shape with $\delta(s = m_R^2) = \frac{\pi}{2}$

$$T(p) = \frac{-\sqrt{s}\,\Gamma(p)}{s - m_R^2 + i\sqrt{s}\,\Gamma(p)} = \frac{1}{\cot\delta(p) - i}\,, \quad \Gamma(p) = g^2\,\frac{p^{2l+1}}{s}. \tag{3}$$

  The fit of $\delta_l(p)$ renders m_R and g or Γ. It is customary to compare g rather than Γ to experiment, since Γ depends on the phase space. Lüsher's approach has been verified on several conventional mesonic resonances.

- The *bound state (B)* in M_1M_2 scattering is realized when $T(p)$ has a pole at $p_B^2 < 0$ or $p_B = i|p_B|$

$$T = \frac{1}{\cot(\delta_l(p_B)) - i} = \infty\,, \quad \cot(\delta(p_B)) = i\,, \quad m_B = E_{H_1}(p_B) + E_{H_2}(p_B)\,. \tag{4}$$

  The location of an s-wave shallow bound state can be obtained by parametrizing δ_0 near threshold and finding p_B which satisfies $\cot(\delta(p_B)) = i$.

DOI: 10.1051/epjconf/201612900018

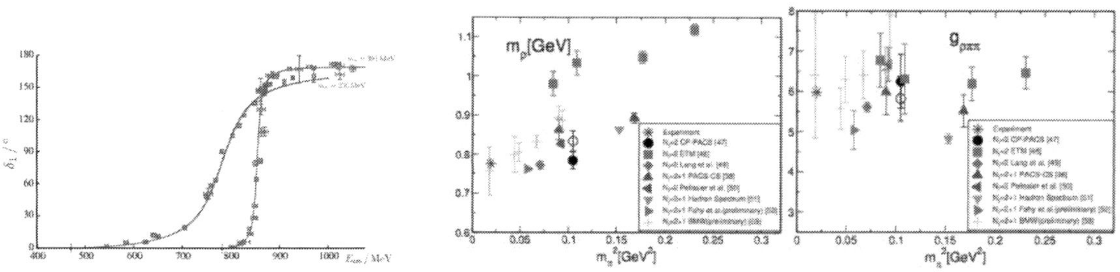

Figure 1. Left: The $\pi\pi$ phase shift in the ρ-meson channel from the simulation by the Hadron Spectrum Collaboration [4, 5]. Right: The compilation [6] of the masses and widths for the ρ meson from recent lattice simulations, where the g parametrizes the width as $\Gamma = g^2 p^3/(6\pi s)$.

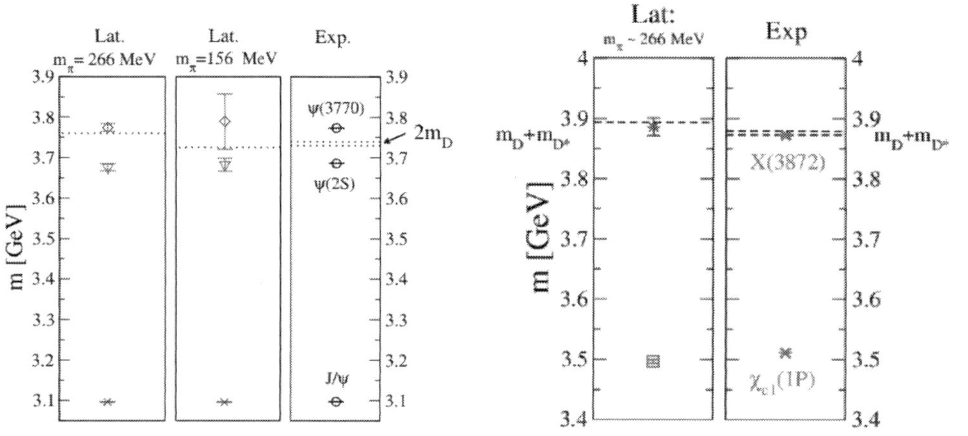

Figure 2. Left: The lattice spectrum of the vector charmonia from [7]: the diamond denotes the resonance mass of $\psi(3770)$, while the triangle denotes the pole mass of the bound state $\psi(2S)$; both are obtained from $D\bar{D}$ scattering matrix. Right: The location of $X(3872)$ with $I = 0$ which emerges as shallow bound state in a simulation of $D\bar{D}^*$ scattering that includes also diquark antidiqark interpolating fields [8].

5 Selected resonances from one-channel scattering

ρ meson: Only one resonance - namely the ρ meson - has been rigorously studied by a number of lattice collaborations, reviwed in [6]. Figure 1 shows an example of extracted $\pi\pi$ phase shift for $L = 1$ and $I = 1$, obtained by the Hadron Spectrum Collaboration [4, 5] for two pion masses. The Breit-Wigner-type fit renders ρ resonance mass and width [4] shown by the brown triangle in Fig. 1, where other lattice results are also shown.

Charmonium resonances: The charmonium spectrum above open charm threshold contains a number of exotic candidates. It is important to verify the approach on the conventional states before addressing the exotic states. Until recently, all charmonia above open-charm threshold were treated ignoring the strong decay to a pair of charmed mesons. The first exploratory simulation aimed at determining the masses as well as the decay widths of these resonances was presented in [7]. The Breit-Wigner-type fit of the $D\bar{D}$ phase shift in p-wave leads to the resonance $\psi(3770)$ mass in agreement with experiment, as shown in Fig. 5. The extracted $\psi(3770) \to D\bar{D}$ decay width $\Gamma = g^2 p/s$

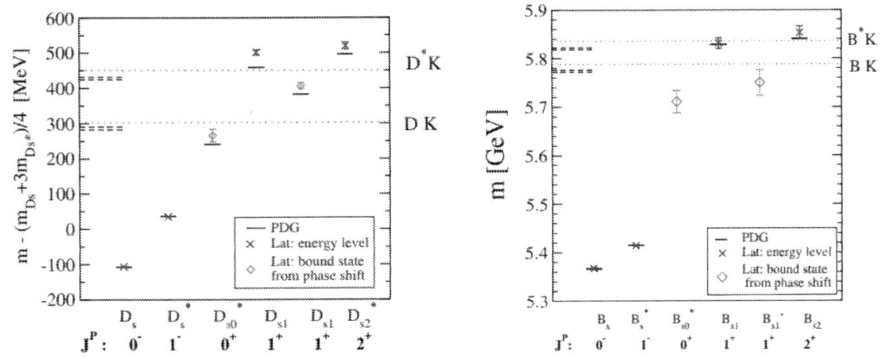

Figure 3. The spectra of D_s [9, 10] and B_s [11] mesons resulting from $N_f = 2 + 1$ lattice simulation on PACS-CS ensembles with $m_\pi \simeq 156$ MeV. The black horizontal lines are experimental masses.

with $g^{lat}_{m_\pi=266\ \mathrm{MeV}} = 19.7 \pm 1.4$ is also close to $g^{exp} = 18.7 \pm 1.4$. The $\psi(2S)$ in Fig. 5 appears as a bound state pole below threshold. In the scalar channel, only the ground state $\chi_{c0}(1P)$ is understood and there is no commonly accepted candidate for its first excitation $\chi_{c0}(2P)$. The lattice results on the $D\bar{D}$ scattering in s-wave [7] leave open puzzles and call for further work on this problem.

6 Selected shallow bound states from one-channel scattering

Scalar and axial mesons D_s and B_s: The quark models expected $D^*_{s0}(2317)$ and $D_{s1}(2460)$ above DK and D^*K thresholds, but they were experimentally found slightly below them. These are one of few shallow bound states in the meson sector, and therefore deserve special attention. The first lattice simulation that takes the effect of these thresholds into account was presented in [9, 10]. It employs DK and D^*K interpolating fields in addition to the $\bar{s}c$. The $D^{(*)}K$ phase shift is extracted and then interpolated in the region close to threshold. A pole of the scattering matrix is found below the thresholds where the relation (4) is satisfied. The poles are related to $D^*_{s0}(2317)$ and $D_{s1}(2460)$ bound states and are close to the masses of experimentally observed states, as shown in Fig. 3. The scalar and axial B_s mesons have not been discovered in experiment yet, and in this case we made a prediction for their masses using analogous simulation of $B^{(*)}K$ scattering. We find poles related to B_{s0} and B_{s1} mesons below BK and B^*K thresholds [11], as shown in Fig. 3.

$X(3872)$: The $X(3872)$ lies experimentally on $D^0\bar{D}^{0*}$ threshold and its existence on the lattice can not be established without taking into account the effect of this threshold. This was first done by simulating $D\bar{D}^*$ scattering in [12], where a pole in $D\bar{D}^*$ scattering matrix was found just below the threshold in $I(J^{PC}) = 0(1^{++})$ channel (4). The pole was associated with $X(3872)$ [12] and was also confirmed by a simulation with HISQ action [13]. The more recent simulation [8] included also the diquark anti-diquark interpolating fields and the location of the $X(3872)$ pole is shown in Fig. 5.

The lattice study [8] investigated which Fock components are essential for appearance of $X(3872)$ with $I = 0$ on the lattice. The energy eigenstate related to $X(3872)$ appears in the simulation only if $D\bar{D}^*$ as well as $\bar{c}c$ interpolating fields are employed. The $X(3872)$ does not appear in absence of $\bar{c}c$ interpolators, even if (localized) interpolators $[\bar{c}\bar{q}]_{3_c}[cq]_{\bar{3}_c}$ or $[\bar{c}\bar{q}]_{6_c}[cq]_{\bar{6}_c}$ are in the interpolator basis. This indicates that $\bar{c}c$ Fock component is most likely more essential for $X(3872)$ than the diquark-antidiquark one.

Pentaquark bound state: The NPLQCD collaboration finds an interesting evidence for a $\eta_c N$ bound state approximately 20 MeV below $\eta_c N$ threshold [14] (N denotes nucleon). To my knowledge, this is the only pentaquark candidate containing $\bar{c}c$ from lattice studies up to now. As the simulation is done at $SU(3)$ flavour symmetric point corresponding to $m_\pi \simeq 800$ MeV, it is not clear yet whether this bound state persists to physical m_π. LHCb has recently found two pentaquark resonances in $J/\psi p$, located about 400 MeV above threshold [15]. The lattice simulation of those is much more challenging due to several open channels and has not been performed yet.

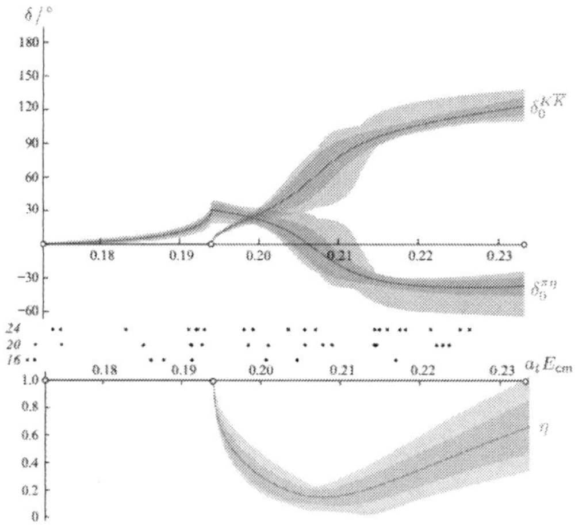

Figure 4. The phases and the inelasticity of the $K\bar{K} - \pi\eta$ coupled channel scattering in the channel where a_0 resonance appears in experiment. The results are obtained from a $N_f = 2+1$ lattice simulation at $m_\pi \simeq 400$ MeV by the Hadron Spectrum Collaboration [16].

7 Towards the coupled-channel scattering and exotic hadrons

Most of the exotic candidates are experimentally found above several thresholds, and can decay in more than one hadronic final state. This problem requires the determination of the scattering matrix for the coupled-channel scattering. This is more challenging for a rigorous treatment than a simulation of one-channel (elastic) scattering.

a_0 resonance: The coupled-channel scattering matrix was extracted from the lattice simulation for the first time only recently [18], where the generalized Lüscher's method was used. Figure 4 shows the results for the scattering phases and inelasticity that parametrize the 2×2 scattering matrix in the a_0 channel [16].

$Z_c^+(3900)$: The lattice search for the manifestly exotic states Z_c^+ with flavour content $\bar{c}c\bar{d}u$ and $I^G(J^{PC}) = 1^+(1^{+-})$ is very challenging since the experimental candidate can strongly decay to several final states $D\bar{D}^*$, $J/\psi\pi$ and $\eta_c\rho$. The corresponding 3×3 scattering matrix has never been determined using the reliable Lüscher's method.

The HALQCD collaboration used the HALQCD approach to extracted the coupled-channel scattering matrix, which is not commonly accepted as an exact treatment, and has not been verified on the conventional resonances. First the potential $V_{\pi J/\psi - \pi J/\psi}(r)$ related to Nambu-Bethe-Salpeter equation

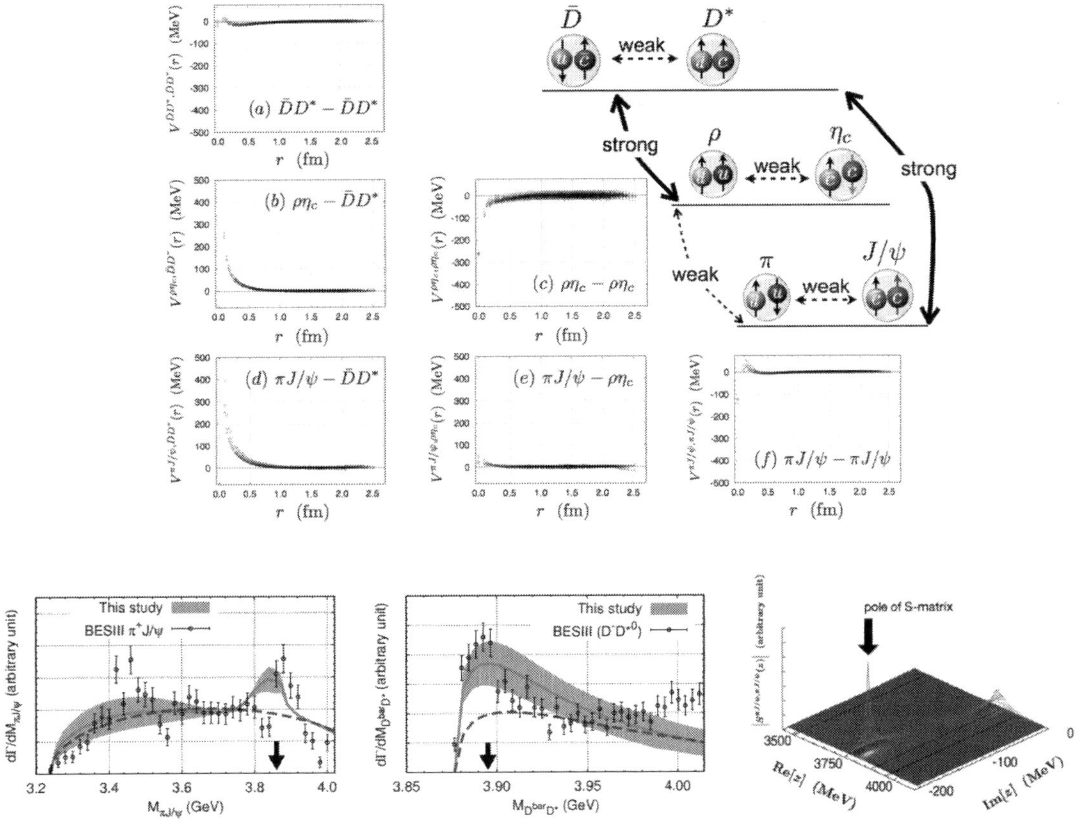

Figure 5. The results related to $Z_c^+(3900)$ from a lattice simulation [17], based on the HALQCD method. Top: potentials $V(r)$ for three channels. Bottom left: red lines present the lattice QCD results for $Y(4260) \to J/\psi\pi\pi$ and $Y(4260) \to D\bar{D}^*\pi$ invariant masses, while blue lines present results if $J/\psi\pi$-$D\bar{D}^*$ coupling is turned off. Bottom right: Poles of the scattering matrix in the complex energy plane.

is determined between the J/ψ and π as a function of their separation r [17]. This is presented in Fig. 5 together with potential for the other two channels, as well as potential corresponding to coupling between different channels according to coupled-channel HALQCD formalism [19]. The off-diagonal potential between channels $\pi J/\psi$ and $D\bar{D}^*$ is larger than other potentials, which seems to indicate a sizable coupled channel effect near $D\bar{D}^*$ threshold. The potentials render 3×3 scattering matrix for three coupled channels [17]. This is then used to determine the three body decay $Y(4260) \to J/\psi\pi\pi$ and $Y(4260) \to D\bar{D}^*\pi$ in a semi-phenomenological way. The result in Fig. 5 indeed shows a peak around Z_c mass. If the coupling between $J/\psi\pi$ and $D\bar{D}^*$ is turned off, the peak disappears as shown by blue dashed line. It indicates that the coupling of $J/\psi\pi$ and $D\bar{D}^*$ channels seems to be crucial for the existence of Z_c. This needs to be verified also by the Lüscher's method.

The first step towards the treatment of this problem with the Lüscher method is to determine the energies of eigenstates for the Z_c system. This has been done in [13, 20]. The number of lattice eigenstates in the relevant energy region was found to be equal to the number of expected two-hadron states in the absence of interactions, and no additional eigenstate was found. This suggests that Z_c is perhaps not a resonance in the usual sense, i.e. it is not associated with a pole above $D\bar{D}^*$ threshold near the real axis. Indeed, the HALQCD collaboration has not found such a pole, as shown in Fig. 5. More analytical work and lattice simulations are required to conclude on this problem.

$X(4140)$: The experimental candidate $X(4140)$ with hidden strangeness was confirmed in $J/\psi\phi$ invariant mass by LHCb recently [21, 22]. The amplitude analysis favors quantum numbers $J^{PC} = 1^{++}$ for it. The rigorous treatment would require extraction of the scattering matrix for at leat two channels $J/\psi\phi$ and $D_s\bar{D}_s^*$. The simplified lattice search was performed in this channel with $D_s\bar{D}_s^*$, $J/\psi\phi$, $[\bar{c}\bar{s}][cs]$ interpolators. The number of lattice eigenstates in the relevant energy region was found to be equal to the number of expected two-hadron states in the absence of interactions. No additional eigenstate was found near 4.1 GeV, which suggests that $X(4140)$ is perhaps not a resonance with a pole above $J/\psi\phi$ threshold near the real axis. The experimental study [22] found that the amplitudes could be fitted within the cusp scenario [23], which might be consistent with absence of a resonance pole near $E_{cm} \simeq m_X - i\Gamma_X/2$.

8 Summary

During the past few years, the lattice QCD provided rigorous results on the hadrons that lie close and above strong-decay thresholds. Verification of the approach on the conventional hadrons is the first important step towards understanding exotic ones. Most of the experimental exotic candidates can strongly decay via several decay channels and lie near thresholds. This presents a challenge for rigorous treatment from first principles, but important encouraging steps in this direction have been accomplished.

Acknowledgments
I would like to thank C.B. Lang, D. Mohler, L. Leskovec, M. Padmanath and R. Woloshyn for pleasant collaboration on these physics problems. Support from Slovenian Research Agency ARRS under project N1-0020 is acknowledged.

References

[1] C. Michael, Nucl. Phys. B **259**, 58 (1985).
[2] B. Blossier, M. DellaMorte, G. von Hippel, T. Mendes, R. Sommer, JHEP **0904**, 094 (2009), 0902.1265.
[3] M. Lüscher, Nucl. Phys. B **364**, 237 (1991).
[4] J.J. Dudek, R.G. Edwards, C.E. Thomas, Phys.Rev. **D87**, 034505 (2013), 1212.0830.
[5] D.J. Wilson, R.A. Briceno, J.J. Dudek, R.G. Edwards, C.E. Thomas, Phys. Rev. **D92**, 094502 (2015), 1507.02599.
[6] T. Yamazaki, PoS **LATTICE2014**, 009 (2015), 1503.08671.
[7] C.B. Lang, L. Leskovec, D. Mohler, S. Prelovsek (2015), 1503.05363.
[8] M. Padmanath, C.B. Lang, S. Prelovsek, Phys. Rev. **D92**, 034501 (2015), 1503.03257.
[9] D. Mohler, C. Lang, L. Leskovec, S. Prelovsek, R. Woloshyn, Phys.Rev.Lett. **111**, 222001 (2013), 1308.3175.
[10] C. Lang, L. Leskovec, D. Mohler, S. Prelovsek, R. Woloshyn, Phys.Rev. **D90**, 034510 (2014), 1403.8103.
[11] C.B. Lang, D. Mohler, S. Prelovsek, R.M. Woloshyn, Phys. Lett. **B750**, 17 (2015), 1501.01646.
[12] S. Prelovsek, L. Leskovec, Phys.Rev.Lett. **111**, 192001 (2013), 1307.5172.

[13] S.h. Lee, C. DeTar, H. Na, D. Mohler (Fermilab Lattice, MILC) (2014), 1411.1389.
[14] S.R. Beane, E. Chang, S.D. Cohen, W. Detmold, H.W. Lin, K. Orginos, A. Parreño, M.J. Savage, Phys. Rev. **D91**, 114503 (2015), 1410.7069.
[15] R. Aaij et al. (LHCb), Phys. Rev. Lett. **115**, 072001 (2015), 1507.03414.
[16] J.J. Dudek, R.G. Edwards, D.J. Wilson (Hadron Spectrum), Phys. Rev. **D93**, 094506 (2016), 1602.05122.
[17] Y. Ikeda, S. Aoki, T. Doi, S. Gongyo, T. Hatsuda, T. Inoue, T. Iritani, N. Ishii, K. Murano, K. Sasaki (2016), 1602.03465.
[18] J.J. Dudek, R.G. Edwards, C.E. Thomas, D.J. Wilson (Hadron Spectrum), Phys. Rev. Lett. **113**, 182001 (2014), 1406.4158.
[19] S. Aoki, B. Charron, T. Doi, T. Hatsuda, T. Inoue et al., Phys.Rev. **D87**, 034512 (2013), 1212.4896.
[20] S. Prelovsek, C.B. Lang, L. Leskovec, D. Mohler, Phys. Rev. **D91**, 014504 (2015), 1405.7623.
[21] R. Aaij et al. (LHCb) (2016), 1606.07895.
[22] R. Aaij et al. (LHCb) (2016), 1606.07898.
[23] E.S. Swanson, Int. J. Mod. Phys. **E25**, 1642010 (2016), 1504.07952.

Overview of LHCb results on beauty and charm spectroscopy

Antimo Palano[1,⋆]

[1] *INFN and University of Bari, Italy, on behalf of the LHCb Collaboration*

Abstract. We present a summary of new experimental results from LHCb experiment on the status of the charm spectroscopy using inclusive approaches and Dalitz plot analyses of B and B_s decays. We also summarize latest results on the spectroscopy of heavy baryons.

1 Introduction

The LHCb experiment [1] is collecting very large samples of $c\bar{c}$ and $b\bar{b}$ events. The present results are mostly based on data data collected in 2011 and 2012 during the Run 1 period, corresponding to an integrated luminosity of ≈ 3.0 fb^{-1}.

2 Charm meson spectroscopy

Two methods are used for studying or discovering new states in LHCb.

- Inclusive approach: the study of inclusive final states $D^0\pi^+$, $D^+\pi^-$ and $D^{*+}\pi^-$ in the reactions

  $$pp \to D_J X,$$

  and of the final states D^0K^+, $D^+K^0_S$ and $D^{*+}K^0_S$ in the reactions

  $$pp \to D_{sJ} X.$$

  In these reactions,

  - All resonances can be produced.
  - The signal to background ratio can be poor.
  - A spin-parity analysis of three-body decays can only distinguish between Natural and Unnatural Parity assignments.
  - No spin analysis is possible for two-body decays.

- Exclusive approach: Dalitz plot analysis of B and B_s decays. In these cases,

  - A full spin-parity analysis is possible.
  - High mass resonances can have low rates or may not be produced.
  - The analysis can be complicated by the presence of multiple interfering contributions.

⋆e-mail: antimo.palano@ba.infn.it

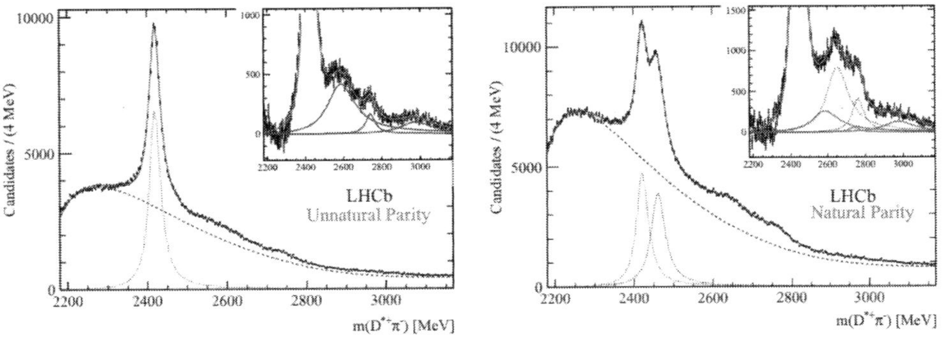

Figure 1. $D^{*+}\pi^-$ mass spectrum with (Left) "Enhanced Unnatural Parity" selections, (Right): "Natural Parity" selections.

2.1 Observation of new D_J mesons

The LHCb experiment has performed an inclusive study of the $D\pi$ and $D^*\pi$ systems using 1 fb^{-1} of data [2]. In the study of the $D^{*+}\pi^-$ system, with $D^{*+} \to D^0\pi^+$, the data are separated in terms of the π^+ helicity angle θ_H . The sample with $|\cos\theta_H| > 0.75$ is defined as "Enhanced Unnatural Parity Sample", while the sample with $|\cos\theta_H| < 0.5$ is defined as "Natural Parity Sample". The two mass spectra are shown in Fig. 1. In the Enhanced Unnatural Parity sample, above the $D_1(2420)$ and $D_2^*(2460)$, three new structures are observed: $D_J(2550)^0$, $D_J(2760)^0$ and $D_J(3000)^0$. In the Natural Parity sample, two further structures are observed: $D_J^*(2600)^0$ and $D_J^*(2760)^0$. The presence of these two Natural Parity states is confirmed by the analysis of the $D^+\pi^-$ and $D^0\pi^+$ data.

2.2 First observation and Dalitz plot analysis of $B^- \to D^+K^-\pi^-$

The $D^+K^-\pi^-$ mass spectrum [3] is shown in Fig. 2 (Left) and contains ≈2000 events in the B^- signal region. A standard Dalitz plot analysis of the $B^- \to D^+K^-\pi^-$ system has been performed. The

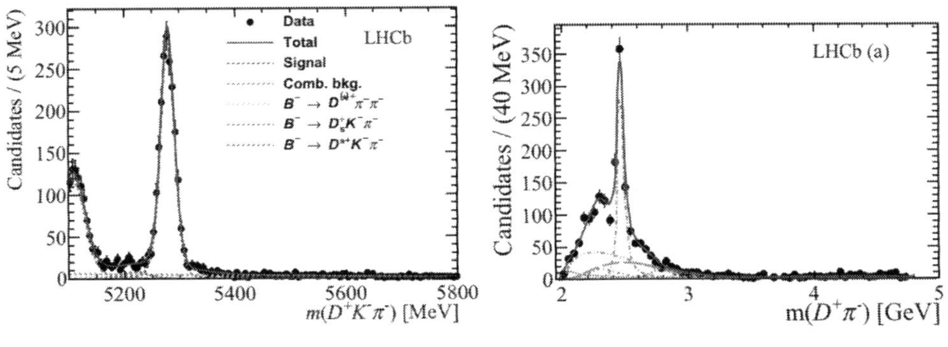

Figure 2. (Left): $D^+K^-\pi^-$ mass spectrum. (Right): $D^+\pi^-$ fit projection with fit result.

best fit is obtained introducing virtual $D_v^*(2007)^0$, B_v^{*0} and nonresonant contributions. A clear spin-2 $D_2^*(2460)$ signal and a $D_J^*(2760)^0$ spin-1 resonance are also observed. The $D^+\pi^-$ fit projection with fit result is shown in Fig. 2(Right). No evidence for a $D_3^*(2760)$ spin-3 resonance is found in this final state.

2.3 Dalitz plot analysis of $B^0 \to \bar{D}^0\pi^+\pi^-$

The $B^0 \to \bar{D}^0\pi^+\pi^-$ [4] signal contains $\approx$9600 events with 97.8% purity. The $B^0 \to \bar{D}^0\pi^+\pi^-$ Dalitz plot is shown in Fig. 3 (Left). The spin-2 $D_2^*(2460)$ signal along the $\bar{D}^0\pi^-$ axis and the spin-1 $\rho(770)$ signal

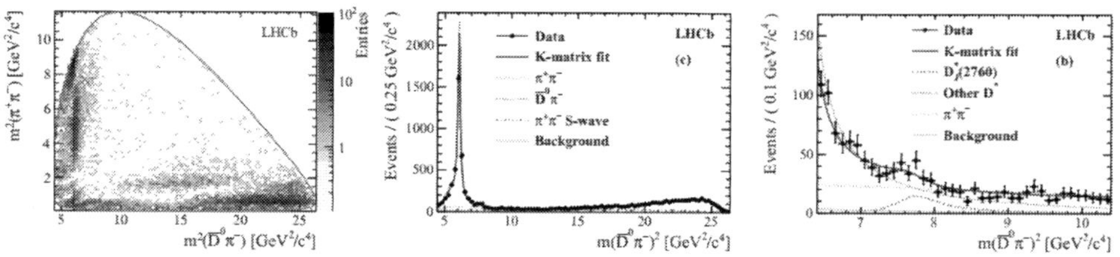

Figure 3. (Left): $B^0 \to \bar{D}^0\pi^+\pi^-$ Dalitz plot. (Center), (Right) $m^2(\bar{D}^0\pi^-)$ fit projection with two different binnings.

along the $\pi^+\pi^-$ axis are observed. The Dalitz plot analysis has been performed using the isobar model and a K-matrix description of the $\pi^+\pi^-$ S-wave. Both methods give a good description of the data. The $m^2(\bar{D}^0\pi^-)$ fit projection is shown in Fig. 3 (Center),(Right). The decay is dominated, in the $\pi^+\pi^-$ system, by S-wave $(16.51\pm0.70\pm1.68\pm1.10)$% and $\rho(770)$ $(36.15\pm1.00\pm2.13\pm0.79)$%. In the $\bar{D}^0\pi^-$ system, the largest contribution comes from the $D_2^*(2460)^-$ resonance $(28.13 \pm 0.72 \pm 1.06 \pm 0.54)$%.

The Dalitz plot analysis requires the presence of an additional $J^P = 3^-$ resonance with a K-matrix model fitted fraction of $(1.58 \pm 0.22 \pm 0.18 \pm 0.07)$%.

2.4 Dalitz plot analysis of $B^0 \to \bar{D}^0K^+\pi^-$

The $B^0 \to \bar{D}^0K^+\pi^-$ [5] signal region contains $\approx$2300 events. The B^0 Dalitz plot is shown in Fig. 4(Left). The fit projections are shown in Fig. 4(Center),(Right). The decay is dominated by in-

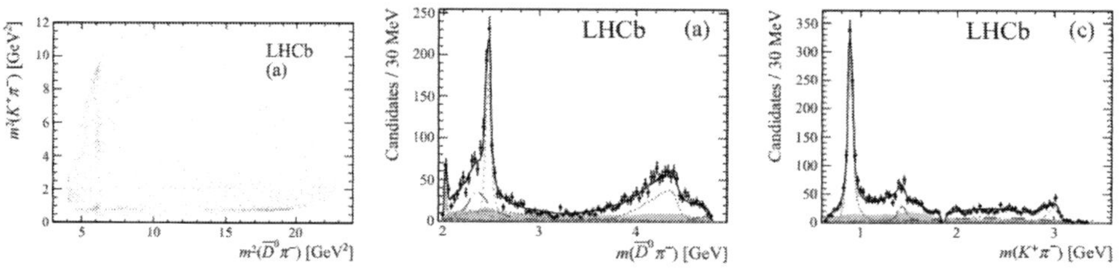

Figure 4. (Left): $B^0 \to \bar{D}^0K^+\pi^-$ Dalitz plot. (Center): $m(\bar{D}^0\pi^-)$ and (Right): $m(K^+\pi^-)$ fit projections.

termediate resonance production of $K^*(892)^0$ (37.4 ± 1.5)%, $D_0^*(2400)^-$ (19.3 ± 2.8)% and $D_2^*(2460)^-$ (23.1 ± 1.2)%. These Dalitz analyses obtain new parameters for the broad $D_0^*(2400)$ resonance which are summarized in Table 1. No evidence is found for additional spin-1 or spin-3 D_J^* resonances.

3 Results on D_{sJ} mesons spectroscopy

3.1 Dalitz plot analysis of $B_s^0 \to \bar{D}^0K^-\pi^+$

Figure 5(Left) shows the $\bar{D}^0K^-\pi^+$ mass spectrum [8]. The B_s^0 signal contains $\approx$11000 signal events. The Dalitz analysis shows that the largest components are: $\bar{K}^*(892)^0$ (28.6 %), $D_{s2}^*(2573)^-$ (25.7 %),

Table 1. $D_0^*(2400)$ resonances parameters from the different Dalitz analyses.

Final state	Method	Mass (MeV)	Width (MeV)
$B^0 \to D^0 K^+ \pi^-$	Free	2360 ± 15	255 ± 26
$B^0 \to \overline{D}^0 \pi^+ \pi^-$	Free	2354 ± 7	230 ± 15
$B^- \to D^+ K^- \pi^-$	(PDG)	2318 ± 29	267 ± 40

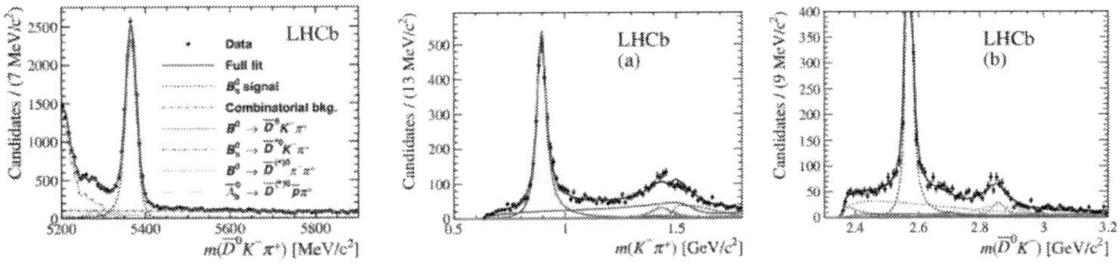

Figure 5. (Left) $\overline{D}^0 K^- \pi^+$ mass spectrum. (Right) $K^- \pi^+$) and $\overline{D}^0 K^-$ mass projections.

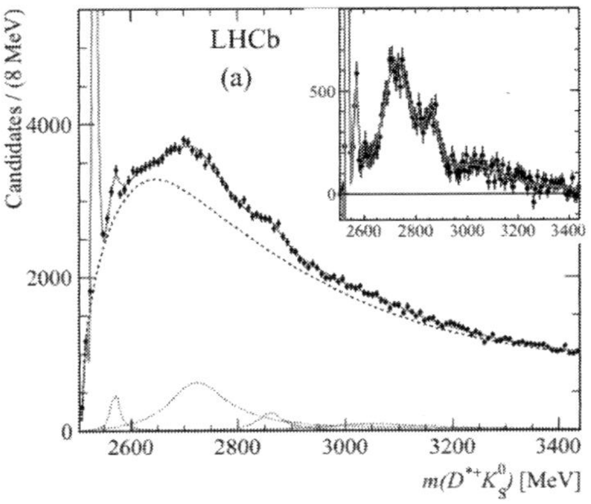

Figure 6. $D^{*+} K_S^0$ with enhanced Natural Parity mass spectrum.

$K\pi$ S-wave (LASS) (21.4 %) $\overline{D}^0 K^-$ nonresonant (12.4 %). A signal present in the 2860 MeV $\overline{D}^0 K^-$ mass region which is described by a superposition of a new spin-1 $(5.0 \pm 1.2 \pm 0.7 \pm 3.3)$ % and a spin-3 $(2.2 \pm 0.1 \pm 0.3 \pm 0.4)$ % resonances. The fitted resonances parameters are:

$$m(D_{s1}^*(2860)^-) = 2859 \pm 12 \pm 6 \pm 23 \text{ MeV}, \Gamma(D_{s1}^*(2860)^-) = 159 \pm 23 \pm 27 \pm 72 \text{ MeV},$$
$$m(D_{s3}^*(2860)^-) = 2860.5 \pm 2.6 \pm 2.5 \pm 6.0 \text{ MeV}, \Gamma(D_{s3}^*(2860)^-) = 53 \pm 7 \pm 4 \pm 6 \text{ MeV}.$$

3.2 Inclusive studies

Using samples of 0.36×10^6 $D^+ K_S^0$ and 3.15×10^6 $D^0 K^+$ inclusive candidates [6], the $D_{s1}^*(2710)^+$ and $D_{sJ}^*(2860)^+$ have been observed with parameters consistent with previous measurements [7].

The inclusive $D^{*+} K_S^0$ mass spectrum with enhanced Natural Parity selection is shown in Fig. 6 [9]. Resonances $D_{s1}^*(2710)^+$ and $D_{sJ}^*(2860)^+$ are observed with parameters consistent with those from

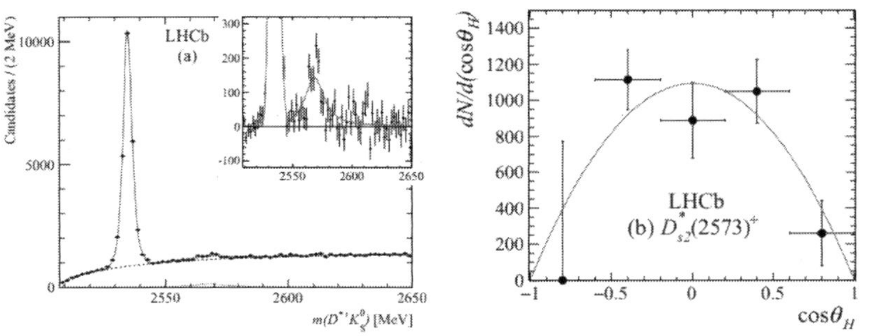

Figure 7. (Left) $D^{*+}K^0_S$ mass spectrum in the threshold region with enhanced Natural Parity selection, (Right) Angular distribution of the $D^*_{s2}(2573)^+ \to D^{*+}K^0_S$.

inclusive and exclusive analyses. Significances are 7.6σ and 7.1σ respectively. A search has been performed for the new spin-1 resonance $D^*_{s1}(2860)^+$ observed in the B_s decay. However, the same χ^2/ndf is obtained in fit that excludes it and its resulting statistical significance is 3.3σ. It is concluded that the data are not sensitive to the presence of the $D^*_{s1}(2860)^+$ resonance.

Figure 7 (Left) shows the threshold region of the $D^{*+}K^0_S$ mass spectrum with enhanced Natural Parity selection. Signals of $D_{s1}(2536)^+$ and $D^*_{s2}(2573)^+$ are observed. $D^*_{s2}(2573)^+$ is observed here for the first time and has a significance of 6.9σ. Its mass is found to be consistent with that obtained from the DK measurements. The $D^*_{s2}(2573)^+$ angular distribution is shown in Fig. 7 (Right) and is consistent with a $J^P = 2^+$ Natural Parity assignment. The relative branching fraction of the decay $D^*_{s2}(2573)^+ \to D^{*+}K^0_S$ is measured as

$$\frac{\mathcal{B}(D^*_{s2}(2573)^+ \to D^{*+}K^0_S)}{\mathcal{B}(D^*_{s2}(2573)^+ \to D^+K^0_S)} = 0.044 \pm 0.005\,(\text{stat}) \pm 0.011\,(\text{syst}). \tag{1}$$

and is in agreement with expectations from quark model calculations which predict a value of 0.058 for this ratio [10].

4 Search for new B^{**} states

The search for new B^{**} states is based on 3 fb^{-1} of data [11]. High purity B^+ (2.5×10^6) and B^0 (1.2×10^6) samples are selected through:

$$B^- \to J/\psi K^-,\ B^{-,0} \to D^{0,-}\pi^+(\pi^+\pi^-)\ [with\ D^0 \to K^-\pi^+(\pi^-\pi^+)],\ B^0 \to J/\psi K^*(892)^0.$$

Figure 8 shows the δm distributions for $B^+\pi^-$ and $B^0\pi^+$ systems. The narrow states are identified as $B_1(5721)^{0,+}$ and $B^*_2(5747)^{0,+}$. New broad states are observed and labelled as $B_J(5840)$ and $B_J(5960)$.

5 New results on baryon spectroscopy

5.1 Observation of excited Λ^0_b states

The quark model predicts orbitally excited Λ^{*0}_b with $J^P = \frac{1}{2}^+$ and $J^P = \frac{3}{2}^+$ decaying to $\Lambda^0_b\pi^+\pi^-$ or $\Lambda^0_b\gamma$. A clean sample of 70.500 $\Lambda_b \to \Lambda_c\pi^-$ decays in 1.0 fb^{-1} of data is reconstructed, with $\Lambda_c \to pK^-\pi^+$. The $\Lambda_b\pi^+\pi^-$ is shown in Fig. 9 [12]. Two narrow states observed with the following parameters

 DOI: 10.1051/epjconf/201612900019

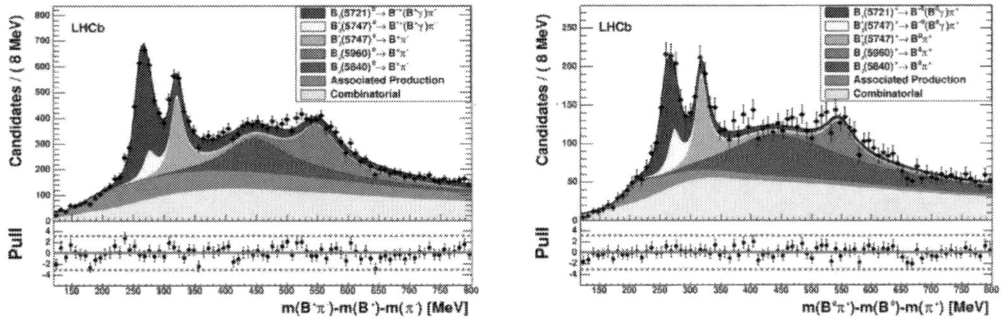

Figure 8. δm distribution for (Left) $B^+\pi^-$ and (Right) $B^0\pi^+$ systems.

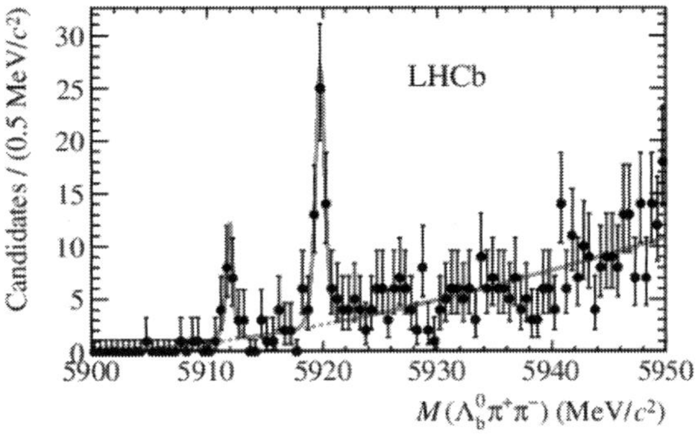

Figure 9. $\Lambda_b\pi^+\pi^-$ mass spectrum.

$M_{\Lambda_b^{*0}(5912)} = 5911.97 \pm 0.12 \pm 0.02 \pm 0.66$ MeV, $\Gamma_{\Lambda_b^{*0}(5912)} < 0.66$ MeV (90% CL),
$M_{\Lambda_b^{*0}(5920)} = 5919.77 \pm 0.08 \pm 0.02 \pm 0.66$ MeV, $\Gamma_{\Lambda_b^{*0}(5920)} < 0.63$ MeV (90% CL).

The $\Lambda_b^{*0}(5912)$ signal contains 17.6 ± 4.8 events and has a significance of 5.2σ; the $\Lambda_b^{*0}(5920)$ signal contains 52.5 ± 8.1 events with a significance of 10.2σ.

5.2 Ξ_b^{*0} Spectroscopy

In 2012, the CMS collaboration [13] observed a new $\Xi_b^{*0} \to \Xi_b^-\pi^+$ state [13]. LHCb has searched for this resonance using $\Xi_b^- \to \Xi_c^0\pi^+$ and $\Xi_c^0 \to pK^-K^-\pi^+$ [14].

Figure 10 shows the $\delta m = m_{cand}(\Xi_b^-\pi^+) - m_{cand}(\Xi_b^-) - m(\pi^+)$ distribution, where a clean $\Xi_b^{*0} \to \Xi_b^-\pi^+$ is observed, confirming the CMS observation.

The resonance parameters are

$m(\Xi_b^{*0}) - m(\Xi_b^-) - m(\pi^+) = 15.727 \pm 0.068 \pm 0.023$ MeV, $\Gamma(\Xi_b^{*0}) = 0.90 \pm 0.16 \pm 0.08$ MeV.

The state is consistent with being the $J^P = 3/2^+$ Ξ_b^{*0} resonance expected in the quark model.

5.3 Observation of two new Ξ_b^- resonances

LHCb experiment has searched for new states in the $\Xi_b^0\pi^-$ mass spectrum, with $\Xi_b^0 \to \Xi_c^+\pi^-$ and $\Xi_c^+ \to pK^-\pi^+$. Figure 11 shows the $\delta m = m_{cand}(\Xi_b^0\pi^-) - m_{cand}(\Xi_b^0) - m(\pi^-)$ distribution where two new states are observed [15].

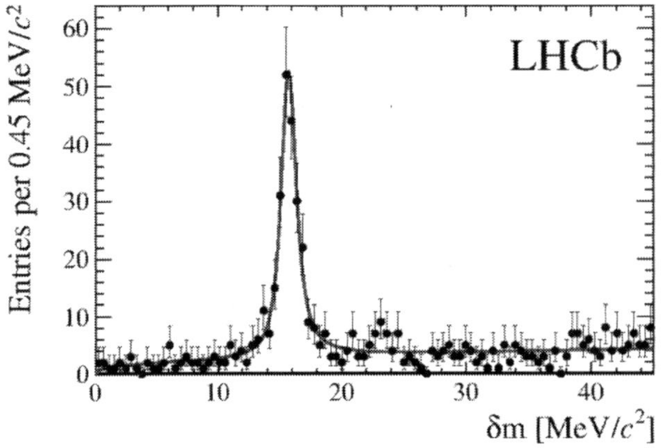

Figure 10. $\delta m = m_{cand}(\Xi_b^- \pi^+) - m_{cand}(\Xi_b^-) - m(\pi^+)$ distribution.

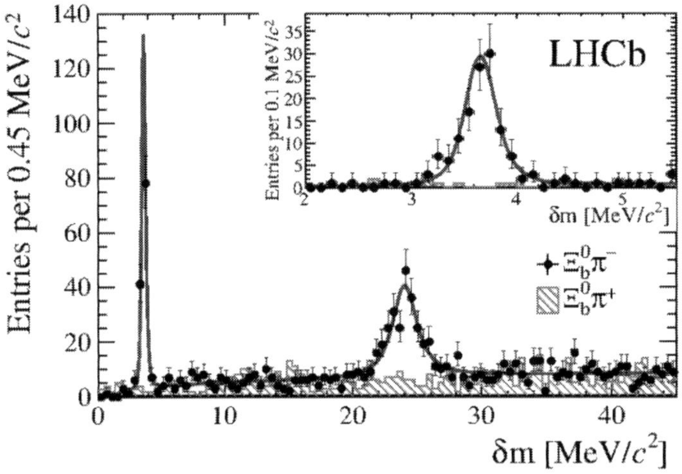

Figure 11. $\delta m = m_{cand}(\Xi_b^0 \pi^-) - m_{cand}(\Xi_b^0) - m(\pi^-)$ distribution.

The two states are consistent with being the $J^P = 1/2^+$ $\Xi_b^{'-}$ and $J^P = 3/2^+$ Ξ_b^{*-} resonances expected from the quark model in this mass region. Their parameters are measured to be

$$m(\Xi_b^{'-}) = 5935.02 \pm 0.02 \pm 0.01 \pm 0.50 \text{ MeV}, \Gamma(\Xi_b^{'-}) < 0.08 \text{ MeV at } 95\% \text{ C.L.},$$
$$m(\Xi_b^{*-}) = 5955.33 \pm 0.12 \pm 0.06 \pm 0.50 \text{ MeV}, \Gamma(\Xi_b^{*-}) = 1.65 \pm 0.31 \pm 0.10 \text{ MeV}.$$

5.4 Search for Ξ_{cc}^+ states

LHCb experiment has performed an initial search for $\Xi_{cc}^+ \to \Lambda_c K^- \pi^+$ [16] using a small data set (0.65 fb^{-1}). No signal has been found. A more accurate search will be performed using the very high statistics data samples actually available.

DOI: 10.1051/epjconf/201612900019

6 Summary

- Several new D_J and D_{sJ} mesons have been observed in both inclusive and exclusive B/B_s decays.
- The observation of these states in B/B_s decays allows the determination of their quantum numbers.
- New high statistics data will be shortly available and new decay channels are being analyzed. These studies will allow to better understand the charm and strange-charm spectrum.
- Progress have been performed in completing the quark model spectrum for heavy baryons. Several new states have been discovered. Large progress are expected in the new future in understanding the heavy baryons using high statistics and high purity new decay channels.

References

[1] R. Aaij *et al.* (LHCb Collaboration), Int. J. Mod. Phys. **A30**, 1530022 (2015).
[2] R. Aaij *et al.* (LHCb Collaboration), JHEP **09**, 145 (2013).
[3] R. Aaij *et al.* (LHCb Collaboration), Phys. Rev. **D91**, 092002 (2015).
[4] R. Aaij *et al.* (LHCb Collaboration), Phys. Rev. **D92**, 032002 (2015).
[5] R. Aaij *et al.* (LHCb Collaboration), Phys. Rev. **D92**, 032002 (2015).
[6] R. Aaij *et al.* (LHCb Collaboration), JHEP **10**, 151 (2012).
[7] K.A. Olive *et al.* (Particle Data Group), Chin. Phys. C, **38**, 090001 (2014).
[8] R. Aaij *et al.* (LHCb Collaboration), Phys. Rev. **D90**, 072003 (2014).
[9] R. Aaij *et al.* (LHCb Collaboration), JHEP **02**, 133 (2016).
[10] S. Godfrey and K. Moats, Phys. Rev. **D93**, 034035 (2016).
[11] R. Aaij *et al.* (LHCb Collaboration), JHEP **04**, 024 (2015).
[12] R. Aaij *et al.* (LHCb Collaboration), Phys. Rev. Lett. **109**, 172003 (2012).
[13] S. Chatrchyan *et al.* (CMS Collaboration), Phys. Rev. Lett. **108**, 252002 (2012).
[14] R. Aaij *et al.* (LHCb Collaboration), JHEP **05**, 161 (2016).
[15] R. Aaij *et al.* (LHCb Collaboration), Phys. Rev. Lett. **114**, 062004 (2015).
[16] R. Aaij *et al.* (LHCb Collaboration), JHEP **12**, 090 (2013).

EPJ Web of Conferences **129**, 00020 (2016) DOI: 10.1051/epjconf/201612900020
QCD@Work 2016

Production of hadron exotics in high energy processes

Wei Wang[1,⋆]

[1] *INPAC, Shanghai Key Laboratory for Particle Physics and Cosmology,
Department of Physics and Astronomy, Shanghai Jiao-Tong University, Shanghai 200240, China*

Abstract. Production of hadron exotics in high energy processes may play a role in unravelling their nature. In high energy limit, the leading power behavior of the cross section of exclusive processes is usually encoded in the constituent counting rule. In this talk I will demonstrate for the process involving hadrons with hidden flavors the constituent counting rule requests modifications. I will use effective field theory to derive the power scalings. After the $e^+e^- \to VP$ example I will discuss the production of hadron exotics including the $X(3872)$ and $Z_c^\pm$ in high energy process. I will also briefly talk about the challenge of the molecular interpretation of $X(3872)$ based on its large production rates in hadron collisions.

1 Introduction

Thanks to worldwide experiments during the last decades at e^+e^- and hadron colliders, our interest in hadron spectroscopy has been revived by the discovery of a plethora of new hadrons. Most of them were discovered in the charmonium and bottomonium sector and many defy a standard quarkonium interpretation. A recent review can be found in Ref. [1].

2 Constituent scaling rule for hadron exotics in exclusive production

2.1 Constituent scaling rule

At high energy with $\sqrt{s} \gg \Lambda_{\rm QCD}$ where s is the center-of-mass energy squared and $\Lambda_{\rm QCD}$ is the confinement scale, exclusive processes can be understood in QCD perturbation theory. The leading power behavior will lead to a constituent scaling rule in the factorization limit and can be formally derived by matching the full theory of QCD to a low-energy effective field theory (EFT).

At high energy, the differential cross-section of a generic process $a+b \to c+d$ obeys the asymptotic behavior [2]:

$$\frac{d\sigma}{dt} \sim s^{2-n} f(\theta_{cm}), \tag{1}$$

with $n = n_a + n_b + n_c + n_d$, and a, b, c, d denoting generic leptons, photons or hadrons. In the above, the s and t are Mandelstam variables, and the θ_{cm} is the scattering angle in the center-of-mass frame.

⋆e-mail: wei.wang@sjtu.edu.cn

Here n_h is the number of constituents in the particle h. A fundamental particle with no substructure like a quark, electron, or a photon has $n_i = 1$. An ordinary meson has $n_i = 2$ and a baryon has $n_i = 3$.

Recently, it has been argued in Refs. [3–7] that the cross section for high energy production of multiquark states should also scale as Eq. (1) with the power of s determined according to their expected valence quark structures. Instead of $n_i = 2$ for an ordinary meson, a meson-meson molecule or a compact tetraquark has $n_i = 4$, and a pentaquark has $n_i = 5$.

If the constituent counting rules were correct, it would provide a very straightforward tool to unravel the quark structures of exotic hadrons. In the following, I will show for hadrons with hidden-flavor quarks it is problematic to naively apply such a constituent counting rule [8]. I will consider a simple example involving only ordinary mesons, $e^+e^- \to VP$, with V and P denoting ordinary light flavor vector and pseudoscalar mesons, respectively. This process requests a helicity flip and thus it implies the following scaling for integrated cross section [2] from Eq. (1):

$$\sigma(e^+e^- \to VP) \propto \frac{1}{s^4}.$$

2.2 SCET power counting

Perturbation theory at high energies allows to express the matrix elements of Heisenberg operators in terms of free local operators and interaction terms. When factorization exists, one may separate the interactions according to the involved scales using the operator product expansion. For a generic matrix element, we have:

$$\begin{aligned}\langle f|O_H(0)|i\rangle &= \langle f|T\left[O\times \exp[i\int d^4x\mathcal{L}_{\rm int}(x)]_{>\mu}\times T\exp[i\int d^4x\mathcal{L}_{\rm int}(x)]_{<\mu}\right]|i\rangle \\ &\sim \langle f|T\left[O'\times \exp[i\int d^4x\mathcal{L}_{\rm int}(x)]_{<\mu}\right]|i\rangle \equiv \langle f|O'_{H,\mu}|i\rangle, \end{aligned} \tag{2}$$

where we have formally integrated out the interactions above the factorization scale μ, and obtained a new set of low-energy effective operators O'. The interaction below μ contains no information on the $1/\sqrt{s}$ and thus the scaling is obtained by counting the pertinent number of constituents in O'.

Now I will use the effective field theory to explicitly derive the leading power contribution to the cross section. The applicable EFT here is the soft-collinear effective theory (SCET) [9, 10]. For an introduction please see the textbook [11]. Instead of directly studying the s dependence, we introduce a dimensionless parameter $\lambda = \Lambda/\sqrt{s}$ and and count the power dependence on λ

$$\sigma \sim \frac{1}{s}\left(\frac{\Lambda}{\sqrt{s}}\right)^n. \tag{3}$$

The Λ is a low-energy scale and may be taken as Λ_{QCD} in the case of a light quark, $m_{c/b}$ in the case of a charm/bottom quark.

At high energy, the energetic quarks are jet-like (collinear) with the typical momentum

$$p = (p^+, p^-, p_\perp) \propto \left(\sqrt{s}, \frac{\Lambda^2}{\sqrt{s}}, \Lambda\right). \tag{4}$$

For an energetic quark, it is convenient to split the quark field into two components:

$$\psi = \xi + \eta, \quad \xi = \frac{\not{n}\not{\bar{n}}}{4}\psi, \quad \eta = \frac{\not{\bar{n}}\not{n}}{4}\psi,$$

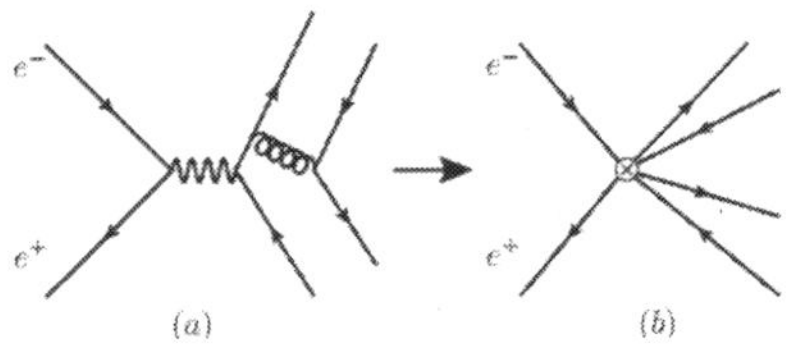

Figure 1. Feynman diagram for $e^+e^- \to VP$ in the full theory (a) and effective field theory (b).

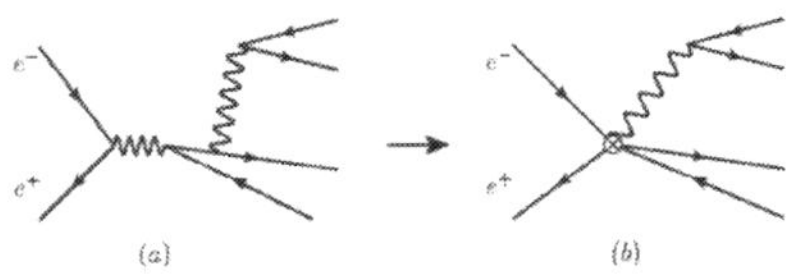

Figure 2. Feynman diagram for the photon contribution to $e^+e^- \to VP$ in the full theory (a) and effective field theory (b). Here the vector meson is a neutral one like ρ^0, ω, ϕ.

where n and $\bar{n}$ is two light-like vectors: $n^2 = \bar{n}^2 = 0$. The quark field scaling can be obtained by considering the two-point correlator:

$$\langle 0|T[\psi(x)\bar{\psi}(y)]|0\rangle = \int \frac{d^4p}{(2\pi)^4} e^{-ip\cdot(x-y)} \frac{i(\not{p}+m)}{p^2-m^2+i\epsilon}. \tag{5}$$

This gives $\xi \propto \lambda$, $\eta \propto \lambda^2$. For a collinear photon or gluon field, one similarly finds $n_+A \propto 1$, $A_\perp \propto \lambda$, $n_-A \propto \lambda^2$. From the normalization, the mesonic state scales as $|M\rangle \propto \lambda^{-1}$.

Matching to the effective field theory is equivalent to integrate out the off-shell intermediate states. For the $e^+e^- \to VP$ shown in Fig. 1, the vector meson is produced by a pair of quark fields with:

$$\langle V|\bar{\psi}\gamma_\perp^\mu\psi|0\rangle = \langle V|(\bar{\xi}\gamma_\perp^\mu\eta + \bar{\eta}\gamma_\perp^\mu\xi)|0\rangle \sim \lambda^2, \tag{6}$$

where the transverse projection receives the additional suppression λ. The leptonic state cancels the scaling of the lepton field. Then the power scaling of amplitude for $e^+e^- \to VP$ is given as

$$\mathcal{M} \propto \langle P|\bar{\xi}\not{n}\xi|0\rangle \times \langle V|\bar{\psi}\gamma_\perp^\mu\psi|0\rangle \propto \lambda^3, \tag{7}$$

and the cross section has

$$\sigma \propto \frac{1}{s}\lambda^6 \propto \frac{\Lambda^6}{s^4}. \tag{8}$$

This recovers the standard perturbative QCD prediction given in Eq. (2).

A vector meson such as ρ^0 can be produced by a photon field $\langle\rho^0|A_\perp^\mu|0\rangle$. A gauge invariant operator can be constructed by including the Wilson line. Then the scaling of decay amplitude

$$\mathcal{M} \propto \langle\pi|\bar{\xi}\not{n}\xi|0\rangle \times \langle\rho^0|A_\perp^\mu|0\rangle \propto \lambda, \tag{9}$$

leads to the cross section

$$\sigma \propto \frac{1}{s}\lambda^2 \propto \frac{\Lambda^2}{s^2}. \tag{10}$$

It is interesting to notice the production through a photon is suppressed by the electromagnetic coupling constant $\alpha_{\rm em} \sim 1/137$ and less important at low energies. But at very high energies this new diagram will dominate and in particular it gives a scaling rule different from the one in Eq. (2) [12].

To understand the above behaviors and accommodate with the constituent scaling rule, one can technically count the valence degrees of freedom of the neutral vector meson as $n_i = 1$, which amounts to count the number of lines (a photon in this case) attached to the effective vertex. The lesson that one can learn from the above example is: not all ingredients undergo the hard momentum transfer. The fall-off scaling is determined by the leading-power (leading-twist) operator at the scale $\mu = \sqrt{s}$ which overlaps with the hadron.

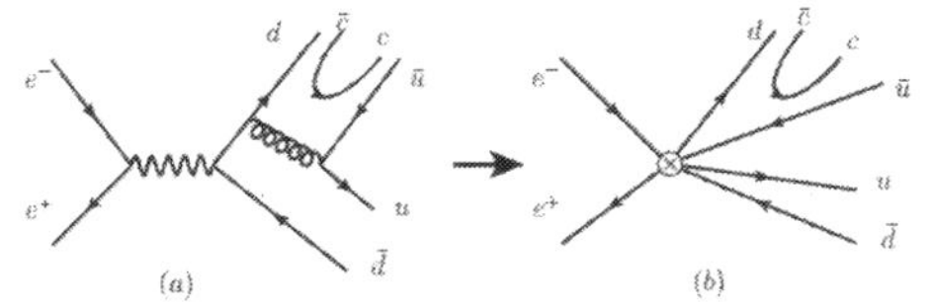

Figure 3. Feynman diagrams for $e^+e^- \to Z_c^{\pm}\pi^{\mp}$.

2.3 Implications

Let us switch to the exclusive production of multiquark states, and consider the $e^+e^- \to Z_c^{\pm}\pi^{\mp}$. The $Z_c^{\pm}(3900)$ was discovered in Refs. [13, 14], but its nature is still controversial. In Ref. [6], it has been argued that its cross section in the $s \to \infty$ limit obeys the scaling

$$\frac{\sigma(e^+e^- \to Z_c^+(\bar{c}c\bar{d}u)\pi^-(\bar{u}d))}{\sigma(e^+e^- \to \mu^+\mu^-)} \overset{?}{\propto} \frac{1}{s^4}, \tag{11}$$

where the $Z_c^+(\bar{c}c\bar{d}u)$ is a tetraquark made of two quarks and two antiquarks. We have put a question mark to the above scaling behavior since we believe that it is problematic at very high energies. We show a production mechanism in Fig. 3 (a). In this diagram, the heavy quark pair $\bar{c}c$ is generated from the QCD vacuum, and such a contribution is suppressed by $O(1/m_c^2)$. Integrating out the off-shell intermediate propagators at the scale $\sqrt{s}$ we find that the Z_c behaves as an ordinary $\bar{q}q$ meson and the s dependence scaling of the cross-section is determined by light quarks of the Z_c:

$$\frac{\sigma(e^+e^- \to Z_c^+\pi^-)}{\sigma(e^+e^- \to \mu^+\mu^-)} \propto \frac{1}{s^2}, \tag{12}$$

which can be derived by counting the number of lines attached to the effective vertex. It is obvious this production mechanism will become dominant at very high energies with $\sqrt{s} \gg m_c$.

Now we come to the production of the $X(3872)$. It is necessary to stress that as pointed out in Ref. [15] the production of the $X(3872)$ meson involves many length scales. The creation of the $\bar{c}c$ pair with small relative momentum requires a hard-scattering process with a short-distance scale m_c. The evolution of the $\bar{c}c$ into a color-singlet hadronic system occurs over a longer length scale $m_c v$ or $m_c v^2$. The evolution of the charmed mesons occurs over an even lower scale m_π. Finally the binding of DD^* into the molecular state X occurs over the very long length scale. It is also interesting to notice that the above treatment for the $X(3872)$ production is different with the mixing scheme [16].

Like the case of Z_c, the leading power operator to produce the $X(3872)$ might have the form:

$$\langle X|\bar{c}\Gamma c|0\rangle, \quad \langle X|\bar{q}q|0\rangle.$$

For instance for $B_c \to X(3872)$ decays, the $\langle X|\bar{c}\Gamma c|0\rangle$ contributes at leading power. A consequence is that this contribution is irrespective of the emitted particles in B_c decays. Thus the ratios of branching fractions of semileptonic and nonleptonic decays for instance $B_c \to X\ell\bar{\nu}$ and $B_c \to J/\psi\rho$ can be precisely predicted and tested by data. For more details one can see Ref. [17].

Actually, the leading power behavior has been incorporated in generalized light-cone distribution amplitude (LCDA), where the leading twist LCDA of $p\bar{p}$ and $\pi\pi$ systems is defined by quark fields. For a detailed review see Ref. [18]. The generalized LCDA has recently been used to calculate heavy meson decays in Refs. [19–25] and agreements with relevant data are found.

3 Prompt $X(3872)$ production at hadron colliders

The $X(3872)$ is a notable exotic candidate and has received great attention [26]. Its tiny width and the vicinity of its mass to the $D\bar{D}^*$ threshold has lead to the speculation as a hadron molecule [27, 28]. A competitive model is compact tetraquark [29], dominated by short-range color interactions.

The fact that the $X(3872)$ hadron can be copiously produced in hadron collisions has lead to the suspect of the molecular interpretation of the $X(3872)$. Ref. [30] has used the Monte Carlo method and calculated the production rates of DD^*. By taking a momentum cutoff set by the binding energy, the authors have found the simulated cross section is smaller than data. Actually, this simulation has used the constituent-molecule duality, namely, the production rate of the constituents in the phase space is equivalent to that of the molecule. This is similar with *local* quark-hadron duality, which often fails for very narrow resonances. To recover the quark-hadron duality, one should include final state interactions which is equivalent to increase the momentum cutoff [31, 32]. For the production of ordinary heavy quarkonium, modern EFT approach is nonrelativistic QCD, in which a hadron is nonperturbatively produced by quark fields. Similarly, for hadronic molecules we have established an approach in which the hadron molecule is produced by its constituents, and the low-energy matrix element has been estimated using an effective theory at hadron level in Refs. [33–35]. This approach avoids the use of *local* constituent-molecule duality, and thus the results should be more reliable.

It has also been argued that "If the $X(3872)$ shares the same molecule nature with light nuclei, one expects similar production cross sections, especially at high values of $p_\perp$" [36]. As we have shown above, unlike the light nuclei like a deutron, the $X(3872)$ contains hidden flavors, and thus the production rate at leading power does not scale according to its low energy nature. This is different with the deutron production, since the deutron has at least six quarks.

4 Conclusions

For a reaction involving hadrons, if factorization exists, short-distance and long-distance degrees of freedom decouple from each other. The long-distance physics depends on hadrons' internal structure but does not affect the short-distance coefficients. The short-distance coefficients are calculable using QCD perturbation theory. Using the $e^+e^- \to VP$ process, I have analyzed the leading power contribution to the cross section and demonstrated that it is not compatible with the naive constituent scaling rule. This has important consequences in the study of the production of hadron exotics, since the misuse of the constituent scaling rule will sometimes lead to misleading results to the nature of exotic hadrons. Such examples include the production of $Z_c^\pm$ and $X(3872)$.

Acknowledgement

I thank the organizers of this workshop, in particular Pietro Colangelo, Fulvia De Fazio, Floriana Giannuzzi, and Stefano Nicotri. I am grateful to Ulf-G. Meißner, Feng-Kun Guo and Qiang Zhao for collaborations. Useful communications with Stanley Brodsky and Richard Lebed are acknowledged. This work is supported in part by the Thousand Talents Plan for Young Professionals, National Natural Science Foundation of China under Grant No.11575110, Natural Science Foundation of Shanghai under Grant No. 15DZ2272100 and No. 15ZR1423100, by the Open Project Program of State Key Laboratory of Theoretical Physics, Institute of Theoretical Physics, Chinese Academy of Sciences, China (No.Y5KF111CJ1), and by Scientific Research Foundation for Returned Overseas Chinese Scholars, State Education Ministry.

References

[1] H. X. Chen, W. Chen, X. Liu and S. L. Zhu, Phys. Rept. **639**, 1 (2016).
[2] G. P. Lepage and S. J. Brodsky, Phys. Rev. D **22**, 2157 (1980).
[3] H. Kawamura, S. Kumano and T. Sekihara, Phys. Rev. D **88**, 034010 (2013).

[4] H. Kawamura and S. Kumano, Phys. Rev. D **89**, no. 5, 054007 (2014).
[5] S. H. Blitz and R. F. Lebed, Phys. Rev. D **91**, no. 9, 094025 (2015).
[6] S. J. Brodsky and R. F. Lebed, Phys. Rev. D **91**, 114025 (2015).
[7] W. C. Chang, S. Kumano and T. Sekihara, Phys. Rev. D **93**, no. 3, 034006 (2016).
[8] F. K. Guo, U. G. Meißner and W. Wang, arXiv:1607.04020 [hep-ph].
[9] C. W. Bauer, S. Fleming and M. E. Luke, Phys. Rev. D **63**, 014006 (2000).
[10] C. W. Bauer, S. Fleming, D. Pirjol and I. W. Stewart, Phys. Rev. D **63**, 114020 (2001).
[11] T. Becher, A. Broggio and A. Ferroglia, Lect. Notes Phys. **896** (2015).
[12] C. D. Lu, W. Wang and Y. M. Wang, Phys. Rev. D **75**, 094020 (2007).
[13] M. Ablikim *et al.* [BESIII Collaboration], Phys. Rev. Lett. **110**, 252001 (2013).
[14] Z. Q. Liu *et al.* [Belle Collaboration], Phys. Rev. Lett. **110**, 252002 (2013).
[15] E. Braaten, Phys. Rev. D **73**, 011501 (2006).
[16] C. Meng, H. Han and K. T. Chao, arXiv:1304.6710 [hep-ph].
[17] W. Wang and Q. Zhao, Phys. Lett. B **755**, 261 (2016).
[18] M. Diehl, Phys. Rept. **388**, 41 (2003).
[19] C. H. Chen and H. n. Li, Phys. Lett. B **561**, 258 (2003).
[20] U. G. Meißner and W. Wang, Phys. Lett. B **730**, 336 (2014).
[21] W. F. Wang, H. C. Hu, H. n. Li and C. D. Lü, Phys. Rev. D **89**, no. 7, 074031 (2014).
[22] W. Wang and R. L. Zhu, Phys. Lett. B **743**, 467 (2015).
[23] W. F. Wang, H. n. Li, W. Wang and C. D. Lü, Phys. Rev. D **91**, no. 9, 094024 (2015).
[24] Y. J. Shi and W. Wang, Phys. Rev. D **92**, no. 7, 074038 (2015).
[25] C. Hambrock and A. Khodjamirian, Nucl. Phys. B **905**, 373 (2016).
[26] S. K. Choi *et al.* [Belle Collaboration], Phys. Rev. Lett. **91**, 262001 (2003).
[27] N. A. Tornqvist, Phys. Lett. B **590**, 209 (2004).
[28] C. Hanhart, Y. S. Kalashnikova, A. E. Kudryavtsev and A. V. Nefediev, Phys. Rev. D **76**, 034007 (2007).
[29] L. Maiani, F. Piccinini, A. D. Polosa and V. Riquer, Phys. Rev. D **71**, 014028 (2005).
[30] C. Bignamini, B. Grinstein, F. Piccinini, A. D. Polosa and C. Sabelli, Phys. Rev. Lett. **103**, 162001 (2009).
[31] P. Artoisenet and E. Braaten, Phys. Rev. D **81**, 114018 (2010).
[32] P. Artoisenet and E. Braaten, Phys. Rev. D **83**, 014019 (2011).
[33] F. K. Guo, U. G. Meißner and W. Wang, Commun. Theor. Phys. **61**, 354 (2014).
[34] F. K. Guo, U. G. Meißner, W. Wang and Z. Yang, Eur. Phys. J. C **74**, no. 9, 3063 (2014).
[35] F. K. Guo, U. G. Meißner, W. Wang and Z. Yang, JHEP **1405**, 138 (2014).
[36] A. Esposito, A. L. Guerrieri, L. Maiani, F. Piccinini, A. Pilloni, A. D. Polosa and V. Riquer, Phys. Rev. D **92**, no. 3, 034028 (2015).

EPJ Web of Conferences **129**, 00021 (2016) DOI: 10.1051/epjconf/201612900021
QCD@Work 2016

LHCb results on exotics and pentaquark states

Marco Pappagallo[1,⋆]

[1] *INFN and Universitá degli studi di Bari*

Abstract. The latest years have seen a resurrection of interest in searches for exotic states. Using the data collected at pp collisions at 7 and 8 TeV by the LHCb experiment, unambiguous new observations of exotic charmonia hadrons produced in B and Λ_b decays are presented. Results of a search for a tetraquark state decaying into $B_s^0\pi^{\pm}$ arealso reported.

1 Introduction

The LHCb detector [1] is a single-arm forward spectrometer covering the pseudorapidity range $2 < \eta < 5$, designed for the study of particles containing b or c quarks. During 2011 and 2012, pp collision data corresponding to integrated luminosities of 1 fb^{-1} at a centre-of-mass energy of 7 TeV and 2 fb^{-1} at 8 TeV have been collected with the LHCb detector.

2 Search for pentaquarks states

2.1 Observation of $J/\psi p$ resonances consistent with pentaquark states in $\Lambda_b \to J/\psi K^- p$ decays

Past claimed observations of pentaquark states have been shown to be spurious but the existence of pentaquark baryon states would not be surprising given the observation of charged charmonium-like meson (e.g. $Z_c(4430)^+$). A large sample (26,000 candidates) of Λ_b baryons decaying to the $J/\psi K^- p$ final state has been reconstructed at LHCb [2]. In Fig. 1(a) the Dalitz plot is shown using the $K^- p$ and $J/\psi p$ invariant masses-squared as independent variables. A distinct horizontal band is observed in the $J/\psi p$ invariant mass distribution near 19.5 GeV/c^2. In order to ascertain if the observed structures are resonant in nature and not due to reflections generated by the Λ^* states, it is necessary to perform a full amplitude analysis, allowing for interference effects between both decay sequences.

The fit uses five decay angles and the $K^- p$ invariant mass m_{Kp} as independent variables. First the data were fit with an amplitude model that contains 14 Λ^* states listed by the Particle Data Group [3]. As this did not give a satisfactory description of the data, one P_c^+ state was added, and when that was not sufficient a second state was included. The two P_c^+ states are found to have masses of 4380 ± 8 ± 29 MeV and 4449.8 ± 1.7 ± 2.5 MeV, with corresponding widths of 205 ± 18 ± 86 MeV and 39 ± 5 ± 19 MeV. The best fit solution has spin-parity J^P values of $(3/2^-, 5/2^+)$. Acceptable solutions are also found for additional cases with opposite parity, either $(3/2^+, 5/2^-)$ or $(5/2^+, 3/2^-)$. The best fit projections are shown in Fig. 1(b). The significances of the lower mass and higher mass states are 9 and 12 standard deviations, respectively.

⋆e-mail: marco.pappagallo@cern.ch

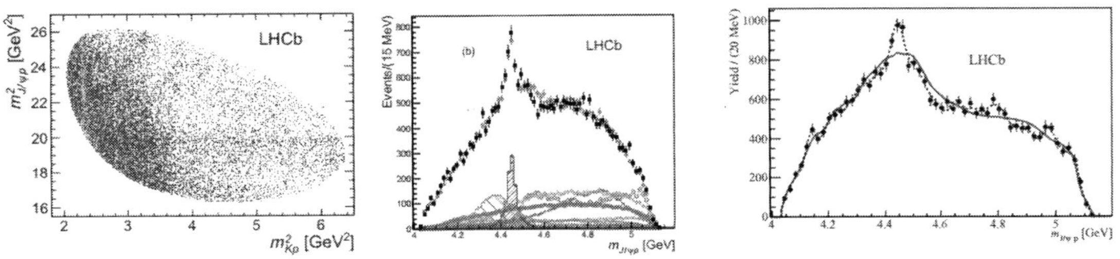

Figure 1. (left) Invariant mass squared of K^-p versus $J/\psi p$ for candidates. (middle) Fit projections for $m_{J/\psi p}$ with two P_c^+ states. The data are shown as solid (black) squares, while the solid (red) points show the results of the fit. The (blue) open squares with the shaded histogram represent the $P_c(4450)^+$ state, and the shaded histogram topped with (purple) filled squares represents the $P_c(4380)^+$ state. (right) Efficiency-corrected and background-subtracted $m_{J/\psi p}$ distribution of the data (black points), with $PDF(m_{J/\psi p}|H_0)$ (solid blue line) and $PDF(m_{J/\psi p}|H_1)$ (dashed black line) superimposed.

Further evidence for the resonant character of the higher mass, narrower, P_c^+ state is obtained by viewing the evolution of the complex amplitude in the Argand diagram [3]. In the amplitude fits discussed above, the $P_c(4450)^+$ is represented by a Breit-Wigner amplitude, where the magnitude and phase vary with $m_{J/\psi p}$ according to an approximately circular trajectory in the (Re A^{P_c}, Im A^{P_c}) plane, where A^{P_c} is the $m_{J/\psi p}$ dependent part of the $P_c(4450)^+$ amplitude. An additional fit to the data has been performed, in which the $P_c(4450)^+$ amplitude is represented as the combination of independent complex amplitudes at six equidistant points in the range $\pm\Gamma_0 = 39$ MeV around $M_0 = 4449.8$ MeV as determined in the default fit. The resulting Argand diagram, shown in Fig. 2(a), is consistent with a rapid counter-clockwise change of the $P_c(4450)^+$ phase when its magnitude reaches the maximum, a behavior characteristic of a resonance. A similar study for the wider state is shown in Fig. 2(b); although the fit does show a large phase change, the amplitude values are sensitive to the details of the Λ^* model and so this latter study is not conclusive.

2.2 Model-independent evidence for $J/\psi p$ contributions to $\Lambda_b \to J/\psi pK^-$ decays

In order to investigate further the presence of pentaquarks in the $\Lambda_b \to J/\psi pK^-$ decay chain, a model-independent analysis has been carried out [4]. It has been demonstrated at more than 9 standard deviations that the $\Lambda_b \to J/\psi pK^-$ decays cannot all be attributed to K^-p resonant or nonresonant contributions (Fig. 1(c)). The analysis requires only minimal assumptions on the mass and spin of the K^-p contributions; no assumptions on their number, their resonant or nonresonant nature, or their lineshapes have been made. Non-K^-p contributions, which must be present in the data, can be either of the exotic hadron type, or due to rescattering effects among ordinary hadrons. This result supports the amplitude model-dependent observation of the $J/\psi p$ resonances presented previously [2].

2.3 Evidence for exotic hadron contributions to $\Lambda_b \to J/\psi p\pi^-$ decays

Observations of the same two pentaquark states in another decay would strengthen their interpretation as genuine exotic baryonic states, rather than kinematical effects related to the so called triangle singularity [5, 6]. The LHCb collaboration has performed an amplitude analysis of $\Lambda_b \to J/\psi p\pi^-$ decays, which are related to $\Lambda_b \to J/\psi pK^-$ decays via Cabibbo suppression [7]. A sample of 1885 $\pm$ 50 candidates has been selected using the same data sample (3 fb^{-1}) and similar seletion tecnique as

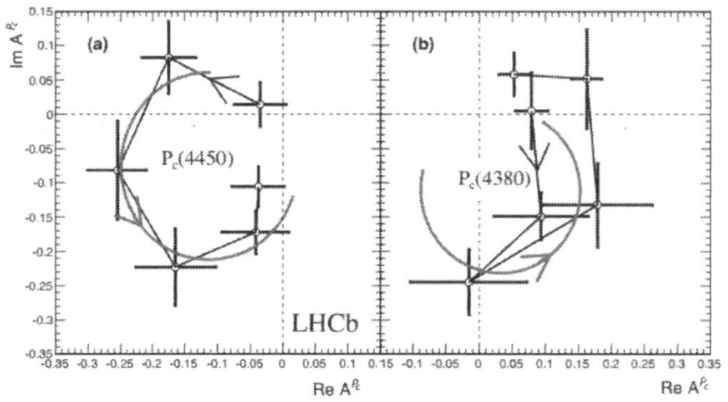

Figure 2. Fitted values of the real and imaginary parts of the amplitudes for the baseline ($3/2^-$, $5/2^+$) fit for a) the $P_c(4450)^+$ state and b) the $P_c(4380)^+$ state, each divided into six $m_{J/\psi p}$ bins of equal width between $-\Gamma_0$ and $+\Gamma_0$ shown in the Argand diagrams as connected points with error bars.

for the $\Lambda_b \to J/\psi p K^-$ decays. Amplitude models for the $\Lambda_b \to J/\psi p \pi^-$ decays are constructed to examine the possibility of exotic hadron contributions from the $P_c(4380)^+$ and $P_c(4450)^+ \to J/\psi p$ states [2] and from the $Z_c(4200)^- \to J/\psi \pi^-$ state, previously reported by the Belle collaboration in $B^0 \to J/\psi K^+ \pi^-$ decays [8].

The 15 well-established N^* resonances [3] are considered in the amplitude model of $p\pi^-$ contributions. The high-mass and high-spin states are not included, since they require $L \geq 3$ in the Λ_b decay and therefore are unlikely to be produced near the upper kinematic limit of $m_{p\pi}$. The data are examined only for the presence of the previously observed $P_c(4380)^+$, $P_c(4450)^+$ states and the claimed $Z_c(4200)^-$ resonance due to the limited number of signal events and the large number of free parameters. In the fits, the mass and width of each exotic state are fixed to the reported central values.

A significantly better description of the data is achieved by either including the two P_c^+ states observed in $\Lambda_b \to J/\psi p K^-$ decays, or the $Z_c(4200)^-$ state (Fig. 3(a)). If both types of exotic resonances are included, the total significance for them is $3.1\,\sigma$. Individual exotic hadron components, or the two P_c^+ states taken together, are not significant as long as the other(s) is (are) present. Within the statistical and systematic errors, the data are consistent with the $P_c(4380)^+$ and $P_c(4450)^+$ production rates expected from their previous observation and Cabibbo suppression.

3 Amplitude analysis of $B^+ \to J/\psi \phi K^+$ and observation of $J/\psi \phi$ structures consistent with exotic states

There has been a great deal of experimental and theoretical interest in $J/\psi\phi$ mass structures in $B^+ \to J/\psi \phi K^+$ decays since the CDF collaboration presented $3.8\,\sigma$ evidence for a near-threshold $X(4140)$ mass peak, with width $\Gamma = 12$ MeV [9]. Much larger widths are expected for charmonium states at this mass because of open flavor decay channels, which should also make the kinematically suppressed $X \to J/\psi\phi$ decays undetectable. Therefore, it has been suggested that the $X(4140)$ peak could be a molecular state, a tetraquark state, a hybrid state, or a rescattering effect.

The LHCb collaboration has performed the first amplitude analysis of $B^+ \to J/\psi \phi K^+$ decays [10, 11]. We have obtained a good description of the data in the 6D phase space composed of invariant masses and decay angles (Fig. 3(b)). The K^{*+} amplitude model extracted from our data is consistent with expectations from the quark model and from the previous experimental results on such

 DOI: 10.1051/epjconf/201612900021

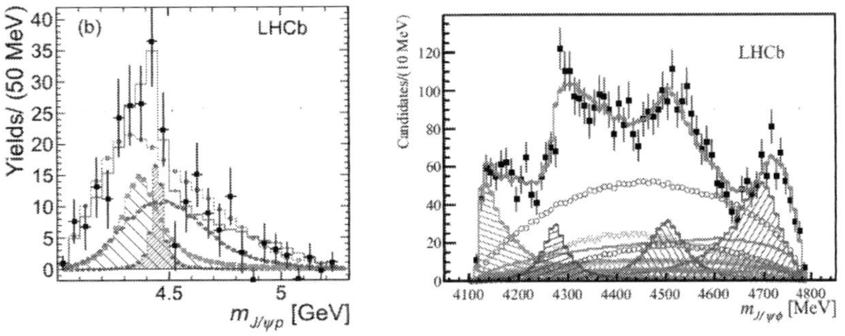

Figure 3. (left) Background-subtracted data and fit projections onto $m(J/\psi p)$ for the $m(p\pi) > 1.8$ GeV region. (right) Distributions of $J/\psi\phi$ invariant masses for the $B^+ \to J/\psi\phi K^+$ candidates (black data points) compared with the results of the default amplitude fit containing eight $K^{*+} \to \phi K^+$ and five $X \to J/\psi\phi$ contributions. The total fit is given by the red points with error bars.

resonances. We determine the J^{PC} quantum numbers of the $X(4140)$ structure to be 1^{++}. This has a large impact on its possible interpretations, in particular ruling out the 0^{++} or 2^{++} $D_s^{*+}D_s^{*-}$ molecular models. The $X(4140)$ width is substantially larger than previously determined. The existence of the $X(4274)$ structure is established and its quantum numbers are determined to be 1^{++}. The high $J/\psi\phi$ mass region is investigated for the first time with good sensitivity and shows very significant structures, which can be described as two 0^{++} resonances: $X(4500)$ and $X(4700)$.

4 Search for structure in the $B_s^0\pi^\pm$ invariant mass spectrum

In february the D0 collaboration has claimed [12] the observation with 5.1 σ significance of a narrow structure, which they dub the $X(5568)$, in the decay sequence $X(5568) \to B_s^0\pi^\pm$, $B_s^0 \to J/\psi\phi$, $J/\psi \to \mu^+\mu^-$, $\phi \to K^+K^-$. The mass and width of the $X(5568)$ state are reported to be $m = 5567.8 \pm 2.9\,^{+0.9}_{-1.9}\,\mathrm{MeV/c^2}$ and $\Gamma = 21.9 \pm 6.4\,^{+5.0}_{-2.5}\,\mathrm{MeV/c^2}$. The fraction of B_s^0 mesons that comes from $X(5568) \to B_s^0\pi^\pm$ decay is reported to be $\rho_X^{\mathrm{D0}} = (8.6 \pm 1.9 \pm 1.4)\,\%$. Recently the significance of the signal has been revisited after loosening the selection of the $B_s^0\pi^\pm$ candidates.

The claimed $X(5568)$ state has been searched in a data sample corresponding to 3 fb^{-1} of pp collision data at $\sqrt{s} = 7$ and 8 TeV recorded by LHCb [13]. The B_s^0 mesons are reconstructed in decays to $D_s^-\pi^+$ and $J/\psi\phi$ final states to obtain a B_s^0 yield approximately 20 times larger than that available to the D0 collaboration. The analysis techniques follow closely those developed for studies of the B^+K^- [14], $B^+\pi^-$ and $B^0\pi^+$ [15] spectra. The B_s^0 candidates are combined with each track that originates from the same primary vertex and that has $p_T > 500\,\mathrm{MeV/c}$.

The $Q \equiv m(B_s^0\pi) - M_{B_s^0} - M_\pi$ distributions are fitted with a function containing components for signal and background (Fig. 4(a)). The signal shape is an S-wave Breit–Wigner function, with mass and width parameters fixed according to the central values obtained by the D0 collaboration. The background is modelled with a polynomial function. The fit has a p-value of 34.0% and no significant $X(5568)$ yield.

The results of the fits to the $B_s^0\pi^\pm$ Q value distributions are used to measure the ratio of cross-sections

$$\rho_X^{\mathrm{LHCb}} \equiv \frac{\sigma(pp \to X(5568) + \text{anything}) \times \mathcal{B}\big(X(5568) \to B_s^0\pi^\pm\big)}{\sigma(pp \to B_s^0 + \text{anything})} = \frac{N(X)}{N(B_s^0)} \times \frac{1}{\epsilon^{\mathrm{rel}}(X)}, \quad (1)$$

DOI: 10.1051/epjconf/201612900021

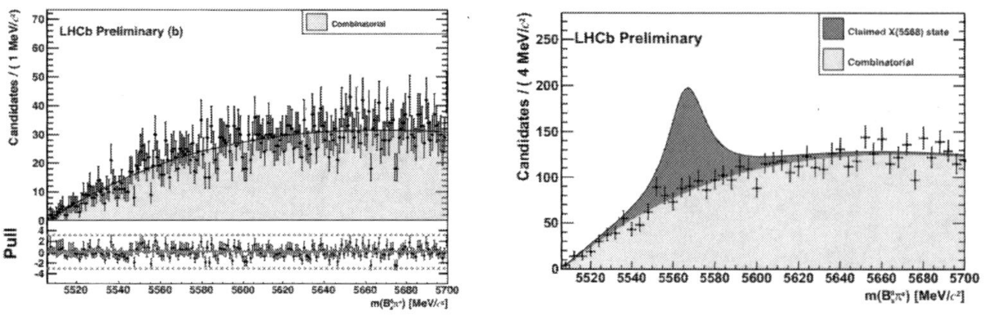

Figure 4. (left) Results of the fit to the Q value distribution, shifted to display the $B_s^0\pi^+$ invariant mass, for $B_s^0\pi$ candidates (both B_s^0 modes combined) with the signal yield fixed to zero. The fits are shown for minimum B_s^0 p_T of 10 GeV/c^2. (right) Distribution of Q value, shifted to display the $B_s^0\pi^+$ invariant mass, for candidates with minimum B_s^0 p_T of 10 GeV/c^2, with a signal component superimposed that corresponds to $\rho_X^{\rm LHCb} = 8.6\,\%$.

where the cross-sections σ are for promptly produced particles within the LHCb acceptance. The relative efficiency $\epsilon^{\rm rel}(X) = \frac{\epsilon(X)}{\epsilon(B_s^0)}$ is determined from simulation and accounts for the reconstruction and selection efficiency of the companion pion as well as the requirement that is within the LHCb acceptance. The dominant sources of systematic uncertainties are due to the variation of the signal width and the effect of the efficiency on the signal shape. Since the signal is not significant, upper limits on $\rho_X^{\rm LHCb}$ are found to be

$$\rho_X^{\rm LHCb}(B_s^0\ p_T > \ 5\ {\rm GeV/c^2}) \ < \ 0.009\,(0.010)\ @\ 90\,(95)\,\%\ {\rm CL}\,,$$
$$\rho_X^{\rm LHCb}(B_s^0\ p_T > 10\ {\rm GeV/c^2}) \ < \ 0.016\,(0.018)\ @\ 90\,(95)\,\%\ {\rm CL}\,.$$

References

[1] A. A. Alves Jr. *et al.* [LHCb Collaboration], JINST **3** (2008) S08005.
[2] R. Aaij *et al.* [LHCb Collaboration], Phys. Rev. Lett. **115** (2015) 072001.
[3] K. A. Olive *et al.* (Particle Data Group), Chin. Phys. C, **38** (2014) 090001.
[4] R. Aaij *et al.* [LHCb Collaboration], Phys. Rev. Lett. **117** (2016) 082002.
[5] T. J. Burns, Eur. Phys. J. **A51** (2015) 152.
[6] E. Wang *et al.*, Phys. Rev. **D93** (2016) 094001.
[7] R. Aaij *et al.* [LHCb Collaboration], Phys. Rev. Lett. **117** (2016) 082003.
[8] K. Chilikin *et al.* [Belle Collaboration], Phys. Rev. **D90** (2014) 112009.
[9] T. Aaltonen *et al.*, [CDF collaboration] Phys. Rev. Lett. **102** (2009) 242002.
[10] R. Aaij *et al.* [LHCb Collaboration], arXiv:1606.07895.
[11] R. Aaij *et al.* [LHCb Collaboration], arXiv:1606.07898.
[12] V. M. Abazov *et al.* [D0 Collaboration], Phys. Rev. Lett. **117** (2016) 022003.
[13] R. Aaij *et al.* [LHCb Collaboration], LHCb-CONF-2016-004 (superseded by Phys. Rev. Lett. **117** (2016) 152003).
[14] R. Aaij *et al.* [LHCb Collaboration], Phys. Rev. Lett. **110** (2013) 151803.
[15] R. Aaij *et al.* [LHCb Collaboration], JHEP **1504** (2015) 024.

EPJ Web of Conferences **129**, 00022 (2016) DOI: 10.1051/epjconf/201612900022
QCD@Work 2016

$X(3872)$ and its charmonium content in Heavy Quark limit

Elif Cincioglu[1], *Juan* Nieves[2], and *Altug* Ozpineci[3,⋆]

[1] *Physics Engineering, Ankara University, Ankara, Turkey*
[2] *IFIC, Centro Mixto Universidad de Valencia-CSIC, Valencia, Spain*
[3] *Physics Department, Middle East Technical University, Ankara, Turkey*

Abstract. $X(3872)$ still presents many puzzles more than a decade after its discovery. Some of its properties, like the isospin violating decays, can easily be accommodated in a molecular model, whereas its other properties, such as radiative decays can be more naturally explained in the quarkonium picture. The best of these schemes can be combined in a picture of $X(3872)$ where it is dominantly a molecular state with some charmonium components. In this work, we present a model based on heavy quark symmetry which describes $X(3872)$ as a superposition of molecular and charmonium components.

1 Introduction

The hadronic spectroscopy has witnessed an exciting period in the recent years. Some of this excitement is due to the observation of mesons that can not be easily explained using the conventional quark-anti-quark picture. One of these mesons is the $X(3872)$ meson which was first observed more than a decade ago by the Belle collaboration [1]. Its existence was later on confirmed by other experiments [2–5]. Its quantum numbers have been confirmed to be $J^{PC} = 1^{++}$ only recently by the LHCb experiment [5]. To explain its properties, various models have been proposed, some of the most common ones are the charm-anti-charm, molecular and tetraquark pictures. In this work, we present a consistent framework based on heavy quark symmetry which tries to combine the pure charmonium and the molecular pictures. Details of the model and its analysis can be found in [6].

In a constituent quark model description of $X(3872)$ it can be identified by the $2P$ state. Various predictions for the mass of the $2P$ state range from 3947.4 MeV [7, 8], to 3906 MeV [9]. All of the predictions on the mass of the $2P$ state are at least 30 MeV above the observed mass of $X(3872)$, which is just below the $D^0 D^{*0}$ threshold. This "coincidence" can easily be explained using a molecular picture for the $X(3872)$. In the molecular picture, the mass of the $X(3872)$ is $m_X = m_{D^0} + m_{D^{*0}} - B$, where $B > 0$ is the binding energy, i.e. in the molecular picture $X(3872)$ is a weakly bound $D\bar{D}^*$ state.

Another puzzling observation about $X(3872)$ is its decays into final states containing a J/ψ and two or three pions. These decays proceed through intermediate ρ and ω mesons. Taking into account the phase space difference, the ratio of the decay amplitudes are obtained to be [10]

$$\left|\frac{A(J/\psi\rho)}{A(J/\psi\omega)}\right| = 0.27 \pm 0.02 \ . \tag{1}$$

⋆e-mail: ozpineci@metu.edu.tr

Considering that $J/\psi\rho$ is an $I = 1$ and $J/\psi\omega$ is an $I = 0$ final state, this ratio points out to a large isospin violation. Such a large isospin violation is hard to reconcile with a pure charmonium state where $X(3872)$ would have $I = 0$ and the observed violation should arise during the decay. In contrast, in the molecular picture, due to the mass difference between the neutral and charged D mesons, $X(3872)$ is a mixture of $I = 0$ and $I = 1$ states. In such a picture, the mixing angle between the $I = 0$ and $I = 1$ components have been estimated to be almost maximal $\theta_{mixing} \simeq 40°$ [11, 12]. Hence the observed isospin violation can easily arise due to this mixing, and not due to the processes that lead to the decay.

When radiative decays of $X(3872)$ are considered, quark-anti-quark picture of $X(3872)$ is favored. $X(3872)$ is observed to decay into $\psi(2S)$ and $\psi(1S)$ by the emission of one photon. The ratio of the branching ratios is measured to be [13, 14]

$$R_{\psi\gamma} = \frac{Br(X \to \psi(2S)\gamma)}{Br(X \to J/\psi\gamma)} = 2.46 \pm 0.64 \pm 0.29 \ . \tag{2}$$

This ratio can be accommodated into the molecular picture if one assumes that $DD^*\psi(2S)$ coupling is about twice as large as DD^*J/ψ coupling. In the charm-anti-charm picture, $R_{\psi\gamma}$ can be explained more naturally since the $X \to \psi(2S)\gamma$ transition is nothing but the $\Delta l = 1$ transition.

Finally, it is observed that in pp collisions, the production rate of $X(3872)$ is $1/20$ times the production rate of $\psi(2S)$ [15]. Such a large rate can not be naturally explained in the pure molecular picture. This problem can be avoided if one assumes that $X(3872)$ is dominantly a molecular state with about 5% charmonium admixture.

In the next section, a leading order effective Lagrangian consistent with the heavy quark symmetry will be presented where $X(3872)$ arises as a mixture of a molecule and charmonium. Section III will be devoted to the $J^{PC} = 1^{++}$ and $J^{PC} = 2^{++}$ spectrum of the model followed by our conclusions.

2 Leading order effective Lagrangian

When the heavy quark mass is taken to be infinite, $m_Q \to \infty$, a new symmetry, the so called heavy quark symmetry, arises due to the decoupling of the spin of the heavy quark from the dynamics. In this limit, D and D^* mesons can be grouped into a doublet field as:

$$\begin{aligned} H_a^{(Q)} &= \frac{1+\not{v}}{2}\left(P_{a\mu}^{*(Q)}\gamma^\mu - P_a^{(Q)}\gamma_5\right), \qquad v\cdot P_a^{*(Q)} = 0, \\ H^{(\bar{Q})a} &= \left(P_\mu^{*(\bar{Q})a}\gamma^\mu - P^{(\bar{Q})a}\gamma_5\right)\frac{1-\not{v}}{2}, \qquad v\cdot P^{*(\bar{Q})a} = 0. \end{aligned} \tag{3}$$

where $v = \lim_{m_Q\to\infty}\frac{p}{m_Q}$ is the heavy quark velocity, a is the isospin index, $P^{*(Q)}$ and $P^{(Q)}$ fields annihilate vector and pseudoscalar mesons containing a heavy quark, and similarly for the fields $P^{*(\bar{Q})}$ and $P^{(\bar{Q})}$. Once the HQSS multiplets are identified, the leading order (LO) heavy quark spin symmetric interactions between the D and D^* mesons can be written as

$$\begin{aligned} \mathcal{L}_{4H} = & \, C_A \operatorname{Tr}\left[\bar{H}^{(Q)a}H_a^{(Q)}\gamma_\mu\right]\operatorname{Tr}\left[H^{(\bar{Q})a}\bar{H}_a^{(\bar{Q})}\gamma^\mu\right] \\ & + C_A^\tau \operatorname{Tr}\left[\bar{H}^{(Q)a}\vec{\tau}^{b}{}_{a}H_b^{(Q)}\gamma_\mu\right]\operatorname{Tr}\left[H^{(\bar{Q})c}\vec{\tau}^{d}{}_{c}\bar{H}_d^{(\bar{Q})}\gamma^\mu\right] \\ & + C_B \operatorname{Tr}\left[\bar{H}^{(Q)a}H_a^{(Q)}\gamma_\mu\gamma_5\right]\operatorname{Tr}\left[H^{(\bar{Q})a}\bar{H}_a^{(\bar{Q})}\gamma^\mu\gamma_5\right] \\ & + C_B^\tau \operatorname{Tr}\left[\bar{H}^{(Q)a}\vec{\tau}^{b}{}_{a}H_b^{(Q)}\gamma_\mu\gamma_5\right]\operatorname{Tr}\left[H^{(\bar{Q})c}\vec{\tau}^{d}{}_{c}\bar{H}_d^{(\bar{Q})}\gamma^\mu\gamma_5\right] \end{aligned} \tag{4}$$

This model predicts that $X(3872)$ should have heavy quark spin symmetric partners with quantum numbers $J^{PC} = 0^{++}$, $J^{PC} = 2^{++}$ (denoted by X_2 in this work) and $J^{PC} = 1^{+-}$ whose binding energy should be similar to the binding energy of $X(3872)$ [16].

To include the charmonium states into this Lagrangian, their multiplet is required. The P-wave heavy quarkonia can be written as a multiplet by means of the four-vector field [17] ($\epsilon_{0123} = +1$):

$$J^{\mu} = \frac{1+\not{v}}{2}\left(\chi_2^{\mu\alpha}\gamma_\alpha + \frac{i}{\sqrt{2}}\epsilon^{\mu\alpha\beta\gamma}\chi_{1\gamma}v_\alpha\gamma_\beta + \frac{1}{\sqrt{3}}\chi_0(\gamma^\mu - v^\mu) + h^\mu\gamma_5\right)\frac{1-\not{v}}{2} \tag{5}$$

with $J_\mu v^\mu = 0$. Here, $\chi_2^{\mu\alpha}, \chi_1^\mu, \chi_0$ and h^μ fields annihilate $\chi_{QJ}(nP)$ and $h_Q(nP)$ quarkonium states, with $J^{PC} = 0^{++}, 1^{++}, 2^{++}$ and 1^{+-}, respectively.

Once the multiplet is identified, there is a single heavy quark spin symmetric interaction term between the quarkonium field J_μ and the $H^{(Q)}$, $H^{(\bar{Q})}$ fields [18],

$$\mathcal{L}_{HHQ\bar{Q}} = \frac{d}{2}\,\mathrm{Tr}[H^{a(\bar{Q})}\bar{J}_\mu H_a^{(Q)}\gamma^\mu] + \frac{d}{2}\,\mathrm{Tr}[\bar{H}^{a(Q)}J_\mu \bar{H}_a^{(\bar{Q})}\gamma^\mu] \tag{6}$$

Note that in the constituent quark model, there is an infinite tower of P-wave quarkonia, all of which can couple to D and D^* mesons. Hence strictly speaking, for each radial excitation, there is such an interaction term in the Lagrangian. But assuming that the constant d has the same order of magnitude for all of them, the P-wave charmonia that will mix dominantly will be the $n = 2$ charmonia due to the closeness of its mass to the threshold.

Once the model is fixed, the spectrum can be identified using the poles of the T matrix in each one of the J^{PC} sectors. The coupling of the states to various channels can be extracted from the residues of the T matrix elements near the threshold through

$$T_{ij} \overset{E\to E_R}{\longrightarrow} \frac{g_i g_j}{E - E_R} \tag{7}$$

where $E_R = M_R - i\frac{\Gamma_R}{2}$ is the position of the pole, M_R the mass of the resonance and Γ_R its width.

Identifying the nature of a resonance from the T matrix elements has its complications. We will define the weights of the molecular component ($\tilde{X}$) and charmonium component ($\tilde{Y}$) through

$$\tilde{X} = -\frac{\Sigma'_{c\bar{c}}(E_R)}{1 - \Sigma'_{c\bar{c}}(E_R)}, \qquad \tilde{Z} = \frac{1}{1 - \Sigma'_{c\bar{c}}(E_R)} \tag{8}$$

where $\Sigma'_{c\bar{c}}(E_R)$ is the derivative with respect to the energy of the self energy of the bare charmonium state due to its coupling to D mesons. Strictly speaking, these parameters correspond to probabilities for bound states, whereas for resonances, their interpretation is more problematic. Nevertheless, even for resonances, they can be used as an indication of the weight of various components in the identified states (for details see [6]).

3 1^{++} and 2^{++} spectrum

The model described in the previous section has nine unknown parameters: four $C_X^{(\tau)}$ ($X = A$ or $X = B$) parameters describing the contact interactions between the D mesons, d parameter describing the coupling of the charmonium to a pair of D mesons, and four masses of the charmonium states. Ideally, these parameters should be fitted to experimental data. Due to the lack of sufficient data to fix all the unknown parameters several simplifications will be made: isospin breaking mass differences will be ignored, and we will concentrate only on the $J^{PC} = 1^{++}$ and $J^{PC} = 2^{++}$ sectors. After these simplifications, the only necessary contact interaction term is $C_0 = C_A + C_B$, reducing the total number of unknowns to four: C_0, d, and the masses of the charmonia in the relevant J^{PC} sectors. Furthermore, the observed mass of $X(3872)$ will be used as an experimental input, further reducing the number of unknowns by one.

In the $J^{PC} = 1^{++}$ sector, for the bare mass of the charmonium, the mass predicted by [8] is used. In the $J^{PC} = 2^{++}$ sector, the χ_{c2} meson is observed to have a mass $m_{\chi_{c2}} = 3906$ MeV. In this sector, this will be used as an experimental input to identify the bare mass of the 2^{++} charmonium. This leaves a single free parameter in both of the J^{PC} sectors. In this work, this parameter is chosen to be the d parameter, and the behavior of the spectrum is analyzed as this parameter is changed.

In Table 1, the spectrum of the $J^{PC} = 1^{++}$ sector is presented. Note that at the mass of the $X(3872)$, the coupling to the charmonium creates an effective attraction proportional to d^2. Hence, in order the keep the mass of the $X(3872)$ fixed, for sufficiently large values of d, the parameter C_0 should increase to positive values, resulting even in a repulsive four particle contact interaction. As expected, as d increases, the molecular component of the $X(3872)$ meson gets smaller and smaller, and for very large values of d, $X(3872)$ is dominantly a charmonium state.

The other pole, which in the $d = 0$ limit is the pure charmonium state, develops a width as the parameter d starts to get nonzero values. As d increases, this pole moves deep into the complex plane, and then goes below threshold. A further increase in the value of d causes this pole to approach the real axis on the second Riemann sheet (SRS). Once this pole reaches the real axis, it collides with its conjugate pole and two poles start to move on the real axis, one moving towards and the other one moving away from the threshold. Since these poles are on the SRS, they will not have any observable consequences if they are deep in the complex plane, or on the real axis, but much below threshold. In Table 1, we only show the pole that might have an observable effect on line shapes, i.e. the pole that moves towards the threshold.

In Table 2, we present the results we obtain in the $J^{PC} = 2^{++}$ sector. In this sector, the mass of the χ_{c2} meson has been fixed on the observed value $m_{\chi_{c2}} = 3927.2 \pm 2.6$ MeV [19]. Note that the closest prediction from the constituent quark model calculations for the mass of the χ_{c2} state is $m_{\chi_{c2}} = 3949$ MeV [9]. As can be seen from Table 2, as d increases, the bare $\mathring{m}_{\chi_{c2}}$ needs to get larger to agree with the experiment. If this bare mass is taken to be the prediction of the constituent quark model, it is seen that the discrepancy between constituent quark model predictions, and the bare mass gets smaller. This implies that the origin of the discrepancy between the constituent quark model predictions and the experimental observation for the χ_{c2} mass is due to the threshold effects.

The other pole, which corresponds to X_2, starts as a stable (dominantly) molecular state for small values of the parameter d. As d increases, it reaches threshold and crosses to the SRS, and for even larger values of d, it moves quickly deep into the complex plane in the SRS, below threshold. Since it is below threshold in the SRS, and deep in the complex plane, it will not have any observable effects, a possible reason for the non observation of this $J^{PC} = 2^{++}$ partner of $X(3872)$.

4 Conclusions

In this talk, the framework, developed in [6], based on heavy quark spin symmetry is presented where both the molecular and the charmonium components of $X(3872)$ can be studied. In the $J^{PC} = 1^{++}$ and $J^{PC} = 2^{++}$ sectors, this model has few parameters. The spectrum and properties of the spectrum of this model is studied. It is shown that a large charmonium component can easily be accommodated within this model depending on the parameter d. Furthermore, in the $J^{PC} = 2^{++}$ sector, the origin of the discrepancy between the constituent quark model predictions and the observation of χ_{c2} can be identified as the coupling of the charmonium to D mesons, an effect that is ignored in constituent quark models (see [20] for a recent analysis on the effects of meson loops on quark models).

Table 1. Properties of the 1^{++} hidden charm poles as a function of d. The position of the $X(3872)$ is fixed at $M_X = 3871.69$ MeV in the FRS. The $\chi_{c1}(2P)$ pole is located in the SRS. Finally, $d^{\rm crit}(\Lambda = 1\,{\rm GeV}) = 0.370$ fm$^{1/2}$.

d [fm$^{1/2}$]	C_{0X} [fm^2]	$g_{D\bar{D}^*}^{X(3872)}$ [GeV$^{-1/2}$]	$\tilde{X}_{X(3872)}$	$(m_{\chi_{c1}}, \Gamma_{\chi_{c1}})$ [MeV]	$g_{D\bar{D}^*}^{\chi_{c1}}$ [GeV$^{-1/2}$]	$\lvert\tilde{X}_{\chi_{c1}}\rvert$	$\tilde{Z}_{\chi_{c1}}$
0.	−0.789	0.90	1	(3906.0,0)	0.	0.	1.
0.05	−0.774	0.89	0.98	(3906.6, 1.9)	$0.01 + 0.16\,i$	0.02	$0.99 + 0.01\,i$
0.1	−0.731	0.87	0.92	(3908.2, 7.9)	$0.03 + 0.31\,i$	0.06	$0.96 + 0.05\,i$
0.15	−0.659	0.83	0.84	(3910.5, 19.2)	$0.07 + 0.44\,i$	0.14	$0.92 + 0.11\,i$
0.20	−0.559	0.78	0.75	(3912.4, 37.8)	$0.14 + 0.56\,i$	0.23	$0.87 + 0.19\,i$
0.25	−0.429	0.73	0.66	(3912.0, 67.0)	$0.24 + 0.65\,i$	0.36	$0.82 + 0.31\,i$
0.30	−0.271	0.68	0.57	(3903.9, 112.8)	$0.38 + 0.73\,i$	0.55	$0.77 + 0.50\,i$
0.35	−0.084	0.63	0.49	(3864.5, 185.2)	$0.63 + 0.85\,i$	> 1	$0.70 + 1.01\,i$
$d^{\rm crit}$	0.000	0.61	0.47	(3798.3, 209.4)	$0.93 + 1.09\,i$	> 1	$0.53 + 2.12\,i$
0.375	0.020	0.61	0.46	(3754.4, 186.4)	$1.21 + 1.37\,i$	> 1	$0.29 + 3.66\,i$
0.3775	0.031	0.61	0.46	(3701.6, 93.5)	$2.19 + 2.39\,i$	> 1	$-0.44 + 12.27\,i$
0.40	0.132	0.59	0.43	(3827.1, 0) at SRS	0.96	$\tilde{X}_{\chi_{c1}} < 0$	2.07
0.45	0.376	0.55	0.37	(3850.9,0) at SRS	0.63	$\tilde{X}_{\chi_{c1}} < 0$	1.52
0.5	0.649	0.51	0.32	(3858.4,0) at SRS	0.51	$\tilde{X}_{\chi_{c1}} < 0$	1.36
1.0	4.963	0.29	0.11	(3869.7, 0) at SRS	0.21	$\tilde{X}_{\chi_{c1}} < 0$	1.08
2.0	22.217	0.15	0.03	(3871.3, 0) at SRS	0.10	$\tilde{X}_{\chi_{c1}} < 0$	1.02
$d \gg d^{\rm crit}$	$\sim \frac{d^2}{\mathring{m}_{\chi_{c1}} - M_X}$	$O(1/d)$	$O(1/d^2)$	$(M_X - O(\frac{1}{d^2}), 0)$ at SRS	$O(1/d)$	$\tilde{X}_{\chi_{c1}} = -O(\frac{1}{d^2})$	$1 + O(\frac{1}{d^2})$

Table 2. Properties of the 2^{++} hidden charm poles as a function of d. The position of the dressed $\chi_{c2}(2P)$ is fixed at $m_{\chi_{c2}}^{\rm exp} = 3927.2$ MeV in the FRS, and we also give the $X(3872)$ meson-molecular probabilities ($\tilde{X}_{X(3872)}$) for each value of d.

d [fm$^{1/2}$]	$\tilde{X}_{X(3872)}$	$g_{D^*\bar{D}^*}^{\chi_{c2}}$ [GeV$^{-1/2}$]	$\tilde{X}_{\chi_{c2}}$	$\mathring{m}_{\chi_{c2}}$ [MeV]	$M_{X_2} - 2M_{D^*} - i\frac{\Gamma_{X_2}}{2}$ [MeV]	$g_{D^*\bar{D}^*}^{X_2}$ [GeV$^{-1/2}$]	$\tilde{X}_{X_2}$
0.	1	0.0	0.0	3927.2	−5.6	0.97	1.
0.05	0.98	0.27	0.01	3927.8	−4.5	0.90	0.996
0.10	0.92	0.51	0.02	3929.6	−1.8	0.67	0.991
0.15	0.84	0.69	0.04	3932.2	−0.0 at SRS	$-0.12\,i$	> 1
0.20	0.75	0.82	0.05	3935.2	−6.4 at SRS	$-0.76\,i$	> 1
0.22	0.71	0.86	0.06	3936.4	−21.2 at SRS	$-1.24\,i$	> 1
0.25	0.66	0.90	0.06	3938.3	$-28.3 - \frac{72.9}{2}\,i$	$0.23 - 0.65\,i$	$0.47 + 0.32\,i$
0.30	0.57	0.95	0.07	3941.2	$-31.2 - \frac{162.8}{2}\,i$	$0.03 + 0.67\,i$	$0.48 - 0.04\,i$
0.35	0.49	0.96	0.07	3943.8	$-59.5 - \frac{312.6}{2}\,i$	$0.30 + 0.71\,i$	$0.52 - 0.39\,i$

Acknowledgments

This research has been supported by the Spanish Ministerio de Economía y Competitividad and European FEDER funds under the contracts FIS2014-51948-C2-1-P, FIS2014-57026-REDT and SEV-2014-0398, by Generalitat Valenciana under contract PROMETEOII/2014/0068 and by TUBITAK under contract 114F234.

References

[1] S. K. Choi et al. (Belle), Phys. Rev. Lett. **91**, 262001 (2003).
[2] D. Acosta et al. (CDF Collaboration), Phys. Rev. Lett. **93**, 072001 (2004).
[3] V. M. Abazov et al. (D0 Collaboration), Phys. Rev. Lett. **93**, 162002 (2004).
[4] B. Aubert et al. (BaBar Collaboration), Phys. Rev. D **71**, 071103 (2005).
[5] R. Aaij et al. (LHCb Collaboration), Eur. Phys. J. C **72**, 1972 (2012).
[6] E. Cincioglu, J. Nieves, A. Ozpineci and A. U. Yilmazer, arXiv:1606.03239 (2016).
[7] P. G. Ortega, J. Segovia, D. R. Entem, and F. Fernandez, Phys. Rev. D **81**, 054023 (2010).
[8] J. Segovia, D. R. Entem, F. Fernandez, and E. Hernandez, Int. J. Mod. Phys. E **22**, 1330026 (2013).
[9] D. Ebert, R. N. Faustov, and V. O. Galkin, Eur. Phys. J. C **71**, 1825 (2011).
[10] C. Hanhart, Yu. S. Kalashnikova, A. E. Kudryavtsev, and A. V. Nefediev, Phys. Rev. D **85**, 011501 (2012).
[11] D. Gamermann, J. Nieves, E. Oset and E. Ruiz Arriola, Phys. Rev. D **81**, 014029 (2010).
[12] C. Hidalgo, J. Nieves, A. Ozpineci and V. S. Zamiralov, Int. J. Mod. Phys. Conf. Ser. **26**, 1460110 (2014).
[13] BaBar collaboration, B. Aubert et al., Phys. Rev. Lett. **102**, 132001 (2009).
[14] R. Aaij et al., LHCb Collaboration, Nucl.Phys. B**886**, 665 (2014).
[15] M. Takizawa and S. Takeuchi, PTEP **2013**, 0903D01 (2013).
[16] J. Nieves and M. P. Valderrama, Phys. Rev. D **86**, 056004 (2012).
[17] R. Casalbuoni, A. Deandrea, N. Di Bartolomeo, R. Gatto, F. Feruglio and G. Nardulli, Phys. Lett. B **302**, 95 (1993).
[18] P. Colangelo, F. De Fazio and T. N. Pham, Phys. Rev. D **69**, 054023 (2004).
[19] K. A. Olive et al. (Particle Data Group), Chin. Phys. C**38**, 090001 (2014).
[20] I. K. Hammer, C. Hanhart and A. V. Nefediev, arXiv:1607.06971 (2016).

EPJ Web of Conferences **129**, 00023 (2016) DOI: 10.1051/epjconf/201612900023
QCD@Work 2016

On properties of the exotic hadrons from QCD sum rules

Wolfgang Lucha[1,*] and Dmitri Melikhov[1,2,3,**]

[1] *Institute for High Energy Physics, Austrian Academy of Sciences, Nikolsdorfergasse 18, A-1050 Vienna, Austria*
[2] *D. V. Skobeltsyn Institute of Nuclear Physics, M. V. Lomonosov Moscow State University, 119991, Moscow, Russia*
[3] *Faculty of Physics, University of Vienna, Boltzmanngasse 5, A-1090 Vienna, Austria*

Abstract. We discuss the specific features of extracting properties of the exotic polyquark hadrons (tetraquarks, pentaquarks) compared to the usual hadrons by the QCD sum-rule approach. In the case of the ordinary hadrons, already the one-loop leading-order $O(\alpha_s^0)$ correlation functions provide the bulk of the hadron observables, e.g., of the form factor; inclusion of radiative corrections $O(\alpha_s)$ modifies already nonzero one-loop contributions. In the case of an exotic hadron, the situation is qualitatively different: discussing strong decays of an exotic tetraquark meson, which provide the main contribution to its width, we show that the *disconnected* leading-order diagrams are not related to the tetraquark properties. For a proper description of the tetraquark decay width, it is mandatory to calculate specific radiative corrections which generate the *connected* diagrams.

1 Properties of individual resonances from QCD correlation functions

One of the most famous applications of the method of QCD sum rules (see [1–3]) is the calculation of the hadronic ground-state properties from the QCD correlation functions involving power corrections.

1.1 Two-point functions

To recall the basics of the method, we briefly review QCD sum rules arising from two-point functions. For obtaining properties of a hadron M, one considers an interpolating current for this hadron, i.e., a current $j(x)$ which produces this hadron from the vacuum:

$$\langle\Omega|j(0)|M\rangle = f_M \neq 0. \tag{1}$$

(For instance, $j(x) = \bar{q}_1(x)Oq_2(x)$ for "normal" mesons, four-quark currents for exotic mesons). One then considers the correlation functions, i.e., the vacuum expectation values of the T-products of the interpolating currents. The simplest object is the two-point function

$$\Pi(p^2) = i\int d^4x\, e^{ipx}\left\langle\Omega\left|T\left(j(x)j^\dagger(0)\right)\right|\Omega\right\rangle. \tag{2}$$

*e-mail: Wolfgang.Lucha@oeaw.ac.at
**e-mail: dmitri_melikhov@gmx.de

Wilson's operator product expansion (OPE) provides the following expansions, for the T-product [4],

$$T\left(j(x)j^\dagger(0)\right) = C_0(x^2,\mu)\hat{1} + \sum_n C_n(x^2,\mu) : \hat{O}_n(x=0,\mu) :, \tag{3}$$

and, for the two-point function,

$$\Pi(p^2) = \Pi_{\rm pert}(p^2,\mu) + \sum_n \frac{C_n}{(p^2)^n}\langle\Omega| : \hat{O}_n(x=0,\mu) : |\Omega\rangle. \tag{4}$$

The basic concept of the method is the difference between the physical QCD vacuum, $|\Omega\rangle$, which has a complicated structure, and the perturbative QCD vacuum, $|0\rangle$: properties of the physical vacuum $|\Omega\rangle$ are characterized by the condensates — nonzero expectation values of gauge-invariant operators over the physical vacuum: $\langle\Omega| : \hat{O}(0,\mu) : |\Omega\rangle \neq 0$ (see [5, 6] for the recent determinations). Hereafter, for notational simplicity, we write $\langle\Omega|\cdots|\Omega\rangle \equiv \langle\cdots\rangle$. The two-point function is an analytic function of p^2:

$$\Pi(p^2) = \int \frac{ds}{s-p^2}\rho(s). \tag{5}$$

One calculates the spectral densities $\rho(s)$ using both OPE and the language of confined hadron states:

$$\rho_{\rm theor}(s) = \left[\rho_{\rm pert}(s,\mu) + \sum_n C_n\delta^{(n)}(s)\langle\Omega|\hat{O}_n(\mu)|\Omega\rangle\right], \qquad \rho_{\rm hadr}(s) = f^2\delta(s-M^2) + \rho_{\rm cont}(s). \tag{6}$$

In order to relate to each other the truncated $\Pi_{\rm OPE}(p^2)$ and the hadronic representation $\Pi_{\rm hadron}(p^2)$, one has to perform a smearing, e.g., by application of a Borel transformation $p^2 \to \tau\left[\frac{1}{s-p^2} \to \exp(-s\tau)\right]$:

$$\Pi(\tau) = \int ds\, e^{-s\tau}\rho(s) = f^2 e^{-M^2\tau} + \int_{s_{\rm phys}}^{\infty} ds\, e^{-s\tau}\rho_{\rm hadr}(s)$$
$$= \int_{(m_1+m_2)^2}^{\infty} ds\, e^{-s\tau}\rho_{\rm pert}(s,\mu) + \Pi_{\rm power}(\tau,\mu). \tag{7}$$

Here, $s_{\rm phys}$ is the physical threshold, and f is the decay constant defined by (1). In order to get rid of the excited-state contributions, one adopts the *duality ansatz* [7–9]: all contributions of excited states are counterbalanced by the perturbative contribution above some *effective continuum threshold* $s_{\rm eff}(\tau)$, which differs from the physical continuum threshold. Applying the duality assumption yields

$$f^2 e^{-M^2\tau} = \int_{(m_1+m_2)^2}^{s_{\rm eff}(\tau)} ds\, e^{-s\tau}\rho_{\rm pert}(s,\mu) + \Pi_{\rm power}(\tau,\mu). \tag{8}$$

As soon as we have an algorithm [10, 11] of fixing the effective threshold $s_{\rm eff}(\tau)$ at our disposal, Eq. (8) provides the decay constant f. To this end, we need either to know the hadron mass M or to set $\tau = 0$.

2 Strong decays from three-point functions in QCD

Having recalled these basic ideas, let us now consider three-point functions, which are the appropriate QCD quantities for various hadron form factors [12, 13]:

$$\Gamma(p,p',q) = \int dx\,dx'\,\langle T(J(x)j(0)j'(x'))\rangle \exp(ipx - ip'x'). \tag{9}$$

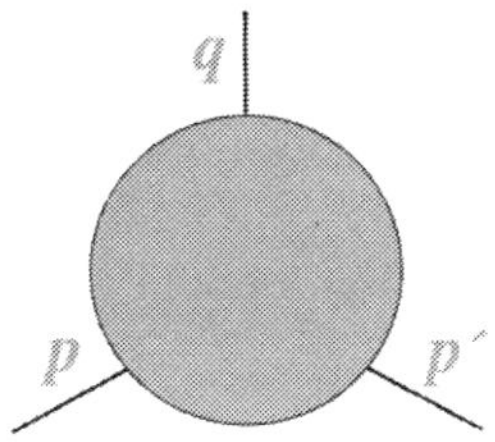

Figure 1. Three-point vertex function (9).

This correlator contains the triple-pole in the Minkowski region, namely,

$$\Gamma(p,p',q) = \frac{ff'}{(p^2-M^2)(p'^2-M'^2)}F(q^2)+\cdots, \tag{10}$$

where the form factor $F(q^2)$ contains a pole at $q^2 = M_q^2$:

$$F(q^2) = \frac{f_q g_{MM'M_q}}{(q^2-M_q^2)}+\cdots. \tag{11}$$

Here, $g_{MM'M_q}$ describes the $M \to M'M_q$ strong transition; f, f', and f_{M_q} are the decay constants of the mesons: $\langle 0|J(0)|M\rangle = f$, $\langle 0|j'(0)|M'\rangle = f'$, and $\langle 0|j(0)|M_q\rangle = f_q$.

The three-point function satisfies the double spectral representation

$$\Gamma(p,p',q) = \int \frac{ds}{s-p^2}\frac{ds'}{s'-p'^2}\Delta(s,s',q^2) \tag{12}$$

A double Borel transformation $p^2 \to \tau$, $p'^2 \to \tau'$ and duality cuts in the p^2 and p'^2 channels [12–14] lead to

$$\exp(-M^2\tau)\exp(-M'^2\tau')F(q^2) = \int^{s_{\rm eff}} ds\exp(-s\tau)\int^{s'_{\rm eff}} ds'\exp(-s'\tau')\Delta_{\rm OPE}(s,s',q^2) \tag{13}$$

Let us discuss this relation for two cases: for normal bilinear and for exotic four-quark currents.

2.1 Normal hadrons

An elastic or a transition form factor of a normal meson corresponds to the choice of the interpolating currents in Eq. (13) in the bilinear form, that is, $J, j, j' = \bar{q}_i\hat{O}q_j$, with appropriate quark fields $q_{i,j}$ and a generic Dirac matrix $\hat{O}$. The perturbative part of the correlation function has the following structure:

$$\Gamma_{\rm OPE}(p^2,p'^2,q^2) = \Gamma_0(p^2,p'^2,q^2)+\alpha_s\Gamma_1(p^2,p'^2,q^2)+\cdots. \tag{14}$$

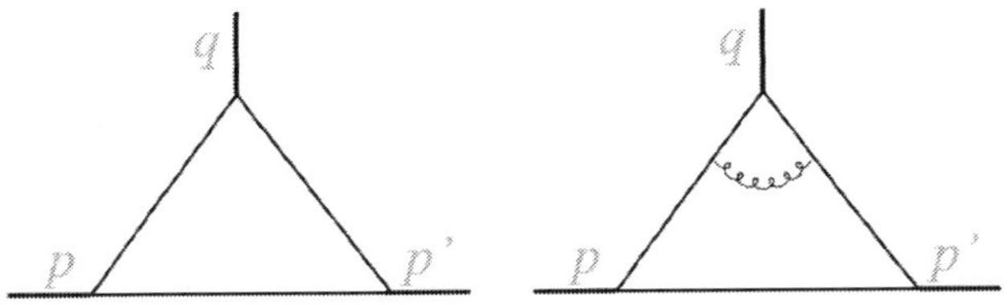

Figure 2. The OPE for the three-point function of the bilinear $\bar{q}q$ interpolating currents for normal mesons. The leading-order diagram and one of the radiative corrections are shown.

Already the one-loop leading-order diagram has a nonzero double-spectral density $\Delta_0(s,s',q^2)$ and therefore provides a nonzero contribution to the form factor (and, respectively, to $g_{MM'M_q}$). Radiative corrections [15–17] give essential contributions, increasingly important at larger q^2 [18, 19]. However, a reasonable estimate may be obtained already from the leading-order three-point function [20–23].

2.2 Three-point function containing one exotic current

The interpolating current for the exotic state (for the sake of clarity, we focus to the case of tetraquarks, i.e., exotic states with four valence quarks) may be chosen in many different ways (see the discussion in the next section). Let us consider an interpolating current $\theta(x)$ of the following rather general form:

$$\theta(x) = \bar{q}_1(x)\hat{O}q_2(x)\bar{q}_3(x)\hat{O}q_4(x), \tag{15}$$

where $\hat{O}$ is an appropriate combination of Dirac and colour matrices and possibly also of (covariant) derivatives. The OPE for the correlation function $\Gamma^{\theta jj} = \langle T(\theta(x)j(y)j(z)) \rangle$ containing one exotic current and two normal currents has the diagrammatic structure shown in Fig. 3 and reads

$$\Gamma^{\theta jj}_{\rm OPE}(p^2, p'^2, q^2) = \Pi(p'^2)\Pi(q^2) + \alpha_s \Gamma_{\rm connected}(p^2, p'^2, q^2). \tag{16}$$

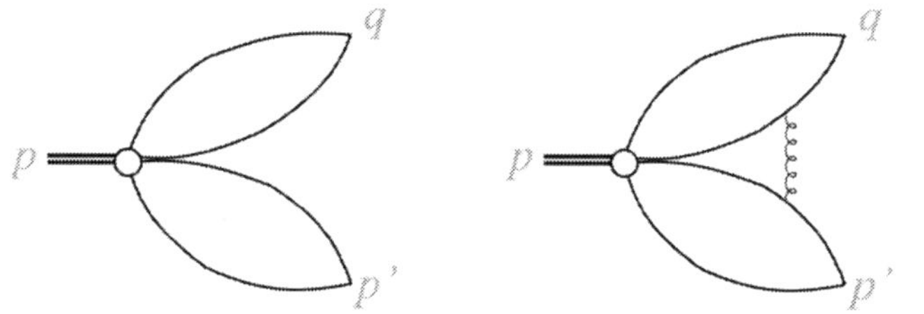

Figure 3. Perturbative expansion of the three-point function containing one exotic tetraquark current and two normal quark–antiquark currents.

As known from the general features of the Bethe-Salpeter equation, the disconnected diagrams are not related to the bound states. This is also clear from the following argument: Performing the Borel transformation $p^2 \to \tau$ (one of the steps of a sum-rule analysis), the Borel image of the disconnected leading-order contribution vanishes. The *connected* diagrams, relevant for the exotic states, emerge at order $O(\alpha_s)$ and higher. Thus, any attempt to extract, e.g., the tetraquark decay amplitude from the leading-order contribution is inconsistent. The common feature of *all* previous investigations of these decays by QCD sum rules (e.g., [24–27]) was the attempt to study tetraquark (and pentaquark) decays relying on the factorizable leading-order contribution, which intrinsically has no relation to tetraquark properties [which fact is evident from both the factorization property $\Gamma(p, p', q) = \Pi(p'^2)\Pi(q^2)$ and the large-N_c behaviour of the QCD diagrams [28–30]]. Therefore, the existing analyses should be strongly revised by calculating and taking into account the nonfactorizable two-loop $O(\alpha_s)$ corrections. In other words, *the "fall-apart" decay of exotic hadrons differs from the decay mechanism of ordinary hadrons and requires an appropriate treatment within QCD sum rules. The leading-order contribution is not related to the exotic-state decay, as is clear from its factorization property $\Gamma(p, p', q) = \Pi(p'^2)\Pi(q^2)$ and from the large-N_c behaviour of the QCD diagrams. The calculation of the radiative corrections is mandatory for a reliable analysis of the properties of the exotic states.*

3 Structure of the exotic states

Obviously, the exotic tetraquark states may have a rather complicated "internal" structure. The most popular scenarios for such a structure are a confined tetraquark state (i.e., a bound state in a confining potential between two colour-triplet diquarks) and a molecular "nuclear-physics-like" bound state in the system of two colourless mesons.

However, an important question about the structure of the exotic state — which to a large extent controls also its production mechanism — is not easy to answer [31]: (i) by a combined colour–spinor Fierz rearrangement of the tetraquark interpolating current $D(x)$ one can express it in either diquark–antidiquark or meson–meson form; (ii) the same quantum numbers of the exotic interpolating current may be obtained by different combinations of its diquark–antidiquark or meson–meson bilinear parts.

3.1 Set of decay constants of exotic states

The simplest characteristic of a meson is its decay constant, i.e., the transition amplitude between the vacuum and the meson, induced by its interpolating current. For a heavy quarkonium state, the decay constant is analogous to its wave function at spatial origin, $\psi(r = 0)$. Whereas an ordinary meson is described by one or few decay constants $\langle 0|\bar{q}_1 O_A q_2|M\rangle = f_A$, exotic states have many decay constants, related to possible different structures of the interpolating four-quark currents; e.g., the interpolating currents may be chosen in "meson–meson" form $\langle 0|M_{12}(x)M_{34}(x)|\theta\rangle = f_{MM}$ with $M_{12} = \bar{q}_1 O q_2$, or in "diquark–diquark" form $\langle 0|\bar{D}^a_{13}(x)D^a_{24}(x)|\theta\rangle = f_{DD}$ with $D^a_{24} = \epsilon^{abd}(q^b_{c4})^T q^d_2$. If, in some limit, a certain class of decay constants vanishes, one is eligible to say that the exotic state is, e.g., a pure molecular or a pure diquark–antidiquark state. In general, the mixing of different "components" seems unavoidable.

An appropriate tool to access the decay constants theoretically is the QCD sum rule for two-point functions, $\Pi^{(\theta\theta)} = \langle\theta(x)\theta(0)\rangle$ [32]. Let us emphasize, however, that *not all contributions to two-point functions of exotic interpolating currents are related to the properties of exotic states. The large-N_c behaviour of the different diagrams should be the guiding principle for the selection of appropriate contributions, similar to selecting the connected contributions to three-point functions.* After selecting the appropriate contributions to the two-point functions $\Pi_{\theta\theta}$, the set of sum rules should be studied and only then the answer about the structure of the observed narrow exotic candidates may be obtained.

3.2 Form factors of the exotic states

The appropriate quantities which describe the structure of bound states — both normal and exotic — are the form factors. For exotic states, these are the form factors related to the connected parts of the three-point functions involving two exotic interpolating currents, $\langle T(\theta\theta j)\rangle$; so far, no OPE calculations for the corresponding quantities are available in the literature.

4 Conclusions and Outlook

- The three-point (vertex) correlation functions for the normal bilinear quark currents show a crucial difference from the three-point correlation functions of exotic polyquark interpolating currents: the latter are described at leading order by disconnected diagrams, not related to strong decays of exotic hadrons. In the exotic-current case, the relevant diagrams start at order $O(\alpha_s)$. This requires taking into account $O(\alpha_s)$ corrections.
- Similar to the case of three-point functions, not all contributions to two-point correlation functions of exotic interpolating currents are related to the properties of the exotic bound state: the N_c scaling may be the guiding principle for selecting the appropriate contributions.
- Our experience in the analysis of the normal hadrons proves that a truncated OPE for the correlation function does not enable one to study at the same time both the *existence* of the isolated ground state and its *properties*. However, if the mass of such a narrow bound state is known, the QCD-sum rule approach allows one to obtain reliable predictions for decay constants [33] and form factors [34]. The observed exotic-state candidates are narrow; therefore, any procedure for the extraction of their parameters from the OPE has the same features and the same challenges as for the normal hadrons.

Acknowledgements

D. M. was supported by the Austrian Science Fund (FWF) under project P29028-N27.

References

[1] M. A. Shifman, A. I. Vainshtein, and V. I. Zakharov, Nucl. Phys. B **147**, 385 (1979).
[2] P. Colangelo and A. Khodjamirian, hep-ph/0010175.
[3] B. L. Ioffe, Prog. Part. Nucl. Phys. **56**, 232 (2006).
[4] V. A. Novikov, M. A. Shifman, A. I. Vainshtein, and V. I. Zakharov, Nucl. Phys. B **249**, 445 (1985).
[5] C. T. H. Davies *et al.*, PoS **ConfinementX** 042 (2012).
[6] C. Dominguez, L. Hernandez, and K. Schilcher, JHEP **1507**, 110 (2015).
[7] B. Blok, M. Shifman, and D.-X. Zhang, Phys. Rev. D **57**, 2691 (1998); **59**, 019901(E) (1999); M. Shifman, hep-ph/0009131.
[8] W. Lucha and D. Melikhov, Phys. Rev. D **73**, 054009 (2006).
[9] W. Lucha, D. Melikhov, and S. Simula, Phys. Lett. B **687**, 48 (2010).
[10] W. Lucha, D. Melikhov, and S. Simula, Phys. Rev. D **76**, 036002 (2007); Phys. Lett. B **657**, 148 (2007).
[11] W. Lucha, D. Melikhov, and S. Simula, Phys. Lett. B **671**, 445 (2009); Phys. Rev. D **79**, 096011 (2009); J. Phys. G **37**, 035003 (2010); D. Melikhov, Phys. Lett. B **671**, 450 (2009).
[12] B. L. Ioffe and A. V. Smilga, Phys. Lett. B **114**, 353 (1982).
[13] V. A. Nesterenko and A. V. Radyushkin, Phys. Lett. B **115**, 410 (1982).
[14] A. P. Bakulev and A. V. Radyushkin, Phys. Lett. B **271**, 223 (1991).
[15] E. Bagan, P. Ball, and P. Godzinsky, Phys. Lett. B **301**, 249 (1993).
[16] P. Colangelo, F. De Fazio, and N. Paver, Phys. Rev. D **58**, 116005 (1998).
[17] V. Braguta and A. Onishchenko, Phys. Lett. B **591**, 255 (2004); **591**, 267 (2004).
[18] V. V. Anisovich, D. Melikhov, and V. Nikonov, Phys. Rev. D **52**, 5295 (1995); **55**, 2918 (1997).
[19] V. Braguta, W. Lucha, and D. Melikhov, Phys. Lett. B **661**, 354 (2008).
[20] W. Lucha and D. Melikhov, Phys. Rev. D **86**, 016001 (2012).
[21] W. Lucha and D. Melikhov, J. Phys. G **39**, 045003 (2012).
[22] I. Balakireva, W. Lucha, and D. Melikhov, J. Phys. G **39**, 055007 (2012); Phys. Rev. D **85**, 036006 (2012).
[23] D. Melikhov and B. Stech, Phys. Rev. D **85**, 051901 (2012); Phys. Lett. B **718**, 488 (2012).
[24] M. Nielsen, F. S. Navarra, and S.-H. Lee, Phys. Rept. **497**, 41 (2010).
[25] F. S. Navarra and M. Nielsen, Mod. Phys. Lett. A **29**, 1430005 (2014).
[26] Z.-G. Wang and T. Huang, Nucl. Phys. A **930**, 63 (2014).
[27] J. M. Dias *et al.*, Phys. Lett. B **758**, 235 (2016).
[28] S. Weinberg, Phys. Rev. Lett. **110**, 261601 (2013).
[29] M. Knecht and S. Peris, Phys. Rev. D **88**, 036016 (2013).
[30] T. D. Cohen and R. F. Lebed, Phys. Rev. D **89**, 054018 (2014); **90**, 016001 (2014).
[31] R. L. Jaffe, Nucl. Phys. A **804**, 25 (2008).
[32] F. Fanomezana, S. Narison, and A. Rabemananjara, Nucl. Part. Phys. Proc. **258–259**, 152 (2015); L. Albuquerque *et al.*, Int. J. Mod. Phys. A **31**, 1650093 (2016).
[33] W. Lucha, D. Melikhov, and S. Simula, J. Phys. G **38**, 105002 (2011); Phys. Lett. B **701**, 82 (2011); Phys. Rev. D **88**, 056011 (2013); Phys. Lett. B **735**, 12 (2014); Phys. Rev. D **91**, 116009 (2015).
[34] W. Lucha, D. Melikhov, H. Sazdjian, and S. Simula, Phys. Rev. D **80**, 114028 (2009).

EPJ Web of Conferences **129**, 00024 (2016) DOI: 10.1051/epjconf/201612900024
QCD@Work 2016

Exotic baryons as a hadronic molecule in the heavy quark region

Yasuhiro Yamaguchi[1,⋆]

[1] *Istituto Nazionale di Fisica Nucleare (INFN), Sezione di Genova, via Dodecaneso 33, 16146 Genova, Italy*

Abstract. We study the hadronic molecules formed by the coupled-channel system of $\bar{D}^{(*)}\Lambda_c$ and $\bar{D}^{(*)}\Sigma_c^{(*)}$, inspired by the two hidden-charm pentaquark states observed by LHCb collaborations in 2015. In these molecules, the coupled channels of $\bar{D}^{(*)}\Sigma_c^{(*)}$ are important because the thresholds of these channels are approximately degenerated by the heavy quark spin symmetry. In addition, we consider the coupling to the $\bar{D}^{(*)}\Lambda_c$ channel whose thresholds are close to the $\bar{D}^{(*)}\Sigma_c^{(*)}$ thresholds, and the coupling to the state with $\ell \neq 0$ mixed by the tensor force. By solving the coupled-channel Schrödinger equations with the one meson exchange potentials, we study the hidden-charm hadronic molecules with $I(J^P) = 1/2(3/2^\pm)$ and $1/2(5/2^\pm)$. We conclude that the J^P assignment of the observed pentaquarks is $3/2^+$ for $P_c^+(4380)$ and $5/2^-$ for $P_c^+(4450)$, which is agreement with the results of the LHCb analysis. In addition, we give predictions for other $J^P = 3/2^\pm$ states.

1 Introduction

In 2015, LHCb collaborations observed the two hidden-charm pentaquarks $P_c^+(4380)$ and $P_c^+(4450)$ in $\Lambda_b^0 \to J/\psi K^- p$ decay [1]. The minimal quark content of them is considered to be $c\bar{c}uud$. The reported masses and widths are $(M, \Gamma) = (4380 \pm 8 \pm 29, 205 \pm 18 \pm 86)$ MeV and $(4449.8 \pm 1.7 \pm 2.5, 39 \pm 5 \pm 19)$ MeV, respectively. In Ref. [1], the J^P of $(P_c^+(4380), P_c^+(4450))$ is $(3/2^-, 5/2^+)$. However, $(3/2^+, 5/2^-)$, and $(5/2^+, 3/2^-)$ are also acceptable. There have been lively discussions about the structure of the pentaquarks, such as the compact pentaquark [2–5] and the hadronic molecule [6–14]. On the other hand, the threshold enhancement by the anomalous triangle singularity is discussed in Refs. [16–18].

Hadronic molecules are expected as the structure of the exotic states near the thresholds. In the strangeness sector, $\Lambda(1405)$ has been discussed as a meson-baryon ($\bar{K}N$ and $\pi\Sigma$) molecule [19–21]. A lot of exotic states have been found near the $D^{(*)}\bar{D}^{(*)}$ thresholds [22], such as $X(3872)$ [23, 24] and the charged charmonium $Z_c(3900)$ [25]. In the bottom quark sector, charged twin resonances Z_b and $Z_b^{(\prime)}$ have been studied as the $B^{(*)}\bar{B}^*$ molecules [26].

The hadronic molecules are dynamically generated by the hadron-hadron interaction. Therefore, the properties of the interaction are important in producing those structures. In the charm quark sector, the SU(4) flavor symmetric interaction has been applied, and has also been used in the hidden-charm molecules [6–8]. However, the large mass of the charm quark would break the SU(4) flavor symmetry.

In the heavy flavor sector, the large mass of the heavy quarks introduces the heavy quark spin symmetry [27–29]. This symmetry appears because the suppression of the spin-dependent interaction between a heavy quark and a light degrees of freedom, a light quark and a gluon. It manifests the mass

⋆e-mail: yasuhiro.yamaguchi@ge.infn.it

Table 1. Various channels of the $\bar{D}^{(*)}\Lambda_c$ and $\bar{D}^{(*)}\Sigma_c^{(*)}$ states for the given total angular momentum and parity J^P. In each channel, the total spin S and orbital angular momentum L are represented as ^{2S+1}L.

J^P	Channels
$3/2^-$	$\bar{D}\Lambda_c(^2D)$, $\bar{D}^*\Lambda_c(^4S,^2D,^4D)$, $\bar{D}\Sigma_c(^2D)$, $\bar{D}\Sigma_c^*(^4S,^4D)$, $\bar{D}^*\Sigma_c(^4S,^2D,^4D)$, $\bar{D}^*\Sigma_c^*(^4S,^2D,^4D,^6D,^6G)$
$3/2^+$	$\bar{D}\Lambda_c(^2P)$, $\bar{D}^*\Lambda_c(^2P,^4P,^4F)$, $\bar{D}\Sigma_c(^2P)$, $\bar{D}\Sigma_c^*(^4P,^4F)$, $\bar{D}^*\Sigma_c(^2P,^4P,^4F)$, $\bar{D}^*\Sigma_c^*(^2P,^4P,^6P,^4F,^6F)$
$5/2^-$	$\bar{D}\Lambda_c(^2D)$, $\bar{D}^*\Lambda_c(^2D,^4D,^4G)$, $\bar{D}\Sigma_c(^2D)$, $\bar{D}\Sigma_c^*(^4D,^4G)$, $\bar{D}^*\Sigma_c(^2D,^4D,^4G)$, $\bar{D}^*\Sigma_c^*(^6S,^2D,^4D,^6D,^4G,^6G)$
$5/2^+$	$\bar{D}\Lambda_c(^2F)$, $\bar{D}^*\Lambda_c(^4P,^2F,^4F)$, $\bar{D}\Sigma_c(^2F)$, $\bar{D}\Sigma_c^*(^4P,^4F)$, $\bar{D}^*\Sigma_c(^4P,^2F,^4F)$, $\bar{D}^*\Sigma_c^*(^4P,^6P,^2F,^4F,^6F,^6H)$

degeneracy of the heavy hadrons with different total spin, such as D and D^* mesons ($\Delta m_{DD^*} \sim 140$ MeV) and Σ_c and Σ_c^* baryons ($\Delta m_{\Sigma_c\Sigma_c^*} \sim 65$ MeV). Therefore, the mass degeneracy induces a coupled-channel system. In the hidden-charm molecules formed by the $\bar{D}^{(*)}$ meson and the $\Sigma_c^{(*)}$ baryons, the $\bar{D}\Sigma_c$, $\bar{D}\Sigma_c^*$, $\bar{D}^*\Sigma_c$ and $\bar{D}^*\Sigma_c^*$ channels will be coupled.

Moreover, we cannot ignore the $\bar{D}^{(*)}\Lambda_c$ channel which was not considered in the early works of the hidden-charm pentaquarks [7, 8, 13, 14]. The $\bar{D}^*\Lambda_c$ threshold is 25 MeV below the $\bar{D}\Sigma_c$ threshold, and therefore, the $\bar{D}^*\Lambda_c$ channel is not irrelevant.

In the $\bar{D}^{(*)}\Lambda_c - \bar{D}^{(*)}\Sigma_c^{(*)}$ system, the tensor force mixing S-wave and D-wave components would be important. In the $\bar{D}N - \bar{D}^*N$ systems, the $\bar{D} - \bar{D}^*$ mixing enhances the effect of the one pion exchange potential (OPEP) having the tensor force producing the strong attraction [30–35]. We expect that this mixing is also helpful to produce the attraction in the hidden-charm molecules.

On the basis of the above discussions, we consider the coupled-channel systems of $\bar{D}^{(*)}\Lambda_c - \bar{D}^{(*)}\Sigma_c^{(*)}$ including states with larger orbital angular momentum, namely D-wave and G-wave for the negative parity state, and F-wave and H-wave for the positive parity state, as summarized in Table 1. This full-channel coupling has never been considered so far. The interaction is obtained by the one meson exchange potential that respects the heavy quark spin symmetry. The bound and resonant states in $I(J^P) = 1/2(3/2^\pm)$ and $1/2(5/2^\pm)$ are studied by solving the coupled-channel Schrödinger equations.

2 Interactions

We employ the meson exchange potential as the interaction between the heavy meson and the heavy baryon. The meson exchange potential is obtained by the Lagrangians satisfying the heavy quark and chiral symmetries. The Lagrangians for a heavy meson and a light meson are given [29, 36–39] as

$$\mathcal{L}_{\pi HH} = g_\pi \mathrm{Tr}\left[H_b \gamma_\mu \gamma_5 A^\mu_{ba} \bar{H}_a\right], \quad \mathcal{L}_{\sigma HH} = g_s \mathrm{Tr}\left[H_a \sigma \bar{H}_a\right], \tag{1}$$

$$\mathcal{L}_{vHH} = -i\beta \mathrm{Tr}\left[H_b v^\mu (\rho_\mu)_{ba} \bar{H}_a\right] + i\lambda \mathrm{Tr}\left[H_b \sigma^{\mu\nu} F_{\mu\nu}(\rho)_{ba} \bar{H}_a\right], \tag{2}$$

for the pion (π), sigma meson (σ), and vector (ρ and ω) meson (v). v^μ is a four-velocity of a heavy quark. The heavy meson field constructed by the pseudoscalar meson P and vector meson P^* are represented [29, 36–39] by

$$H_a = \frac{1+\not{v}}{2}\left[P^*_{a\mu}\gamma^\mu - P_a\gamma_5\right], \quad \bar{H}_a = \gamma_0 H_a^\dagger \gamma_0, \tag{3}$$

Table 2. Masses of the exchanged mesons and coupling constants of the interaction Lagrangians for the heavy mesons and heavy baryons [15, 32, 40].

	m_α [MeV]	Meson	Baryon
π	137.27	$g_\pi = 0.59$	$g_1 = (3/\sqrt{8})g_4 = 1.0$
ρ	769.9	$\beta = 0.9$	$\beta_S = -2\beta_B = 12.0/g_V$
		$\lambda = 0.59\,[\text{GeV}^{-1}]$	$\lambda_S = -2\sqrt{2}\lambda_I = 19.2/g_V\,[\text{GeV}^{-1}]$
ω	781.94	$\beta = 0.9$	$\beta_S = -2\beta_B = 12.0/g_V$
		$\lambda = 0.59\,[\text{GeV}^{-1}]$	$\lambda_S = -2\sqrt{2}\lambda_I = 19.2/g_V\,[\text{GeV}^{-1}]$
σ	550.0	$g_s = -0.76$	$\ell_S = -2\ell_B = 7.30$

where the subscripts a, b are for the light flavor. A_μ and $F_{\mu\nu}(\rho)$ are the axial vector current and the field tensor given by

$$A_\mu = \frac{i}{2}\left(\xi^\dagger\partial_\mu\xi - \xi\partial_\mu\xi^\dagger\right), \quad F_{\mu\nu}(\rho) = \partial_\mu\rho_\nu - \partial_\nu\rho_\mu , \tag{4}$$

where $\xi = \exp(i\hat{\pi}/2f_\pi)$ with the pion decay constant $f_\pi = 92.3$ MeV.

The Lagrangians for a heavy baryon and a light meson [38, 40] are given by

$$\mathcal{L}_{\pi BB} = \frac{3}{2}g_1 i v_\kappa \epsilon^{\mu\nu\lambda\kappa}\text{tr}\left[\bar{S}_\mu A_\nu S_\lambda\right] + g_4\text{tr}\left[\bar{S}^\mu A_\mu B_{\bar{3}}\right] + \text{H.c.}, \quad \mathcal{L}_{\sigma BB} = \ell_B\text{tr}\left[\bar{B}_{\bar{3}}\sigma B_{\bar{3}}\right] + \ell_S\text{tr}\left[\bar{S}_\mu\sigma S^\mu\right], \tag{5}$$

$$\mathcal{L}_{vBB} = -i\beta_B\text{tr}\left[\bar{B}_{\bar{3}}v^\mu\rho_\mu B_{\bar{3}}\right] - i\beta_S\text{tr}\left[\bar{S}_\mu v^\alpha\rho_\alpha S^\mu\right] + \lambda_S\text{tr}\left[\bar{S}_\mu F^{\mu\nu}S_\nu\right] + i\lambda_I\epsilon^{\mu\nu\lambda\kappa}v_\mu\text{tr}\left[\bar{S}_\nu F_{\lambda\kappa}B_{\bar{3}}\right] + \text{H.c.}. \tag{6}$$

The superfield S_μ for $\Sigma_Q^{(*)}$ is represented by

$$S_\mu = B^*_{6\mu} + \frac{\delta}{\sqrt{3}}\left(\gamma_\mu + v_\mu\right)\gamma_5 B_6, \quad \bar{S}_\mu = \gamma_0 S^\dagger_\mu\gamma_0 . \tag{7}$$

The phase factor is chosen by $\delta = -1$ as discussed in Ref. [40]. The coupling constants of the Lagrangians are summarized in Table 2.

From the effective Lagrangians introduced above, we obtain the meson exchange potentials as

$$V_\pi^{ij}(r) = G_\pi^{ij}\left[\vec{O}_1^i\cdot\vec{O}_2^j C(r;m_\pi) + S_{O_1^iO_2^j}(\hat{r})T(r;m_\pi)\right], \quad V_\sigma^{ij}(r) = G_\sigma^{ij}C(r;m_\sigma), \tag{8}$$

$$V_v^{ij}(r) = G_v^{ij}C(r;m_v) + F_v^{ij}\left[-2\vec{O}_1^i\cdot\vec{O}_2^j C(r;m_v) + S_{O_1^iO_2^j}(\hat{r})T(r;m_v)\right]. \tag{9}$$

i and j stand for the indices of the channels. G_α^{ij} ($\alpha = \pi, \rho, \omega, \sigma$) is the constant of the (i, j) component given by the coupling constants of the Lagrangians. O_1^i and O_2^j are the (transition) spin operator of the heavy meson and heavy baryon vertices, respectively [31–33, 40]. $S_{O_1^iO_2^j}(\hat{r})$ is the tensor operator $S_{O_1^iO_2^j}(\hat{r}) = 3\vec{O}_1^i\cdot\hat{r}\vec{O}_2^j\cdot\hat{r} - \vec{O}_1^i\cdot\vec{O}_2^j$. The potential for the isovector mesons, π and ρ, is multiplied by the isospin factor, $\sqrt{6}$ for $\bar{D}^{(*)}\Lambda_c - \bar{D}^{(*)}\Sigma_c^{(*)}$ and -2 for $\bar{D}^{(*)}\Sigma_c^{(*)} - \bar{D}^{(*)}\Sigma_c^{(*)}$ with $I = 1/2$. The functions $C(r;m_\alpha)$ and $T(r;m_\alpha)$ are given by

$$C(r;m_\alpha) = \int\frac{d^3q}{(2\pi)^3}\frac{m_\alpha^2}{\vec{q}^2 + m_\alpha^2}e^{i\vec{q}\cdot\vec{r}}F_\alpha(\Lambda,\vec{q}), \tag{10}$$

$$S_{O_1^iO_2^j}(\hat{r})T(r;m_\alpha) = \int\frac{d^3q}{(2\pi)^3}\frac{-\vec{q}^2}{\vec{q}^2 + m_\alpha^2}S_{O_1^iO_2^j}(\hat{q})e^{i\vec{q}\cdot\vec{r}}F_\alpha(\Lambda,\vec{q}), \tag{11}$$

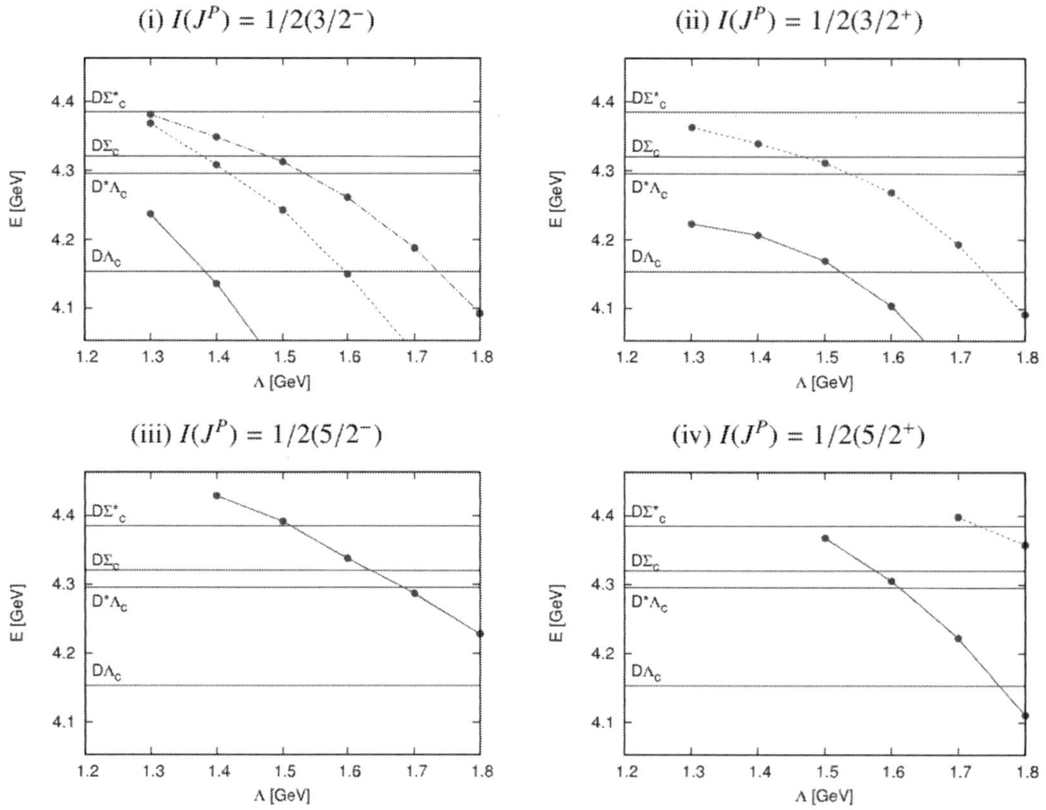

Figure 1. Cutoff Λ dependence of the obtained energies E in $J^P = 3/2^{\pm}$ and $5/2^{\pm}$ and isospin $I = 1/2$ from Ref. [15]. The black horizontal lines show the meson-baryon thresholds ($\bar{D}^{(*)}\Lambda_c$ and $\bar{D}^{(*)}\Sigma_c^{(*)}$).

with the dipole-type form factor $F(\Lambda, \vec{q})$ for spatially extended hadrons [14, 31–33, 40]

$$F_\alpha(\Lambda, \vec{q}) = \left(\frac{\Lambda^2 - m_\alpha^2}{\Lambda^2 + \vec{q}^{\,2}}\right)^2 , \tag{12}$$

with the cutoff parameters Λ.

3 Numerical results

We solve the Schrödinger equations with the meson exchange potential between the heavy meson and the heavy baryon for the coupled-channels. The molecular states of $\bar{D}^{(*)}\Lambda_c^{(*)} - \bar{D}^{(*)}\Sigma^{(*)}$ with $J^P = 3/2^{\pm}, 5/2^{\pm}$ and isospin $I = 1/2$ are studied. The obtained energies with the cutoff dependence are shown in Fig. 1. The cutoff parameter Λ is determined to reproduce the observed mass of the hidden-charm pentaquarks. We then use the spectrum of the narrow resonance $P_c^+(4450)$. We find that the state appearing near $P_c^+(4450)$, is the $J^P = 5/2^-$ state with the resonance energy 4428.6 MeV in $\Lambda = 1400$ in our calculation. Therefore, the cutoff Λ is chosen by $\Lambda = 1400$. We conclude that the J^P assignment of $P_c^+(4450)$ is $J^P = 5/2^-$. In according to the analysis of the LHCb collaboration [1], the J of the observed pentaquarks are $3/2$ and $5/2$, and these states have the opposite parity. Therefore,

Table 3. Obtained masses with full channel coupling (Full), without $\bar{D}^{(*)}\Lambda_c$ (w/o $\bar{D}^{(*)}\Lambda_c$) and without large orbital angular momentum ℓ (w/o $\ell > 0$ or w/o $\ell > 1$) in $\Lambda = 1400$ MeV from Ref. [15].

J^P	Channels	Mass [MeV]	J^P	Channels	Mass [MeV]
$3/2^-$	Full	4136.0, 4307.9, 4348.7	$3/2^+$	Full	4206.7, 4339.7
	w/o $\bar{D}^{(*)}\Lambda_c$	4278.4, 4400.4		w/o $\bar{D}^{(*)}\Lambda_c$	—
	w/o $\ell > 0$	4220.4, 4376.6		w/o $\ell > 1$	4275.3
$5/2^-$	Full	4428.6	$5/2^+$	Full	—
	w/o $\bar{D}^{(*)}\Lambda_c$	—		w/o $\bar{D}^{(*)}\Lambda_c$	—
	w/o $\ell > 0$	—		w/o $\ell > 0$	—

the J^P of the $P_c^+(4380)$ will be $3/2^+$. In Fig. 1, the upper state of $J^P = 3/2^+$ has the mass, 4339.7 MeV, which is near the mass of $P_c^+(4380)$. Therefore, this obtained state of $J^P = 3/2^+$ corresponds to $P_c^+(4380)$.

In addition, new resonances are found in $\Lambda = 1400$ MeV. As shown in Fig. 1, we obtain three states in $J^P = 3/2^-$, whose masses are 4136.0 MeV, 4307.9 MeV and 4348.7 MeV, respectively. In the $J^P = 3/2^+$ state, we find two states, where the upper state corresponds to $P_c^+(4380)$ as discussed above, and the lower state is new one with the mass, 4206.7 MeV.

Finally, we discuss the effects of the channel couplings in the meson-baryon molecules. In the obtained results shown in Fig. 1, the full-channel coupling of $\bar{D}^{(*)}\Lambda_c - \bar{D}^{(*)}\Sigma_c^{(*)}$ is considered. However, the mass spectra are changed when the some of the channels are ignored. In Table 3, we compare the results with full channel coupling, without $\bar{D}^{(*)}\Lambda_c$, and without large orbital angular momentum $\ell > 0$ ($\ell > 1$) for the negative (positive) parity. We find that the masses increase by tens of MeV and some of states disappear when the $\bar{D}^{(*)}\Lambda_c$ channel or the states with large orbital angular momentum are ignored. Specifically the $J^P = 5/2^-$ state which corresponds to $P_c^+(4450)$ disappears when the full channel coupling calculation is not carried out.

4 Summary

We studied the hidden-charm pentaquarks as meson-baryon molecules, $\bar{D}^{(*)}\Lambda_c - \bar{D}^{(*)}\Sigma_c$. We considered the channel coupling analysis with $\bar{D}^{(*)}\Sigma_c^{(*)}$ manifested by the heavy quark spin symmetry, $\bar{D}^{(*)}\Lambda_c$ near the $\bar{D}^{(*)}\Sigma_c^{(*)}$ thresholds, and the states with larger orbital angular momentum mixed by the tensor force. The analysis of this molecular systems with $\bar{D}^{(*)}\Lambda_c - \bar{D}^{(*)}\Sigma_c^{(*)}$ had not been performed so far. As for the meson-baryon interaction, the meson exchange potential was employed which was obtained by the effective Lagrangians that respects the heavy quark and chiral symmetries. By solving the coupled-channel Schrödinger equations, we studied the bound and resonant states in $I(J^P) = 1/2(3/2^\pm)$ and $1/2(5/2^\pm)$ with $I = 1/2$. We concluded that the J^P assignments of $P_c^+(4380)$ and $P_c^+(4450)$ are $3/2^+$ and $5/2^-$, respectively. We also found new states in $J^P = 3/2^\pm$. In this calculations, we found that the coupling to the $\bar{D}^{(*)}\Lambda_c$ channel and to the state with large orbital angular momentum is not ignored, which produced the attraction.

Acknowledgments

This work was supported in part by the INFN Fellowship Programme.

References

[1] R. Aaij *et al.* [LHCb Collaboration], Phys. Rev. Lett. **115**, 072001 (2015).

[2] W. L. Wang, F. Huang, Z. Y. Zhang and B. S. Zou, Phys. Rev. C **84**, 015203 (2011).
[3] S. G. Yuan, K. W. Wei, J. He, H. S. Xu and B. S. Zou, Eur. Phys. J. A **48**, 61 (2012).
[4] G. N. Li, X. G. He and M. He, JHEP **1512**, 128 (2015).
[5] G. Yang and J. Ping, arXiv:1511.09053 [hep-ph].
[6] J. Hofmann and M. F. M. Lutz, Nucl. Phys. A **763**, 90 (2005).
[7] J. J. Wu, R. Molina, E. Oset and B. S. Zou, Phys. Rev. Lett. **105**, 232001 (2010).
[8] J. J. Wu, R. Molina, E. Oset and B. S. Zou, Phys. Rev. C **84**, 015202 (2011).
[9] C. Garcia-Recio, J. Nieves, O. Romanets, L. L. Salcedo and L. Tolos, Phys. Rev. D **87**, 074034 (2013).
[10] C. W. Xiao, J. Nieves and E. Oset, Phys. Rev. D **88**, 056012 (2013).
[11] M. Karliner and J. L. Rosner, Phys. Rev. Lett. **115**, no. 12, 122001 (2015).
[12] T. Uchino, W. H. Liang and E. Oset, Eur. Phys. J. A **52**, no. 3, 43 (2016).
[13] R. Chen, X. Liu and S. L. Zhu, Nucl. Phys. A **954**, 406 (2016).
[14] Y. Shimizu, D. Suenaga and M. Harada, Phys. Rev. D **93**, no. 11, 114003 (2016).
[15] Y. Yamaguchi and E. Santopinto, arXiv:1606.08330 [hep-ph].
[16] F. K. Guo, U. G. Meißner, W. Wang and Z. Yang, Phys. Rev. D **92**, 071502 (2015).
[17] X. H. Liu, Q. Wang and Q. Zhao, Phys. Lett. B **757**, 231 (2016).
[18] F. K. Guo, U. G. Meißner, J. Nieves and Z. Yang, arXiv:1605.05113 [hep-ph].
[19] E. Oset and A. Ramos, Nucl. Phys. A **635**, 99 (1998).
[20] B. Krippa, Phys. Rev. C **58**, 1333 (1998).
[21] T. Hyodo and D. Jido, Prog. Part. Nucl. Phys. **67**, 55 (2012).
[22] N. Brambilla *et al.*, Eur. Phys. J. C **71** (2011) 1534.
[23] S. K. Choi *et al.* [Belle Collaboration], Phys. Rev. Lett. **91**, 262001 (2003).
[24] E. S. Swanson, Phys. Rept. **429**, 243 (2006).
[25] M. Ablikim *et al.* [BESIII Collaboration], Phys. Rev. Lett. **110**, 252001 (2013).
[26] A. Bondar *et al.* [Belle Collaboration], Phys. Rev. Lett. **108**, 122001 (2012).
[27] N. Isgur and M. B. Wise, Phys. Lett. B **232**, 113 (1989).
[28] N. Isgur and M. B. Wise, Phys. Lett. B **237**, 527 (1990).
[29] A. V. Manohar and M. B. Wise, *Heavy quark physics* (Camb. Monogr. Part. Phys. Nucl. Phys. Cosmol. **10**, 2000) 1.
[30] T. D. Cohen, P. M. Hohler and R. F. Lebed, Phys. Rev. D **72**, 074010 (2005).
[31] S. Yasui and K. Sudoh, Phys. Rev. D **80**, 034008 (2009).
[32] Y. Yamaguchi, S. Ohkoda, S. Yasui and A. Hosaka, Phys. Rev. D **84**, 014032 (2011).
[33] Y. Yamaguchi, S. Ohkoda, S. Yasui and A. Hosaka, Phys. Rev. D **85**, 054003 (2012).
[34] Y. Yamaguchi, S. Ohkoda, S. Yasui and A. Hosaka, Phys. Rev. D **87**, 074019 (2013).
[35] A. Hosaka, T. Hyodo, K. Sudoh, Y. Yamaguchi and S. Yasui, arXiv:1606.08685 [hep-ph].
[36] M. B. Wise, Phys. Rev. D **45**, no. 7, R2188 (1992).
[37] A. F. Falk and M. E. Luke, Phys. Lett. B **292**, 119 (1992).
[38] T. M. Yan, H. Y. Cheng, C. Y. Cheung, G. L. Lin, Y. C. Lin and H. L. Yu, Phys. Rev. D **46**, 1148 (1992) Erratum: [Phys. Rev. D **55**, 5851 (1997)].
[39] R. Casalbuoni, A. Deandrea, N. Di Bartolomeo, R. Gatto, F. Feruglio and G. Nardulli, Phys. Rept. **281**, 145 (1997).
[40] Y. R. Liu and M. Oka, Phys. Rev. D **85**, 014015 (2012).

EPJ Web of Conferences **129**, 00025 (2016) DOI: 10.1051/epjconf/201612900025
QCD@Work 2016

Charmed Mesons and Charmonia: Three-Meson Strong Couplings

Wolfgang Lucha[1,⋆], Dmitri Melikhov[2,3,⋆⋆], Hagop Sazdjian[4,⋆⋆⋆], and Silvano Simula[5,⋆⋆⋆⋆]

[1] *Institute for High Energy Physics, Austrian Academy of Sciences, Nikolsdorfergasse 18, A-1050 Vienna, Austria*
[2] *D. V. Skobeltsyn Institute of Nuclear Physics, M. V. Lomonosov Moscow State University, 119991, Moscow, Russia*
[3] *Faculty of Physics, University of Vienna, Boltzmanngasse 5, A-1090 Vienna, Austria*
[4] *IPN, CNRS/IN2P3, Universitè Paris-Sud 11, F-91406 Orsay, France*
[5] *INFN, Sezione di Roma Tre, Via della Vasca Navale 84, I-00146 Roma, Italy*

Abstract. Revisiting the strong couplings of three mesons, each of which involves at least one charm quark, proves clear disaccord between quark-model and QCD sum-rule results.

1 Three-meson strong couplings from relativistic constituent-quark model

We extract the strong couplings of three mesons among which there is, at least, one of the charmonia η_c and J/ψ from the residues of poles in adequate transition form factors for timelike momentum transfer, which, in turn, are inferred from a relativistic dispersion approach based on a constituent-quark model.

2 Definitions of strong couplings, transition form factors, decay constants

Preparatively, let us introduce all the quantities necessary for the formulation of the relation sought, for *pseudoscalar* mesons P, with mass M_P, and *vector* mesons V, with mass M_V and polarization vector ε_μ.

- With momentum transfer $q \equiv p_1 - p_2$, the *strong couplings* $g_{PP'V}$ and $g_{PV'V}$ determine the amplitudes

$$\langle P'(p_2)\,V(q)|P(p_1)\rangle = -\frac{g_{PP'V}}{2}\,(p_1+p_2)^\mu\,\varepsilon^*_\mu(q)\,, \qquad \langle V'(p_2)\,V(q)|P(p_1)\rangle = -g_{PV'V}\,\epsilon_{\varepsilon^*(q)\,\varepsilon^*(p_2)\,p_1\,p_2}\,.$$

- The *transition form factors* $\mathcal{F}(q^2) = F_+^{P\to P'}(q^2)$, $V^{P\to V}(q^2)$ or $A_0^{P\to V}(q^2)$ enter in the two-meson matrix elements of the vector quark current $j_\mu \equiv \bar{q}_1\,\gamma_\mu\,q_2$ and the axial-vector quark current $j^5_\mu \equiv \bar{q}_1\,\gamma_\mu\,\gamma_5\,q_2$

$$\langle P'(p_2)|j_\mu|P(p_1)\rangle = F_+^{P\to P'}(q^2)\,(p_1+p_2)_\mu + \cdots\,, \qquad \langle V(p_2)|j_\mu|P(p_1)\rangle = \frac{2\,V^{P\to V}(q^2)}{M_P+M_V}\,\epsilon_{\mu\,\varepsilon^*(p_2)\,p_1\,p_2}\,,$$
$$\langle V(p_2)|j^5_\mu|P(p_1)\rangle = \mathrm{i}\,q_\mu\,(\varepsilon^*(p_2)\,p_1)\,\frac{2\,M_V}{q^2}\,A_0^{P\to V}(q^2) + \cdots\,. \qquad (1)$$

⋆e-mail: Wolfgang.Lucha@oeaw.ac.at
⋆⋆e-mail: dmitri_melikhov@gmx.de
⋆⋆⋆e-mail: sazdjian@ipno.in2p3.fr
⋆⋆⋆⋆e-mail: simula@roma3.infn.it

- The vector and pseudoscalar *decay constants* $f_{V,P}$ govern the meson–vacuum matrix elements of $j_\mu^{(5)}$:

$$\langle 0|j_\mu|V(q)\rangle = f_V\, M_V\, \varepsilon_\mu(q)\,, \qquad \langle 0|j_\mu^5|P(q)\rangle = \mathrm{i}\, f_P\, q_\mu\,. \tag{2}$$

In terms of the above quantities, the contributions of the *poles* residing at the masses M_{V_R} or M_{P_R} of the relevant vector and pseudoscalar resonances, V_R and P_R, to the form factors introduced in Eqs. (1) read

$$F_+^{P\to P'}(q^2) = \frac{g_{PP'V_R}\, f_{V_R}}{2\, M_{V_R}}\,\frac{1}{1-q^2/M_{V_R}^2} + \cdots\,, \qquad V^{P\to V}(q^2) = \frac{(M_V+M_P)\, g_{PVV_R}\, f_{V_R}}{2\, M_{V_R}}\,\frac{1}{1-q^2/M_{V_R}^2} + \cdots\,,$$

$$A_0^{P\to V}(q^2) = \frac{g_{PP_RV}\, f_{P_R}}{2\, M_V}\,\frac{1}{1-q^2/M_{P_R}^2} + \cdots\,.$$

3 Dispersion formalism relying on relativistic constituent-quark framework

For the actual theoretical computation of the three transition form factors $\mathcal{F}(q^2)$ lying at the core of our strong-couplings study, past experience leads us to trust in the relativistic *constituent-quark* model [1]. Adhering to this conviction requires us to match our currents to those built up by constituent quarks, Q.

- For heavy-quark currents, this can be easily accomplished by use of corresponding form factors $g_{V,A}$:

$$j_\mu = g_V\, \bar{Q}_1\, \gamma_\mu\, Q_2 + \cdots\,, \qquad j_\mu^5 = g_A\, \bar{Q}_1\, \gamma_\mu\, \gamma_5\, Q_2 + \cdots\,.$$

- For light-quark currents, partial axial-current conservation, e.g., renders this rather cumbersome [2]. Following Ref. [3], we use for our model parameter values the constituent-quark masses of Table 1 and

$$g_V = g_A = 1\,.$$

Table 1. Numerical values of the constituent-quark mass parameters m_Q entering in the present investigation [3].

Quark flavour Q	u	d	s	c
Quark mass m_Q (GeV)	0.23	0.23	0.35	1.45

Then, by application of the relativistic dispersion formalism, we are put in the position to represent the leptonic decay constants $f_{P,V}$ in the form of dispersion integrals of *spectral densities* $\rho_{P,V}(s)$ and the transition form factors $\mathcal{F}(q^2)$ by double dispersion integrals of *double spectral densities* $\Delta_{\mathcal{F}}(s_1, s_2, q^2)$,

$$f_{P,V} = \int \mathrm{d}s\, \phi_{P,V}(s)\, \rho_{P,V}(s)\,, \qquad \mathcal{F}(q^2) = \int \mathrm{d}s_1\, \phi_1(s_1) \int \mathrm{d}s_2\, \phi_2(s_2)\, \Delta_{\mathcal{F}}(s_1, s_2, q^2)\,, \tag{3}$$

involving the *wave functions* of pseudoscalar and vector mesons taking part in the studied reactions [4]

$$\phi_{P,V}(s) = \frac{\pi}{s^{3/4}}\sqrt{\frac{s^2-(m_1^2-m^2)^2}{2\,[s-(m_1-m)^2]}}\, w_{P,V}\!\left(\frac{(s-m_1^2-m^2)^2-4\,m_1^2\,m^2}{4\,s}\right), \qquad \int \mathrm{d}k\, k^2\, w_{P,V}^2(k^2) = 1\,.$$

The spectral densities may be derived from one-loop Feynman graphs, of the kind shown in Fig. 1. For the radial meson wave functions $w_{P,V}(k^2)$, it has become customary to assume simple Gaussian shapes:

$$w_{P,V}(k^2) \propto \exp\left(-\frac{k^2}{2\,\beta_{P,V}^2}\right). \tag{4}$$

The necessary input parameter values, drawn from a variety of related sources, are collected in Table 2.

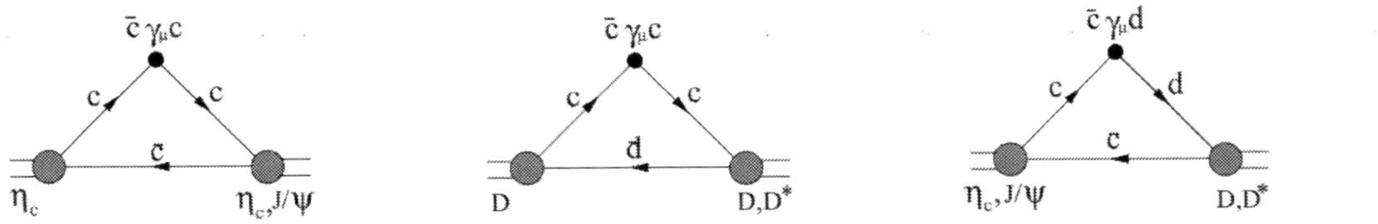

Figure 1. $M_1 > M_2$ meson–meson transitions for $M_{1,2} = \eta_c, J/\psi, D, D^*$, mediated by the *constituent*-quark vector currents $\bar{c}\gamma_\mu c$ or $\bar{c}\gamma_\mu d$: Feynman graphs yielding the one-loop contributions to the spectral density $\Delta_{\mathcal{F}}(s_1, s_2, q^2)$.

Table 2. Numerical values of the relevant parameters of the charmed mesons $D_{(s)}^{(*)}$ and the charmonia η_c and J/ψ: meson mass M, leptonic decay constant f defined in Eq. (2), and slope β fixing the width of the Gaussian (4) [5].

Meson	D	D^*	D_s	D_s^*	η_c	J/ψ
M (GeV)	1.87	2.010	1.97	2.11	2.980	3.097
f (MeV)	206 ± 8	260 ± 10	248 ± 2.5	311 ± 9	394.7 ± 2.4	405 ± 7
β (GeV)	0.475	0.48	0.545	0.54	0.77	0.68

In order to extract the strong couplings under discussion, we determine the momentum dependence for the transition form factors $\mathcal{F}(q^2)$ sufficiently far off their resonances R, where $R = V$ for $\mathcal{F} = F_+^{P\to P'}$ and $\mathcal{F} = V^{P\to V}$ but $R = P$ for $\mathcal{F} = A_0^{P\to V}$, *interpolate* our results by means of the simple parametrization

$$\mathcal{F}(q^2) = \frac{\mathcal{F}(0)}{\left(1 - q^2/M_R^2\right)\left(1 - \sigma_1\, q^2/M_R^2 + \sigma_2\, q^4/M_R^4\right)}\,, \qquad \operatorname{Res}\mathcal{F}(M_R^2) = \frac{\mathcal{F}(0)}{1 - \sigma_1 + \sigma_2}\,, \tag{5}$$

governed by the three parameters $\sigma_{1,2}$ and $\mathcal{F}(0)$, and extrapolate this momentum dependence of $\mathcal{F}(q^2)$ to the resonance region $q^2 \approx M_R^2$. From the resulting *residues* of the meson poles at $q^2 = M_R^2$, the strong couplings are found by factorizing off all known quantities such as meson masses and decay constants. In case a particular strong coupling shows up in residues of resonance poles of more than one transition form factor, for such multipresent strong coupling an optimized estimate is obtained by a combined fit.

4 Strong couplings among three η_c or J/ψ mesons: $\eta_c\eta_c J/\psi$ and $\eta_c J/\psi J/\psi$

An illustration of such *multipresence* is given by the strong coupling $g_{\eta_c\eta_c\psi}$ [6], with appearance in both

- the residue of $F_+^{\eta_c\to\eta_c}(M_\psi^2)$ arising from the vector current $\bar{c}\gamma_\mu c$ coupling, with strength f_ψ, to J/ψ and
- the residue of $A_0^{\eta_c\to\psi}(M_{\eta_c}^2)$ from the axial-vector current $\bar{c}\gamma_\mu\gamma_5 c$ that couples, with strength f_{η_c}, to η_c:

$$\operatorname{Res} F_+^{\eta_c\to\eta_c}(M_\psi^2) = g_{\eta_c\eta_c\psi}\,\frac{f_\psi}{2\,M_\psi}\,, \qquad \operatorname{Res} A_0^{\eta_c\to\psi}(M_{\eta_c}^2) = g_{\eta_c\eta_c\psi}\,\frac{f_{\eta_c}}{2\,M_\psi}\,.$$

After such detailed preliminaries, the way how to proceed should be pretty plain: We determine the meson wave-function parameters $\beta_{\eta_c,\psi}$ of both charmonia such that the dispersion representation (3) of their decay constants, $f_{\eta_c,\psi}$, reproduces the observed values. With these meson vertices at our disposal, we deduce the strong coupling of interest, for each meson–meson transition where this strong coupling enters (for the case of the $\eta_c\eta_c J/\psi$ coupling, see Fig. 2), from the spectral representation (3) of the form factor corresponding to the respective transition. Thus, our η_c and J/ψ PPV and PVV couplings are [6]

$$g_{\eta_c\eta_c\psi} = 25.8 \pm 1.7\,, \qquad g_{\eta_c\psi\psi} = (10.6 \pm 1.5)\ \text{GeV}^{-1}\,.$$

 DOI: 10.1051/epjconf/201612900025

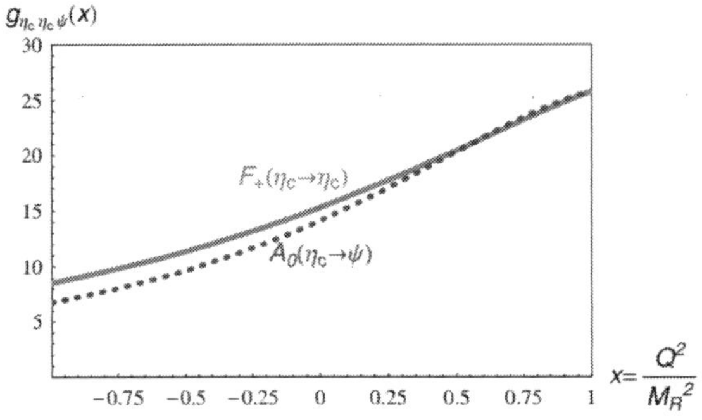

Figure 2. Off-shell strong coupling $g_{\eta_c\eta_c\psi}(x)$ as function of $x \equiv \frac{q^2}{M_R^2}$ for $\eta_c > \eta_c$ (red) or $\eta_c > J/\psi$ (blue) transitions.

5 Strong three-meson couplings of the charmonia J/ψ or η_c to $D_{(s)}$ and $D^*_{(s)}$

Along the same route as in Sec. 4 — and by taking into account, in addition, constituent-quark currents involving one d or s quark — we may likewise discuss the strong couplings of J/ψ or η_c to non-strange ($D^{(*)}$) or strange ($D_s^{(*)}$) charmed mesons. Combining the options sketched in Fig. 3, our findings are [6]

$$\begin{aligned} g_{DD\psi} &= 26.04 \pm 1.43\,, & g_{D^*D\psi} &= (10.7 \pm 0.4)\ \mathrm{GeV}^{-1}\,,\\ g_{DD^*\eta_c} &= 15.51 \pm 0.45\,, & g_{D^*D^*\eta_c} &= (9.76 \pm 0.32)\ \mathrm{GeV}^{-1}\,,\\ g_{D_sD_s\psi} &= 23.83 \pm 0.78\,, & g_{D_sD_s^*\psi} &= (9.6 \pm 0.8)\ \mathrm{GeV}^{-1}\,,\\ g_{D_sD_s^*\eta_c} &= 14.15 \pm 0.52\,, & g_{D_s^*D_s^*\eta_c} &= (8.27 \pm 0.37)\ \mathrm{GeV}^{-1}\,. \end{aligned} \qquad (6)$$

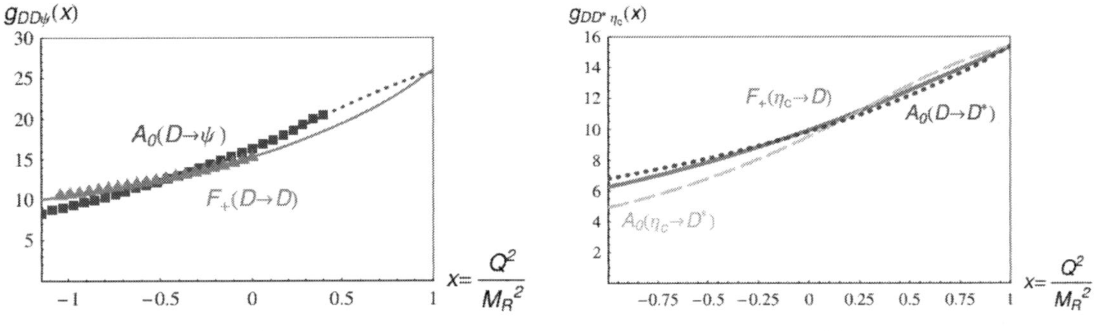

(a) (b)

Figure 3. Off-shell strong couplings $g_{DD\psi}$ and $g_{DD^*\eta_c}$ to $D^{(*)}$ mesons (with resonances indicated by circumflexes): dependences on $x \equiv \frac{q^2}{M_R^2}$ of (a) $g_{D\hat{D}\psi}(x) = \frac{2M_\psi}{f_D}(1-x)A_0^{D>\psi}(q^2)$ (blue) and $g_{DD\hat{\psi}}(x) = \frac{2M_\psi}{f_\psi}(1-x)F_+^{D>D}(q^2)$ (red), got with (lines) and without (symbols) interpolation, and of (b) $g_{D\hat{D}^*\eta_c}(x)$ (red), $g_{DD^*\hat{\eta}_c}(x)$ (blue) and $g_{\hat{D}D^*\eta_c}(x)$ (green).

6 Observations, comparison with findings of different origin, conclusions

Our application of a relativistic dispersion technique, relying on the constituent-quark model, to strong three-meson couplings of quarkonia among each other and to $D^{(*)}_{(s)}$ mesons yields some crucial insights:

1. The interpolation of our numerical transition-form-factor results found at low q^2 by means of the ansatz (5) yields values of the resonance-mass fit parameter M_R close to the experimental meson masses; this can be interpreted as confirmation of the presence of the poles expected at $q^2 \approx M_R^2$.
2. The replacement of the d quark by the s quark (or vice versa) in the quark currents mediating any transition under study enables us to arrive at some estimate of the size of SU(3)-breaking effects. Inspecting Eq. (6), we get a change of the strong couplings under consideration by roughly 10%.
3. Table 3 confronts, for the strong couplings between *charmonia and* $D_{(s)}^{(*)}$ *mesons*, the predictions of our dispersive constituent-quark formalism with (available) corresponding figures from QCD sum rules [7–9]; surprisingly, the latter prove to be smaller than our results [6] by a factor of two.

Table 3. Strong couplings of three mesons which involve one J/ψ: findings of the present relativistic constituent quark-model framework [6], confronted with available results [7–9] extracted from the QCD sum-rule approach.

Coupling	$g_{DD\psi}$	$g_{DD^*\psi}$ (GeV^{-1})	$g_{D_sD_s\psi}$	$g_{D_sD_s^*\psi}$ (GeV^{-1})
Quark model [6]	26.04 ± 1.43	10.7 ± 0.4	23.83 ± 0.78	9.6 ± 0.8
QCD sum rules	11.6 ± 1.8 [7]	4.0 ± 0.6 [7]	11.96 ± 1.34 [8]	4.30 ± 1.53 [9]

References

[1] W. Lucha, F. F. Schöberl, and D. Gromes, Phys. Rep. **200** (1991) 127; F. Cardarelli, E. Pace, G. Salmè, and S. Simula, Phys. Lett. B **357** (1995) 267, arXiv:nucl-th/9507037; R. N. Faustov and V. O. Galkin, Z. Phys. C **66** (1995) 119.

[2] D. Melikhov and B. Stech, Phys. Rev. D **74** (2006) 034022, arXiv:hep-ph/0606203; W. Lucha, D. Melikhov, and S. Simula, Phys. Rev. D **74** (2006) 054004, arXiv:hep-ph/0606281.

[3] D. Melikhov and B. Stech, Phys. Rev. D **62** (2000) 014006, arXiv:hep-ph/0001113.

[4] D. Melikhov, Eur. Phys. J. direct **C4** (2002) 2, arXiv:hep-ph/0110087.

[5] C. T. H. Davies *et al.*, Phys. Rev. D **82** (2010) 114504, arXiv:1008.4018 [hep-lat]; W. Lucha, D. Melikhov, and S. Simula, Phys. Lett. B **701** (2011) 82, arXiv:1101.5986 [hep-ph]; D. Bečirević *et al.*, JHEP **1202** (2012) 042, arXiv:1201.4039 [hep-lat]; G. C. Donald *et al.*, Phys. Rev. D **86** (2012) 094501, arXiv:1208.2855 [hep-lat]; W. Lucha, D. Melikhov, and S. Simula, Phys. Lett. B **735** (2014) 12, arXiv:1404.0293 [hep-ph]; K. A. Olive *et al.* (PDG), Chin. Phys. C **38** (2014) 090001.

[6] W. Lucha, D. Melikhov, H. Sazdjian, and S. Simula, Phys. Rev. D **93** (2016) 016004, **93** (2016) 019902(E), arXiv:1506.09213 [hep-ph].

[7] R. D. Matheus *et al.*, Int. J. Mod. Phys. E **14** (2005) 555.

[8] B. Osório Rodrigues, M. E. Bracco, and M. Chiapparini, Nucl. Phys. A **929** (2014) 143; arXiv: 1309.1637 [hep-ph].

[9] B. Osório Rodrigues *et al.*, Eur. Phys. J. A **51** (2015) 28; arXiv:1501.03088 [hep-ph].

EPJ Web of Conferences **129**, 00026 (2016) DOI: 10.1051/epjconf/201612900026
QCD@Work 2016

Heavy-Meson Decay Constants: QCD Sum-Rule Glance at Isospin Breaking

Wolfgang Lucha[1,⋆], Dmitri Melikhov[1,2,3,⋆⋆], and Silvano Simula[4,⋆⋆⋆]

[1] *Institute for High Energy Physics, Austrian Academy of Sciences, Nikolsdorfergasse 18, A-1050 Vienna, Austria*
[2] *D. V. Skobeltsyn Institute of Nuclear Physics, M. V. Lomonosov Moscow State University, 119991, Moscow, Russia*
[3] *Faculty of Physics, University of Vienna, Boltzmanngasse 5, A-1090 Vienna, Austria*
[4] *INFN, Sezione di Roma Tre, Via della Vasca Navale 84, I-00146 Roma, Italy*

Abstract. QCD sum rules for decay constants of heavy mesons with u or d quark yield for B mesons much less isospin breaking than lattice QCD but good agreement for D mesons.

1 Hadronic properties scrutinized from the point of view of QCD sum rules

QCD sum rules [1] are analytic relations between hadron features and QCD parameters emerging upon evaluation of correlation functions of appropriate interpolating operators at both the hadronic level and the QCD level by insertion of complete sets of hadron states, exploitation of Wilson's operator product expansion, Borel transformation, and the quark–hadron duality assumption above effective thresholds.

In order to both raise the *accuracy* of QCD sum-rule predictions and estimate the *systematic errors* [2], some time ago we developed [3] a modified formalism which takes into account the dependence of the effective thresholds on the Borel parameters. This enabled the *improved extraction* of heavy-meson decay constants f [4], notably of the inequalities $f_{B^*_{(s)}} < f_{B_{(s)}}$ [5], later on confirmed by lattice QCD [6].

QCD sum rules provide in *analytic* form the *dependences* of the properties of the hadrons currently under study on the basic parameters of QCD. This allows us to study the impact of the isospin breaking provoked by the slight difference in the masses of the light u and d quarks on the decay constants of the heavy–light mesons, by working out the explicit relation of the decay constants to the light quark mass.

2 Heavy-meson decay constants: extraction from QCD sum-rule approach

Hence, let us try to get hold of the dependence of our QCD sum-rule approach on the light-quark mass. We analyse pseudoscalar (P_q) and vector (V_q) mesons (generically labelled $M_q \equiv P_q, V_q$) of mass M_{M_q}, composed of a heavy quark $Q = b, c$ and a light quark $q = u, d$ with masses m_Q and m_q, respectively. In terms of suitable interpolating heavy–light quark-current operators, that is, the axial-vector (A) current $A_\mu(x) \equiv \bar{q}(x)\,\gamma_\mu\,\gamma_5\,Q(x)$ and the vector (V) current $V_\mu(x) \equiv \bar{q}(x)\,\gamma_\mu\,Q(x)$, the *decay constants* f_{M_q} of the mesons [with momentum p and, in the case of vector mesons, polarization vector $\varepsilon_\mu(p)$] are defined by

⋆e-mail: Wolfgang.Lucha@oeaw.ac.at
⋆⋆e-mail: dmitri_melikhov@gmx.de
⋆⋆⋆e-mail: simula@roma3.infn.it

$$\langle 0|\,A_\mu(0)\,|P_q(p)\rangle = \mathrm{i}\, f_{P_q}\, p_\mu\,, \qquad \langle 0|\,V_\mu(0)\,|V_q(p)\rangle = f_{V_q}\, M_{V_q}\, \varepsilon_\mu(p)\,.$$

The decay constants may be inferred from the correlators of two of such currents $J_\mu(x) = A_\mu(x), V_\mu(x)$:

$$\Pi^{(J)}_{\mu\nu}(p) = \mathrm{i}\int \mathrm{d}^4x \exp(\mathrm{i}\,p\,x)\,\langle 0|\,T\big(J_\mu(x)\,J^\dagger_\nu(0)\big)\,|0\rangle\,, \qquad J = \mathrm{A}, \mathrm{V}\,.$$

The QCD sum rules providing, for the mesons $M_q = P_q, V_q$, their masses M_{M_q} and decay constants f_{M_q} receive both purely perturbative contributions, represented by dispersion integrals of spectral densities $\rho_J(s, m_Q, m_q, \alpha_s)$, given by series expansions in powers of the strong coupling α_s, and non-perturbative contributions $\widehat{\Pi}^{(J)}_N(\tau, m_Q, m_q)$, parametrized by vacuum condensates; generically, they assume the form

$$f^2_{M_q}\, M^{2N}_{M_q} \exp\left(-M^2_{M_q}\,\tau\right) = \int_{(m_Q+m_q)^2}^{s_{\rm eff}(\tau)} \mathrm{d}s\, s^N \exp(-s\,\tau)\,\rho_J(s, m_Q, m_q, \alpha_s) + \widehat{\Pi}^{(J)}_N(\tau, m_Q, m_q)\,, \qquad J = \mathrm{A}, \mathrm{V}\,.$$

The integer N relates to the Lorentz nature of the employed interpolating operator: the axial-vector and vector currents $A_\mu(x)$ and $V_\mu(x)$ imply $N = 1$, whereas, for pseudoscalar currents, we would get $N = 2$.

The innovative change introduced, in Refs. [3], to the concept of QCD sum rules simply consists in realizing (and accepting) that, in general, the effective threshold $s_{\rm eff}$ will depend on the Borel variable τ (the parameter of dimension of inverse mass squared arising upon Borel transformation): $s_{\rm eff} = s_{\rm eff}(\tau)$. Our favourite possibility for acquiring information on the actual τ behaviour of $s_{\rm eff}(\tau)$ is to fit our QCD sum-rule predictions of the masses of the mesons under study to their experimentally measured values. It suffices to model $s_{\rm eff}(\tau)$ by polynomial ansätze of rather low order n, with expansion coefficients $s^{(n)}_j$:

$$s^{(n)}_{\rm eff}(\tau) = \sum_{j=0}^{n} s^{(n)}_j\, \tau^j\,.$$

By its central value and half-width, the spread of results for linear ($n = 1$), quadratic ($n = 2$), and cubic ($n = 3$) $s^{(n)}_{\rm eff}(\tau)$ behaviour enables an estimate of some decay constant and its systematic uncertainty [3].

We should achieve our goals by just imitating the application of our algorithm [3] to the analysis of heavy-meson decay constants in related studies [4, 5] based on cutting-edge QCD contributions [7–9]. In order to optimize our predictions' perturbative performance [8], we use quark masses defined by the modified minimal-subtraction ($\overline{\rm MS}$) renormalization scheme. Table 1 shows all numerical ingredients.

Table 1. Numerical parameter values entering in QCD sum rules for heavy-meson decay constants [4, 8, 10, 11].

Operator-product-expansion parameter		Numerical input value
u-quark $\overline{\rm MS}$ mass	$m_u(2\ \mathrm{GeV})$	$(2.3^{+0.7}_{-0.5})$ MeV
d-quark $\overline{\rm MS}$ mass	$m_d(2\ \mathrm{GeV})$	$(4.8^{+0.5}_{-0.3})$ MeV
s-quark $\overline{\rm MS}$ mass	$m_s(2\ \mathrm{GeV})$	(93.8 ± 2.4) MeV
c-quark $\overline{\rm MS}$ mass	$m_c(m_c)$	(1275 ± 25) MeV
b-quark $\overline{\rm MS}$ mass	$m_b(m_b)$	(4247 ± 34) MeV
Strong coupling	$\alpha_s(M_Z)$	0.1185 ± 0.0006
Light-quark condensate	$\langle \bar{q}\,q\rangle(2\ \mathrm{GeV})$	$-[(267 \pm 17)\ \mathrm{MeV}]^3$
Strange-quark condensate	$\langle \bar{s}\,s\rangle(2\ \mathrm{GeV})$	$(0.8 \pm 0.3) \times \langle \bar{q}\,q\rangle(2\ \mathrm{GeV})$
Gluon condensate	$\left\langle \frac{\alpha_s}{\pi}\,G\,G\right\rangle$	$(0.024 \pm 0.012)\ \mathrm{GeV}^4$

3 Summary of procedures, results, insights, conclusions, and outlook [12]

Allowing the light-quark mass m_q to assume any value within the interval $0 \leq m_q \leq m_s$ and defining on this domain a function $f_M(m_q)$ by the outcomes of our QCD sum rule for the decay constant of a meson M with light quark of just this mass m_q, we minimize uncertainties by analyzing the difference quotient

$$\frac{f_M(m_d) - f_M(m_u)}{m_d - m_u}\,, \qquad M|_{m_q=m_u,m_d,m_s} = M_{u,d,s}\,, \tag{1}$$

or the *slope* of $f_M(m_q)$ in the vicinity of $\overline{m} \equiv (m_u+m_d)/2$, requiring us to specify some m_q dependences:

- For the light-quark condensate, $\langle \bar{q}\, q\rangle$, we assume a linear m_q dependence from its value $\langle \bar{u}\, u\rangle \approx \langle \bar{d}\, d\rangle$ at $\overline{m} \equiv (m_u+m_d)/2 = 3.5^{+0.7}_{-0.2}$ MeV [10] up to the strange-quark condensate, $\langle \bar{s}\, s\rangle$, at $m_s = 93.8$ MeV.
- For the mass $M_M(m_q)$ of any heavy–light $[\bar{Q}\, q]$ meson, we allow for a linear m_q dependence from the experimental [10] nonstrange-meson mass $M_{M_{ud}}$ up to the associated strange-meson mass M_{M_s} [10]:

$$M_M(m_q) = M_{M_{ud}} + \frac{m_q - \overline{m}}{m_s - \overline{m}}\,(M_{M_s} - M_{M_{ud}})\,.$$

Figure 1 shows the decay-constant function $f_M(m_q)$ for all $M = B^{(*)}, D^{(*)}$, whereas Fig. 2 illustrates nicely the reduction of the systematic errors, when normalizing $f_M(m_q)$ to their values $f_M(0)$ at $m_q = 0$.

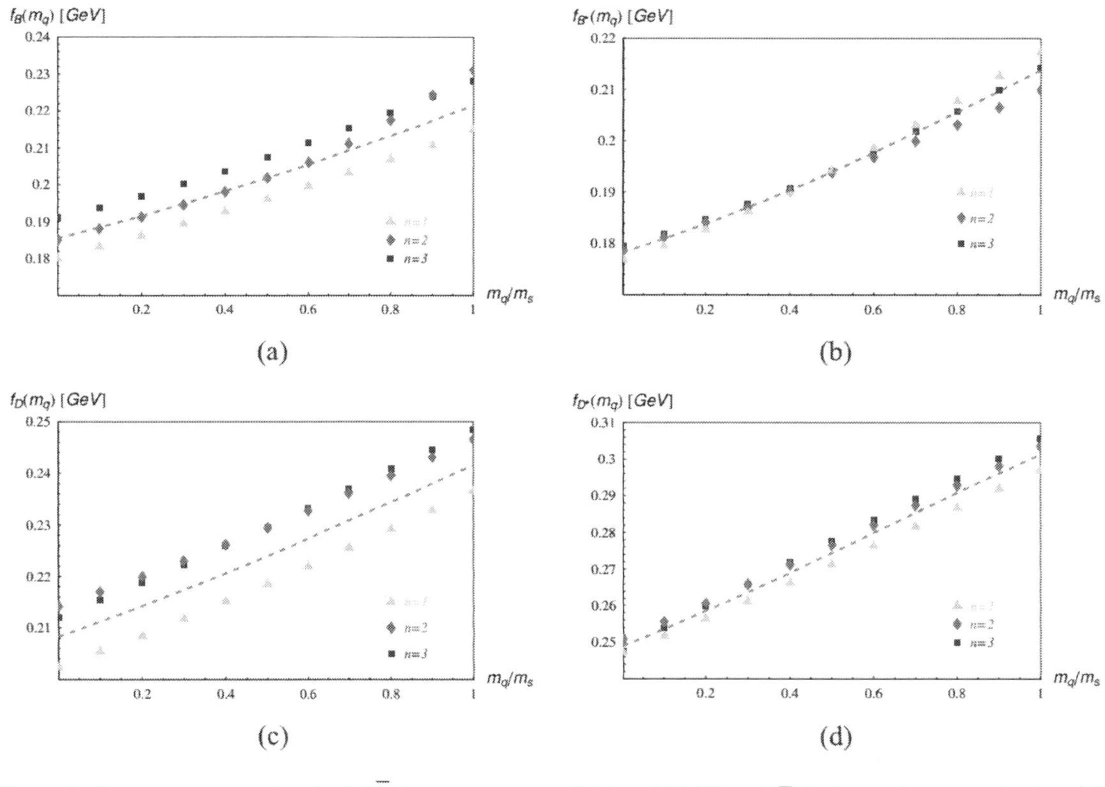

Figure 1. Decay constants f_M of both $[\bar{b}\, q]$ beauty mesons B (a) and B^* (b) and $[\bar{c}\, q]$ charmed mesons D (c) and D^* (d): dependence on the light quark mass m_q predicted by QCD sum rules [3] employing, for the effective threshold $s_{\rm eff}(\tau)$, polynomial ansatzes of order $n = 1$ (green triangles ▲), $n = 2$ (red diamonds ◆), and $n = 3$ (blue squares ■). The dashed magenta line indicates the position of the center of the band spanned by these individual-n extractions.

 DOI: 10.1051/epjconf/201612900026

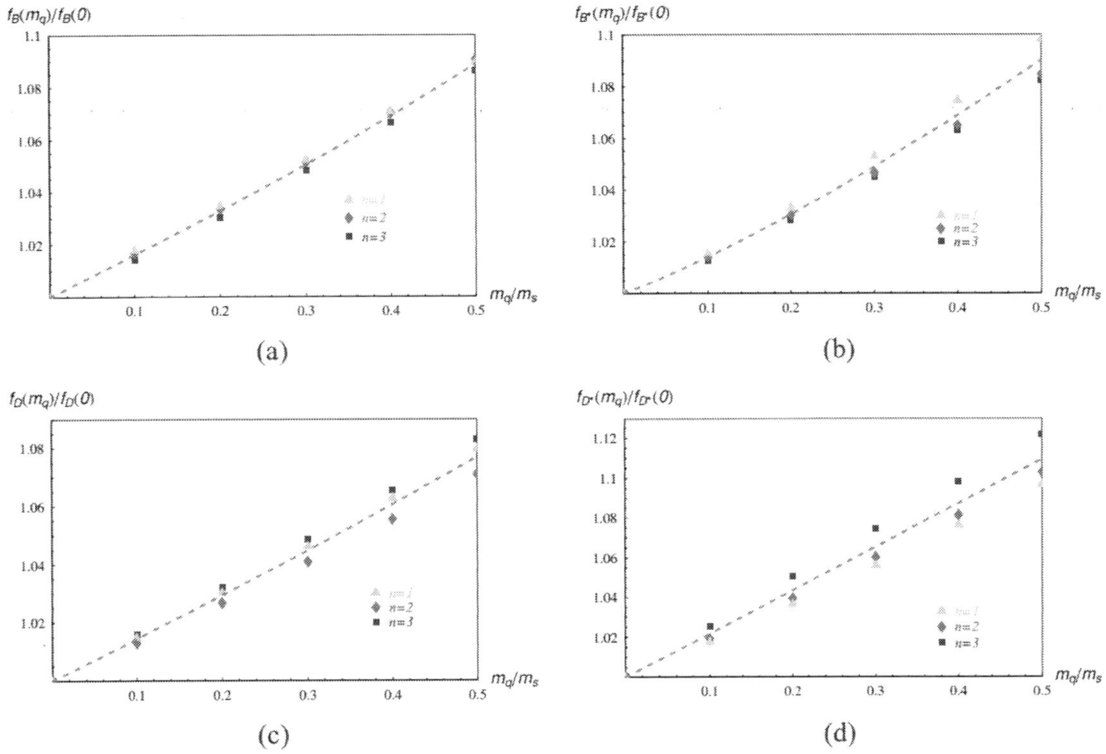

Figure 2. Decay-constant ratios, $f_M(m_q)/f_M(0)$, of the bottom mesons B (a) and B^* (b) and the charmed mesons D (c) and D^* (d): dependence on the light-quark mass m_q from QCD sum rules [3] relying, for the effective threshold $s_{\rm eff}(\tau)$, on polynomial ansätze of order $n = 1$ (green triangles ▲), $n = 2$ (red diamonds ◆) or $n = 3$ (blue squares ■). The dashed magenta line indicates the position of the center of the band spanned by these individual-n extractions.

Albeit, as physical quantities, the decay constants f_{M_q} must not depend on the renormalization scale μ, inevitable truncations of perturbative expansions induce artificial μ dependences of any QCD sum-rule results, raising the errors. We find f_{M_q} from the ranges [4, 5] $1 < \mu$ (GeV) < 3 for charmed mesons and $3 < \mu$ (GeV) < 5 for beauty mesons. For each of the mesons analyzed, $f_M(m_q)$ may be parametrized as

$$\frac{f_M(m_q)}{f_M(0)} = 1 + r_\ell \, \frac{m_q}{m_s} \log \frac{m_q}{m_s} + r_1 \, \frac{m_q}{m_s} + \cdots \,, \qquad M = B^{(*)}, D^{(*)} \,. \tag{2}$$

Table 2. Difference of the decay constants f_{M_q} for d- vs. u-quark mesons signalling isospin breaking, normalized to $f_M(0)$, for $M = B^{(*)}, D^{(*)}$, and numerical values of the coefficients $r_{\ell,1}$ in the parametrization (2) of these ratios.

Meson M_q	r_ℓ	r_1	$\dfrac{f_{M_d} - f_{M_u}}{f_M(0)}$
B	0.011 ± 0.008	0.181 ± 0.003	$(4.1 \pm 0.4) \times 10^{-3}$
B^*	0.024 ± 0.003	0.19 ± 0.02	$(3.6 \pm 0.3) \times 10^{-3}$
D	0.006 ± 0.015	0.156 ± 0.010	$(3.8 \pm 0.4) \times 10^{-3}$
D^*	0.002 ± 0.010	0.22 ± 0.010	$(5.7 \pm 1.2) \times 10^{-3}$

Table 2 reveals, for each meson M_q, the values of the coefficients $r_{\ell,1}$ and the decay-constant difference $f_{M_d}-f_{M_u}$, which satisfies $(f_{M_d}-f_{M_u})/(m_d-m_u) > 0$, tantamount to $f_{M_d} > f_{M_u}$ for all $M = B^{(*)}, D^{(*)}$ [12]. Upon taking into account all the uncertainties induced by the QCD parameters and the systematic ones inherent to the QCD sum-rule technique, we arrive at the following $f_{B^{(*)},D^{(*)}}$ decay-constant differences:

$$f_{B^0} - f_{B^\pm} = (0.79 \pm 0.14)\ \text{MeV}\,, \qquad f_{B^{*0}} - f_{B^{*\pm}} = (0.65 \pm 0.10)\ \text{MeV}\,,$$
$$f_{D^\pm} - f_{D^0} = (0.78 \pm 0.13)\ \text{MeV}\,, \qquad f_{D^{*\pm}} - f_{D^{*0}} = (1.41 \pm 0.42)\ \text{MeV}\,.$$

Comparing with related lattice-QCD outcomes [13], our predictions for the decay-constant differences are, for the B mesons, lower than their lattice-QCD counterparts but agree, in the case of the D mesons, with lattice-QCD results. Our findings are of similar size as those got by lattice QCD for the K mesons. In Ref. [12], we present a more detailed analysis, including the uncertainties of the light-quark masses.

Acknowledgement. D. M. was supported by the Austrian Science Fund (FWF): Project P29028-N27.

References

[1] M. A. Shifman, A. I. Vainshtein, and V. I. Zakharov, Nucl. Phys. B **147** (1979) 385.

[2] W. Lucha, D. Melikhov, and S. Simula, Phys. Rev. D **76** (2007) 036002, arXiv:0705.0470 [hep-ph]; Phys. Lett. B **657** (2007) 148, arXiv:0709.1584 [hep-ph]; Phys. At. Nucl. **71** (2008) 1461; Phys. Lett. B **671** (2009) 445, arXiv:0810.1920 [hep-ph]; D. Melikhov, Phys. Lett. B **671** (2009) 450, arXiv:0810.4497 [hep-ph].

[3] W. Lucha, D. Melikhov, and S. Simula, Phys. Rev. D **79** (2009) 096011, arXiv:0902.4202 [hep-ph]; J. Phys. G **37** (2010) 035003, arXiv:0905.0963 [hep-ph]; Phys. Lett. B **687** (2010) 48, arXiv:0912.5017 [hep-ph]; Phys. At. Nucl. **73** (2010) 1770, arXiv:1003.1463 [hep-ph]; W. Lucha, D. Melikhov, H. Sazdjian, & S. Simula, Phys. Rev. D **80** (2009) 114028, arXiv:0910.3164 [hep-ph].

[4] W. Lucha, D. Melikhov, and S. Simula, J. Phys. G **38** (2011) 105002, arXiv:1008.2698 [hep-ph]; Phys. Lett. B **701** (2011) 82, arXiv:1101.5986 [hep-ph]; Phys. Rev. D **88** (2013) 056011, arXiv:1305.7099 [hep-ph]; Phys. Lett. B **735** (2014) 12, arXiv:1404.0293 [hep-ph]; EPJ Web Conf. **80** (2014) 00043, arXiv:1407.5512 [hep-ph].

[5] W. Lucha, D. Melikhov, and S. Simula, EPJ Web Conf. **80** (2014) 00046, arXiv:1410.6684 [hep-ph]; arXiv:1411.3890 [hep-ph]; AIP Conf. Proc. **1701** (2016) 050007, arXiv:1411.7844 [hep-ph]; Phys. Rev. D **91** (2015) 116009, arXiv:1504.03017 [hep-ph]; PoS (EPS-HEP2015) 532, arXiv:1508.07595 [hep-ph].

[6] B. Colquhoun *et al.*, HPQCD Coll., Phys. Rev. D **91** (2015) 114509, arXiv:1503.05762 [hep-lat].

[7] K. G. Chetyrkin and M. Steinhauser, Phys. Lett. B **502** (2001) 104, arXiv:hep-ph/0012002; Eur. Phys. J. C **21** (2001) 319, arXiv:hep-ph/0108017.

[8] M. Jamin and B. O. Lange, Phys. Rev. D **65** (2002) 056005, arXiv:hep-ph/0108135.

[9] P. Gelhausen, A. Khodjamirian, A. A. Pivovarov, and D. Rosenthal, Phys. Rev. D **88** (2013) 014015, arXiv:1305.5432 [hep-ph]; **89** (2014) 099901(E); **91** (2015) 099901(E).

[10] K. A. Olive *et al.*, Particle Data Group, Chin. Phys. C **38** (2014) 090001.

[11] S. Aoki *et al.*, FLAG Working Group, Eur. Phys. J. C **74** (2014) 2890, arXiv:1310.8555 [hep-lat].

[12] W. Lucha, D. Melikhov, and S. Simula, HEPHY-PUB 973/16 (2016), in preparation.

[13] G. M. de Divitiis *et al.*, RM123 Coll., J. High Energy Phys. 04 (2012) 124, arXiv:1110.6294 [hep-lat]; R. J. Dowdall *et al.*, HPQCD Coll., Phys. Rev. Lett. **110** (2013) 222003, arXiv:1302.2644 [hep-lat]; A. Bazavov *et al.*, Fermilab Lattice and MILC Collaborations, Phys. Rev. D **90** (2014) 074509, arXiv:1407.3772 [hep-lat]; N. H. Christ *et al.*, RBC and UKQCD Collaborations, Phys. Rev. D **91** (2015) 054502, arXiv:1404.4670 [hep-lat].

EPJ Web of Conferences **129**, 00027 (2016) DOI: 10.1051/epjconf/201612900027
QCD@Work 2016

Factorization, resummation and sum rules for heavy-to-light form factors

Yu-Ming Wang[1,2,⋆]

[1] *Fakultät für Physik, Universität Wien, Boltzmanngasse 5, 1090 Vienna, Austria*
[2] *School of Physics, Nankai University, 300071 Tianjin, China*

Abstract. Precision calculations of heavy-to-light form factors are essential to sharpen our understanding towards the strong interaction dynamics of the heavy-quark system and to shed light on a coherent solution of flavor anomalies. We briefly review factorization properties of heavy-to-light form factors in the framework of QCD factorization in the heavy quark limit and discuss the recent progress on the QCD calculation of $B \to \pi$ form factors from the light-cone sum rules with the B-meson distribution amplitudes. Demonstration of QCD factorization for the vacuum-to-B-meson correlation function used in the sum-rule construction and resummation of large logarithms in the short-distance functions entering the factorization theorem are presented in detail. Phenomenological implications of the newly derived sum rules for $B \to \pi$ form factors are further addressed with a particular attention to the extraction of the CKM matrix element $|V_{ub}|$.

1 Introduction

Heavy-to-light form factors serve as fundamental inputs of describing many exclusive heavy hadron decays which are of great phenomenological interest to the ongoing and forthcoming collider experiments. Extensive efforts have been devoted to develop systematic theoretical frameworks for the precision calculation of heavy-to-light form factors in QCD (see, for instance[1–7]). In addition to the nonperturbative hadronic distribution amplitudes (DA), the symmetry-conserving "soft" form factors have to be introduced in QCD factorization for $B \to \pi$ form factors due to the emergence of rapidity divergences in the corresponding convolution integrals. An alternative approach to compute $B \to \pi$ form factors is QCD light-cone sum rules (LCSR) constructed from the vacuum-to-B-meson correlation functions with the aid of the dispersion relations and parton-hadron duality approximation [8–11]. In the following, we will first establish QCD factorization for the vacuum-to-B-meson correlation function at one loop employing the method of regions and perform the resummation of large logarithms in the hard and jet functions with the renormalization-group approach in section 2, where the resummation improved LCSR for $B \to \pi$ form factors and a new determination of the CKM matrix element $|V_{ub}|$ are also presented.

⋆e-mail: yu-ming.wang@univie.ac.at

2 The LCSR for the $B \to \pi$ form factors at $O(\alpha_s)$

The construction of LCSR for the $B \to \pi$ form factors can be achieved with the following correlation function [12]:

$$\begin{aligned}\Pi_\mu(n \cdot p, \bar{n} \cdot p) &= \int d^4x\, e^{ip\cdot x} \langle 0|T\left\{\bar{d}(x)\, \not{n}\, \gamma_5\, u(x), \bar{u}(0)\, \gamma_\mu\, b(0)\right\}|\bar{B}(p+q)\rangle \\ &= \Pi(n \cdot p, \bar{n} \cdot p)\, n_\mu + \widetilde{\Pi}(n \cdot p, \bar{n} \cdot p)\, \bar{n}_\mu\,, \end{aligned} \tag{1}$$

where $p + q = m_B\, v$ and the two light-cone vectors satisfy the relations $n \cdot v = \bar{n} \cdot v = 1$. To facilitate the perturbative calculation of short-distance functions in the factorization formula of (1), we need to establish the power counting scheme for the external momentum p_μ

$$n \cdot p \simeq \frac{m_B^2 + m_\pi^2 - q^2}{m_B} = 2E_\pi\,, \qquad \bar{n} \cdot p \sim O(\Lambda_{\rm QCD})\,. \tag{2}$$

Applying the light-cone operator-produce-expansion (OPE) at space-like p^2 yields the tree-level factorization formula

$$\begin{aligned}\widetilde{\Pi}(n \cdot p, \bar{n} \cdot p) &= \tilde{f}_B(\mu)\, m_B \int_0^\infty d\omega' \, \frac{\phi_B^-(\omega')}{\omega' - \bar{n} \cdot p - i\,0} + O(\alpha_s)\,, \\ \Pi(n \cdot p, \bar{n} \cdot p) &= O(\alpha_s)\,. \end{aligned} \tag{3}$$

in the heavy quark limit, where the B-meson light-cone DA in the coordinate space is defined as

$$\langle 0|\bar{d}_\beta(\tau\, \bar{n})\, [\tau\bar{n}, 0]\, b_\alpha(0)|\bar{B}(p+q)\rangle = -\frac{i\tilde{f}_B(\mu)\, m_B}{4}\left\{\frac{1+\not{v}}{2}\left[2\,\tilde{\phi}_B^+(\tau) + \left(\tilde{\phi}_B^-(\tau) - \tilde{\phi}_B^+(\tau)\right)\not{n}\right]\gamma_5\right\}_{\alpha\beta}. \tag{4}$$

Applying the standard definitions of $B \to \pi$ form factors and the pion decay constant

$$\begin{aligned}\langle \pi(p)|\bar{u}\gamma_\mu b|\bar{B}(p_B)\rangle &= f_{B\pi}^+(q^2)\left[p_B + p - \frac{m_B^2 - m_\pi^2}{q^2}\, q\right]_\mu + f_{B\pi}^0(q^2)\, \frac{m_B^2 - m_\pi^2}{q^2}\, q_\mu\,, \\ \langle \pi(p)|\bar{d}\, \not{n}\, \gamma_5\, u|0\rangle &= -i\, n \cdot p\, f_\pi\,, \end{aligned} \tag{5}$$

it is straightforward to derive the hadronic dispersion relation for the correlator (1)

$$\begin{aligned}&\Pi_\mu(n \cdot p, \bar{n} \cdot p) \\ &= \frac{f_\pi\, n \cdot p\, m_B}{2\,(m_\pi^2 - p^2)}\left\{\bar{n}_\mu\left[\frac{n \cdot p}{m_B}\, f_{B\pi}^+(q^2) + f_{B\pi}^0(q^2)\right] + n_\mu\, \frac{m_B}{n \cdot p - m_B}\left[\frac{n \cdot p}{m_B}\, f_{B\pi}^+(q^2) - f_{B\pi}^0(q^2)\right]\right\} \\ &+ \int_{\omega_s}^{+\infty} d\omega'\, \frac{1}{\omega' - \bar{n} \cdot p - i0}\left[\rho^h(\omega', n \cdot p)\, n_\mu + \tilde{\rho}^h(\omega', n \cdot p)\, \bar{n}_\mu\right], \end{aligned} \tag{6}$$

where the threshold parameter ω_s in the pion channel scales as Λ^2/m_b. Matching the hadronic and OPE representations of the vacuum-to-B-meson correlation function leads to

$$\begin{aligned}f_{B\pi}^+(q^2) &= \frac{\tilde{f}_B(\mu)\, m_B}{f_\pi\, n \cdot p}\exp\left[\frac{m_\pi^2}{n \cdot p\, \omega_M}\right]\int_0^{\omega_s} d\omega'\, e^{-\omega'/\omega_M}\, \phi_B^-(\omega') + O(\alpha_s)\,, \\ f_{B\pi}^0(q^2) &= \frac{n \cdot p}{m_B}\, f_{B\pi}^+(q^2) + O(\alpha_s)\,. \end{aligned} \tag{7}$$

Now we are in a position to demonstrate QCD factorization for $\Pi_\mu(n \cdot p, \bar{n} \cdot p)$ at one loop using the diagrammatical approach. Technically, this amounts to justifying the equivalence between the soft

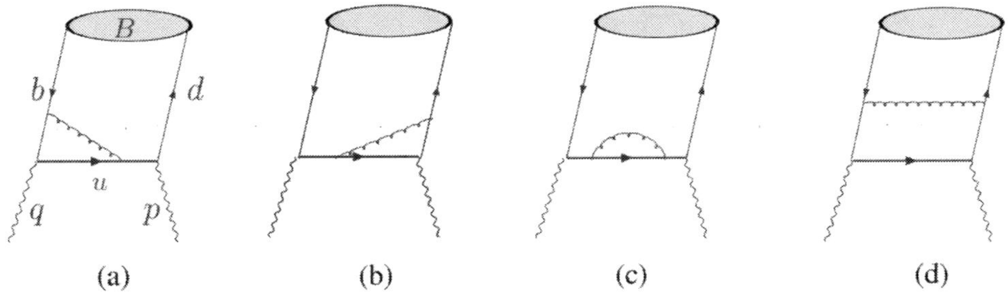

Figure 1. Next-to-leading-order QCD correction to the correlation function $\Pi_\mu(n \cdot p, \bar{n} \cdot p)$.

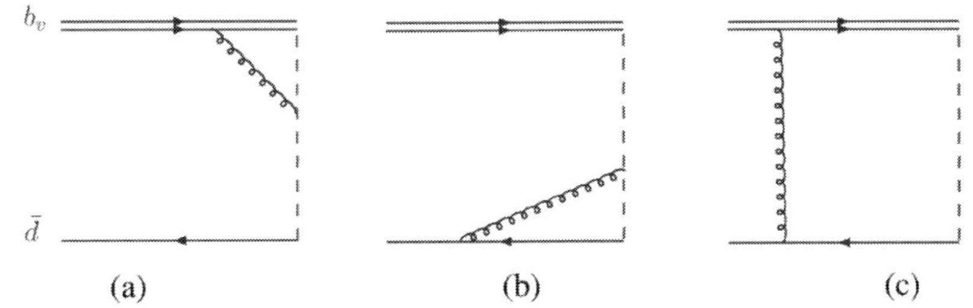

Figure 2. One-loop effective diagrams for the infrared subtraction.

contribution to the one-loop QCD diagrams displayed in Figure 1 and the infrared subtraction shown in Figure 2 in the heavy quark limit and extracting the hard and hard-collinear contributions to the correlation function at leading power in Λ/m_b. We will apply the standard strategies to establish the diagrammatical factorization for QCD Green functions at one-loop accuracy [13, 14]:

- Identify the leading regions of the QCD amplitudes with the power counting scheme (2);
- Evaluate the loop integrals with the method of regions and extract the "bare" perturbative kernels;
- Apply the ultraviolet renormalization and perform the infrared subtraction;
- Substitute the momentum-space light-cone projector of the B-meson.

Taking the weak vertex diagram displayed in Figure 1(a) as an example, the corresponding QCD amplitude can be written as [12]

$$\begin{aligned} \Pi^{(1)}_{\mu,\, weak} &= \frac{g_s^2 C_F}{2\,(\bar{n} \cdot p - \omega)} \int \frac{d^D l}{(2\pi)^D} \frac{1}{[(p-k+l)^2 + i0][(m_b v + l)^2 - m_b^2 + i0][l^2 + i0]} \\ &\quad \bar{d}(k)\, \not{n}\, \gamma_5\, \not{\bar{n}}\, \gamma_\rho\, (\not{p} - \not{k} + \not{l})\, \gamma_\mu\, (m_b\, \not{v} + \not{l} + m_b)\, \gamma^\rho\, b(v)\,. \end{aligned} \tag{8}$$

Employing the power counting scheme (2) one can readily identify that the leading power contributions to $\Pi^{(1)}_{\mu,\, weak}$ come from the hard, hard-collinear and soft regions. Expanding the QCD amplitude $\Pi^{(1)}_{\mu,\, weak}$ in terms of the powers of Λ/m_b in the soft region and keeping the leading-order terms yield

$$\begin{aligned} \Pi^{(1),s}_{\mu,\, weak} &= \frac{g_s^2 C_F}{2\,(\bar{n} \cdot p - \omega)} \int \frac{d^D l}{(2\pi)^D} \frac{1}{[\bar{n} \cdot (p-k+l) + i0][v \cdot l + i0][l^2 + i0]} \\ &\quad \bar{d}(k)\, \not{n}\, \gamma_5\, \not{\bar{n}}\, \gamma_\mu\, b(v)\,. \end{aligned} \tag{9}$$

On the other hand, the one-loop correction to the B-meson DA can be computed with the Wilson-line Feynman rules

$$\begin{aligned}\Phi^{\alpha\beta,(1)}_{b\bar{d},a}(\omega,\omega') &= i g_s^2 C_F \int \frac{d^D l}{(2\pi)^D} \frac{1}{[\bar{n}\cdot l + i0][v\cdot l + i0][l^2 + i0]} \\ &\times[\delta(\omega' - \omega - \bar{n}\cdot l) - \delta(\omega' - \omega)]\,[\bar{d}(k)]_\alpha\,[b(v)]_\beta\,, \end{aligned} \tag{10}$$

from which the infrared subtraction term can be deduced as $\Phi^{(1)}_{b\bar{d},a} \otimes T^{(0)} = \Pi^{(1),s}_{\mu,weak}$, verifying QCD factorization for the vacuum-to-B-meson correlation function at one loop diagrammatically. By proceeding in a similar way, the hard-colliner contribution from the weak vertex diagram can be extracted as follows:

$$\begin{aligned}\Pi^{(1),hc}_{\mu,weak} &= \frac{g_s^2 C_F}{2(\bar{n}\cdot p - \omega)} \int \frac{d^D l}{(2\pi)^D} \frac{2\, m_b\, n\cdot(p+l)}{[n\cdot(p+l)\,\bar{n}\cdot(p-k+l) + l_\perp^2 + i0][m_b\, n\cdot l + i0][l^2 + i0]} \\ &\bar{d}(k)\ \not{n}\,\gamma_5\ \not{\bar{n}}\,\gamma_\mu\, b(p_b)\,. \end{aligned} \tag{11}$$

which can be further computed with the loop integrals collected in the Appendix A of [12]. Finally, expanding the QCD amplitude (8) in the hard region gives rise to

$$\begin{aligned}\Pi^{(1),h}_{\mu,\,weak} &= \frac{\alpha_s C_F}{4\pi}\, \tilde{f}_B(\mu)\, m_B\, \frac{\phi^-_{b\bar{d}}(\omega)}{\bar{n}\cdot p - \omega} \left\{ \bar{n}_\mu \left[\frac{1}{\epsilon^2} + \frac{1}{\epsilon}\left(2\ln\frac{\mu}{n\cdot p} + 1\right) + 2\ln^2\frac{\mu}{n\cdot p} + 2\ln\frac{\mu}{m_b} \right.\right. \\ &\left.\left. - \ln^2 r - 2\,\mathrm{Li}_2\left(-\frac{\bar{r}}{r}\right) + \frac{2-r}{r-1}\ln r + \frac{\pi^2}{12} + 3 \right] + n_\mu \left[\frac{1}{r-1}\left(1 + \frac{r}{\bar{r}}\ln r\right) \right] \right\}, \end{aligned} \tag{12}$$

with $r = n\cdot p/m_b$ and $\bar{r} = 1 - r$.

Along the same vein, one can evaluate the leading power contributions to the remaining diagrams in Figure 1 and the resulting factorization formulae for the correlation function are given by

$$\begin{aligned}\Pi &= \tilde{f}_B(\mu)\, m_B \sum_{k=\pm} C^{(k)}(n\cdot p,\mu) \int_0^\infty \frac{d\omega}{\omega - \bar{n}\cdot p}\, J^{(k)}\left(\frac{\mu^2}{n\cdot p\,\omega}, \frac{\omega}{\bar{n}\cdot p}\right) \phi_B^{(k)}(\omega,\mu)\,, \\ \widetilde{\Pi} &= \tilde{f}_B(\mu)\, m_B \sum_{k=\pm} \widetilde{C}^{(k)}(n\cdot p,\mu) \int_0^\infty \frac{d\omega}{\omega - \bar{n}\cdot p}\, \widetilde{J}^{(k)}\left(\frac{\mu^2}{n\cdot p\,\omega}, \frac{\omega}{\bar{n}\cdot p}\right) \phi_B^{(k)}(\omega,\mu)\,, \end{aligned} \tag{13}$$

where the hard and hard-collinear functions at one loop read [12]

$$\begin{aligned}C^{(+)} &= \tilde{C}^{(+)} = 1, \qquad C^{(-)} = \frac{\alpha_s C_F}{4\pi}\frac{1}{\bar{r}}\left[\frac{r}{\bar{r}}\ln r + 1\right], \\ \tilde{C}^{(-)} &= 1 - \frac{\alpha_s C_F}{4\pi}\left[2\ln^2\frac{\mu}{n\cdot p} + 5\ln\frac{\mu}{m_b} - \ln^2 r - 2\,\mathrm{Li}_2\left(-\frac{\bar{r}}{r}\right) + \frac{2-r}{r-1}\ln r + \frac{\pi^2}{12} + 5\right], \end{aligned} \tag{14}$$

and

$$\begin{aligned}J^{(+)} &= \frac{1}{r}\tilde{J}^{(+)} = \frac{\alpha_s C_F}{4\pi}\left(1 - \frac{\bar{n}\cdot p}{\omega}\right)\ln\left(1 - \frac{\omega}{\bar{n}\cdot p}\right), \qquad J^{(-)} = 1\,, \\ \tilde{J}^{(-)} &= 1 + \frac{\alpha_s C_F}{4\pi}\left[\ln^2\frac{\mu^2}{n\cdot p(\omega - \bar{n}\cdot p)} - 2\ln\frac{\bar{n}\cdot p - \omega}{\bar{n}\cdot p}\ln\frac{\mu^2}{n\cdot p(\omega - \bar{n}\cdot p)} \right. \\ &\left. - \ln^2\frac{\bar{n}\cdot p - \omega}{\bar{n}\cdot p} - \left(1 + \frac{2\bar{n}\cdot p}{\omega}\right)\ln\frac{\bar{n}\cdot p - \omega}{\bar{n}\cdot p} - \frac{\pi^2}{6} - 1\right]. \end{aligned} \tag{15}$$

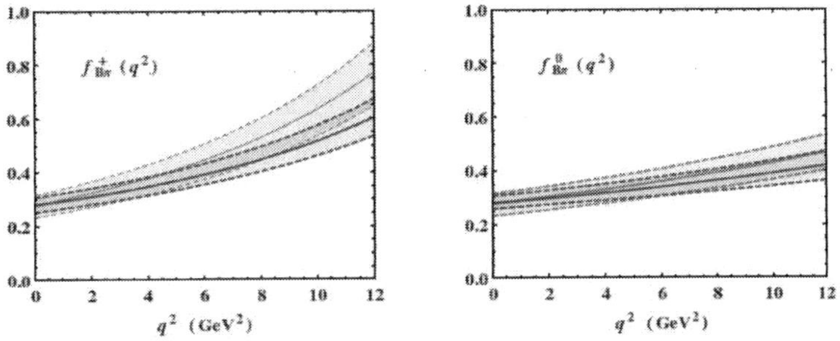

Figure 3. The momentum-transfer dependence of the $B \to \pi$ form factors $f_{B\pi}^{+,0}(q^2)$ from the NLL resummation improved LCSR (17) with the B-meson DA (pink curves) and from the NLO QCD sum rules with the pion DA (blue curves).

The hard coefficients presented in (14) are in compatible with the perturbative matching coefficients of the weak current $\bar{q}\,\gamma_\mu\, b$ from QCD to soft-collinear effective theory (SCET) [15], and the hard-collinear functions displayed in (15) coincide with the corresponding jet functions computed with the SCET Feynman rules [11].

It is evident that there is no common value of μ that can avoid the parametrically large logarithms of order $\ln(m_b/\Lambda)$ in the hard functions, the jet functions and the B-meson DA. The summation of these large logarithms in the hard coefficient functions can be accomplished by solving the following renormalization-group equations

$$\begin{aligned} \frac{d}{d\ln\mu}\tilde{C}^{(-)}(n\cdot p,\mu) &= \left[-\Gamma_{\rm cusp}(\alpha_s)\ln\frac{\mu}{n\cdot p}+\gamma(\alpha_s)\right]\tilde{C}^{(-)}(n\cdot p,\mu)\,, \\ \frac{d}{d\ln\mu}\,\tilde{f}_B(\mu) &= \tilde{\gamma}(\alpha_s)\,\tilde{f}_B(\mu)\,, \end{aligned} \tag{16}$$

where the three-loop cusp anomalous dimension and the two-loop γ ($\tilde{\gamma}$) are needed to achieve the next-to-leading-logarithmic (NLL) resummation. Applying the standard procedure for the construction of QCD sum rules leads to

$$\begin{aligned} & f_\pi\, e^{-m_\pi^2/(n\cdot p\,\omega_M)}\left\{\frac{n\cdot p}{m_B}\,f_{B\pi}^{+}(q^2)\,,\;\; f_{B\pi}^{0}(q^2)\right\} \\ &= \left[U_2(\mu_{h2},\mu)\,\tilde{f}_B(\mu_{h2})\right]\int_0^{\omega_s} d\omega'\;\, e^{-\omega'/\omega_M}\Big[\left[U_1(n\cdot p,\mu_{h1},\mu)\,\widetilde{C}^{(-)}(n\cdot p,\mu_{h1})\right]\,\phi_{B,\rm eff}^{-}(\omega',\mu) \\ &\quad + r\,\phi_{B,\rm eff}^{+}(\omega',\mu)\pm\frac{n\cdot p-m_B}{m_B}\,\left(\phi_{B,\rm eff}^{+}(\omega',\mu)+C^{(-)}(n\cdot p,\mu)\,\phi_B^{-}(\omega',\mu)\right)\Big]\,, \end{aligned} \tag{17}$$

where the explicit expressions of $\phi_{B,\rm eff}^{\pm}(\omega',\mu)$ can be found in [12].

Having at our disposal the NLL resummation improved LCSR presented in (17), it is straightforward to plot the q^2 dependence of the $B \to \pi$ form factors $f_{B\pi}^{+,0}(q^2)$ in Figure 3 where the theoretical predictions from the sum rules with the pion DA [4] are also shown for a comparison. The observed discrepancy of the q^2 shape for the vector form factor $f_{B\pi}^{+}(q^2)$ predicted from the two distinct sum rules could be attributed to the systematical uncertainties generated by the different parton-hadron duality ansatz and to the yet unaccounted higher order/power corrections in the perturbative calculations of the corresponding correlation functions. Expressing the differential branching fraction of $B \to \pi \mu \nu_\mu$ in terms of the form factor $f^+_{B\pi}(q^2)$

$$\frac{d\Gamma}{dq^2}(B \to \pi \mu \nu_\mu) = \frac{G_F^2 |V_{ub}|^2}{24\pi^3} |\vec{p}_\pi|^3 \, |f^+_{B\pi}(q^2)|^2 . \tag{18}$$

and employing the experimental measurements for the integrated decay rate [16, 17] lead to

$$|V_{ub}| = \left(3.05^{+0.54}_{-0.38}|_{\rm th.} \pm 0.09|_{\rm exp.}\right) \times 10^{-3} , \tag{19}$$

which is consistent with the exclusive determination of $|V_{ub}|$ from the leptonic $B \to \tau \nu_\tau$ decay [18].

3 Conclusion

To summarize, we demonstrated QCD factorization for the vacuum-to-B-meson correlation function used in the construction of the $B \to \pi$ form factors at one-loop accuracy explicitly with the method of regions and achieved the resummation of large logarithms in the short-distance functions at NLL by solving the renormalization-group equations in the momentum space. Phenomenological applications of the newly derived sum rules with the B-meson DA were also discussed in brief, focusing on the extraction of exclusive $|V_{ub}|$ from the semi-leptonic $B \to \pi \mu \nu_\mu$ decay.

References

[1] M. Beneke and T. Feldmann, Nucl. Phys. B **592** (2001) 3 [hep-ph/0008255].
[2] P. Ball and R. Zwicky, Phys. Rev. D **71**, 014015 (2005) [hep-ph/0406232].
[3] G. Duplancic, A. Khodjamirian, T. Mannel, B. Melic and N. Offen, JHEP **0804**, 014 (2008) [arXiv:0801.1796 [hep-ph]].
[4] A. Khodjamirian, T. Mannel, N. Offen and Y.-M. Wang, Phys. Rev. D **83**, 094031 (2011) [arXiv:1103.2655 [hep-ph]].
[5] Y. M. Wang, Y. Li and C. D. Lü, Eur. Phys. J. C **59**, 861 (2009) [arXiv:0804.0648 [hep-ph]].
[6] H. n. Li, Y. L. Shen and Y. M. Wang, Phys. Rev. D **85**, 074004 (2012) [arXiv:1201.5066 [hep-ph]].
[7] C. D. Lü, Y. M. Wang, H. Zou, A. Ali and G. Kramer, Phys. Rev. D **80**, 034011 (2009) [arXiv:0906.1479 [hep-ph]].
[8] A. Khodjamirian, T. Mannel and N. Offen, Phys. Lett. B **620**, 52 (2005) [hep-ph/0504091].
[9] A. Khodjamirian, T. Mannel and N. Offen, Phys. Rev. D **75**, 054013 (2007) [hep-ph/0611193].
[10] F. De Fazio, T. Feldmann and T. Hurth, Nucl. Phys. B **733**, 1 (2006) [hep-ph/0504088].
[11] F. De Fazio, T. Feldmann and T. Hurth, JHEP **0802**, 031 (2008) [arXiv:0711.3999 [hep-ph]].
[12] Y. M. Wang and Y. L. Shen, Nucl. Phys. B **898**, 563 (2015) [arXiv:1506.00667 [hep-ph]].
[13] Y. M. Wang and Y. L. Shen, JHEP **1602**, 179 (2016) [arXiv:1511.09036 [hep-ph]].
[14] Y. M. Wang, arXiv:1606.03080 [hep-ph].
[15] C. W. Bauer, S. Fleming, D. Pirjol and I. W. Stewart, Phys. Rev. D **63**, 114020 (2001) [hep-ph/0011336].
[16] J. P. Lees *et al.* [BaBar Collaboration], Phys. Rev. D **86**, 092004 (2012) [arXiv:1208.1253 [hep-ex]].
[17] A. Sibidanov *et al.* [Belle Collaboration], Phys. Rev. D **88**, 032005 (2013) [arXiv:1306.2781 [hep-ex]].
[18] B. Kronenbitter *et al.* [Belle Collaboration], Phys. Rev. D **92**, 051102 (2015) [arXiv:1503.05613 [hep-ex]].

EPJ Web of Conferences **129**, 00028 (2016) DOI: 10.1051/epjconf/201612900028
QCD@Work 2016

QCD sum rules for 70-plet baryons

Valeri Zamiralov[1,★]

[1] *Skobeltsyn Institute of Nuclear Physics, Lomonosov Moscow State University, 119991 Moscow, Russia*

Abstract. QCD sum rules for the 70-plet baryons are proposed, using unitary pattern of these rules. As it was previously shown, all correlators of QCD sum rules for the 56-plet depend only on two generalized functions of the F and D types, characteristic for the unitary symmetry. The same is shown to be valid for the 70-plet baryons.

1 Introduction

Magnetic moments of the octet $1/2^-$ baryons attracted some attention recently in quark models as well as in the QCD sum rule formalism. In the quark model these baryons are assigned usually to the 70-plet representation of SU(6), in contrast to the $1/2^+$ baryons of the 56-plet (see [1, 2] and the references therein).

Within the QCD sum rules formalism one treats mostly $1/2^+$ baryons starting from pioneering works of Ioffe et al. [3]. The baryon interpolating currents reflect baryon wave functions of the non-relativistic 56-plet quark model (NRQM). The same reasoning is maintained also for the $1/2^-$ baryons (see, e.g., ref.[4]).

In a series of papers we have shown that QCD sum rules for couplings and magnetic moments of the octet $1/2^+$ baryons reproduce unitary symmetry results in terms of F and D quantities ([5] and references therein). In fact, any QCD sum rule for magnetic moments of the octet baryons exhibits the unitary structure. We take a little old-fashioned QCD sum rule [6] as a simple example. If the correlator is

$$\Pi_B = i\int d^4x e^{ipx}\langle 0|T\{\eta_B(x)\bar{\eta}_B(0)\}|0\rangle_\gamma \ , \tag{1}$$

T being time ordering operator, γ means external electromagnetic field while η_B is interpolating current with the quantum numbers of the B baryon, then, say, for Σ^0 it could be written (we put to shorten formulae $m_s = 0$ disregarding differences between vacuum expectation values of various flavors) as:

$$\begin{aligned} \frac{1}{4}\tilde{\beta}_\Sigma^2\Big[\mu(\Sigma^0) + AM^2\Big]e^{-m_\Sigma^2/M^2} &= (e_u + e_d)\Big\{\frac{M^6}{4L^{4/9}} - \frac{a^2L^{4/9}}{144}[2-(\kappa-2\xi)] \\ &- \frac{\chi a^2}{24L^{4/27}}(M^2 - \frac{m_0^2}{8L^{4/9}}) + \frac{bM^2}{96L^{4/9}}\Big\} + e_s\Big\{-\frac{a^2L^{4/9}}{24} + \frac{bM^2}{192L^{4/9}}\Big\} \\ &= e_u\Pi_1(u,d,s;M^2) + e_d\Pi_1(d,u,s;M^2) + e_s\Pi_2(u,d,s;M^2). \end{aligned} \tag{2}$$

Here M^2 is the Borel parameter, the vev's are $a = -(2\pi)^2\langle\bar{q}q\rangle$=0.45 GeV^3, $b = \langle g_s^2G^2\rangle$=0.5 GeV^4, m_0^2 = 0.8 GeV^2, $(\kappa - 2\xi)$=1.1, $\chi = -3.3$, while L=$\ln(M^2/\Lambda_{QCD}^2)/\ln(\mu^2/\Lambda_{QCD}^2)$ with Λ_{QCD}=100

★e-mail: zamir@depni.sinp.msu.ru

MeV and normalization point μ=500 MeV. In the limit of degenerated masses of quarks and vacuum expectation values (vev) it reduces to the relation [7]:

$$\mu_{\Sigma^0} = (e_u + e_d)F + e_s(D - F). \tag{3}$$

Note that this relation goes to the NRQM formula with $F = 2/3$ and $D = 1$ and $e_q \rightarrow \mu_q$, and from the other side it reproduces the unitary formula by putting values of e_q's for all octet baryons but Λ. The Λ quantities are reproduced easily with our relation [8].

2 NRQM and quark-diquark model for 70-plet

With the 70-plet octet baryons we loose this correspondence. QCD sum rules remain essentially the same [9, 10] (only with γ_5 inside, to change baryon parity) while NRQM gives $\mu^{70}_{\Sigma^0} = 1/3(\mu_u + \mu_d + \mu_s)$ (see [1] and references therein). We propose a way to arrive at the desired QCD sum rules in accord with the NRQM limit [11] with the help of quark-diquark model which has led previously to Eq.(3). Let us take quark wave functions of the 70-plet baryon [1] (and we take Σ^0) as (subindices 1,2 mean $\uparrow$,$\downarrow$, correspondingly):

$$|\Sigma^{*0}\rangle = |\Sigma^0\rangle_i + |\Sigma^0\rangle_{ii};$$

$$|\Sigma^0\rangle_i = \frac{1}{2\sqrt{3}}|(u_1s_1d_2 + d_1s_1u_2 - s_1u_1d_2 - s_1d_1u_2 + u_1d_2s_1 + d_1u_2s_1$$
$$-u_1s_2d_1 - d_1s_2u_1 + s_2u_1d_1 + s_2d_1u_1 - u_2d_1s_1 - d_2u_1s_1)\psi'(0^+)\rangle;$$

$$|\Sigma^0\rangle_{ii} = \frac{1}{6}|(u_1s_1d_2 + d_1s_1u_2 + s_1u_1d_2 + s_1d_1u_2 + u_1d_2s_1 + d_1u_2s_1 \tag{4}$$
$$+u_1s_2d_1 + d_1s_2u_1 + s_2u_1d_1 + s_2d_1u_1 + u_2d_1s_1 + d_2u_1s_1$$
$$-2u_1d_1s_2 - 2d_1u_1s_2 - 2s_1u_2d_1 - 2s_1d_2u_1 - 2u_2s_1d_1 - 2d_2s_1u_1)\psi''(0^+)\rangle;$$

($\psi'(0^+)$ and $\psi''(0^+)$ are spatial wave functions [1]). To obtain Λ wave functions, and also to control the formulae, we use relations with the exchange of quarks:

$$|\Sigma^{*0}\rangle|_{ds} + |\Sigma^{*0}\rangle|_{us} = -|\Sigma^{*0}\rangle, \quad |\Sigma^{*0}\rangle|ds - |\Sigma^{*0}\rangle|_{us} = -\sqrt{3}|\Lambda^*\rangle.$$

We follow previous reasoning of our quark-diquark model assuming that photon distinguishes between the diquark of (quasi)similar quarks (qq) and single quark Q [7]. That is, we introduce 4 matrix elements of the new operator $\hat{\omega}_q$:

$$\langle q_\uparrow q_\uparrow, Q_\downarrow|\hat{\omega}_q|q_\uparrow q_\uparrow, Q_\downarrow\rangle = w_{11}, \qquad \langle q_\uparrow q_\downarrow, Q_\uparrow|\hat{\omega}_q|q_\uparrow q_\downarrow, Q_\uparrow\rangle = w_{12},$$
$$\langle q_\uparrow q_\uparrow, Q_\downarrow|\hat{\omega}_Q|q_\uparrow q_\uparrow, Q_\downarrow\rangle = v_{11}, \qquad \langle q_\uparrow q_\downarrow, Q_\uparrow|\hat{\omega}_Q|q_\uparrow q_\downarrow, Q_\uparrow\rangle = v_{12}. \tag{5}$$

For the 56-plet one obtains for $\Sigma^0(ud, s)$ with the modified magnetic moment operator $e_q\hat{\omega}_q\sigma^q_3$ (disregarding $\hat{\omega}_q$ one just returns to NRQM):

$$\mu_{\Sigma^0} = (1/6)[4(e_u + e_d)w_{11} + e_s(4v_{11} - 2v_{12})], \tag{6}$$

which, with $w_{11} = 3/2F$, $w_{12} = D$, $v_{11} = D$, $v_{12} = 3F - D$ [7], yields Eq.(3). Instead, for the 70-plet one obtains:

$$\mu^{70}_{\Sigma^0} = 1/3[(e_u + e_d)w_{11} - e_s(v_{11} - 2v_{12})] = (e_u + e_d + 4e_s)\frac{1}{2}F - e_sD. \tag{7}$$

But now, with $F = 2/3$ and $D = 1$, we return to the 70-plet NRQM result $\mu_{\Sigma^0} = 1/3(\mu_u + \mu_d + \mu_s)$, while putting values of e_q's we arrive at magnetic moments of 70-plet baryons in terms of F and D constants [11]:

$$\mu_{N^{*+}} = \mu_{\Sigma^{*+}} = \frac{1}{3}D, \qquad \mu_{\Sigma^{*-}} = \mu_{\Xi^{*-}} = (-F + \frac{1}{3}D),$$
$$\mu_{N^{*0}} = \mu_{\Xi^{*0}} = -2\mu_{\Sigma^{*0}} = 2\mu_{\Lambda^{*0}} = (F - \frac{2}{3}D). \tag{8}$$

We also note that the transition magnetic moment is equal to zero for the 70-plet, $\langle\Lambda^*|\hat{\mu}^{70}|\Sigma^{*0}\rangle=0$. It can be proved also with the relation in ref.[8].

3 QCD sum rules for 70-plet magnetic moments

We take it as a base for the construction of QCD sum rules for the magnetic moments (here for $1/2^+$),

$$\Pi^{70}(u,d,s) = e_u\tilde{\Pi}_1(u,d,s) + e_d\tilde{\Pi}_1(d,u,s) + e_s\tilde{\Pi}_2(u,d,s),$$

with $\tilde{\Pi}_1(u,d,s) = 1/2\Pi_1(u,d,s)$, $\tilde{\Pi}_2 = \Pi_1(d,u,s)+\Pi_2(u,d,s)$. These formulae yield correct NRQM limit and maintain F, D pattern of the QCD sum rules. We proceed with these QCD sum rules in more detail, and take as an example the QCD sum rules for the baryon magnetic moments [6]. The Eq.(2) according to our formulae Eq.(7) transforms itself for the 70-plet into:

$$\begin{aligned}\frac{1}{4}\tilde{\beta}^2_{\Sigma^*}[\mu(\Sigma^{*0}) + A'M^2]e^{-m^2_{\Sigma^*}/M^2} \quad = \quad & \frac{1}{2}(e_u + e_d + 4e_s) \\ & \Big\{\frac{M^6}{4L^{4/9}} - \frac{a^2L^{4/9}}{144}[2-(\kappa-2\xi)] - \frac{\chi a^2}{24L^{4/27}}\left(M^2 - \frac{m_0^2}{8L^{4/9}}\right) \\ + \quad & \frac{bM^2}{96L^{4/9}}\Big\} - e_s\Big\{-\frac{a^2L^{4/9}}{24} + \frac{bM^2}{192L^{4/9}}\Big\}.\end{aligned} \tag{9}$$

It is seen that sum rules (2) and (9) differ drastically. Model predictions for both sum rules and also some results for $1/2^-$ baryons are given in Table 1.

4 Conclusions

As a result we have shown the way to construct QCD sum rules for the magnetic moments of the 70-plet octet baryons which have the correct NRQM limit. Octet baryons in the 70-plet are analyzed in a way similar to those of the 56-plet. Magnetic moments are written in terms of the D and F quantities characteristic for octet coupling. The main formulae for the magnetic moments are written in such a way as to obtain the NRQM results as well as unitary symmetry ones. In NRQM in the 70-plet octet Tansition magnetic moment is shown to be zero. Borel QCD sum rules are constructed for the magnetic moments of the 70-plet octet. Comparison with some other results in the field are made.

I am grateful to T.M. Aliev for partecipating in part of this work, and to the Workshop Organizers for the warm hospitality.

Table 1. Magnetic moments of 70-plet and 56-plet $1/2^{\pm}$ baryons.

$\mu(B)$	56-plet $1/2^+$ [6]	This Work Eq.(9)	$1/2^-$[12] Lattice	NCQM [2]	$1/2^-$ [10]	Eq.(8) F=2.5 D=4.5
$\mu(p)$	2.72	0.83	-1.8	1.894	1.4	1.5
$\mu(n)$	-1.65	0.22	-1.0	-1.284	-0.54	-0.5
$\mu(\Sigma^+)$	2.52	0.70	-0.6	1.814	1.8	1.5
$\mu(\Sigma^-)$	-1.13	-1.13	1.0	-0.689	-1.1	-1.0
$\mu(\Sigma^0)$		0.11	0.1	0.820	0.4	0.25
$\mu(\Lambda)$	-0.50	-0.11	-0.1		-0.26	-0.25
$\mu(\Xi^0)$	-0.89	-0.43	-0.5	-0.990	-0.55	-0.5
$\mu(\Xi^-)$	-1.18	-1.18	0.8	-0.315	-1.2	-1.0

References

[1] H. Dahiya and M. Gupta, Phys. Rev. D 67, 114015 (2003).
[2] N. Sharma, A. Martinez Torres, K.P. Khemchandani and H. Dahiya, Eur. Phys. J. A49, 11 (2013).
[3] B.L. Ioffe, Nucl. Phys. B188, 317 (1981).
[4] D. Jido, N. Kodama and M. Oka, Phys. Rev. D54, 4532 (1996).
[5] T.M. Aliev, A. Ozpineci, M. Savci, and V. Zamiralov, Phys. Rev. D80, 01610105 (2009).
[6] C. B. Chiu, J. Pasupathy, and S.J. Wilson, Phys. Rev. D33, 1961 (1986).
[7] L. Gelmi, V.S. Zamiralov, and S.N. Lepshokov, MSU Bulletin. Ser.3. Phys. Astron. 2, 33 (1989).
[8] A. Ozpineci, S. Yakovlev, and V. Zamiralov, At. Nucl. Phys. 68, 304 (2005); Mod. Rev. Lett. 20, 243 (2005).
[9] Y. Kondo, O. Morimatsu, T. Nishikawa, and Y. Kanada-En'yo, Phys. Rev. D 75, 034010 (2007).
[10] T.M. Aliev, and M. Savci, arXiv:1402.4609, 2014.
[11] T.M. Aliev and V.S. Zamiralov, Adv. High Energy Phys. 2015 (2015) 406875.
[12] F.X. Lee and A. Alexandru, Proc. Sci. LATTICE2010, 148 (2010).

ALICE @ LHC: Status and Highlights

Alberica Toia[1,2,*]

[1] *Goethe University of Frankfurt*
[2] *GSI Helmholzzentrum fuer Schwerionenforschung*

Abstract. ALICE is the dedicated heavy-ion experiment at the Large Hadron Collider at CERN. After a two-year long shutdown, the LHC restarted its physics programme in June 2015 with proton-proton collisions at $\sqrt{s}$ = 13 TeV and Pb-Pb collisions at $\sqrt{s_{\mathrm{NN}}}$ = 5.02 TeV, the highest centre-of-mass energy ever reached in laboratory. Recent results and future perspective for ALICE will be presented.

1 Introduction

The harvest of the results collected in heavy ion collisions during the first Run of LHC operations demonstrate the formation of a deconfined phase of matter, the quark-gluon plasma (QGP), at high temperature and energy density [1]. The higher centre-of-mass energy achieved in Run 2, as well as the increased luminosity, offer the possibility to further constrain the fundamental properties of the QGP. This paper will briefly review some of the most recent results of the ALICE experiment and their impact in the characterization of the QGP.

2 Measuring the collectivity of the system

2.1 The energy density

The charged particle multiplicity density is a measure of the initial energy density and provides information on the dynamics of the soft particle production and its relation to the initial collision geometry. ALICE measured the pseudorapidity density of charged particles measured at central rapidity ($|\eta| < 0.5$), for pp and Pb–Pb collisions at the LHC top energies, divided by the average number of participating nucleon pairs, $\langle N_{\mathrm{part}}\rangle/2$, calculated with the MC-Glauber [2]. It is shown in Fig.1 together with a compilation of results at different collider energies. The new measurement at an increased energy allows to probe the increased importance of hard scatterings proportional to the number of binary collisions . The data confirms the trend established by lower energy data of a much stronger rise than the one measured for pp and even for p–Pb collisions, indicating that the particle production in AA is not solely related to the multiple collisions undergone by the participants. The multiplicity shows a strong dependence on the centrality (Fig.1 right) and a smooth trend towards the value measured in minimum bias p–Pb collisions. The centrality dependence of $\mathrm{d}N_{\mathrm{ch}}/\mathrm{d}\eta$ is nearly identical to that measured at $\sqrt{s_{\mathrm{NN}}}$ = 2.76 TeV, which showed to be very similar to those measured at RHIC, pointing to a similar mechanism of particle production independent of energy.

*e-mail: alberica.toia@cern.ch

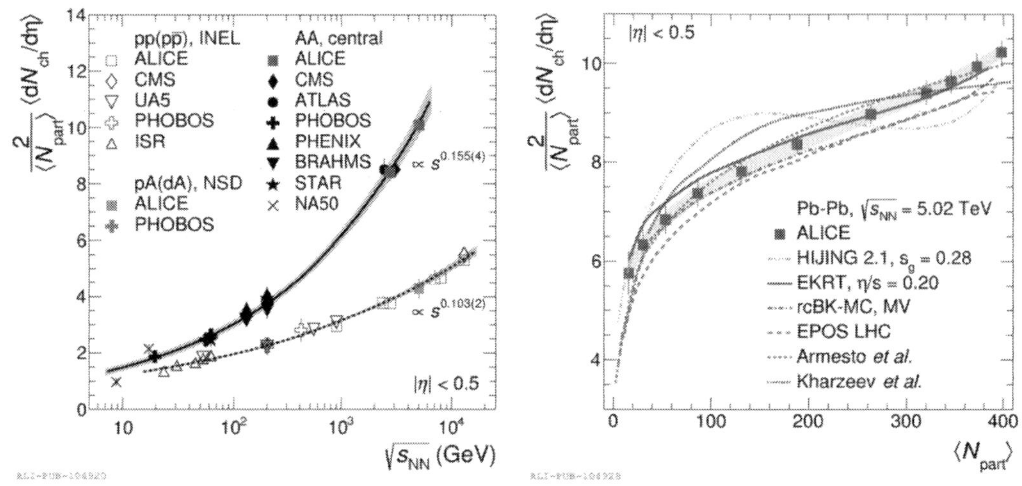

Figure 1. Left: Values of $\frac{2}{\langle N_{\rm part}\rangle}\langle dN_{\rm ch}/d\eta\rangle$ for pp, p$\bar{\rm p}$ and central AA collisions as a function of $\sqrt{s_{\rm NN}}$. The s-dependences of the AA and pp (p$\bar{\rm p}$) collision data are well described by the functions $s_{\rm NN}^{0.155}$ (solid line) and $s_{\rm NN}^{0.103}$ (dashed line), respectively. Right: The $\frac{2}{\langle N_{\rm part}\rangle}\langle dN_{\rm ch}/d\eta\rangle$for Pb–Pb collisions at $\sqrt{s_{\rm NN}} = 5.02$ TeV in the centrality range 0–80%, as a function of $N_{\rm part}$ in each centrality class.

2.2 The viscosity

An important measurement to test the collectivity of the system is the anisotropic flow, which assess if and to which extend the initial spatial anisotropy of the overlap region of colliding nuclei is translated into an anisotropy in momentum space via interactions of produced particles. It is therefore sensitive to the initial collision geometry and transport mechanism and therefore provides a measurement of the collectivity of the medium.

Figure 2 presents the centrality dependence of v_2, v_3 and v_4, integrated over $0.2 < p_T < 5.0$ GeV/c, for 2.76 and 5.02 TeV Pb–Pb collision, with the corresponding ratios on the lower panels. The data is compared to the predictions from hydrodynamic models, which combine the changes in initial spatial anisotropy and the hydrodynamic response [3]. The predictions are compatible with the measured anisotropic flow coefficients. At the same time, a different hydrodynamic calculation, which employs both constant and temperature dependent viscosity can also describe the increase in anisotropic flow measurements of v_2. In particular, among the different scenarios, the increase from the two energies, rather moderate for v_2 and v_3 and more pronounced for v_4, but independent on centrality, seem to favor a constant viscosity going from 2.76 to 5.02 TeV Pb–Pb collisions.

The measurements of p_T-differential flow are more sensitive to initial conditions and viscosity, and are expected to provide important information to constrain further details of the theoretical calculations, e.g. determination of radial flow and freeze-out conditions. Figure 2 presents the flow coefficient v2 as a function of transverse momentum for the 0-5% and 30-40% centrality classes. The results are similar to the corresponding measurements at 2.76 TeV, indicating that the increase of the p_T-integrated flow can be attributed to an increase of mean transverse momentum. While in the mid-central class v2 remains higher than v3, with no crossing of the different order flow coefficients, in the top-most central events v3 becomes larger than v2 at p_T >2 GeV/c, while v4 is compatible with v2, within uncertainties.

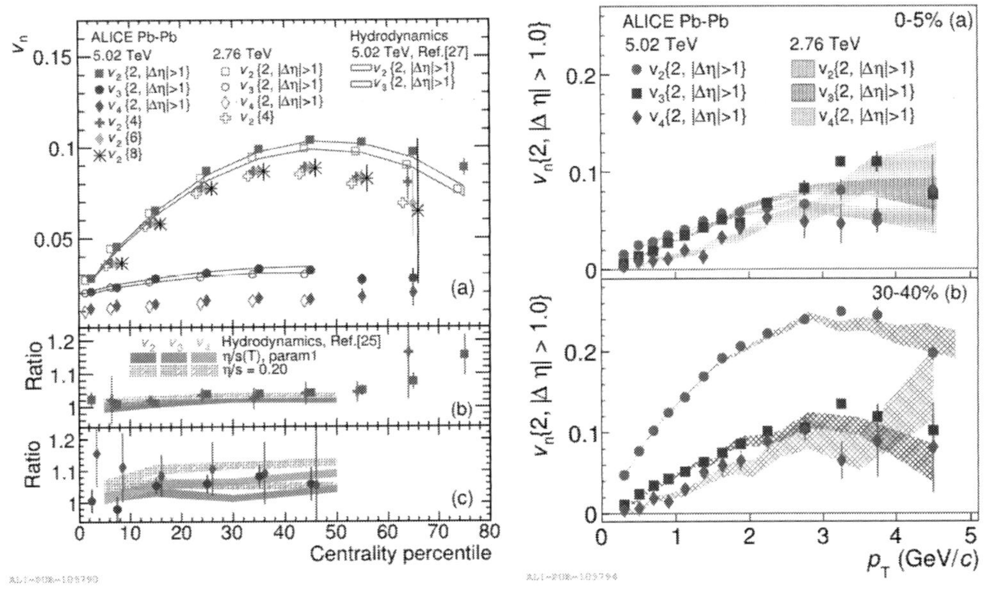

Figure 2. Left: Anisotropic flow v_n integrated over $0.2 < p_T < 5.0$ GeV/c, as a function of event centrality. The ratios between Pb-Pb collisions at 5.02 TeV and 2.76 TeV, are presented in the lower panels. Various hydrodynamic calculations are also presented. Right: vn(p_T) for 0-5% and 30-40% centrality classes.

3 Understanding the energy loss

Hard processes are those which occur with high momentum transfer, where the production rate in pp collisions can be calculated with pQCD techniques using the universality of PDF and FF and the factorization theorem. In AA collisions the produciton rate is expected to scale with the number of elementary nucleon-nucleon collisions N_{coll}, typically calculated with Glauber models. Any deviation from this scaling is experimentally quantified by the nuclear modification factor, R_{AA}, defined as the ratio of the yield in Pb–Pb to that in pp, scaled by N_{coll}. Experimental observables connected to hard processes are hadrons with high p_T, jets, or hadrons from open heavy flavour.

3.1 The energy loss for leading hadrons

ALICE has measured the transverse momentum spectra of primary charged particles at various collision energies. In Run 2 the top LHC energy of 13 TeV was reached in pp and 5.02 TeV for Pb–Pb collsions. Figure 3 presents the measured p_T spectrum and its comparison with calculations with various MC models, i.e. PYTHIA 6 (Perugia-2011), PYTHIA 8 (Monash-2013) and EPOS LHC [4]. These event generators, benefitting from the tuning performed on the LHC data in Run 1, describe the p_T spectrum reasonably well, although not in detail. As expected, and reasonably reproduced by MC models, the spectrum is significantly harder at $\sqrt{s} = 13$ TeV than at $\sqrt{s} = 7$ TeV. We have also measured the p_T spectra in Pb–Pb collisions, for various centrality classes and we have compared it to the spectrum measured at the reference energy of 5.02 TeV in pp scaled by the corresponding number of binary collisions calculated for each centrality class with a Glauber Monte Carlo model. We want to point out the reconstruction and track selection is very much improved with respect to the Run 1 analyses leading to a reduced systematic uncertainties. Moreover a much larger statistics is recorded

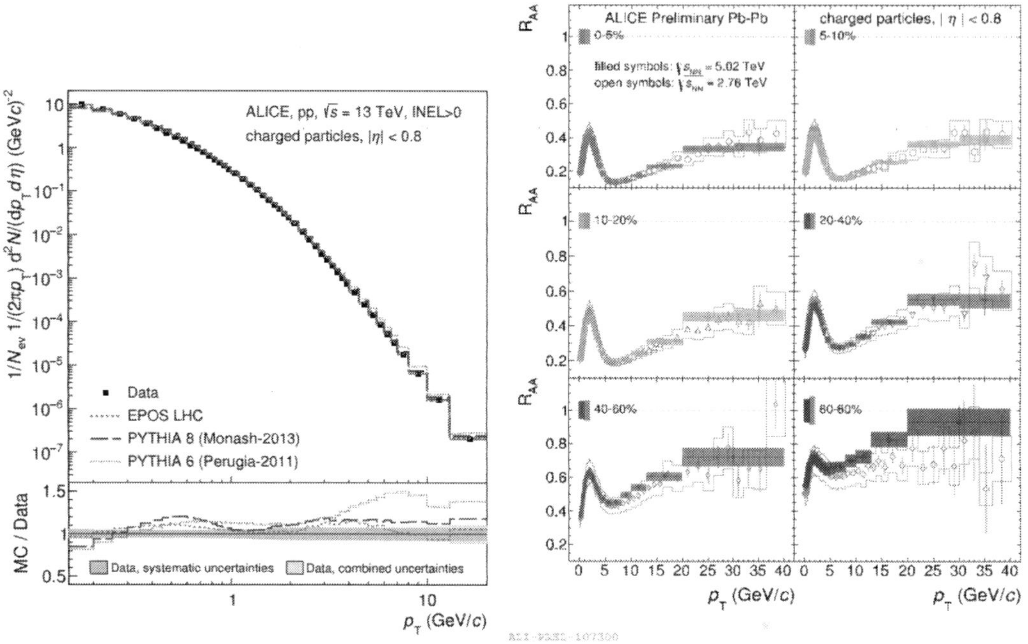

Figure 3. Left:Invariant charged-particle yield as a function of p_T for ppcollisions at $\sqrt{s} = 13$ TeV, compared to Monte Carlo calculations. Right: Nuclear modifications calculated for different classes of centrality. The filled symbols represent the measurement at $\sqrt{s_{NN}} = 5.02$ TeV while the open symbols stand for the measurement at $\sqrt{s_{NN}} = 2.76$ TeV.

for Pb–Pb and pp which is currently under reconstruction. When comparing Pb–Pb to pp spectra a clear strong modification of the spectral shape is visible. It has a minimum at 6 GeV followed by a slow increase. The suppression has a strong centrality dependence, and looks very similar to that one measured at 2.76 TeV. Considering that the pp spectrum is harder at the higher energy, this similarity of the R_{AA} may point to a hotter and denser medium produced at the higher centre-of-mass energy. While these results are in agreement with the expectation for models describing energy loss in the medium, improved measurements and detailed comparison with the lower energy will allow to cancel some of the uncertainties and therefore put stronger constrain on the transport properties of the medium.

3.2 The flavour dependence of energy loss

Because of their large mass, charm and beauty quark can be produced only in parton scattering with large momentum transfer. while their measurement in pp collisions provides an important test of the factorization hypothesis and the theoretical models based on pQCD, in AA collisions it allows to test the quenching properties of the medium. A milder energy loss is expected for hadrons with charm and beauty than the light hadrons which mostly come from gluon fragmentation. Also the so-called dead cone effect predicts a suppressed gluon radiation at small angles for quarks with large mass at low-p_T. The different energy loss, larger for light hadrons than for charm than for beauty, is therefore expeted to result in a hierarchy in the R_{AA}, smaller for light hadrons than for charm than for beauty. ALICE has measured the R_{AA} for (various) D mesons, which is similar at high p_T to the one of light quarks [5]. While this can be explained by the softer fragmentation and p_T spectrum of gluons with

DOI: 10.1051/epjconf/201612900029

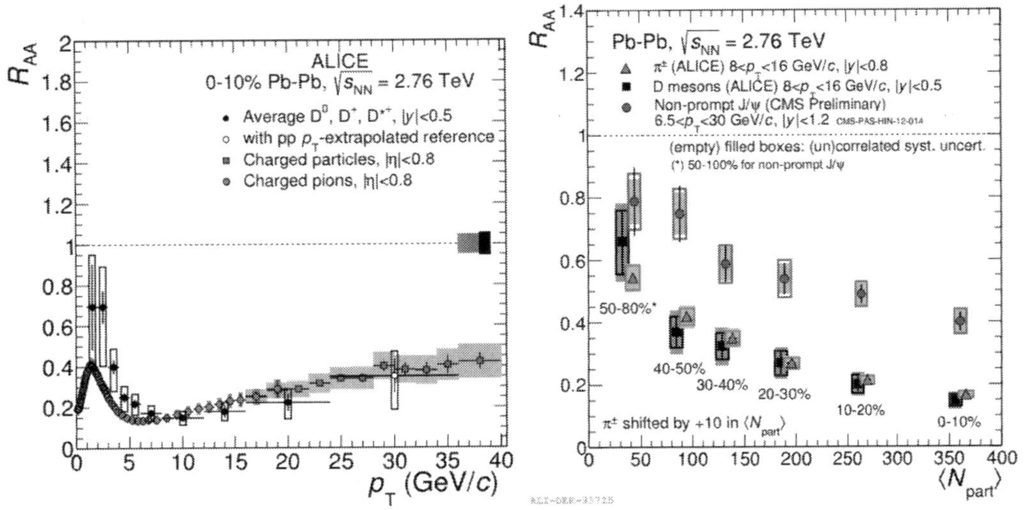

Figure 4. Left: Prompt D-meson R_{AA}(average of D^0, D^+ and D^{*+}) as a function of p_T compared to the nuclear modification factors of pions and charged particles in the 0-10% centrality classes. Right: Comparison of the D meson R_{AA}, to the charged pion R_{AA} integrated over $8 < p_T \lesssim 16$ GeV/c and of the preliminary R_{AA} of non-prompt J/ψ mesons in $6.5 < p_T \lesssim 30$ GeV/c measured by the CMS experiment.

respect to those of c quarks, there are indications at low p_T of $R_{AA}(D) > R_{AA}(\pi)$. The hierarchy between charm and beauty is tested instead with the comparison between D and secondary J/ψ (from B decays) measured by CMS for central collisions, where the expected hierarchy of $R_{AA}(c) < R_{AA}(b)$ is observed.

4 The quarkonium saga

One of the very early signature for the discovery of the QGP was suggested to be the suppression of quarkonia, tigthly bound states of charm or beauty quark and its anti-quark. Debye screening of the color fields in the medium would prevents binding in the QGP because the color charge of one quark is masked by the surrounding medium. A sequential melting of the quarkonium states would be observed as a result of their different binding energies On the other hand it became soon clear that a completely different scenario may occur, under the name of recombination, which becomes more and more relevant with increasing collision energy which in turn increases the cc pair multiplicity: the J/ψ, melted in the QGP or eventually never really formed, may (re)generate from independently produced cc quarks at freeze out, therefore leading to an enhancement of J/ψ (or less suppression). While the data collected so far indicate that the suppression measured at $\sqrt{s_{NN}} = 2.76$ GeV is smaller, has a weaker centrality dependence, and a stronger momentum dependence than what measured at RHIC at $\sqrt{s_{NN}} = 200$ GeV, which is very suggestive of the recombination scenario, the energy increase of Run 2 span will provide both a higher color screening, leading to more suppression, and a higher charm cross section, leading to more regeneration.

The statistics collected in Pb–Pb collisions at $\sqrt{s_{NN}} = 5.02$ TeV is large and allows a more precise study than in Run1, in particular more narrow centrality bins and larger p_T-reach [6]. The data collected in pp collisions at the same energy provides a reference, which is in very good agreement with the interpolation values used until now. Figure 5 shows a clear suppression with almost no centrality

 DOI: 10.1051/epjconf/201612900029

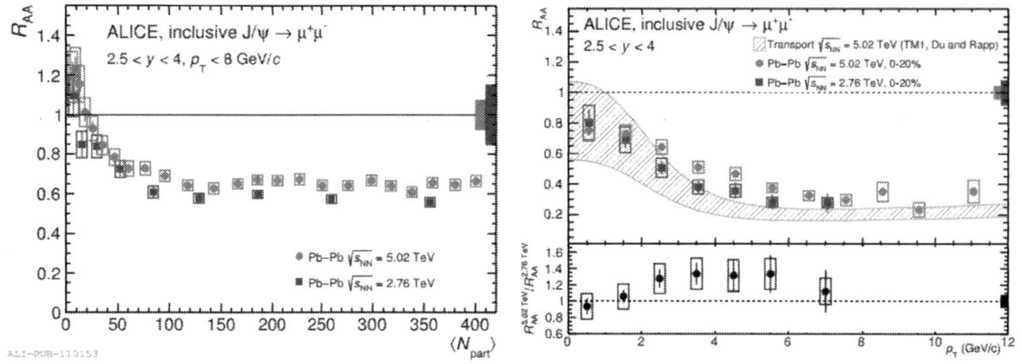

Figure 5. Left: The nuclear modification factor for inclusive J/ψ production, as a function of centrality, at $\sqrt{s_{NN}}$ = 5.02 TeV, compared to published results at $\sqrt{s_{NN}}$ = 2.76 TeV. Right: The p_T dependence of the inclusive J/ψ R_{AA} at $\sqrt{s_{NN}}$ = 5.02 TeV, compared to the corresponding result at $\sqrt{s_{NN}}$ = 2.76 TeV and to the prediction of a transport model (TM1), in the centrality interval 0–20%. The p_T dependence of the ratio is also shown.

dependence above N_{part} 100. There is a systematic difference of approximately 20% with respect to the data at $\sqrt{s_{NN}}$ = 2.76 TeV, which however is within the total uncertainty of the measurements. The data are compared to various models, among which the statistical model, where J/ψ are created at chemical freeze-out, as well as to transport models, one (TM1), where J/ψ are created following a thermal rate equation and which impements dissociation and regeneration in QGP and in the hadronic phase; another one (TM2) which describe the medium evolution with hydrodynamics, and finally to the 'co-mover' model where J/ψ are dissociated via interactions with partons/hadrons in the same y-range. While in general the models suffer of large uncertainties due to the choice of input parameters in particular cc cross section, for most calculations a better agreement is found when considering their upper limit which, for transport models corresponds to the absence of nuclear shadowing, which can be clearly considered as an extreme assumption. The double ratio allows for (some) error cancellation, especially in the models. However data are, within uncertainties, compatible with the theoretical models. We also study the p_T dependence, now extended up to 12 GeV/c, also shown in Fig. 5. The R_{AA} shows less suppression at low with respect to high p_T, with stronger p_T dependence for central events as expected from models with strong regeneration component. There is a hint for an increase of R_{AA} with $\sqrt{s_{NN}}$ visible in $2 < p_T < 6$ GeV/c, while they are consistent elsewhere. In general models do not have the sensitivity to discriminate the different scenarios for the energy increase.

5 The control experiment: benchmarks and surprises

The studies of hard probes are typically based on the comparison between the data from pp and Pb–Pb collisions, and attribute the suppression of the yields observed in Pb–Pb to energy loss in the medium created in the final state. A control experiment is necessary to disentangle strong final state effects from initial state effects, due to the presence of the nuclei themselves and not the QGP. This control experiment may come from either from the measurement of electroweak probes, that do not experience strong interaction withe the medium, or from the measurements in p–Pb collisions where only cod nuclear matter is created.

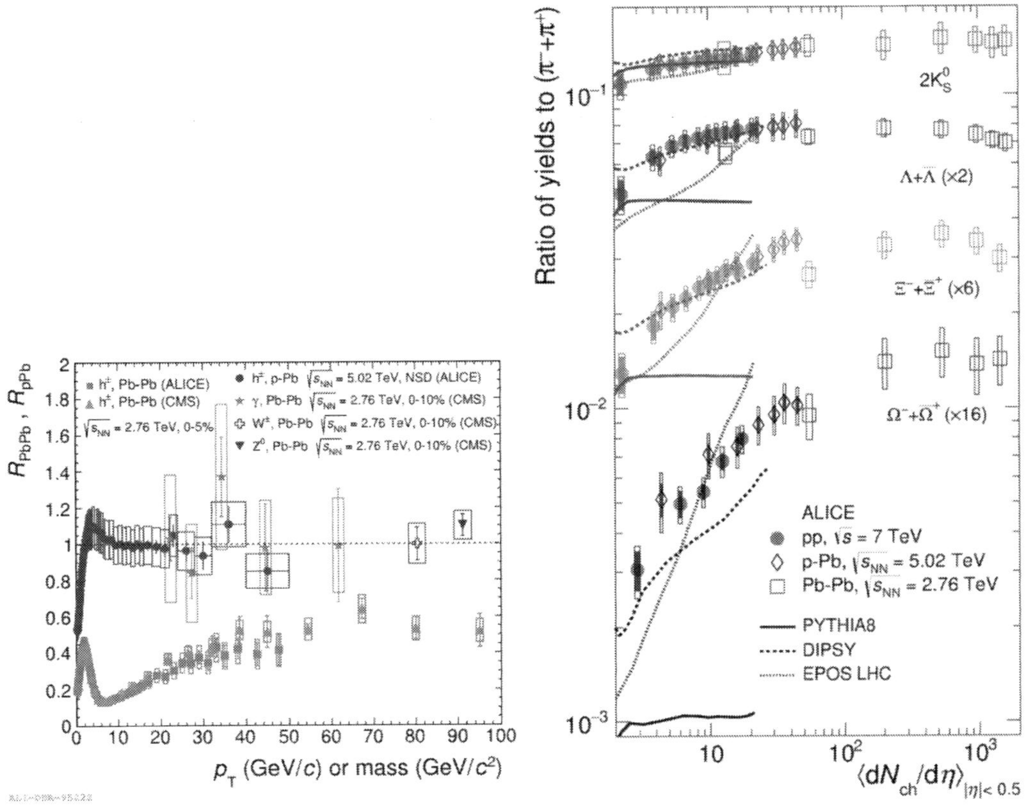

Figure 6. Left: R_{AA} of charged particles of a function of p_T measured in minimum-bias p–Pb collisions at $\sqrt{s_{NN}}$= 5.02 TeV in comparison to data in central Pb–Pb collisions at $\sqrt{s_{NN}}$= 2.76 TeV for charged particle, direct photon, Z^0 and $W^{\pm}$ production. Right: p_T-integrated yield ratios of strange and multi-strange hadrons to pions as a function of $\langle dN_{ch}/d\eta \rangle$measured in the rapidity interval $|y| < 0.5$ in pp, p–Pband Pb–Pbcollisions, compared to calculations from MC models of pp collisions.

5.1 Benchmarking hard probes

ALICE measured the p_T of charged hadrons in p–Pb collisions and the nuclear modification factor R_{pA} consistent with unity [7]. No suppression is observed also for the $R_{pA}$of heavy flavor [8] and jets [9]. This may indicate the absence of cold nuclear matter effects, however calculations with models that include moderate initial or final state effects can reproduce the data. These benchmark measurements in p–Pb collisions, together with the measurement of electroweak probes, measured by CMS, such as the direct photons [10], and the weak bosons $W^{\pm}$ [11] and Z^0 [12] (see Fig.6) confirm the validity of the N_{coll}binary scaling, and therefore provide the experimental demonstration that the suppression observed in Pb–Pb collisions is due to parton energy loss in a hot QGP.

5.2 Collectivity in small systems

However, while the measurement of benchmark processes has served as an important baseline for the understanding and the interpretation of the Pb–Pb data, several measurements, especially in the low and intermediate p_T region, could not be explained by an incoherent superposition of pp collisions, but

 DOI: 10.1051/epjconf/201612900029

instead are compatible with the presence of coherent and collective effects, whose strength increases with multiplicity. ALICE has performed various studies of multiplicity dependence of particle production, in pp, p–Pb, and Pb–Pb collisions. Of particular interest are the measurements of strange and multi-strange hadrons, whose hierarchical enhanced production in AA collisions was one of the earliest proposed indicators for the formation of the QGP state and, although observed in collisions of heavy nuclei at SPS, RHIC and LHC, is not yet understood at fundamental level. The right panel of Fig. 6 shows the ratios of strange hadron to pion yields as a function of $\langle dN_{ch}/d\eta \rangle$in pp, p–Pb, and Pb–Pb collisions [13]. A significant enhancement of strange to non-strange hadron production is observed with increasing particle multiplicity, already in pp collisions. The relative increase is more pronounced for Ω than for Ξ, indicating that the strangeness content may control the rate of increase with multiplicity. The behaviour pp collisions resembles that of p–Pb collisions, both in the values of the ratios and in their evolution with the event activity, and, at the highest multiplicities, reach the values measure in Pb–Pb collisions, which were shown to be compatible with the Gran Canonical limits. This suggests that the origin of strangeness production in hadronic collisions is driven by the characteristics of the event activity rather than by the initial-state collision system or energy.

6 Summary

Heavy ion collisions produce the most extreme state of matter ever created in the laboratory The AA physics program is rich but still full of open questions. Quantitative understanding requires not only a firm baseline from pp measurements, but also constrains of cold nuclear matter effects from p–Pb data. While baseline measurements provide clear proof that effects observed in Pb–Pb collisions are genuine effects related to parton energy loss in the hot deconfined QCD matter, various measureuments have been performed, suggestive of the existence of collective effects at high multiplicities also in small systems. ALICE has collected an excellent set of data for pp p–Pb and Pb–Pb collisions both in Run1 and Run2 which allow a significant progree in the precision measurements of the QGP properties.

References

[1] B. Muller, J. Schukraft, and B. Wyslouch, Ann. Rev. Nucl. Part. Sci. 62 (2012) 361-386.
[2] ALICE Collaboration, Phys. Rev. Lett. 116 (2016) 222302.
[3] ALICE Collaboration, Phys. Rev. Lett. 116 (2016) 132302.
[4] ALICE Collaboration, Phys. Lett. B 753 (2016) 319-329.
[5] ALICE Collaboration, JHEP 03 (2016) 081.
[6] ALICE Collaboration, arXiv:1606.08197.
[7] ALICE Collaboration, EPJC 74 (2014) 3054.
[8] ALICE Collaboration, Phys. Rev. Lett. 113 (2014) 232301.
[9] ALICE Collaboration, Eur. Phys. J. C 76 (2016) 5, 271.
[10] CMS Collaboration, Phys. Lett. B 710 (2012) 256.
[11] CMS Collaboration, Phys. Lett. B 715 (2012) 66.
[12] CMS Collaboration, Phys. Rev. Lett. 106 (2011) 212301.
[13] ALICE Collaboration, arXiv: 1606.07424.

Medium effects on heavy-flavour observables in high-energy nuclear collisions

Andrea Beraudo[1,*]

[1] *INFN - Sezione di Torino - Via Pietro Giuria 1, I-10125 Torino*

Abstract. The peculiar role of heavy-flavour observables in relativistic heavy-ion collisions is discussed. Produced in the early stage, c and b quarks cross the hot medium arising from the collision, interacting strongly with the latter, until they hadronize. Depending on the strength of the interaction heavy quarks may or not approach kinetic equilibrium with the plasma, tending in the first case to follow the collective flow of the expanding fireball. The presence of a hot deconfined medium may also affect heavy-quark hadronization, being possible for them to recombine with the surrounding light thermal partons, so that the final heavy-flavour hadrons inherit part of the flow of the medium. Here we show how it is possible to develop a complete transport setup allowing one to describe heavy-flavour production in high-energy nuclear collisions, displaying some major results one can obtain. Finally, the possibility that the formation of a hot deconfined medium even in small systems (high-multiplicity p-Au and d-Au collisions, so far) may affect also heavy-flavour observables is investigated.

1 Introduction

Quantum Chromo-Dynamics is characterized by a non trivial phase-diagram in which, depending on the temperature and the baryon density, the active degrees of freedom may be coloured quarks and gluons or "white" hadrons. Relativistic heavy-ion collisions are the only experimental tool to study, under controlled conditions, such a phase diagram. In particular, at the highest energies, nuclear collisions allows one to explore the transition from a plasma of deconfined quarks and gluons (QGP) to a gas of hadrons and resonances in the region of high temperature and (almost) vanishing baryon density. In such a regime, which is the one experienced also by the universe during its thermal evolution, the transition is actually a smooth crossover and is accompanied by the spontaneous breaking of chiral symmetry (process responsible for most of the baryonic mass of our universe).

The matter produced in heavy-ion collisions undergoes the following evolution. There is an initial pre-equilibrium stage, during which the energy stored in the strong colour-fields is converted into particles (quarks and gluons). There is evidence that, quite rapidly ($\tau_0 \lesssim 1$ fm/c), the system reaches local thermodynamical equilibrium spending some time ($\sim$ 5–10 fm/c) in the QGP phase, during which it expands and cools until getting to a regime in which the active degrees of freedom are again mesons and baryons. When the collision rate becomes too small to maintain kinetic equilibrium hadrons decouple and follow a free-streaming until being detected. Hence, information on the possible onset

[*] e-mail: beraudo@to.infn.it

of deconfinement and on the properties of the produced medium has to be obtained quite indirectly, starting from hadronic observables.

Soft probes, i.e. low-p_T hadrons (representing the bulk of particle production), provide evidence for the collective behaviour of the produced system, which displays a hydrodynamics expansion driven by pressure gradients. The success of hydrodynamics in explaining peculiar features of soft-particle production (in particular the azimuthal asymmetry of the spectra) suggests that the medium formed in heavy-ion collisions is characterized by a mean-free-path much smaller than its size, $\lambda_{\rm mfp} \ll L$, and that during most of its expansion it remains close to local thermal equilibrium. *Hard probes*, on the contrary, are those particles (high-p_T partons, heavy quarks and quarkonia) produced in hard pQCD processes in the very first instants and which, before reaching the detectors, cross the fireball formed in the collision of the two nuclei, performing a sort of tomography of the latter. In particular, the suppression of high-p_T particle (and jet) production, loosely speaking referred to as jet-quenching, provides evidence that the medium formed in heavy-ion collisions is very opaque: partons propagating through it lose a non-negligible fraction of their energy due to collisions and medium-induced gluon radiation. Within this framework, heavy-flavour (HF) particles play a peculiar role. Soft observables can be satisfactory described by hydrodynamics, assuming to deal with a system close to local thermal equilibrium as a working hypothesis. The study of jet quenching involves simply modeling the energy-degradation of *external* probes. On the other hand, the description of HF observables requires dealing with a more general situation and in particular developing a setup (based on *transport theory*) able to describe how particles initially shot into the medium would asymptotically approach kinetic equilibrium with the latter. Notice that, at high-p_T, the interest in HF particles is no longer related to their possible thermalization, but to the study of the mass and colour-charge dependence of parton energy-loss and jet quenching: we will not address this issue here.

Why are c and b-quarks considered heavy, based on their mass M? First of all because $M \gg \Lambda_{\rm QCD}$, setting a hard scale which ensures that their initial production is well described by pQCD. Secondly, because $M \gg T$, entailing that their thermal abundance in the plasma would be negligible: in heavy-ion collisions, with an expanding fireball with a quite short lifetime $\sim$ 10 fm/c, the final multiplicity is set by the initial hard production. Finally, because $M \gg gT$, gT being the typical momentum exchange in the collisions with the plasma particles, implying that many soft scatterings are necessary to change significantly the momentum and trajectory of a heavy quark: this will be of relevance in order to developed a simplified transport setup.

The present contribution is organized as follows. In Sec. 2 we give an overview on medium modification of HF observables in heavy-ion collisions, illustrating the complex setup needed to get a meaningful theory-to-experiment comparison and discussing the information one can obtain from the present experimental data. In Sec. 3 we move to smaller systems, discussing whether also in p-A or d-A collisions (when selecting high-multiplicity events) one can expect final-state effects associated to the formation of a hot deconfined medium and whether this can affect also HF observables. Finally, in Sec. 4, we draw our conclusions and discuss the perspectives of future studies.

2 Heavy Flavour in heavy-ion collisions

A realistic study of HF observables in high-energy nuclear collisions requires developing a multi-step setup, which here we briefly summarize and describe in detail in the following. First of all, one needs to simulate the initial hard production of the $c\bar{c}$ and $b\bar{b}$ pairs in the elementary nucleon-nucleon collisions: for this automated pQCD tools are available, which one can generalize to include initial-state effects, such as nuclear parton distribution functions (nPDF's) and transverse-momentum broadening in nuclear matter. Secondly, one must have at his disposal a realistic description of the fireball in which the heavy quarks propagate, provided by hydrodynamic calculations validated against soft-hadron data. The core of the setup is represented by the modeling of the heavy-quark propagation in the hot plasma: one has to develop a transport code working in the case of an evolving medium, taking advantage of the previous information. Once the heavy quarks reach a fluid cell below the deconfinement temperature they hadronize and one has to model also this stage: can it be described simply via fragmentation functions like in the vacuum or are other mechanisms, such as recombination with light thermal partons, possible? Clearly, medium modification of hadronization in heavy-ion collisions is an item of interest in itself; however, it is at the same time a source of systematic uncertainty if one wants to extract information on the properties of the deconfined medium, since it also contribute to the modification of the HF spectra with respect to p-p collisions. Finally, since experimental data sometimes does not refer directly to D and B-mesons, one must also simulate the decay into the relevant channels (e.g. $D \to X \nu_e e$, $D \to X \nu_\mu \mu$, $B \to X J/\Psi$...) accessible to the experiments.

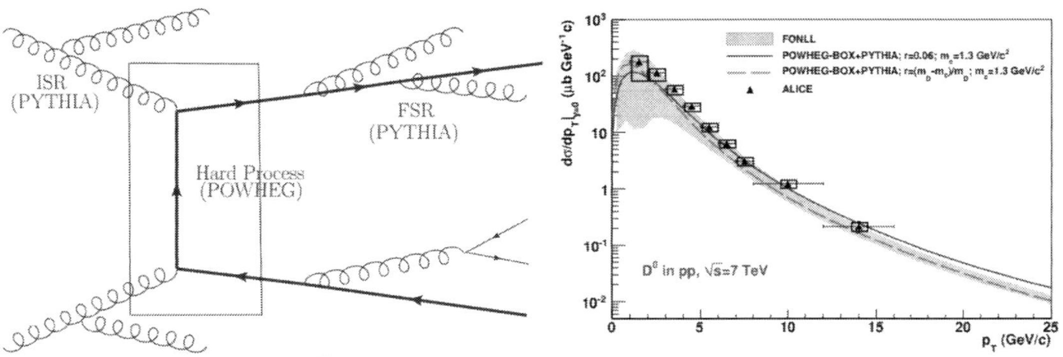

Figure 1. Left panel: a typical $Q\overline{Q}$ event generated by the POWHEG-BOX code [2], with the hard process interfaced with a parton-shower stage. Right panel: POWHEG-BOX predictions for D^0 production in p-p collisions at 7 TeV (different fragmentation functions are considered) compared to ALICE data [4]; the green band is the result of the FONLL [1] calculation.

Due to the large mass, heavy-quark production is a hard short-distance process occurring at the level of the individual binary nucleon-nucleon collisions. It is then necessary to validate the simulation of the initial hard production against p-p data. The state of the art for its theoretical description is represented by the Fixed Order Next-to-Leading Log (FONLL) calculation, developed to provide accurate results both at low and at high-p_T through a proper resummation of large transverse-momentum logarithms [1]. However, FONLL simply provides an inclusive single-particle spectrum, while for practical applications it is more convenient to work with an event generator. It this regard, the POWHEG-BOX package [2] represents a convenient tool. The latter generates hard events at NLO accuracy and interface them with a parton-shower stage, simulating initial and final-state radiation (see left panel of Fig. 1). A comparison of FONLL and POWHEG-BOX results (with $m_c = 1.3$ GeV, $\mu_R = \mu_F = m_T$ and HQET fragmentation functions [3]) with ALICE data for D^0 production in p-p collisions at the LHC [4] is displayed in the right panel of Fig. 1. Overall, the data tend to stay at the upper edge of the (quite large, in the case of charm) theoretical uncertainty band, arising from the choice of the quark mass and of the factorization and renormalization scales. Once validated in the p-p case and supplemented with nPDF's [5], the above pQCD tools can be employed also to simulated the initial $Q\overline{Q}$ production in heavy-ion collisions. For each event the full kinematic information on the produced $Q\overline{Q}$ pairs can be stored and used in the subsequent steps of the simulations.

One needs then to implement a numerical transport calculation describing the heavy-quark propagation in the plasma. The starting point of all transport calculation is the Boltzmann equation for the evolution of the heavy-quark phase-space distribution

$$\frac{d}{dt}f_Q(t,\boldsymbol{x},\boldsymbol{p}) = C[f_Q] \quad \text{with} \quad C[f_Q] = \int d\boldsymbol{q}[w(\boldsymbol{p}+\boldsymbol{q},\boldsymbol{q})f_Q(t,\boldsymbol{x},\boldsymbol{p}+\boldsymbol{q}) - w(\boldsymbol{p},\boldsymbol{q})f_Q(t,\boldsymbol{x},\boldsymbol{p})], \tag{1}$$

where the collision integral $C[f_Q]$ is expressed in terms of the $\boldsymbol{p} \to \boldsymbol{p}-\boldsymbol{q}$ transition rate $w(\boldsymbol{p},\boldsymbol{q})$. The direct solution of the Boltzmann equation is numerically demanding; however, as long as $q \ll p$ (q being typically of order gT) one can expand the collision integrand in powers of the momentum exchange. Truncating the expansion to second order corresponds to the Fokker-Planck (FP) approximation, which leads to the (relativistic) Langevin equation. The latter in its discretized form

$$\Delta\vec{p}/\Delta t = -\eta_D(p)\vec{p} + \vec{\xi}(t), \tag{2}$$

provides a recipe to update the heavy quark momentum through the sum of a deterministic friction force and a random noise term specified by its temporal correlator

$$\langle \xi^i(\boldsymbol{p}_t)\xi^j(\boldsymbol{p}_{t'})\rangle = b^{ij}(\boldsymbol{p}_t)\delta_{tt'}/\Delta t \qquad b^{ij}(\boldsymbol{p}) \equiv \kappa_\|(p)\hat{p}^i\hat{p}^j + \kappa_\perp(p)(\delta^{ij}-\hat{p}^i\hat{p}^j). \tag{3}$$

After evaluating the transport coefficients $\kappa_{\|/\perp}(p)$ (representing the average longitudinal/transverse squared momentum exchanged with the plasma per unit time) and $\eta_D(p)$ (the latter being fixed by the Einstein relation, so that particles asymptotically reach kinetic equilibrium) one has then to solve Eq. (2) throughout the whole medium evolution [6, 7], described by hydrodynamic calculations [8, 9].

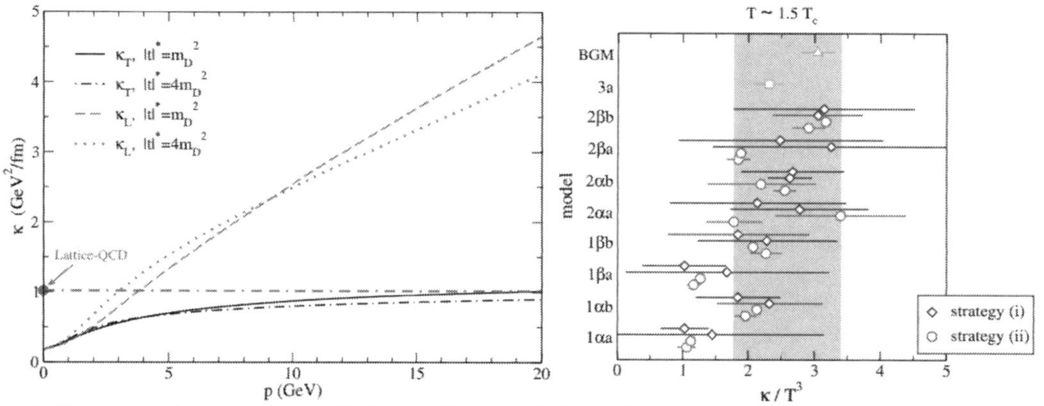

Figure 2. Heavy-quark momentum diffusion coefficients. Left panel: weak-coupling results for charm at $T = 400$ MeV, with different choices of the cutoff separating hard and soft collisions [6]. Right panel: lattice-QCD result for a static quark ($\kappa_T = \kappa_L \equiv \kappa$) in a gluonic plasma [10]. The uncertainty band arise from extracting real-time information from imaginary-time correlators.

In the Langevin approach physics enters through the transport coefficients, which one aims at deriving through a first-principle calculation. In the left panel of Fig. 2, referring to charm, we show the results of a weak-coupling evaluation, accounting for $2 \to 2$ collisions, with resummation of medium effects in the case of soft-gluon exchange [6]. A non-perturbative estimate can be provided in principle by lattice-QCD simulation, so far limited to the case of a static, infinitely heavy, quark. In this limit one can neglect the momentum dependence of the strength of the noise, so that

$$\langle \xi^i(t)\xi^j(t')\rangle = \delta^{ij}\delta(t-t')\kappa \quad \longrightarrow \quad \kappa = \frac{1}{3}\int_{-\infty}^{+\infty} dt\langle \xi^i(t)\xi^i(0)\rangle_{\rm HQ} \underset{p\to 0}{\sim} \frac{1}{3}\int_{-\infty}^{+\infty} dt\langle F^i(t)F^i(0)\rangle_{\rm HQ}. \tag{4}$$

The momentum-diffusion coefficient κ can be extracted from a force-force correlator in an ensemble of states containing one heavy quark: for a static colour source the only force playing a role is the chromo-electric field. Notice that, on the lattice, correlators can be evaluated only for imaginary-times, while, as displayed in Eq. (4), κ, as any other transport coefficient, is a real-time quantity: this leads to the large systematic uncertainty band in the right panel of Fig. 2, displaying the lattice-QCD estimate given in [10].

Finally, heavy quarks, when reaching a fluid cell below a critical temperature, are allowed to hadronize. Various models have been developed to describe how the medium can modify this stage, all based on the idea that a heavy quark can easily find one or more light partons nearby to recombine with, producing a colour-singlet object, which can be taken to be directly a HF hadron (coalescence [11, 12]), a heavy resonance [13] or a string to further fragment [7].

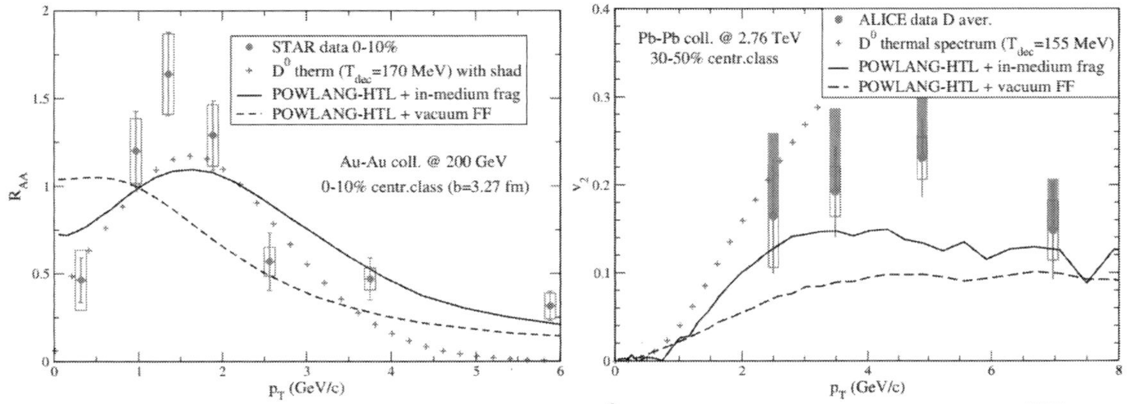

Figure 3. Left panel: the nuclear modification factor of D^0 spectra in 0-10% Au-Au collision at $\sqrt{s_{\rm NN}} = 200$ GeV compared to STAR data [14]. Right panel: the elliptic flow of D^0 mesons in 30-50% Pb-Pb collisions at $\sqrt{s_{\rm NN}} = 2.76$ TeV compared to ALICE data [15]. Theory results include the full kinetic-equilibrium limit (green crosses), transport calculations interfaced with vacuum fragmentation functions (dashed curves) and transport calculations followed by in-medium hadronization (continuous curves).

Let us now present some selected outcomes of our transport calculations [7], pointing out some quite general qualitative findings common with the analysis performed by other groups [11–13]. In Fig. 3 we consider the *nuclear modification factor* of the p_T-spectra and the *elliptic flow*

$$R_{\rm AA} \equiv \frac{(dN/dp_T)^{\rm AA}}{\langle N_{\rm coll}\rangle (dN/dp_T)^{\rm pp}} \quad \text{and} \quad v_2 \equiv \langle \cos[2(\phi - \psi_{\rm RP})]\rangle \tag{5}$$

of D^0 mesons in Au-Au and Pb-Pb collisions at RHIC and LHC ($\langle N_{\rm coll}\rangle$ being the average number of binary nucleon-nucleon collisions). It is interesting to start addressing the limit of very (infinitely) large transport coefficients. In this case HF particles would approach local kinetic equilibrium with the fireball and would be thermally emitted from a decoupling hypersurface, identified (rather schematically) by a temperature below which interactions are supposed to become too rare to further modify the spectra. The usual Cooper-Frye formula gives:

$$E\frac{dN}{d\vec{p}} = \frac{1}{(2\pi)^3}\int_{\Sigma_{\rm dec}} p^\mu d\Sigma_\mu \exp\left[-\frac{p\cdot u(x)}{T_{\rm dec}}\right]. \tag{6}$$

As a consequence, high-energy particles would be thermally suppressed; on the other hand, the collective flow of the fluid u^μ would tend to drag very soft particles to larger p_T. This gives rise to the bump in the $R_{\rm AA}$ seen in the left panel of Fig. 3 (green crosses) and displayed also by the experimental data. At the same time, in non-central collision the flow of the fluid is characterized by an azimuthal anisotropy, due to the initial larger pressure gradient along the reaction-plane (RP): the azimuthal distribution of particles flowing with the fluid would inherit this anisotropy, displaying a strong elliptic flow, as shown in the right panel of Fig. 3 and supported by the data. Does the above considerations entail that in heavy-ion collisions charm actually reaches kinetic equilibrium with the medium during its partonic phase? This is not necessarily the case, as shown by the curves in Fig. 3 referring to Langevin calculations with finite (weak-coupling) transport coefficients. Transport in the QGP phase alone does not appear able to transfer to the charmed particles enough radial and elliptic flow (dashed curves with vacuum fragmentation functions). On the other hand, modeling HF hadronization through recombination with thermal partons from the medium leads to results in better agreement with the data, thanks to the additional flow acquired from the light companion (continuous curves).

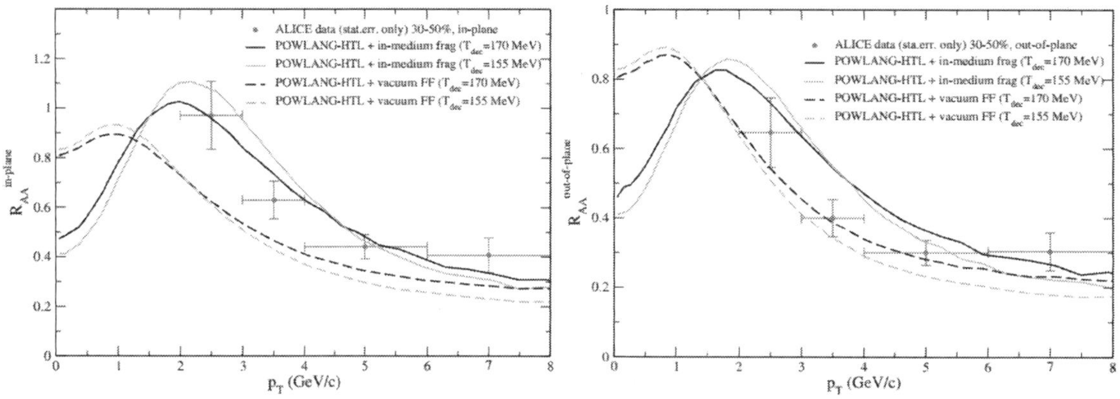

Figure 4. The in-plane (left panel) and out-of-plane (right panel) nuclear modification factor of D^0 mesons in 30-50% Pb-Pb collisions at $\sqrt{s_{\rm NN}}=2.76$ TeV. Transport results with different hadronization schemes and decoupling temperatures are compared to ALICE data [16].

In Fig. 4 we compare the nuclear modification factors of D^0 mesons in non-central Pb-Pb events at the LHC emitted along (in-plane) or orthogonally (out-of-plane) to the direction of the impact parameter of the collision. We explore the sensitivity to different decoupling temperatures and hadronization schemes. Curves including in-medium hadronization are again characterized by a bump at moderate p_T: experimental data at lower p_T are clearly necessary in order to confirm such an occurrence.

3 Heavy Flavour in small systems: room for medium effects?

One of the most surprising findings in the experimental search for the Quark-Gluon Plasma is certainly the signature of possible collective effects, suggestive of the formation of a hot strongly-interacting medium, recently observed in collisions involving small projectiles, like p-Pb at the LHC and d-Au (and now also ^{3}He-Au) at RHIC, in particular when selecting events characterized by a high multiplicity of produced particles. Various observables support the above picture, in particular transverse-momentum spectra and azimuthal correlations, suggesting that soft hadrons take part in a common collective expansion, with a radial, elliptic and triangular flow arising from the response of a strongly-interacting medium to the initial pressure gradients and fluctuating geometry. On the contrary, high-p_T hadrons and jets look essentially unmodified with respect to the p-p benchmark. Due to the peculiar role of HF particles, produced in initial hard events but tending to (at least partially) thermalize with the medium in heavy-ion collisions, it is clearly of interest to study their behaviour also in small systems, looking for possible modifications in the final spectra introduced by a hot deconfined plasma. Due to the large systematic error bands, current experimental data are not able to provide a clear answer: the nuclear modification factor of HF electrons (from charm and beauty decays) in

 DOI: 10.1051/epjconf/201612900030

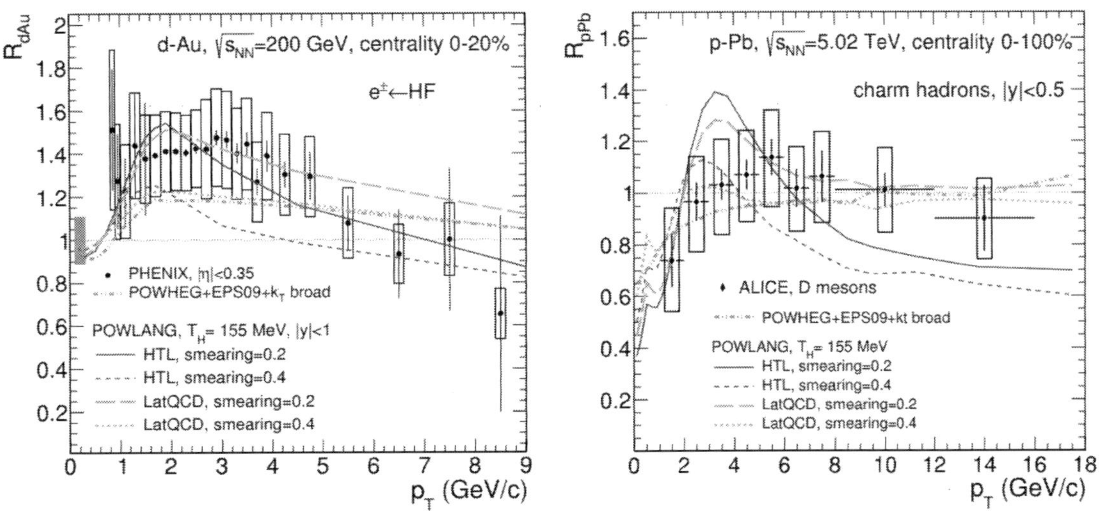

Figure 5. Left panel: the nuclear modification factor of HF decay electrons in 0-20% most central d-Au collisions at RHIC measured by PHENIX [17]. Right panel: the nuclear modification factor of D-mesons in 0-100% p-Pb collisions at the LHC measured by the ALICE collaboration [18]. The various curve refers to different choices of transport coefficients and initial conditions.

d-Au collisions at RHIC measured by PHENIX [17] seems to display a moderate enhancement in the intermediate-p_T region; on the other hand, within the present uncertainties, the D-meson R_{pA} measured by ALICE in p-Pb collisions at the LHC looks compatible with unity (i.e. no medium effect). In [19] we applied our HF transport setup to such small systems. At variance with the nucleus-nucleus case, the initial energy-density profile is not related to the impact parameter of the collision, but to event-by-event fluctuations. Hence, we had to start the hydrodynamic evolution of the medium with Glauber Monte-Carlo initial conditions, so to capture at least the fluctuations arising from the random event-by-event nucleon positions. As discussed in detail in [19], the final results arise from the interplay of several initial/final-state effects: nuclear PDF's, with gluon shadowing tending to reduce charm production at low-p_T; transverse-momentum broadening in nuclear matter (before the initial $Q\overline{Q}$ production), tending to move particles to moderate p_T; interaction with the deconfined plasma produced in the collision, tending to suppress heavy-quark spectra at high-p_T and making them inherit part of the flow of the fireball; in-medium hadronization, which also transfer part of the (radial and elliptic) flow of the fireball to the final HF hadrons. Overall, the final outcomes are the ones shown in Fig. 5, in which the various curves reflect the different initial conditions and transport coefficients, giving an assessment of the theoretical uncertainty.

4 Conclusions and perspectives

The comparison between transport calculations and experimental data provides robust indications that heavy quarks strongly interact with the hot deconfined medium formed in high-energy nuclear collisions. Due to such interactions, besides soft hadrons, also heavy-flavour particles carry signatures of the collective flow of the background medium, although the limited lifetime of the latter together with its expansion may prevent them to reach full kinetic equilibrium with the plasma. A number of theoretical questions and experimental challenges remain to be addressed. Charm measurements down to $p_T \to 0$ will be important to get information on the flow and thermalization of charm and its total cross-section in heavy-ion collisions. D_s and Λ_c measurements will give a definite answer on possible changes of HF hadrochemistry due to the recombination of the heavy quarks with light thermal partons at hadronization, which seems currently supported by D-meson R_{AA} and v_2 data. Furthermore, D_s and Λ_c reconstruction will contribute to a quantitative estimate of the total charm cross-section, a measurement of relevance to study charmonium suppression in heavy-ion collisions on solid ground. In the near future the issue of HF production in small systems (e.g. high-multiplicity p-A collisions), already started in the last few years, will certainly deserve further investigation in order to point out and give a quantitative assessment of initial/final-state effects arising e.g. from the nuclear PDF's and from the possible formation of a hot deconfined medium. Besides the interest for the heavy-ion community, theoretical results for charm production in high-energy p-A collisions validated against experimental data will be of relevance for experiments like IceCube, to get a robust estimate of the charm-decay contribution to the background of high-energy atmospheric neutrinos.

The most important challenge, which would represent a major experimental achievement, remains however the extraction of the HF transport coefficients in hot-QCD: are current experimental data sufficient to put tight constraints on them? In case so, how do they compare with first-principle theoretical calculations? Notice that the concept itself of transport coefficient relies on the idea that the dynamics of the heavy quarks in the medium can be given an effective Langevin description, requiring simply the knowledge of the average squared-momentum exchange per unit time with the plasma. Although current experimental results, once compared to theory calculations, already gives some qualitative guidance, a solid answer to this issue requires waiting for beauty measurements via exclusive B-meson hadronic decays. Beauty is in fact a better probe of the medium due to its large mass: the latter makes the initial hard production (described by pQCD) under better control, ensures the effective equivalence between the Langevin and the Boltzmann equation and, furthermore, makes the extraction of HF transport coefficients from lattice-QCD simulations conceived for static quarks more meaningful.

References

[1] Cacciari *et al.*, JHEP **9805** (1998) 007 and JHEP **0103** (2001) 006.
[2] S. Alioli, P. Nason, C. Oleari and E. Re, JHEP **1006** (2010) 043.
[3] E. Braaten, K.M. Cheung, S. Fleming and T. C. Yuan, Phys. Rev. D 51 (1995) 4819.
[4] ALICE Collaboration (B. Abelev *et al.*), JHEP 1201 (2012) 128.
[5] K.J. Eskola, H. Paukkunen and C.A. Salgado, JHEP **0904**, 065 (2009).
[6] W.A. Alberico *et al.*, Eur. Phys. J. C 71 (2011) 1666 and Eur. Phys. J. C 73 (2013) 2481.
[7] A Beraudo *et al.*, Eur. Phys. J. C 75 (2015) 121.
[8] M. Luzum and P. Romatschke, Phys. Rev. C 78 (2008) 034915.
[9] L. Del Zanna *et al.*, Eur. Phys. J. C 73 (2013) 2524.
[10] A. Francis *et al.*, Phys. Rev. D 92 (2015) 116003.
[11] P.B. Gossiaux, R. Bierkandt and J. Aichelin, Phys. Rev. C 79 (2009) 044906.
[12] H. van Hees et al., Phys. Rev. Lett. 100 (2008) 192301.
[13] M. He, R.J. Fries and R. Rapp, Phys.Rev. C 86 (2012) 014903.
[14] STAR Collaboration (L. Adamczyk *et al.*), Phys. Rev. Lett. 113 (2014) 142301.
[15] ALICE Collaboration (B. Abelev *et al.*), Phys. Rev. Lett. 111 (2013) 102301.
[16] ALICE Collaboration (B. Abelev *et al.*), Phys. Rev. C 90 (2014) 034904.
[17] PHENIX Collaboration (A. Adare *et al.*), Phys. Rev. Lett. 109 (2012) 242301.
[18] ALICE Collaboration (B. Abelev *et al.*), Phys. Rev. Lett. 113 (2014) 232301.
[19] A Beraudo *et al.*, JHEP **1603** (2016) 123.

EPJ Web of Conferences **129**, 00031 (2016) DOI: 10.1051/epjconf/201612900031
QCD@Work 2016

Classical gluon production amplitude in heavy-ion collisions

Giovanni Antonio Chirilli[1,2,⋆]

[1] *Institut für Theoretische Physik, University of Regensburg, D-93040 Regensburg, Germany*

Abstract. The initial condition for the formation of the Quark-Gluon Plasma, the distribution of quarks and gluons produced in the initial stages of nuclear collisions, is the fundamental building block of heavy-ion theory. In the framework of saturation physics and Wilson lines formalism we calculate the scattering amplitude, beyond the leading order, of the classical gluon produced in heavy-ion collisions.
known as the initial condition of the Quark-Gluon Plasma formation, is the fundamental building block of heavy-ion theory.

1 Introduction

To understand the initial stages of nuclear collisions and the formation of quark gluon plasma, one has to study the dynamics of the collisions from first principle in Quantum Chromodynamic (QCD). To this end, we calculate the scattering amplitude, beyond the leading order, of the classical gluon produced in Nucleus-Nucleus collisions.

As a first approximation, we study the collision of two large nuclei with atomic number $1 \ll A_1 \ll A_2$. The smaller nucleus can be treated as a diluted system of constituents when compared to the large nucleus. In particular, the saturation scale of nucleus A_1, which is proportional to its atomic number, $Q^2_{s1} \sim \alpha_s^2 A_{s1}^{1/3}$ is much smaller then the saturation scale of the large nucleus, $Q^2_{2s} \sim \alpha_s^2 A_2^{1/3}$. The atomic number of the nucleus enters in the resummation parameter. Indeed, if we consider a particle propagation through a large nucleus with atomic number A, with good probability, will encounter, along its path, $A^{1/3}$ nucleons. In the two-gluon approximation, the particle propagating through the nucleus, interacts with each nucleon with the exchange of two gluons. Hence, for large nuclei, there are contributions to the scattering amplitude that are proportional to $\alpha_s^2 A^{1/3}$ and since A is a large number, these contributions are the leading ones. In our case we have $\alpha_s^2 A_2^{1/3} \sim 1$ for the larger nucleus (the target) which in the boosted Lorentz frame reduces to a shock-wave (represented as a red stripe in all diagrams below), and $\alpha_s^2 A_1^{1/3} \leq 1$ for the smaller one (the projectile).

2 From proton-Nucleus to Heavy-light Ion Collisions

The leading-order (LO) contribution to the classical gluon production amplitude in heavy-light ions collisions coincides with proton-Nucleus (pA) collisions. The diagrams for pA collisions are shown in Fig. 1. Using light-cone coordinates $x^\pm = \frac{1}{\sqrt{2}}(x^0 \pm x^3)$, the proton propagates along x^+ direction and

⋆e-mail: giovanni.chirilli@physik.uni-regensburg.de

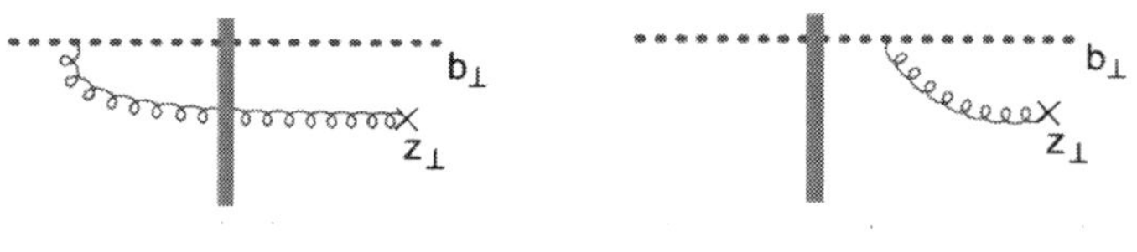

Figure 1. Feynman diagrams for pA collisions. Dashed lines represent Wilson lines.

the target (large nucleus) in x^- direction. Transverse coordinate are $\vec{x}_\perp = (x^1, x^2)$. We will perform the calculation in the $A^+ = 0$ light-cone gauge. Note that the blue dashed lines represent Wilson lines. The scattering amplitude for pA collision is [1]

$$A(\vec{z}_\perp, \vec{b}_\perp) = \frac{ig}{\pi} \frac{\vec{\epsilon}_\perp^{\,\lambda *} \cdot (\vec{z}_\perp - \vec{b}_\perp)}{|\vec{z}_\perp - \vec{b}_\perp|^2} \left[U^{ab}_{\vec{z}_\perp} - U^{ab}_{\vec{b}_\perp} \right] \left(V_{\vec{b}_\perp} t^b \right) \tag{1}$$

where $\vec{\epsilon}_\perp^{\,*}$ is the polarization vector of the emitted gluon, $U^{ab}_{\vec{z}_\perp}$ is the Wilson line in the adjoint representation at transverse position $\vec{z}_\perp$, while $V_{\vec{b}_\perp}$ is the Wilson line in the fundamental representation. The gluon production cross section is proportional to the amplitude square and is given by

$$\frac{d\sigma}{d^2k_T\, dy} = \frac{1}{2\,(2\pi)^3} \int d^2z\, d^2z'\, d^2b\, e^{-i\vec{k}_\perp \cdot (\vec{z}_\perp - \vec{z}'_\perp)} \left\langle A(\vec{z}_\perp, \vec{b}_\perp)\, A^*(\vec{z}'_\perp, \vec{b}_\perp) \right\rangle . \tag{2}$$

We are interested in calculating the next-to-leading order (NLO) contribution to the gluon production cross-section which is proportional to $\alpha_s^3 A_1^{2/3}$. The diagrams that contribute at this order may be divided in three classes as shown in Fig. 2 [2]. It is possible to reduce dramatically the number of

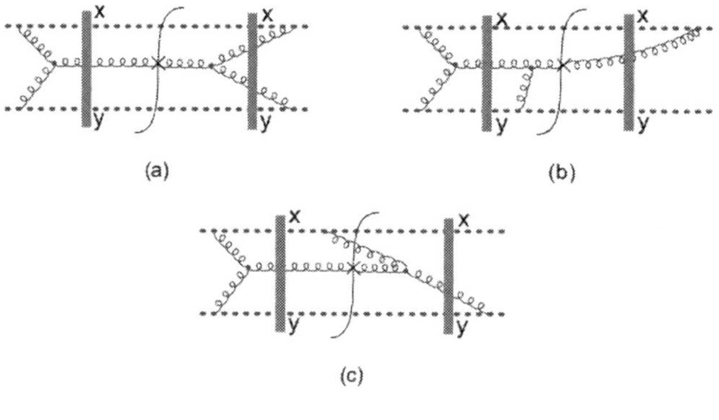

Figure 2. Here we show the three different classes of diagrams at order g^6: (a) Square of order g^3 amplitudes; (b) Interference between order g^5 and order g amplitudes; (c) Interference between order g^4 and order g^2 amplitudes.

diagrams if we sum up diagrams having gluons on one side of the cut with the corresponding diagrams having gluons crossing the cut, thus reducing the two diagrams in a single diagram with a retarded gluon propagator (see Fig.3)

$$\frac{-id^{\mu\nu}(k)}{k^2 + i\epsilon} + 2\pi\theta(-k^+)\delta(k^2)d^{\mu\nu}(k) = \frac{-id^{\mu\nu}(k)}{k^2 + i\epsilon k^+} \tag{3}$$

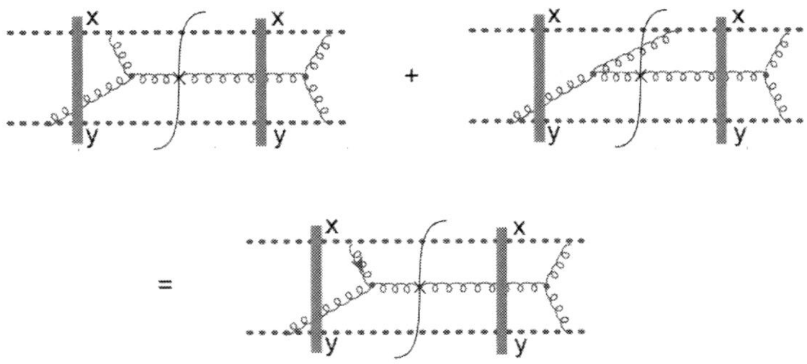

Figure 3. Summing up diagrams with gluons on one side of the cut with the corresponding diagram with gluons crossing the cut, we get a single diagram with a retarded gluon propagator.

where $d^{\mu\nu}(k) = g^{\mu\nu} - \frac{\eta^\mu k^\nu + \eta^\nu k^\mu}{\eta \cdot k}$. A consequence of this observation is that, the only diagrams we will have to consider are those at order g^3 and those at order g^5 (at amplitude level), but with each gluon propagator being retarded propagator.

A further reduction of the number of diagrams to be calculated is given by the following two observations. First, we notice that the sum of diagrams of the type in Fig. 5 cancel. Second, using the

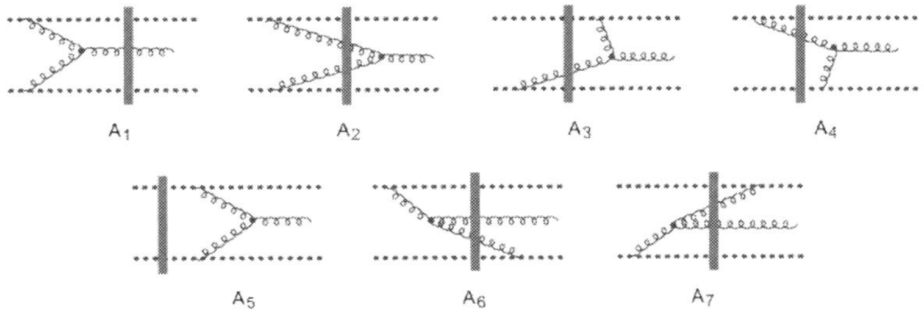

Figure 4. Diagrams A have single 3-gluon vertex with gluons connecting to both quarks. All gluons are retarded and the time flows to the right.

cancellation described in Fig. 5 we can combine diagrams as shown in Fig. 6 thus reducing further the number of diagrams (see Fig. 7).

The result for the calculation of diagrams in Fig. 4 and in Fig. 7 is [2]

$$\sum_{i=1}^{7} A_i + \sum_{i=1}^{12}{}' B_i + \sum_{i=1}^{12}{}' C_i = -\frac{g^3}{4\pi^4} \int d^2x_1\, d^2x_2\, \delta[(\vec{z}_\perp - \vec{x}_{1\perp}) \times (\vec{z}_\perp - \vec{x}_{2\perp})]$$
$$\times \left[\frac{\vec{\epsilon}_\perp^{\,\lambda *} \cdot (\vec{x}_{2\perp} - \vec{x}_{1\perp})}{|\vec{x}_{2\perp} - \vec{x}_{1\perp}|^2} \frac{\vec{x}_{1\perp} - \vec{b}_{1\perp}}{|\vec{x}_{1\perp} - \vec{b}_{1\perp}|^2} \cdot \frac{\vec{x}_{2\perp} - \vec{b}_{2\perp}}{|\vec{x}_{2\perp} - \vec{b}_{2\perp}|^2} - \frac{\vec{\epsilon}_\perp^{\,\lambda *} \cdot (\vec{x}_{1\perp} - \vec{b}_{1\perp})}{|\vec{x}_{1\perp} - \vec{b}_{1\perp}|^2} \frac{\vec{z}_\perp - \vec{x}_{1\perp}}{|\vec{z}_\perp - \vec{x}_{1\perp}|^2} \cdot \frac{\vec{x}_{2\perp} - \vec{b}_{2\perp}}{|\vec{x}_{2\perp} - \vec{b}_{2\perp}|^2} \right.$$
$$\left. + \frac{\vec{\epsilon}_\perp^{\,\lambda *} \cdot (\vec{x}_{2\perp} - \vec{b}_{2\perp})}{|\vec{x}_{2\perp} - \vec{b}_{2\perp}|^2} \frac{\vec{x}_{1\perp} - \vec{b}_{1\perp}}{|\vec{x}_{1\perp} - \vec{b}_{1\perp}|^2} \cdot \frac{\vec{z}_\perp - \vec{x}_{2\perp}}{|\vec{z}_\perp - \vec{x}_{2\perp}|^2} \right] \times$$

$$
\begin{aligned}
&\times f^{abc}\left[U^{bd}_{\vec{x}_{1\perp}} - U^{bd}_{\vec{b}_{1\perp}}\right]\left[U^{ce}_{\vec{x}_{2\perp}} - U^{ce}_{\vec{b}_{2\perp}}\right]\left(V_{\vec{b}_{1\perp}} t^d\right)_1 \left(V_{\vec{b}_{2\perp}} t^e\right)_2 + \frac{ig^3}{4\pi^3} f^{abc}\left(V_{\vec{b}_{1\perp}} t^d\right)_1 \left(V_{\vec{b}_{2\perp}} t^e\right)_2 \\
&\times \int d^2x \left[U^{bd}_{\vec{b}_{1\perp}}\left(U^{ce}_{\vec{x}_\perp} - U^{ce}_{\vec{b}_{2\perp}}\right)\left(\frac{\vec{\epsilon}^{\,\lambda *}_\perp\cdot(\vec{z}_\perp - \vec{x}_\perp)}{|\vec{z}_\perp - \vec{x}_\perp|^2}\,\frac{\vec{x}_\perp - \vec{b}_{1\perp}}{|\vec{x}_\perp - \vec{b}_{1\perp}|^2}\cdot\frac{\vec{x}_\perp - \vec{b}_{2\perp}}{|\vec{x}_\perp - \vec{b}_{2\perp}|^2}\right.\right. \\
&-\frac{\vec{\epsilon}^{\,\lambda *}_\perp\cdot(\vec{z}_\perp - \vec{b}_{1\perp})}{|\vec{z}_\perp - \vec{b}_{1\perp}|^2}\,\frac{\vec{z}_\perp - \vec{x}_\perp}{|\vec{z}_\perp - \vec{x}_\perp|^2}\cdot\frac{\vec{x}_\perp - \vec{b}_{2\perp}}{|\vec{x}_\perp - \vec{b}_{2\perp}|^2} - \left.\frac{\vec{\epsilon}^{\,\lambda *}_\perp\cdot(\vec{z}_\perp - \vec{b}_{1\perp})}{|\vec{z}_\perp - \vec{b}_{1\perp}|^2}\,\frac{\vec{x}_\perp - \vec{b}_{1\perp}}{|\vec{x}_\perp - \vec{b}_{1\perp}|^2}\cdot\frac{\vec{x}_\perp - \vec{b}_{2\perp}}{|\vec{x}_\perp - \vec{b}_{2\perp}|^2}\right) \\
&-\left(U^{bd}_{\vec{x}_\perp} - U^{bd}_{\vec{b}_{1\perp}}\right) U^{ce}_{\vec{b}_{2\perp}}\left(\frac{\vec{\epsilon}^{\,\lambda *}_\perp\cdot(\vec{z}_\perp - \vec{x}_\perp)}{|\vec{z}_\perp - \vec{x}_\perp|^2}\,\frac{\vec{x}_\perp - \vec{b}_{1\perp}}{|\vec{x}_\perp - \vec{b}_{1\perp}|^2}\cdot\frac{\vec{x}_\perp - \vec{b}_{2\perp}}{|\vec{x}_\perp - \vec{b}_{2\perp}|^2}\right. \\
&\left.\left.-\frac{\vec{\epsilon}^{\,\lambda *}_\perp\cdot(\vec{z}_\perp - \vec{b}_{2\perp})}{|\vec{z}_\perp - \vec{b}_{2\perp}|^2}\,\frac{\vec{z}_\perp - \vec{x}_\perp}{|\vec{z}_\perp - \vec{x}_\perp|^2}\cdot\frac{\vec{x}_\perp - \vec{b}_{1\perp}}{|\vec{x}_\perp - \vec{b}_{1\perp}|^2} - \frac{\vec{\epsilon}^{\,\lambda *}_\perp\cdot(\vec{z}_\perp - \vec{b}_{2\perp})}{|\vec{z}_\perp - \vec{b}_{2\perp}|^2}\,\frac{\vec{x}_\perp - \vec{b}_{1\perp}}{|\vec{x}_\perp - \vec{b}_{1\perp}|^2}\cdot\frac{\vec{x}_\perp - \vec{b}_{2\perp}}{|\vec{x}_\perp - \vec{b}_{2\perp}|^2}\right)\right] \\
&-\frac{ig^3}{4\pi^2} f^{abc}\left(V_{\vec{b}_{1\perp}} t^d\right)_1 \left(V_{\vec{b}_{2\perp}} t^e\right)_2\left[(U^{bd}_{\vec{z}_\perp} - U^{bd}_{\vec{b}_{1\perp}})\,U^{ce}_{\vec{b}_{2\perp}}\,\frac{\vec{\epsilon}^{\,\lambda *}_\perp\cdot(\vec{z}_\perp - \vec{b}_{1\perp})}{|\vec{z}_\perp - \vec{b}_{1\perp}|^2}\ln\frac{1}{|\vec{z}_\perp - \vec{b}_{2\perp}|\Lambda} - U^{bd}_{\vec{b}_{1\perp}}\,(U^{ce}_{\vec{z}_\perp}\right. \\
&\left.-U^{ce}_{\vec{b}_{2\perp}})\,\frac{\vec{\epsilon}^{\,\lambda *}_\perp\cdot(\vec{z}_\perp - \vec{b}_{2\perp})}{|\vec{z}_\perp - \vec{b}_{2\perp}|^2}\ln\frac{1}{|\vec{z}_\perp - \vec{b}_{1\perp}|\Lambda}\right] - \frac{ig^3}{4\pi^3}\int d^2x\left[U^{ab}_{\vec{x}_\perp} - U^{ab}_{\vec{z}_\perp}\right] f^{bde}\left(V_{\vec{b}_{1\perp}} t^d\right)_1 \left(V_{\vec{b}_{2\perp}} t^e\right)_2 \\
&\times\frac{\vec{\epsilon}^{\,\lambda *}_\perp\cdot(\vec{z}_\perp - \vec{x}_\perp)}{|\vec{z}_\perp - \vec{x}_\perp|^2}\,\frac{\vec{x}_\perp - \vec{b}_{1\perp}}{|\vec{x}_\perp - \vec{b}_{1\perp}|^2}\cdot\frac{\vec{x}_\perp - \vec{b}_{2\perp}}{|\vec{x}_\perp - \vec{b}_{2\perp}|^2}\,\mathrm{Sign}(b_2^- - b_1^-)\,.
\end{aligned}
\tag{4}
$$

Another class of diagrams that contribute at g^5 order is shown in Fig. 8 and the result is [2]

$$
\begin{aligned}
&\sum_{i=1}^{6}{}' D_i \\
&= -\frac{g^3}{8\pi^4}\int d^2x_1\, d^2x_2\, \delta[(\vec{z}_\perp - \vec{x}_{1\perp})\times(\vec{z}_\perp - \vec{x}_{2\perp})]\left[\frac{\vec{\epsilon}^{\,\lambda *}_\perp\cdot(\vec{x}_{2\perp} - \vec{x}_{1\perp})}{|\vec{x}_{2\perp} - \vec{x}_{1\perp}|^2}\,\frac{\vec{x}_{1\perp} - \vec{b}_{2\perp}}{|\vec{x}_{1\perp} - \vec{b}_{2\perp}|^2}\cdot\frac{\vec{x}_{2\perp} - \vec{b}_{2\perp}}{|\vec{x}_{2\perp} - \vec{b}_{2\perp}|^2}\right. \\
&\left.-\frac{\vec{\epsilon}^{\,\lambda *}_\perp\cdot(\vec{x}_{1\perp} - \vec{b}_{2\perp})}{|\vec{x}_{1\perp} - \vec{b}_{2\perp}|^2}\,\frac{\vec{z}_\perp - \vec{x}_{1\perp}}{|\vec{z}_\perp - \vec{x}_{1\perp}|^2}\cdot\frac{\vec{x}_{2\perp} - \vec{b}_{2\perp}}{|\vec{x}_{2\perp} - \vec{b}_{2\perp}|^2} + \frac{\vec{\epsilon}^{\,\lambda *}_\perp\cdot(\vec{x}_{2\perp} - \vec{b}_{2\perp})}{|\vec{x}_{2\perp} - \vec{b}_{2\perp}|^2}\,\frac{\vec{x}_{1\perp} - \vec{b}_{2\perp}}{|\vec{x}_{1\perp} - \vec{b}_{2\perp}|^2}\cdot\frac{\vec{z}_\perp - \vec{x}_{2\perp}}{|\vec{z}_\perp - \vec{x}_{2\perp}|^2}\right] \\
&\times f^{abc}\left[U^{bd}_{\vec{x}_{1\perp}} - U^{bd}_{\vec{b}_{2\perp}}\right]\left[U^{ce}_{\vec{x}_{2\perp}} - U^{ce}_{\vec{b}_{2\perp}}\right]\left(V_{\vec{b}_{1\perp}}\right)_1\left(V_{\vec{b}_{2\perp}} t^e t^d\right)_2 \\
&+\frac{ig^3}{4\pi^3}\int d^2x\, f^{abc}\, U^{bd}_{\vec{b}_{2\perp}}\left[U^{ce}_{\vec{x}_\perp} - U^{ce}_{\vec{b}_{2\perp}}\right]\left(V_{\vec{b}_{1\perp}}\right)_1\left(V_{\vec{b}_{2\perp}} t^e t^d\right)_2\left(\frac{\vec{\epsilon}^{\,\lambda *}_\perp\cdot(\vec{z}_\perp - \vec{x}_\perp)}{|\vec{z}_\perp - \vec{x}_\perp|^2}\,\frac{1}{|\vec{x}_\perp - \vec{b}_{2\perp}|^2}\right. \\
&\left.-\frac{\vec{\epsilon}^{\,\lambda *}_\perp\cdot(\vec{z}_\perp - \vec{b}_{2\perp})}{|\vec{z}_\perp - \vec{b}_{2\perp}|^2}\,\frac{\vec{z}_\perp - \vec{x}_\perp}{|\vec{z}_\perp - \vec{x}_\perp|^2}\cdot\frac{\vec{x}_\perp - \vec{b}_{2\perp}}{|\vec{x}_\perp - \vec{b}_{2\perp}|^2} - \frac{\vec{\epsilon}^{\,\lambda *}_\perp\cdot(\vec{z}_\perp - \vec{b}_{2\perp})}{|\vec{z}_\perp - \vec{b}_{2\perp}|^2}\,\frac{1}{|\vec{x}_\perp - \vec{b}_{2\perp}|^2}\right) \\
&+\frac{ig^3}{4\pi^2} f^{abc}\, U^{bd}_{\vec{b}_{2\perp}}\left[U^{ce}_{\vec{z}_\perp} - U^{ce}_{\vec{b}_{2\perp}}\right]\left(V_{\vec{b}_{1\perp}}\right)_1\left(V_{\vec{b}_{2\perp}} t^e t^d\right)_2\,\frac{\vec{\epsilon}^{\,\lambda *}_\perp\cdot(\vec{z}_\perp - \vec{b}_{2\perp})}{|\vec{z}_\perp - \vec{b}_{2\perp}|^2}\ln\frac{1}{|\vec{z}_\perp - \vec{b}_{2\perp}|\Lambda}\,.
\end{aligned}
\tag{5}
$$

3 Gauge invariance

As we stated above, the calculation was carried out in the light-cone gauge. It is well known that the light-cone gauge posses an ambiguity about the light-cone singularity $k^+ = 0$. In Ref. [3] we showed that this ambiguity is related to the residual sub-gauge choice, and that different prescriptions

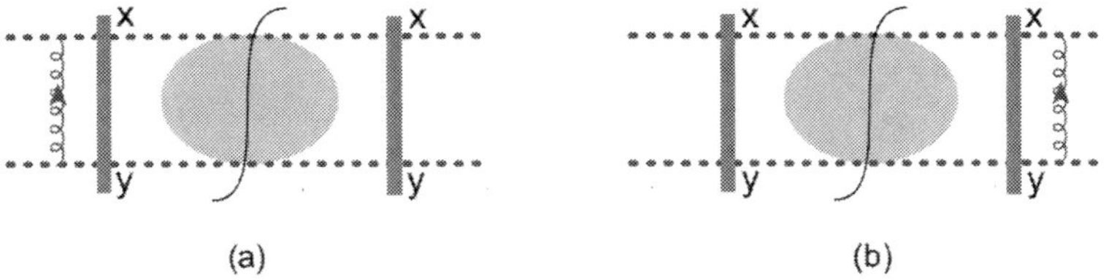

Figure 5. The shaded region represents a general interaction. The two diagrams, with retarded gluon propagators with time flow in the direction of the arrow, cancel.

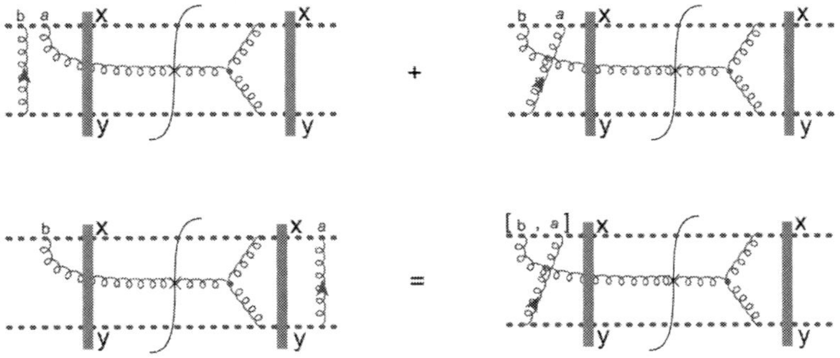

Figure 6. The sum of these three diagrams is equal to the diagram with the color factor associated with the top quark replaced by a commutator. The commutator is represented by a bracket, [,].

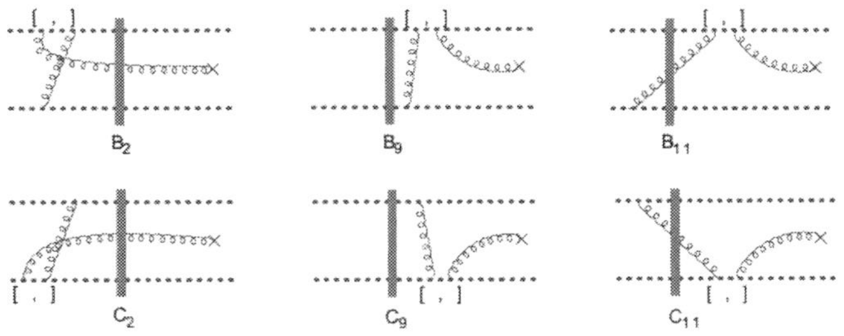

Figure 7. Diagrams B and Diagrams C consist of a single gluon emission with one gluon traveling between the quarks. The color factor associated with the top quark has been replaced with a commutator. The gluons connecting the quarks are retarded with the momentum flowing to the top quark line for diagrams B, and retarded with momentum flowing to the bottom quark line for diagrams C.

for the k^+ singularity coincide with different sub-gauge conditions. It turned out that the quark lines reduce to Wilson lines if we regulate the light-cone singularity with Principle-Value (PV). The sub-gauge condition that leads to PV-prescription is $\partial_\perp \cdot A_\perp(x^- = +\infty) + \partial_\perp \cdot A_\perp(x^- = -\infty) = 0$ and the light-cone propagator is

$$D^{\mu\nu}_{PV}(x,y) \equiv \int \frac{d^4k}{(2\pi)^4} e^{-ik\cdot(x-y)} \frac{-i}{k^2+i\epsilon}\left[g^{\mu\nu}_\perp - \left(k^\mu_\perp \eta^\nu + k^\nu_\perp \eta^\mu\right)\mathrm{PV}\left\{\frac{1}{k^+}\right\} - 2\eta^\mu\eta^\nu k^- \mathrm{PV}\left\{\frac{1}{k^+}\right\}\right] \quad (6)$$

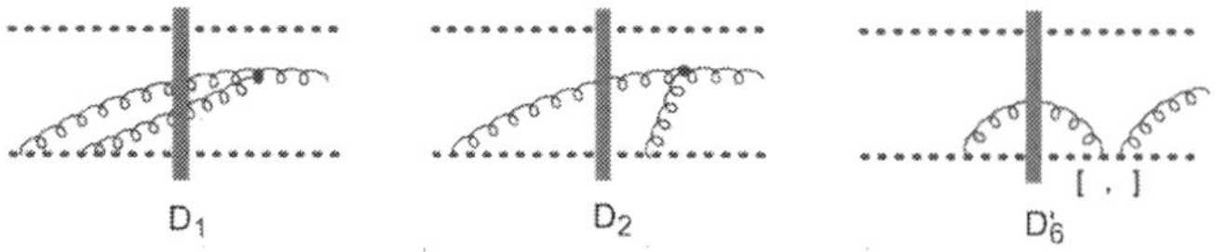

Figure 8. Diagrams involving only one quark at the amplitude level. All gluons are retarded with momentum flowing to the right.

where $\mathrm{PV}\left\{\frac{1}{k^+}\right\} \equiv \frac{1}{2}\left(\frac{1}{k^++i\epsilon} + \frac{1}{k^+-i\epsilon}\right)$. Alternatively, one may choose sub-gauge condition $\vec{\partial}_\perp \cdot \vec{A}_\perp(x^- = +\infty) = 0$ and the propagator is

$$D_1^{\mu\nu}(x,y) \equiv \int \frac{d^4k}{(2\pi)^4} e^{-ik\cdot(x-y)} \frac{-i}{k^2+i\epsilon}\left[g^{\mu\nu} - \frac{k^\mu\eta^\nu}{k^+-i\epsilon} - \frac{k^\nu\eta^\mu}{k^++i\epsilon}\right]. \tag{7}$$

For sub-gauge condition at $x^- = -\infty$ *i.e* $\vec{\partial}_\perp \cdot \vec{A}_\perp(x^- = -\infty) = 0$ the propagator is

$$D_2^{\mu\nu}(x,y) \equiv \int \frac{d^4k}{(2\pi)^4} e^{-ik\cdot(x-y)} \frac{-i}{k^2+i\epsilon}\left[g^{\mu\nu} - \frac{k^\mu\eta^\nu}{k^++i\epsilon} - \frac{k^\nu\eta^\mu}{k^+-i\epsilon}\right]. \tag{8}$$

The results we obtained in Ref. [2] is shown to be invariant under the above sub-gauge choices.

4 Conclusion and Outlook

The results presented here represent a first step toward obtaining the single inclusive gluon production cross-section for heavy-light ions collisions at the classical level beyond the LO. The results we obtained are reasonably compact considering the number of diagrams that contribute at this order. For the success of this calculation it was important to organize the diagrams in different classes and to combine many of them to fewer *effective* diagrams.

It is certainly desirable to compare our result to the similar one obtained in Ref. [4], but such comparison will not be possible until all the diagrams of order g^5 (at amplitude level) will be calculated. Moreover, since our result is at the classical level and there are no quantum fluctuations, it also desirable to check its conformal invariance in the transverse coordinate space. However, at this point this check cannot be performed since diagrams at order g^5 at the amplitude level have still to be calculated.

The author is grateful to the organizers of QCD@work 2016 for warm hospitality.

References

[1] Y. V. Kovchegov and A. H. Mueller, *Gluon production in current nucleus and nucleon nucleus collisions in a quasi-classical approximation, Nucl. Phys.* **B529** (1998) 451-479 [hep-ph/9802440].

[2] G. A. Chirilli, Y. V. Kovchegov, D. E. Wertepny, *Classical Gluon Production Amplitude for Nucleus-Nucleus Collisions: First Saturation Correction in the Projectile, JHEP* **1513** (2015) 015 [arxiv:1501.03106].

[3] G. A. Chirilli, Y. V. Kovchegov and D. E. Wertepny, JHEP **1512**, 138 (2015) doi:10.1007/JHEP12(2015)138 [arXiv:1508.07962 [hep-ph]].

[4] I. Balitsky, *Scattering of shock waves in QCD, Phys. Rev.* **D70** (2004) 114030 [hep-ph/0409314].

EPJ Web of Conferences **129**, 00032 (2016) DOI: 10.1051/epjconf/201612900032
QCD@Work 2016

A perturbative approach to the confinement-deconfinement phase transition*

Urko Reinosa[1,**]

[1] *CPHT, Ecole Polytechnique, CNRS, Université Paris-Saclay, F-91128 Palaiseau, France*

Abstract. We summarize recent progress in describing the confinement-deconfinement transition from a novel perturbative approach.

The low energy regime of Quantum Chromodynamics is not accessible to standard perturbative methods based on the Faddeev-Popov gauge fixing procedure, because the corresponding running coupling becomes large at these scales and even diverges at a finite scale known as $\Lambda_{\rm QCD}$. To remedy these shortcomings, various non-perturbative approaches have been invented, from lattice QCD to non-perturbative continuum methods, and much progress has been achieved over the years. Here, we would like to report on yet another possible route to the low energy regime of QCD. It is based on the speculation that perturbation theory might be applicable after all, provided one takes into account modifications of the Faddeev-Popov procedure. These modifications are needed because, in most practical cases, the gauge fixing is ambiguous in the infrared due to the presence of so-called Gribov copies. Various ways of dealing with the Gribov ambiguity in the Landau gauge have been proposed in the literature but we shall focus more specifically on the proposal of [1], based on the addition of a gluon mass term to the usual Landau gauge-fixed action.[1] This modified action remains renormalizable and, interestingly, some of the corresponding RG trajectories yield a running coupling that remains bounded and effectively small, even at low energies [3] which justifies the use of perturbation theory in this regime. In particular, at one-loop, the approach has been shown to reproduce lattice results for correlation functions to a good accuracy.

We shall further test the validity of this novel perturbative approach by studying the phase structure of QCD in the heavy-quark limit, for which the lattice has provided a wealth of data to compare with, including the case of a real chemical potential. In this regime, the relevant transition is the confinement-deconfinement transition, which we study by evaluating the Polyakov loop effective potential $V(\ell,\bar{\ell})$. The appropriate extremization of the latter, see below, allows us to access the actual values of the Polyakov loops (P denotes the path-oredering and $\bar{P}$ the anti-path ordering)

$$\ell \equiv \frac{1}{3}{\rm tr}\left\langle P\exp\left(ig\int_0^\beta d\tau A_0^a t^a\right)\right\rangle \quad\text{and}\quad \bar{\ell} \equiv \frac{1}{3}{\rm tr}\left\langle \bar{P}\exp\left(-ig\int_0^\beta d\tau A_0^a t^a\right)\right\rangle, \tag{1}$$

as functions of the temperature $T = 1/\beta$. These quantities are directly related to the free energies of a static quark and a static anti-quark respectively, $\ell = e^{-\beta F_q}$ and $\bar{\ell} = e^{-\beta F_{\bar{q}}}$, and are thus order parameters

*Based on various works in collaboration with Julien Serreau, Matthieu Tissier and Nicolás Wschebor.

**e-mail: urko.reinosa@polytechnique.edu

[1] We refer to [2] and references therein for details on the so-called (refined) Gribov-Zwanziger approach.

for the confinement-deconfinement transition. In fact, they are strict order parameters only in the limit of infinite quark masses, where they probe the breaking of the center symmetry of Yang-Mills theory at finite temperature, but they remain nevertheless good signatures of the transition for large but finite quark masses.

1 Polyakov loop effective potential at one-loop order

By definition, the Polyakov loop effective potential is a gauge-invariant function. However, computing it perturbatively requires one to fix the gauge. Here it is important to realize that not all gauges are equivalent within a given approximation method. In particular, if one aims at obtaining a clear signal of the transition, which is governed by center symmetry in the pure glue case, one should work in a gauge that does not break this symmetry explicitly. One possibility is to use the so-called Landau-DeWitt gauge, see for instance [4], which leads to the following Faddeev-Popov contribution to the QCD action:

$$S_{\rm FP} = \int_x \left\{(\bar{D}_\mu \bar{c})^a (D_\mu c)^a + ih^a(\bar{D}_\mu a_\mu)^a\right\}. \tag{2}$$

Here $a_\mu \equiv A_\mu - \bar{A}_\mu$ with $\bar{A}_\mu$ a given background field configuration and $\bar{D}^{ab}_\mu \equiv \delta^{ab}\partial_\mu + gf^{acb}\bar{A}^c_\mu$ is the corresponding covariant derivative. Since the Landau-DeWitt gauge suffers from the Gribov ambiguity, we modify the Faddeev-Popov action by adding the term

$$S_m = \int_x \frac{1}{2} m^2 a^a_\mu a^a_\mu, \tag{3}$$

along the lines of [1, 3, 5]. In the SU(3) case, the value of the mass that allows to reproduce lattice data for the correlators at zero temperature is roughly $m \simeq 510\,\mathrm{MeV}$ and we shall consider this value in what follows.

A priori, the background $\bar{A}$ is arbitrary. However, within a given approximation scheme such as the perturbative scheme considered here, it might be convenient to choose a specific $\bar{A}$ in order to unveil certain properties of the system. In particular, at finite temperature T, it is natural to choose temporal and uniform background fields and, without loss of generality, one can take them along the Cartan subalgebra of $su(3)$: $g\bar{A}_\mu(x) = T\delta_{\mu 0}(r_3\lambda_3 + r_8\lambda_8)/2$. Moreover, in order to keep a good handle on center symmetry, it is convenient to choose so-called self-consistent backgrounds, such that $\bar{A} = \langle A\rangle_{\bar{A}}$ where $\langle ...\rangle_{\bar{A}}$ denotes the expectation value in the Landau-DeWitt gauge (including the additional mass term) with background $\bar{A}$. It can be shown that such backgrounds play the role of order parameters for center symmetry in the pure Yang-Mills case [6] and that they are obtained by minimizing a center-symmetric potential (see below for the appropriate extremization procedure in the presence of quarks), the so-called background effective potential $V(r_3, r_8)$. Its one-loop expression (including quarks) in the modified gauge-fixing defined by Eqs. (2) and (3) can be found in [7]. Correspondingly, one also obtains the tree-level expressions for the Polyakov loops in terms of the background components:

$$\ell = \frac{1}{3}\left[e^{-i\frac{r_8}{\sqrt{3}}} + 2e^{i\frac{r_8}{2\sqrt{3}}}\cos(r_3/2)\right] \quad \text{and} \quad \bar{\ell} = \frac{1}{3}\left[e^{i\frac{r_8}{\sqrt{3}}} + 2e^{-i\frac{r_8}{2\sqrt{3}}}\cos(r_3/2)\right]. \tag{4}$$

By inverting this mapping and by plugging it back into the one-loop expression for $V(r_3, r_8)$, one arrives at a one-loop expression for the Polyakov loop effective potential. More precisely, one obtains $V(\ell, \bar{\ell}) = V_{\rm gauge}(\ell, \bar{\ell}) + \sum_f V_f(\ell, \bar{\ell})$, with

$$V_{\rm f}(\ell, \bar{\ell}) = -\frac{T}{\pi^2}\int_0^\infty dq\, q^2 \left\{\ln\left[1 + 3\ell\, e^{-\beta(\varepsilon^f_q - \mu)} + 3\bar{\ell}\, e^{-2\beta(\varepsilon^f_q - \mu)} + e^{-3\beta(\varepsilon^f_q - \mu)}\right] + (\mu \to -\mu)\right\} \tag{5}$$

and

$$V_{\rm gauge}(\ell,\bar\ell) = \frac{T}{\pi^2}\int_0^\infty dq\,q^2\left\{\frac{3}{2}\ln\Big[1+e^{-8\beta\varepsilon_q}-(9\ell\bar\ell-1)(e^{-\beta\varepsilon_q}+e^{-7\beta\varepsilon_q})\right.$$
$$+(27\ell^3+27\bar\ell^3-27\ell\bar\ell+1)(e^{-2\beta\varepsilon_q}+e^{-6\beta\varepsilon_q})$$
$$-(81\ell^2\bar\ell^2-27\ell\bar\ell+2)(e^{-3\beta\varepsilon_q}+e^{-5\beta\varepsilon_q})$$
$$\left.+(162\ell^2\bar\ell^2-54\ell^3-54\bar\ell^3+18\ell\bar\ell-2)e^{-4\beta\varepsilon_q}\Big]-\frac{1}{2}(m\to 0)\right\}, \quad (6)$$

where $\varepsilon_q \equiv \sqrt{q^2+m^2}$ and $\varepsilon_q^f \equiv \sqrt{q^2+M_f^2}$, with M_f the mass of the quark of flavour f. In what follows, we discuss the phase structure that emerges from these one-loop expressions, both at imaginary and at real chemical potential, and compare it with lattice predictions.

2 Phase structure at imaginary chemical potential

The case of an imaginary chemical potential has attracted a lot of attention in recent years for the fermion determinant that appears under the QCD functional integral is real and positive. This means that there is no sign problem and lattice simulations provide accurate results in this case. Although this situation does not correspond to physical QCD, one hopes at gaining some knowledge about the structural properties of the theory and even at extracting some information about the case of a real chemical potential, through analytic continuation.

The fermion determinant being real, it is simple to argue that the Polyakov loops defined in Eq. (1) should be complex conjugate of each other, as expectation values of complex conjugate quantities in the presence of a real weight under the functional integral. This means that we can restrict the study of $V(\ell,\bar\ell)$ to pairs $(\ell,\bar\ell)$ that belong to $\Sigma \equiv \{(u,v)\in\mathbb{C}^2\,|\,v=u^*\}$. In other words, this means that we can consider $V(\ell)\equiv V(\ell,\bar\ell=\ell^*)$. This fact, together with the tree-level expressions (4), is compatible with our findings in [7], where in the case of an imaginary chemical potential, we argued that the background components r_3 and r_8 could be taken real. We note however that, even though the pair (r_3,r_8) can take any value in $\mathbb{R}\times\mathbb{R}$, ℓ and $\bar\ell$ are only allowed to take certain values in the complex plane, see figure 1. We shall thus restrict the analysis of $V(\ell)$ to this region.

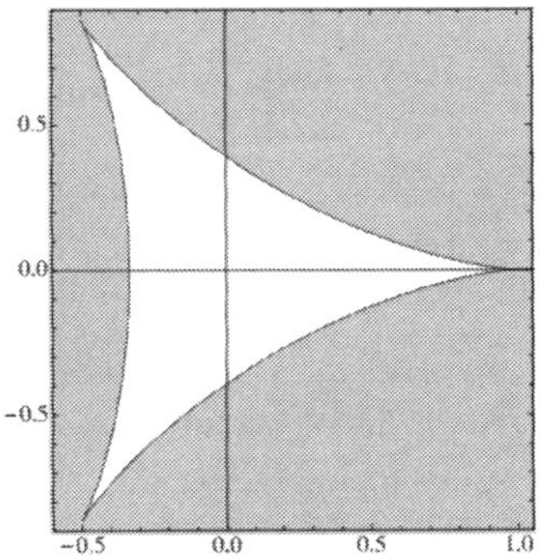

Figure 1. Allowed values (white region) for the Polyakov loop in the complex plane for ℓ, in the case of an imaginary chemical potential.

As already mentioned, the fermion determinant is not only real but positive. This in turn allows to argue that the physical value (1) of the Polyakov loop is obtained at the absolute minimum of $V(\ell)$, more details will be given elsewhere. With this rule in mind, we can now discuss the phase structure that emerges from our one-loop potential.

DOI: 10.1051/epjconf/201612900032

Let us first start with the infinite quark mass limit, that is the pure Yang-Mills case. We find a first order phase transition at a temperature $T_d \simeq 185\,\text{MeV}$, see [5], the unique, Z_3-symmetric minimum of $V(\ell)$ at $\ell = 0$ for $T < T_d$, turning into a triplet of Z_3-breaking minima with $\ell \neq 0$ for $T > T_d$. The order of the transition is in agreement with the lattice predictions, even though the value for T_d is somewhat below the lattice value $T_d \simeq 270\,\text{MeV}$, see for instance [8]. However, one should keep in mind that the present calculation is just a one-loop result and one needs to assess how large the corrections are. We have computed the two-loop corrections to the background effective potential and we have again found a first order phase transition with $T_d \simeq 254\,\text{Mev}$, see Ref. [6], a value much closer to the lattice result or to results from functional methods [9].

Including the quarks, first at zero chemical potential, we find that the nature of the transition depends on the values of the quark masses. For large enough u, d and s quark masses, the transition remains of first order. But as the masses are lowered, one encounters a line of second order phase transitions in the $(M_u = M_d, M_s)$ plane (the so-called Columbia plot), see figure 2. For even smaller masses, the transition becomes a crossover. The agreement with lattice results is not only qualitative

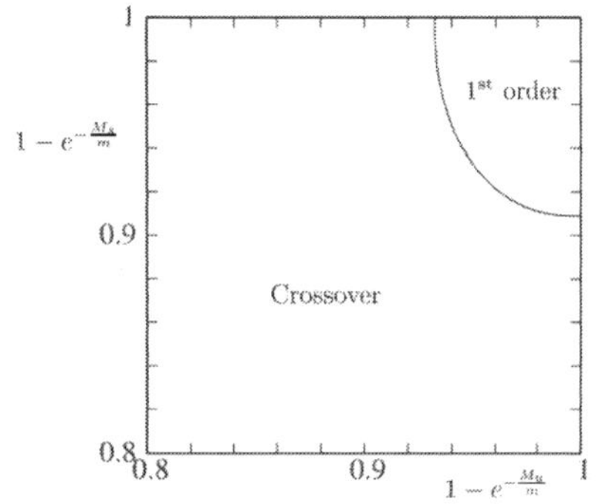

Figure 2. Top right corner of the Columbia plot as obtained from our perturbative approach at one-loop order.

but also quantitative as shown in table 1 where we show the ratio of the critical mass to the critical temperature for $N_f = 1, 2, 3$ degenerate flavours. It is also worth noting that, at this order of approximation, this ratio does not dependent on the specific value of the gluon mass parameter m. We mention

Table 1. Ratio of the critical quark mass to the critical temperature for $N_f = 1, 2, 3$ degenerate flavours, as obtained in the present approach (second column) and compared to results from the lattice [10] (third column) and a matrix model [11] (fourth column).

N_f	M_c/T_c	$(M_c/T_c)^{\text{latt.}}$	$(M_c/T_c)^{\text{matr.}}$
1	6.74	7.22(5)	8.04
2	7.59	7.91(5)	8.85
3	8.07	8.32(5)	9.33

finally that, as long as $\mu = 0$, ℓ remains real (and positive), in agreement with its interpretation in terms of the quark free energy. Since $\bar{\ell} = \ell^*$, this implies also that $\ell = \bar{\ell}$, in line with the fact that the quark and anti-quark free energies need to be equal since charge conjugation is manifest at $\mu = 0$.

As we turn on the chemical potential along the imaginary axis $i\mathbb{R}$, we observe that the Polyakov loop ℓ acquires a phase which rotates as $\mu_i \equiv \text{Im}\,\mu$ is increased. If the temperature is large enough, this occurs with a first order jump at $\mu_i/T = \pi/3$. We have thus recovered the well known Roberge-Weiss transition, see figure 3. The dependence of the phase structure on the quark masses is again in qualitative agreement with that observed on the lattice. In particular, we observe that the critical points of the Columbia plot at $\mu = 0$ enter the imaginary chemical potential phase diagram as the

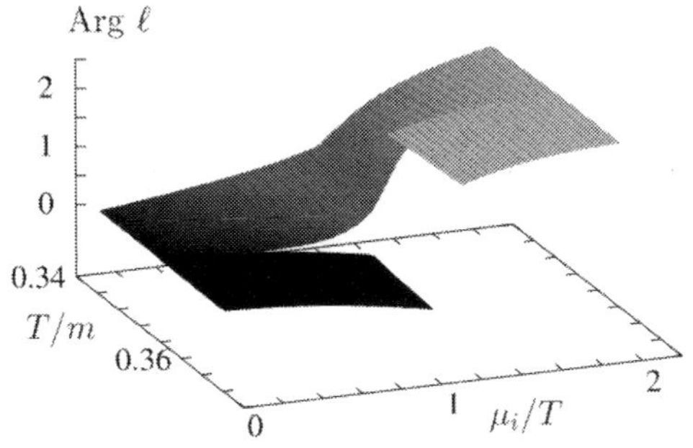

Figure 3. Argument of the Polyakov loop as a function of the temperature and imaginary chemical potential. For large enough temperatures, the phase undergoes a first order phase transition at $\mu_i/T = \pi/3$.

mass is decreased below the critical mass. By decreasing the mass even further, two of such critical points end up meeting at a tricritical point whose scaling is essentially governed by the corresponding mean-field exponent:

$$\frac{M_c(\mu)}{T_c(\mu)} = \frac{M_{\text{tric.}}}{T_{\text{tric.}}} + K\left[\left(\frac{\pi}{3}\right)^2 + \left(\frac{\mu}{T}\right)^2\right]^{2/5}, \tag{7}$$

as observed also on the lattice [12]. The agreement with the lattice also extends to non-universal quantities such as the ratio $M_{\text{tric.}}/T_{\text{tric.}}$ or the prefactor K. We find $M_{\text{tric.}}/T_{\text{tric.}} = 6.15$ and $K = 1.85$, to be compared with the lattice result of Ref. [10], $(M_{\text{tric.}}/T_{\text{tric.}})^{\text{latt.}} = 6.66$ and $K^{\text{latt.}} = 1.55$ for 3 degenerate quark flavours.

3 Phase structure at real chemical potential

In the case of a real chemical potential, the fermion determinant becomes complex. This is at the root of the sign problem in lattice QCD and, as we discuss below, leads to a (milder) sign problem in continuum approaches.

First of all, it is not true anymore that ℓ and $\bar{\ell}$ are complex conjugate of each other and it is not consistent to restrict the analysis of the Polyakov potential to $V(\ell, \bar{\ell} = \ell^*)$. In fact, using some symmetries under the QCD functional integral, one shows that ℓ and $\bar{\ell}$ are both real and independent a priori. This means that the analysis of $V(\ell, \bar{\ell})$ can be restricted to pairs $(\ell, \bar{\ell}) \in \mathbb{R} \times \mathbb{R}$. This change of paradigm from the case of imaginary chemical potential is represented in figure 4. Once again, it is consistent with our findings in the case of the background field effective potential $V(r_3, r_8)$ for which we found that, in the case of a real chemical potential, r_3 needed to be taken real, while r_8 needed to be taken imaginary. We mention once more that, if r_3 and r_8 are free to take any value in $\mathbb{R}$ and $i\mathbb{R}$ respectively, this is not the case for the pair $(\ell, \bar{\ell})$ which explores only a subregion of $\mathbb{R} \times \mathbb{R}$ to which one should restrict the analysis of $V(\ell, \bar{\ell})$.

The previous discussion does not really qualify as a sign problem. In fact, it does not originate in the fermion determinant not being positive but only in the fermion determinant not being real. Moreover, it is clear what needs to be done, that is taking ℓ and $\bar{\ell}$ real (and independent a priori) as follows from some of the symmetries of the QCD action. We mention also that the same symmetries allow us to argue that $V(\ell, \bar{\ell})$ is real if ℓ and $\bar{\ell}$ are taken real, a welcomed property if we expect to extract any meaningful thermodynamical information from the potential.

The true sign problem reemerges instead in that it is now not completely clear how to extract the physics from the potential $V(\ell, \bar{\ell})$. Even though the physical point of the system is certainly an extremum of $V(\ell, \bar{\ell})$ for it corresponds to the limit of zero sources, it is not clear which extremum one should choose in the case where various extrema exist. In fact the characterization of the zero source limit as the absolute minimum of $V(\ell, \ell^*)$ in the case of an imaginary chemical potential does

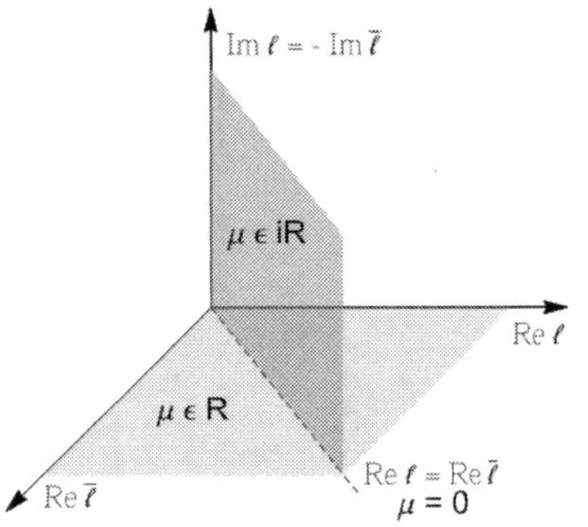

Figure 4. The two different planes where the Polyakov loop effective potential should be considered depending on whether $\mu \in i\mathbb{R}$ or $\mu \in \mathbb{R}$. These two planes are submanifolds of $\mathbb{C} \times \mathbb{C}$, but for the sake of representing them in a three dimensional figure, we considered the space $\{(\ell, \bar{\ell}) \in \mathbb{C}^2 \,|\, \text{Im}\,\ell = -\text{Im}\,\bar{\ell}\}$.

not apply to the case of a real chemical potential. This is rooted in the non-positivity of the fermion determinant and has thus the same origin than the lattice sign problem.

To proceed, we thus need to choose a recipe for selecting the physical extremum. To gain some insight, we can go back to the $\mu = 0$ case for which the physical extremum is characterized as a minimum in the vertical plane of figure 4 and lies at the intersection with the horizontal plane. Viewed from the perspective of this latter plane, it appears as a saddle point, actually the deepest saddle point. For any other $\mu \in \mathbb{R}$, we then choose to identify the physical extremum with the deepest saddle point, but we emphasize that we do not have a rigorous mathematical justification of this choice, even though it sounds plausible in terms of the minimisation of the Gibbs energy among all possible extrema. Given this recipe, our result for the Columbia plot for increasing μ is depicted in the left plot of figure 5. We find that the critical surface is pushed towards larger masses, as is also observed on the lattice.

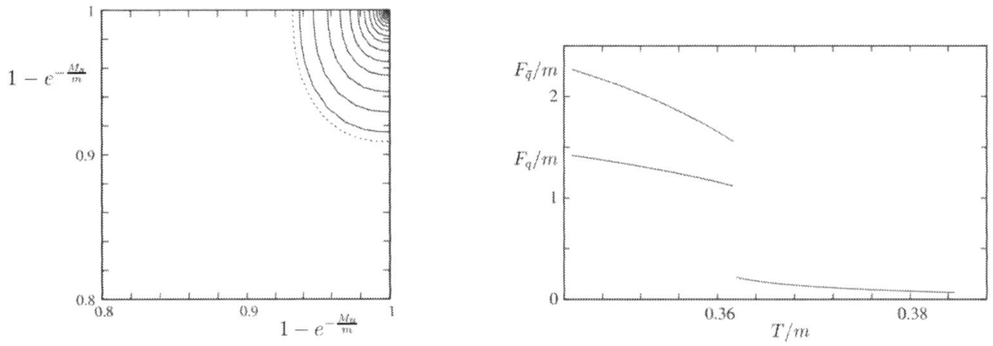

Figure 5. Left: Columbia plot at non-zero real μ. Projected critical surface on the $\mu = 0$ plane. Right: Quark and anti-quark free-energies as functions of the temperature and across the (first order) transition.

We mention finally that our choice of an imaginary component r_8 and thus of real and independent Polyakov loops ℓ and $\bar{\ell}$ is crucial if one wants to obtain different quark and anti-quark free-energies, see the right plot of figure 5, in line with the explicit breaking of charge conjugation symmetry that a non-zero μ implies. In other approches, r_8 is taken equal to 0 without much justification, which leads to $\ell = \bar{\ell}$ and thus to $F_q = F_{\bar{q}}$.

 DOI: 10.1051/epjconf/201612900032

References

[1] J. Serreau and M. Tissier, Phys. Lett. B **712** 97 (2012).
[2] D. Dudal, J. A. Gracey, S. P. Sorella, N. Vandersickel and H. Verschelde, Phys. Rev. D **78**, 065047 (2008).
[3] M. Tissier and N. Wschebor, Phys. Rev. D **82**,101701 (2010).
[4] J. Braun, H. Gies and J. M. Pawlowski, Phys. Lett. B **684**, 262 (2010).
[5] U. Reinosa, J. Serreau, M. Tissier and N. Wschebor, Phys. Lett. B **742**, 61 (2015).
[6] U. Reinosa, J. Serreau, M. Tissier and N. Wschebor, Phys. Rev. D **93**, 105002 (2016).
[7] U. Reinosa, J. Serreau and M. Tissier, Phys. Rev. D **92**, 025021 (2015).
[8] B. Lucini and M. Panero, Phys. Rept. **526**, 93 (2013).
[9] L. Fister and J. M. Pawlowski, Phys. Rev. D **88**, 045010 (2013).
[10] M. Fromm, J. Langelage, S. Lottini and O. Philipsen, JHEP **1201**, 042 (2012).
[11] K. Kashiwa, R. D. Pisarski and V. V. Skokov, Phys. Rev. D **85**, 114029 (2012).
[12] P. de Forcrand and O. Philipsen, Phys. Rev. Lett. **105**, 152001 (2010).

Non prompt D-meson measurements with ALICE at the LHC

Marianna Mazzilli on behalf of the ALICE Collaboration[1,*]

[1] *Dipartimento Interateneo di Fisica 'M.Merlin' and Sezione INFN, Bari, Italy*

Abstract. The production of hadrons with open heavy flavour (charm and beauty) in high-energy nucleus-nucleus collisions is a powerful tool to study the properties of the deconfined phase of strongly interacting matter known as the Quark-Gluon Plasma (QGP). The production of charm and beauty quarks occurs in hard partonic scattering processes in the early stage of the collisions. ALICE is the LHC experiment devoted to the study of heavy-ion physics. It is able to reconstruct charmed mesons in exclusive decays (e.g. $D^0 \rightarrow K^-\pi^+$) and beauty hadrons in semi-inclusive decays (e.g. B→eX, B→J/ψ X). At LHC energies a significant component of the inclusive D-meson yield originates from the decay of beauty-flavoured hadrons, whose knowledge is essential to determine the production of prompt D mesons coming from charm quarks. A precise determination of the non-prompt fraction combined with the determination of the inclusive D-meson yield would allow a measurement of beauty production. A data-driven method that exploits the different shapes of the distributions of the transverse-plane impact parameter to the primary vertex of prompt and feed-down D mesons in p-Pb collisions is used in ALICE. An alternative approach based on the D-meson decay length for Pb–Pb collisions is under study.

1 Introduction

The ALICE experiment at the LHC investigates the properties of the Quark-Gluon Plasma, the hot and dense state of strongly-interacting matter produced in high-energy heavy-ion collisions. Charm and beauty quarks are among the most powerful probes to study the QGP in heavy-ion collisions: they are produced in hard parton scattering processes occurring in the early stages of the collision, traverse the QCD medium and interact with its constituents experiencing the whole evolution of the medium. Charm production in pp, p-Pb and Pb–Pb collisions is studied in ALICE, e.g. by reconstructing exclusive D-meson hadronic decays. The yield obtained from the fit to the invariant mass distribution of D-meson candidates includes contributions from both prompt D mesons (i.e. produced at the primary interaction vertex from the charm quark, either directly or through decays of excited open charm and charmonium states) and feed-down (i.e. originating from beauty-hadron decays). The fraction of prompt D mesons in the raw yield, $f_{\rm prompt}$, can be extracted using different approaches. The first method employs perturbative QCD (pQCD) calculations, namely FONLL [1][2], in order to obtain the fraction of prompt and feed-down D-meson cross sections (FONLL-based method). A data-driven approach for the extraction of $f_{\rm prompt}$ has also been used in ALICE [3]. The method exploits the different shapes of the distributions of the transverse-plane impact parameter to the primary vertex

* e-mail: marianna.mazzilli@ba.infn.it

DOI: 10.1051/epjconf/201612900033

(d_0) of prompt and feed-down D mesons. An alternative approach based on the pseudo-proper decay length is under study.

2 D-meson reconstruction

D^0, D^+, D^{*+} and D_s^+ mesons, and their charge conjugates, are reconstructed via their hadronic decay channels $D^0 \rightarrow K^-\pi^+$ (with a branching ratio, BR, of 3.88$\pm$ 0.05%), $D^+ \rightarrow K^+\pi^-\pi^+$ (BR = 9.13 $\pm$ 0.19%), $D^{*+} \rightarrow D^0\pi^+$ (BR = 67.7 $\pm$ 0.5%) followed by $D^0 \rightarrow K^-\pi^+$, and $D_s^+ \rightarrow \phi\pi^+ \rightarrow K^-K^+\pi^+$ (BR = 2.24 $\pm$ 0.1%) [4]. The D^0, D^+, and D_s^+ mesons decay weakly with mean proper decay lengths ($c\tau$) of about 123, 312 and 150 μm [4], respectively. The analysis strategy is based on the reconstruction of secondary vertices separated by a few hundred μm from the interaction point. Both primary and secondary vertices are reconstructed with the Inner Tracking System (ITS) [7]. The selection of the D-meson decay topology is mainly based on the displacement of the tracks from the interaction vertex, the separation of the primary and secondary vertices, and the pointing of the reconstructed D-meson momentum to the primary vertex [5,6]. Further reduction of the combinatorial background is obtained by applying particle identification (PID) with the Time Projection Chamber (TPC) and the Time Of Flight detector (TOF) for pion and kaon tracks [7]. The applied PID strategy provides a reduction of the combinatorial background by a factor of about three at low transverse momentum (p_T) while preserving an efficiency of 95% for the D^0, D^+ and D^{*+} signals and of 85% for the D_s^+ signal. The D-meson raw yields are extracted from fits to the D^0, D^+ and D_s^+ candidate invariant-mass distributions and to the mass difference $\Delta M = M(K\pi\pi) - M(K\pi)$ distributions for D^{*+} candidates. In the fit function, the signal is modeled with a Gaussian and the background is described by an exponential term for D^0, D^+ and D_s^+ candidates and by a threshold function multiplied by an exponential for the D^{*+} case. For all four D-meson species, the mean values of the Gaussian functions in all transverse momentum and rapidity intervals were found to be compatible within uncertainties with the PDG world average values [4]. Figure 1 shows the D^0 and D^+ candidate invariant mass and D^{*+} mass-difference distribution in Pb–Pb collisions in three p_T intervals for the centrality[1] class 0–10% [5].

The D-meson raw yields extracted in each p_T and y interval include contributions from both prompt and from feed-down D mesons. The raw yields were corrected by a factor (f_{prompt}) in order to obtain the prompt D-meson cross sections:

$$\frac{d^2\sigma^D}{dp_T dy} = \frac{1}{2}\frac{1}{\Delta p_T} \cdot \frac{f_{prompt} \cdot N^{D+\bar{D},raw}(p_T, y)}{\Delta y} \cdot \frac{1}{(Acc \times \epsilon)_{prompt}(p_T)} \cdot \frac{1}{BR \cdot L_{int}} \tag{1}$$

In the formula, $N^{D+\bar{D},raw}$ is the raw yield (sum of particles and antiparticles). The factor 1/2 accounts for the fact that the measured yields include particles and antiparticles while the cross sections are given for particles only; f_{prompt} is the fraction of prompt D mesons in the raw yield; $(Acc\times\epsilon)_{prompt}$ is the product of acceptance and efficiency for prompt D-mesons, where ϵ accounts for the D-meson decay track reconstruction and selection and for D-meson candidate selection with secondary vertex and PID cuts; Δp_T and Δy are the widths of the transverse momentum and rapidity intervals; BR is the branching ratio of the considered decay channel, and L_{int} is the integrated luminosity. The acceptance and efficiency correction factors were obtained from Monte Carlo simulations including detailed descriptions of the geometry of the apparatus and of the detector response.

[1]Centrality is a variable that describes the geometry of a Pb–Pb collision in terms of percentiles of the total hadronic Pb–Pb cross section. Lowest percentiles correspond to lowest values of the impact parameter between the two colliding nuclei.

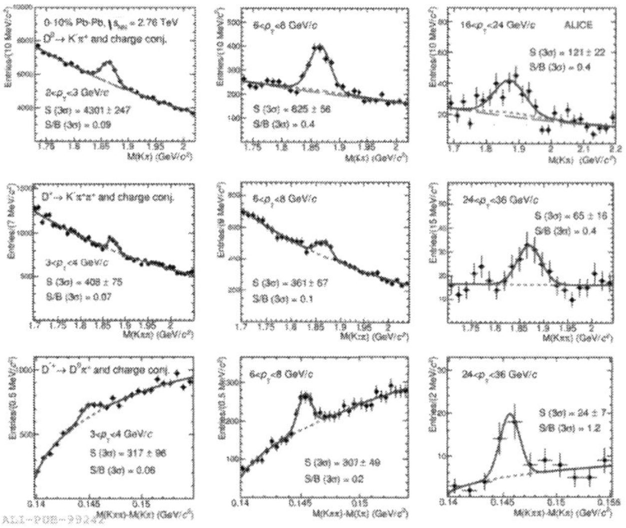

Figure 1. Invariant mass distributions for D^0 (top row) and D^+ (central row) candidates, and mass difference distribution for D^{*+} candidates $\Delta M = M(K\pi\pi) - M(K\pi)$ (bottom row) for the centrality class 0-10% in Pb–Pb collisions at $\sqrt{s_{NN}} = 2.76$ TeV. The distributions are reported in three p_T intervals for each meson (left, middle and right column).

3 Prompt D-meson fraction

3.1 FONLL-based method

The FONLL-based method estimates f_{prompt} as:

$$f_{prompt} = 1 - \frac{N^{D+\bar{D},feed-down}}{N^{D+\bar{D},raw}} = 1 - R_{AA}^{feed-down} \cdot \langle T_{AA} \rangle \left(\frac{d\sigma}{dp_T} \right)^{FONLL}_{feed-down,|y|<0.5} \frac{\Delta p_T \Delta y (Acc \times \epsilon)_{feed-down} \cdot BR \cdot L_{int}}{N^{D+\bar{D}}_{raw}/2} \quad (2)$$

The procedure uses the B-meson production cross section in pp collisions estimated with FONLL calculations [1,6,8,10], the B→D+X decay kinematics from the EvtGen package [9] and the efficiencies for D-meson reconstruction from beauty-hadron decays obtained from Monte Carlo simulations. Different hypotheses on the nuclear modification factor $R_{AA}^{feed-down}$, defined as the ratio of the p_T-differential yield of D mesons from B decays in Pb–Pb collisions to the corresponding cross section in pp collisions scaled by the average nuclear overlap function ($\langle T_{AA} \rangle$), are also introduced. The resulting f_{prompt} values and their uncertainties are shown in the right-hand panels of Fig.2 for D^0, D^+ and D^{*+} mesons in the $|y_{lab}| < y_{fid}(p_T)$ interval ($y_{fid} \sim$ 0.5 - 0.8 depending on the p_T). The central values of f_{prompt} range between 0.81 and 0.96 depending on D-meson species and p_T with no significant rapidity dependence.

3.2 Data-driven method

The data-driven method exploits the different shapes of the distributions of the transverse-plane impact parameter to the primary vertex (d_0) of prompt and feed-down D mesons and estimates the prompt fraction (f_{prompt}) via an unbinned likelihood fit of the d_0 distribution of $D^0(D^+)$-meson candidates

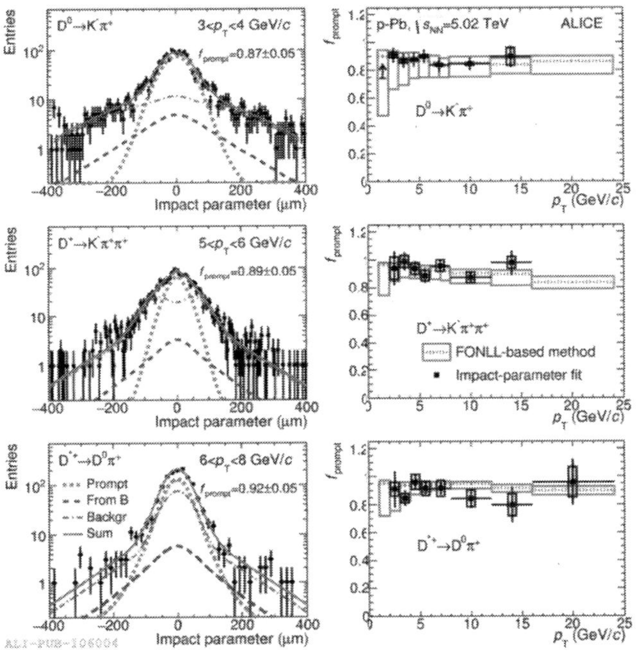

Figure 2. Left: Examples of fits to D^0 (top), D^+ (middle) and D^{*+} (bottom) impact-parameter distributions for p -Pb collisions at $\sqrt{s_{NN}} = 5.02$ TeV in three p_T intervals: $3 < p_T < 4$ GeV/c, $5 < p_T < 6$ GeV/c and $6 < p_T < 8$ GeV/c. The curves show the fit functions describing the prompt, feed-down and background contributions, as well as their sum. Right: fraction of prompt D^0 (top), D^+ (middle) and D^{*+} (bottom) raw yield as a function of p_T compared to the FONLL-based approach (red lines). The results from the data-driven method are shown as square markers with the error bars (boxes) representing the statistical (systematic) uncertainty. The arrow in the interval $1 < p_T < 2$ GeV/c represents the minimum value within a 95% confidence level. The central values of $f_{\rm prompt}$ from the FONLL-based approach are shown by the dashed line and their uncertainty by the red boxes.

with invariant mass $|M - M_D| < 1.5(2)\sigma$ (where σ is the width of the Gaussian function describing the D-meson signal in the invariant-mass fits) and of D^{*+}-meson candidates with a mass difference $|\Delta M - \Delta M_{D^{*+}}| < 2.5\sigma$. The fit function used is [3]:

$$F(d_0) = S \cdot [(1 - f_{\rm prompt})F^{\rm feed-down}(d_0) + f_{\rm prompt}F^{\rm prompt}(d_0)] + B \cdot F^{\rm backgr}(d_0) \qquad (3)$$

In this function, S and B are, respectively, the signal and background raw yields in the selected invariant-mass range; $F^{\rm prompt}(d_0)$, $F^{\rm feed-down}(d_0)$ and $F^{\rm backgr}(d_0)$ are functions describing the impact parameter distributions of prompt D mesons, feed-down D mesons, and background, respectively. The function $F^{\rm prompt}$ is a detector resolution term and is modeled with a Gaussian and a symmetric exponential term, describing the tails of the impact-parameter distribution of prompt D mesons. $F^{\rm feed-down}$ is the convolution of the detector resolution term with a symmetric double-exponential function ($F^{\rm feed-down}_{\rm true}$) describing the intrinsic impact parameter distribution of secondary D mesons from B-meson decays, which is determined by the decay length and decay kinematics of B mesons. $F^{\rm backgr}$ is the sum of a Gaussian and a symmetric exponential term centered at zero. For D^+ mesons, the background impact-parameter distribution has a double-peak structure with a depletion around zero induced by the selections applied. The shape is thus modeled with two Gaussians and two symmetric exponential terms. The parameters of $F^{\rm backgr}$ are fixed by fitting the impact parameter distribution of background candidates in the side-bands of the signal peak in the invariant-mass distributions. Figure 2 (left) shows examples of fits to the impact-parameter distributions of D^0, D^+ and D^{*+} mesons in the transverse-momentum intervals $3 < p_T < 4$ GeV/c, $5 < p_T < 6$ GeV/c and $6 < p_T < 8$ GeV/c, respectively. The prompt fraction of D^0, D^+ and D^{*+} mesons measured with this method is shown in Fig. 2 (right). For the interval $1 < p_T < 2$ GeV/c, given the limited precision of the impact-parameter fit, a lower limit could be estimated only for D^0 mesons at a 95% confidence level on the basis of the statistical and systematic uncertainties. For the same reason, in the highest p_T interval, $16 < p_T < 24$ GeV/c, the prompt fraction could be determined with the data-driven method only for D^{*+} mesons. The prompt fraction measured with the impact-parameter fits is found to be compatible with the FONLL-based estimation within uncertainties. For D^0 mesons, the data-driven approach provides a more precise determination of the prompt fraction [3], while for D^{*+} and D^+ mesons smaller uncertainties are obtained with the FONLL-based method.

3.3 Method based on pseudo-proper decay length

A new method is under study for the separation of the prompt D-meson fraction from the non-prompt component in Pb–Pb collisions. This new approach is based on a fit to the distribution of the pseudo-proper decay length variable, x, which is obtained by considering the scalar variable L_{xy}, defined as the signed projection of the D^0 flight distance onto its transverse momentum vector:

$$L_{xy} = \vec{L} \cdot \frac{\vec{p}_T^{\,D}}{|\vec{p}_T^{\,D}|} \tag{4}$$

The pseudo proper decay length is then defined as:

$$x = \frac{L_{xy} \cdot M^D \cdot c}{|\vec{p}_T^{\,D}|} \tag{5}$$

Here $\vec{L}$ is the vector pointing from the primary vertex to the D^0 decay vertex and M^D the mass of the D^0 taken from [4]. This approach would achieve the separation of the two components exploiting the different shapes of the pseudo-proper decay distributions of prompt and feed-down D mesons. The pseudo-proper decay method was successfully employed for the measurement of prompt and non-prompt J/ψ in pp [11] and Pb-Pb collisions [12].

References

[1] M. Cacciari, S. Frixione, N. Houdeau, M. L. Mangano, P. Nason, G. Ridolfi, JHEP **10** (2012) 137.
[2] M. Cacciari, M. Greco, and P. Nason, JHEP **05** (1998) 007.
[3] J. Adam *et al.* [ALICE Collaboration], CERN-EP-2016-127, arXiv:1605.07569.
[4] K. A. Olive *et al.* [Particle Data Group Collaboration], Chin. Phys. **C38** (2014) 090001.
[5] J. Adam *et al.* [ALICE Collaboration], JHEP **03** (2016) 081.
[6] B. Abelev *et al.* [ALICE Collaboration], JHEP **01** (2012) 128.
[7] B. Abelev *et al.* [ALICE Collaboration], Int. J. Mod. Phys. A29 (2014) 1430044.
[8] B. Abelev *et al.* [ALICE Collaboration], JHEP **07** (2012) 191.
[9] D. J. Lange, Nucl. Inst. and Meth. in Phys. Res. Section A 462 no. 1 - 2, (2001) 152 - 155.
[10] LHCb Collaboration, R. Aaij *et al.*, Nucl. Phys. B871 (2013) 1-20.
[11] B. Abelev *et al.* [ALICE Collaboration], JHEP **11** (2012) 065.
[12] J. Adam *et al.* [ALICE Collaboration], JHEP **07** (2015) 051.

J/ψ production at central rapidity in p–Pb collisions at $\sqrt{s_{\rm NN}}$ = 5.02 TeV with ALICE

Giuseppe Trombetta[1,⋆] on behalf of the ALICE Collaboration

[1] *Dipartimento Interateneo di Fisica 'M. Merlin' and Sezione INFN, Bari, Italy*

Abstract. The ALICE detector is capable of reconstructing J/ψ at central rapidity through the e^+e^- decay channel, down to zero transverse momentum, and has measured its production cross section as well as the fraction of non-prompt J/ψ, produced by the decay of beauty-flavoured hadrons. The latter quantity was evaluated in both pp and Pb–Pb systems down to $p_{\rm T}$ = 1.3 GeV/c. The results obtained by ALICE from the measurements of the inclusive J/ψ yield in p–Pb collisions at $\sqrt{s_{\rm NN}}$ = 5.02 TeV, as well as the statistical techniques and the status of the analysis concerning the non-prompt J/ψ measurement will be presented in this paper.

1 Introduction

The observation of a suppression of the J/ψ yield in ultra-relativistic heavy-ion collisions relative to that expected from a bare superposition of elementary pp collisions has long been interpreted as a signature of the formation of a deconfined state of hadronic matter, known as Quark-Gluon Plasma (QGP) [1]. Various Cold Nuclear Matter (CNM) effects, such as nuclear shadowing or partonic energy loss, are however expected to affect J/ψ production in addition to the modifications due to the presence of the QGP.

The study of p–Pb collisions, where the formation of a QGP medium is not expected, represents a necessary baseline for characterizing the CNM effects affecting J/ψ production and is crucial for improving our understanding of Pb–Pb collision results. Moreover, the determination of the fraction of secondary non-prompt J/ψ originating from the decay of beauty-flavoured hadrons provides a measurement of the inclusive b-quark production, and its study in p–Pb systems allows the evaluation of CNM effects on beauty production.

2 J/ψ identification with the ALICE detector

The ALICE experiment at the LHC consists of a central barrel embedded in a large solenoidal magnet, covering the pseudo-rapidity range $|\eta| < 0.9$, and a muon spectrometer in the forward direction [2]. At central rapidity (y), the J/ψ resonance is detected through its di-electronic decay channel J/$\psi \to e^+e^-$

⋆e-mail: giuseppe.trombetta@ba.infn.it

by means of the Time Projection Chamber (TPC) and Inner Tracking System (ITS) detectors, which provide excellent tracking performances within $|y| < 0.9$. The TPC is the main tracking detector of the barrel, consisting of a large cylindrical drift chamber which allows also charged particle identification through specific energy loss (dE/dx) measurements. The ITS is a cylindrically-shaped tracker made up of six layers of silicon detectors which provide precise track and vertex reconstruction close to the interaction point, and in particular a good secondary vertex separation for non-prompt J/ψ extraction. Thanks to the unique acceptance and low-momentum electron identification potential of these detectors, ALICE is capable of reconstructing J/ψ down to zero transverse momentum (p_T) and to separate, on a statistical basis, their non-prompt component down to $p_T = 1.3$ GeV/c, in a momentum region complementary to other LHC experiments.

3 Inclusive J/ψ measurement

ALICE has measured the inclusive J/ψ production as a function of p_T in p–Pb collisions at $\sqrt{s_{NN}} =$ 5.02 TeV in the mid-rapidity region corresponding to $-1.37 < y_{cms} < 0.43$ in the center of mass frame [3]. The analysis was based on a sample of about 10^8 Minimum Bias p–Pb events collected by ALICE in 2013 and corresponding to an integrated luminosity $L_{int} = 51\ \mu b^{-1}$.
J/ψ candidate selection is performed by combining opposite charged tracks within the ALICE central barrel acceptance. Electron identification is performed by requiring each track to have a specific energy loss signal in the TPC gas compatible with that of an electron and by rejecting tracks compatible with the pion and proton assumptions. Dedicated kinematic and quality selection criteria are furthermore applied to each track in order to reduce combinatorial background from low-momentum electrons and to reject the background electrons from photon conversions.
The raw J/ψ yields are obtained by counting the number of entries within the di-electron signal invariant mass range $2.92 < m_{e^+e^-} < 3.16$ GeV/c^2 after the subtraction of the background, which is evaluated from the invariant mass distribution of mixed-event electron pairs.
In order to extract the p_T-differential cross section $d^2\sigma/dydp_T$, the raw yields measured in five transverse momentum intervals are corrected by the product of the acceptance times efficiency ($A \times \epsilon$),

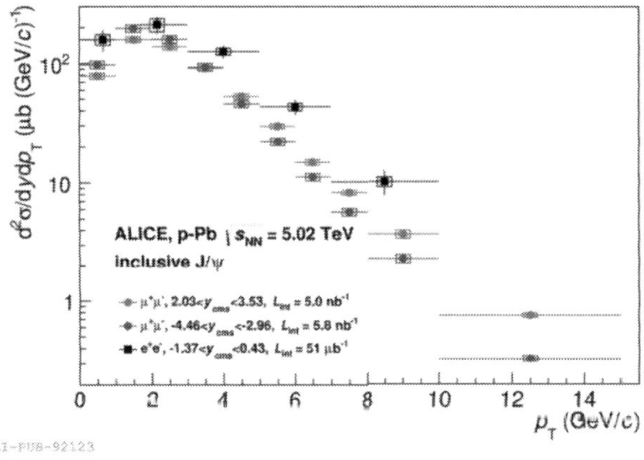

Figure 1. p_T-differential J/ψ cross sections measured by ALICE in different rapidity regions [3]. Vertical bars stand for statistical uncertainties while open and shaded boxes show the uncorrelated and correlated systematic uncertainties, respectively.

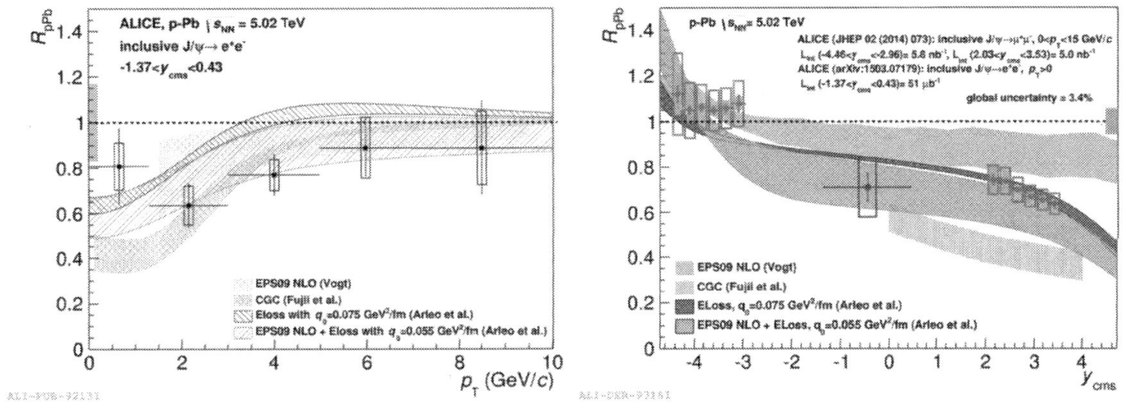

Figure 2. J/ψ nuclear modification factor as a function of p_{T} (left) and rapidity (right). Vertical bars stand for statistical uncertainties while open and shaded boxes show the uncorrelated and correlated systematic uncertainties, respectively. Results are compared to various theoretical models including pure shadowing [4] (with EPS09 parametrization), CGC [5] and coherent energy loss [6] predictions.

which is evaluated by means of a Monte Carlo (MC) simulation taking into account a realistic description of the ALICE experimental set-up.
The measured differential J/ψ cross section is shown in Figure 1 along with similar measurements performed at forward rapidities through the J/ψ di-muon decay channel in the ALICE muon spectrometer.
Systematic uncertainties affecting the cross section measurements are mainly due to the signal extraction procedure, the di-electron reconstruction efficiency and to the choice of the J/ψ p_{T} and y distributions used as input in the MC simulation.
The modifications affecting the J/ψ production due to the presence of the nuclear medium are evaluated by means of the nuclear modification factor R_{pPb}, which is obtained as the ratio of the differential cross sections of proton-nucleus and proton-proton collisions, scaled by the Pb mass number A_{Pb}:

$$R_{\mathrm{pPb}} = \frac{\mathrm{d}^2\sigma^{\mathrm{J}/\psi}_{\mathrm{pPb}}/\mathrm{d}y\mathrm{d}p_{\mathrm{T}}}{A_{\mathrm{Pb}} \cdot \mathrm{d}^2\sigma^{\mathrm{J}/\psi}_{\mathrm{pp}}/\mathrm{d}y\mathrm{d}p_{\mathrm{T}}}.$$

Figure 2 shows the p_{T}-differential (left) and y-differential (right) nuclear modification factors of inclusive J/ψ compared to predictions from various theoretical models based on calculations for J/ψ production. The main uncertainties on the measured R_{pPb} values come from the $\mathrm{d}^2\sigma^{\mathrm{J}/\psi}_{\mathrm{pp}}/\mathrm{d}y\mathrm{d}p_{\mathrm{T}}$ reference cross sections at $\sqrt{s} = 5.02$ TeV, which were obtained by means of an interpolation/extrapolation procedure, as described in [3].
A J/ψ suppression at low p_{T}, which tends to vanish at high p_{T}, is observed from data. Calculations including cold nuclear matter effects such as shadowing and coherent energy loss reproduce within uncertainties the observed p_{T} dependence suppression for $p_{\mathrm{T}} > 1.5$ GeV/c. Predictions based on the Color Glass Condensate (CGC) framework appear in fair agreement with the p_{T}-differential data at mid-rapidity, but clearly underestimate the measurements performed in the full p_{T} range at forward rapidity.

4 Non-prompt J/ψ fraction measurement

At LHC energies, a significant contribution to J/ψ production comes from non-prompt J/ψ which are produced after the weak decay of beauty-hadrons at an experimentally resolvable distance from the primary interaction vertex.

In ALICE, the non-prompt component of the J/ψ yield can be statistically separated by a maximum likelihood fit procedure which relies on the measurement of the pseudo-proper decay length x of each J/ψ candidate, which is defined as

$$x = \frac{\vec{L} \cdot \vec{p}_{\mathrm{T}}^{\,\mathrm{J}/\psi}}{p_{\mathrm{T}}^{\mathrm{J}/\psi}} \cdot \frac{c \cdot m^{\mathrm{J}/\psi}}{p_{\mathrm{T}}^{\mathrm{J}/\psi}}$$

with $\vec{L}$ being the vector from the primary vertex to the J/ψ decay vertex.

The measurement of the f_{B} fraction of non-prompt J/ψ is performed by means of an un-binned likelihood fit to the two-dimensional distribution of invariant mass $m_{e^+e^-}$ and x distributions of the di-electron pairs. The likelihood probability density function (p.d.f.) is computed for each candidate as the product of different p.d.f. describing, for both the prompt and non-prompt component, the x and $m_{e^+e^-}$ distributions of both signal and background pairs. The f_B fraction is included as a free parameter for the likelihood fit.

ALICE has published the measurement of the non-prompt J/ψ component at central rapidity as a function of p_{T} in pp collisions at $\sqrt{s} = 7$ TeV [7] and as a function of p_{T} and multiplicity in Pb–Pb collisions at $\sqrt{s_{\mathrm{NN}}} = 2.76$ TeV [8].

The analysis in the p–Pb system is being performed on the same data sample described for the inclusive cross section measurement. J/ψ candidates are considered down to transverse momenta as low as $p_{\mathrm{T}} \simeq 1.3$ GeV/c in order to ensure a good resolution of secondary vertices as well as a better separation of the prompt and non-prompt components with respect to x. Prompt and non-prompt J/ψ x distributions resulting from dedicated MC simulations are shown in Figure 3.

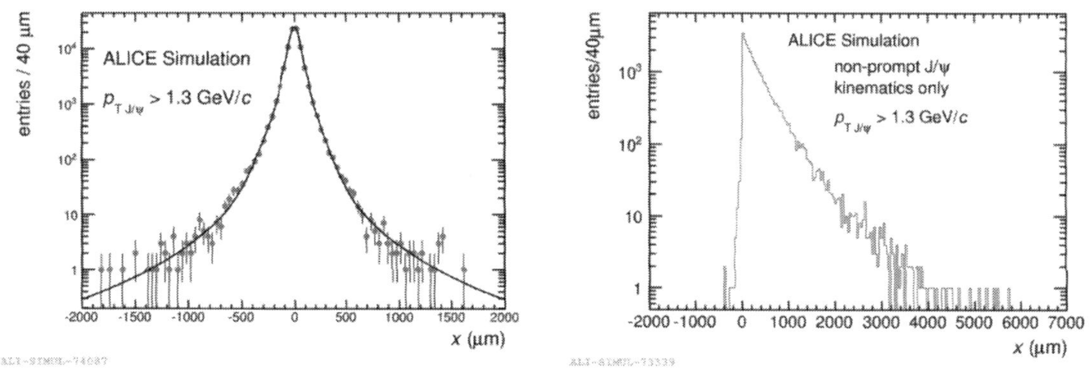

Figure 3. Pseudo-proper decay length distributions of prompt (left) and non-prompt (right) J/ψ, with $p_{\mathrm{T}} > 1.3$ GeV/c, resulting from MC simulations. Since prompt J/ψ production occurs at the primary interaction vertex, their distribution describes the effect of detector resolution on the the x observable.

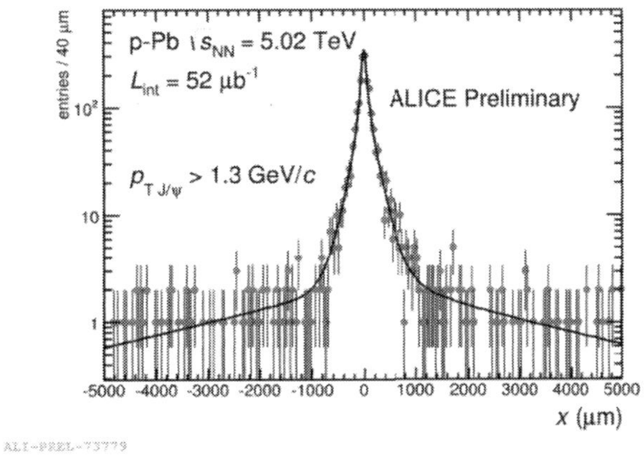

Figure 4. Pseudo-proper decay length distribution of di-electron pairs within the J/ψ invariant mass region (2.4 $< m_{e^+e^-} <$ 4.0 GeV/c^2). Solid line represents the scaled x projection of the likelihood fit function, which includes the prompt, non-prompt and background components.

The resulting maximized likelihood function, projected over the x distributions of J/ψ candidates with $p_T >$ 1.3 GeV/c from the p–Pb sample, is reported in Figure 4.
The raw value of f_B is extracted from the function and is corrected in order to take into account the different reconstruction efficiencies related to prompt and non-prompt J/ψ within the considered trasverse momentum interval.
The analysis for the measurement of the non-prompt component is currently being finalized. The main contributions to the systematic uncertainties affecting this measurement are expected to be related to the evaluation of the functional terms involved in the fit. The result can be furthermore combined with theoretical predictions in order to extract the non-prompt J/ψ and $b\bar{b}$-quark pair production cross sections down to the lowest accessible transverse momenta.

References

[1] T. Matsui and H. Satz, Phys. Lett. B **178** (1986) 416.
[2] K. Aamodt *et al.* [ALICE Collaboration], JINST **3** (2008) S08002.
[3] J. Adam *et al.* [ALICE Collaboration], JHEP **1506** (2015) 055.
[4] J. L. Albacete *et al.*, Int. J. Mod. Phys. E **22** (2013) 1330007.
[5] H. Fujii and K. Watanabe, Nucl. Phys. A **915** (2013) 1.
[6] F. Arleo, R. Kolevatov, S. Peigné and M. Rustamova, JHEP **1305** (2013) 155.
[7] B. Abelev *et al.* [ALICE Collaboration], JHEP **1211** (2012) 065.
[8] J. Adam *et al.* [ALICE Collaboration], JHEP **1507** (2015) 051.

EPJ Web of Conferences **129**, 00035 (2016) DOI: 10.1051/epjconf/201612900035
QCD@Work 2016

Vortex structure in superfluid color-flavor locked quark matter

Mark G. Alford[1,⋆], *S. Kumar* Mallavarapu[1,⋆⋆], *Tanmay* Vachaspati[2,⋆⋆⋆], and *Andreas* Windisch[1,⋆⋆⋆⋆]

[1] *Physics Department, Washington University, St. Louis, MO 63130, USA*
[2] *Physics Department, Arizona State University, Tempe, AZ 85287, USA*

Abstract. The core region of a neutron star may feature quark matter in the color-flavor-locked (CFL) phase. The CFL condensate breaks the baryon number symmetry, such that the phenomenon of superfluidity arises. If the core of the star is rotating, vortices will form in the superfluid, carrying the quanta of angular momentum. In a previous study we have solved the question of stability of these vortices, where we found numerical proof of a conjectured instability, according to which superfluid vortices will decay into an arrangement of so-called semi-superfluid fluxtubes. Here we report first results of an extension of our framework that allows us to study multi-vortex dynamics. This will in turn enable us to investigate the structure of semi-superfluid string lattices, which could be relevant to study pinning phenomena at the boundary of the core.

1 Introduction

In its densest form, matter appears in the color-flavor locked (CFL) phase [1]. The CFL condensate breaks the baryon number symmetry, which renders this phase a superfluid. If this form of matter is present in the core region of a rotating neutron star, vortices will carry the angular momentum of the spinning core. In a recent study [2], we addressed the question of stability of these superfluid vortices, using a Ginzburg-Landau effective theory. There, from a topological point of view, stable vortex solutions are expected, since the first homotopy group of the vacuum manifold is non-trivial [3]. However, the global vortex solution does *not* possess the lowest energy, and it has been conjectured that the vortex undergoes a decay into a configuration of a triplet of well-separated semi-superfluid strings [4]. In our study [2], we not only observed and numerically confirmed this decay, but we also mapped out the stability-/metastability boundary in the parameter space of the couplings. We furthermore identified an analytically constructed mode that proved to be sufficient to trigger the decay of a global vortex. Let us briefly review our findings here. Assuming $m_u = m_d = m_s = 0$, the Ginzburg-Landau Lagrangian of the effective theory reads

$$\mathcal{L} = \mathrm{Tr}\left[-\frac{1}{4}F_{ij}F^{ij} + D_i\Phi^\dagger D^i\Phi + m^2\Phi^\dagger\Phi - \lambda_2(\Phi^\dagger\Phi)^2\right] - \lambda_1(\mathrm{Tr}[\Phi^\dagger\Phi])^2 + \frac{3m^4}{4\lambda}\,, \tag{1}$$

⋆e-mail: alford@physics.wustl.edu
⋆⋆e-mail: kumar.s@wustl.edu
⋆⋆⋆e-mail: tvachasp@asu.edu
⋆⋆⋆⋆e-mail: windisch@physics.wustl.edu

where $D_i = \partial_i - igA_i$ is the covariant derivative, $F_{ij} = \partial_i A_j - \partial_j A_i - ig\left[A_i, A_j\right]$ is the gauge field-strength tensor, and the A_i represent the gauge fields (gluons). The coupling λ is a linear combination of the original self-couplings of the condensate,

$$\lambda \equiv 3\lambda_1 + \lambda_2. \tag{2}$$

The matter field Φ represents the CFL condensate, and has a $3_c \times 3_f$ complex matrix structure. A general entry of the Φ matrix is thus characterized by a color index α and a flavor index a, $\phi_{\alpha a}$. In the broken phase, the vacuum expectation value (vev) is given by

$$A_i = 0\,, \quad \Phi = \bar{\phi}\mathbf{1}_{3\times 3}\,, \quad \bar{\phi} = \sqrt{\frac{m^2}{2\lambda}}\,. \tag{3}$$

A (global) *superfluid vortex* can then be written as

$$A_i = 0\,, \quad \Phi_{\rm sf} = \bar{\phi}\beta(r)e^{i\theta}\,\mathbf{1}_{3\times 3}\,, \tag{4}$$

where $\beta(r)$ is the radial profile obtained from solving

$$\beta'' + \frac{\beta'}{r} - \frac{\beta}{r^2} - m^2\beta(\beta^2 - 1) = 0\,, \tag{5}$$

with the boundary conditions $\beta \to 0$ as $r \to 0$, and $\beta \to 1$ as $r \to \infty$. A (red) *semi-superfluid flux tube*, on the other hand, can be written as

$$\Phi_{\rm ssft}(r,\theta) = \bar{\phi}\begin{pmatrix} f(r)e^{i\theta} & 0 & 0 \\ 0 & g(r) & 0 \\ 0 & 0 & g(r) \end{pmatrix}, \tag{6}$$

$$A_\theta^{\rm ssft}(r) = -\frac{1}{gr}(1 - h(r))\begin{pmatrix} -\frac{2}{3} & 0 & 0 \\ 0 & \frac{1}{3} & 0 \\ 0 & 0 & \frac{1}{3} \end{pmatrix}, \tag{7}$$

$$A_r^{\rm ssft} = 0\,. \tag{8}$$

and the solutions for the green and blue flux tubes follow from swapping the diagonal elements of the matrix. The profile functions $f(r)$, $g(r)$ and $h(r)$ obey a set of coupled differential equations, see equations (10)-(14) in [2]. As discussed in our paper, far from the core of the vortex, the energy density of one semi-superfluid flux tube is one ninth of the energy density of a global vortex,

$$\begin{aligned} \epsilon_{\rm sf} &= 3\bar{\phi}^2/r^2, \\ \epsilon_{\rm ssft} &= \tfrac{1}{3}\bar{\phi}^2/r^2, \end{aligned} \tag{9}$$

which is the cause of the instability. In the case of vanishing gauge coupling, we identified an unstable mode analytically,

$$\delta\Phi^{(8)} \;=\; \epsilon\,\hat{n}\cdot\nabla\psi(r,\theta)\,T_8\,. \tag{10}$$

This mode establishes a distortion of the red and green components of the vortex by a small amount ϵ in direction of the unit vector $\hat{n}$, while shifting the blue component by an amount of 2ϵ in the opposite direction. This perturbation of the global vortex can be plugged into the Hamiltonian density, and the change in energy density evaluates to

$$\delta E_8 = \epsilon^2(\lambda_2 - \lambda)\frac{\pi m^4}{\lambda^2}\int_0^\infty r{\rm d}r\beta'^2\beta^2\,. \tag{11}$$

DOI: 10.1051/epjconf/201612900035

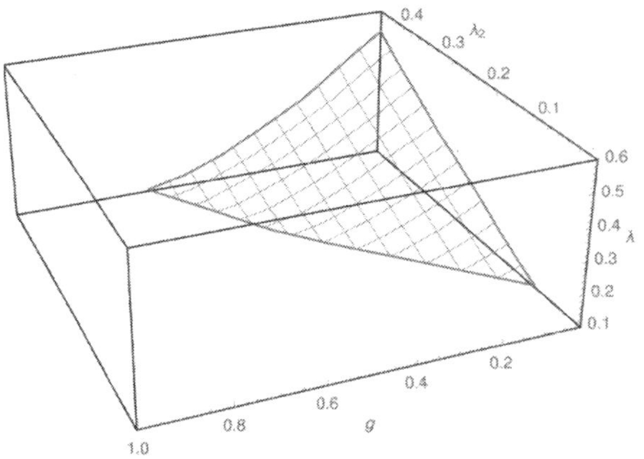

Figure 1. The parameter space of the couplings g, λ and λ_2. Points behind the surface form the region of meta-stability, all other points constitute the region of instability.

For $\lambda_2 > \lambda$, this clearly lowers the energy and is thus an unstable direction. Depending on the gauge coupling g and the condensate self-couplings λ_1 and λ_2, there are regions in parameter space where the global vortex solution is unstable and decays immediately, but we could also identify regions of meta-stability, see Figures 1 and 2. In Figure 2, the solid line corresponds to the instability boundary in parameter space as derived from the perturbation (10), which seems to hold approximately for small gauge couplings.

The dashed line in Figure 2 corresponds to the region in parameter space where $\lambda_1 = \lambda_2$. This has been identified as the physically relevant regime at ultra-high density, where the coupling is sufficiently small to allow for mean-field calculations, [5, 6],

$$\lambda_1 = \lambda_2 = \frac{\lambda}{4} = \frac{36}{7} \frac{\pi^4}{\zeta(3)} \left(\frac{T_c}{\mu}\right)^2 . \tag{12}$$

If this result can be extrapolated to densities where the system is strongly coupled, our study indicates that there are no regions of metastability for the physically relevant case. This is supported by the fact that the region of metastability dies away quickly with increasing coupling g.

2 Initial condition and boundary treatment for multi-vortex dynamics

So far we have explored the behavior of a single global vortex, as well as arrangements of three well-separated semi-superfluid flux tubes. Currently we are investigating two possible extensions to our numerical framework. On the one hand, we introduced strange-quark mass asymmetry, where we followed the work of [7]. Our findings will be presented in a future publication. On the other hand, we study the dynamics of multiple vortices on the lattice, which introduces several complications. In these proceedings, we discuss the first steps in the simulation of multiple vortices on the lattice. To set up many vortices on the lattice, we first obtain a radial profile for a winding-one global vortex by solving (5) using a relaxation method on a large grid with high precision. In the absence of a gauge field at initial time (that is, the link variables are unity), we construct the Φ-field in the presence of N

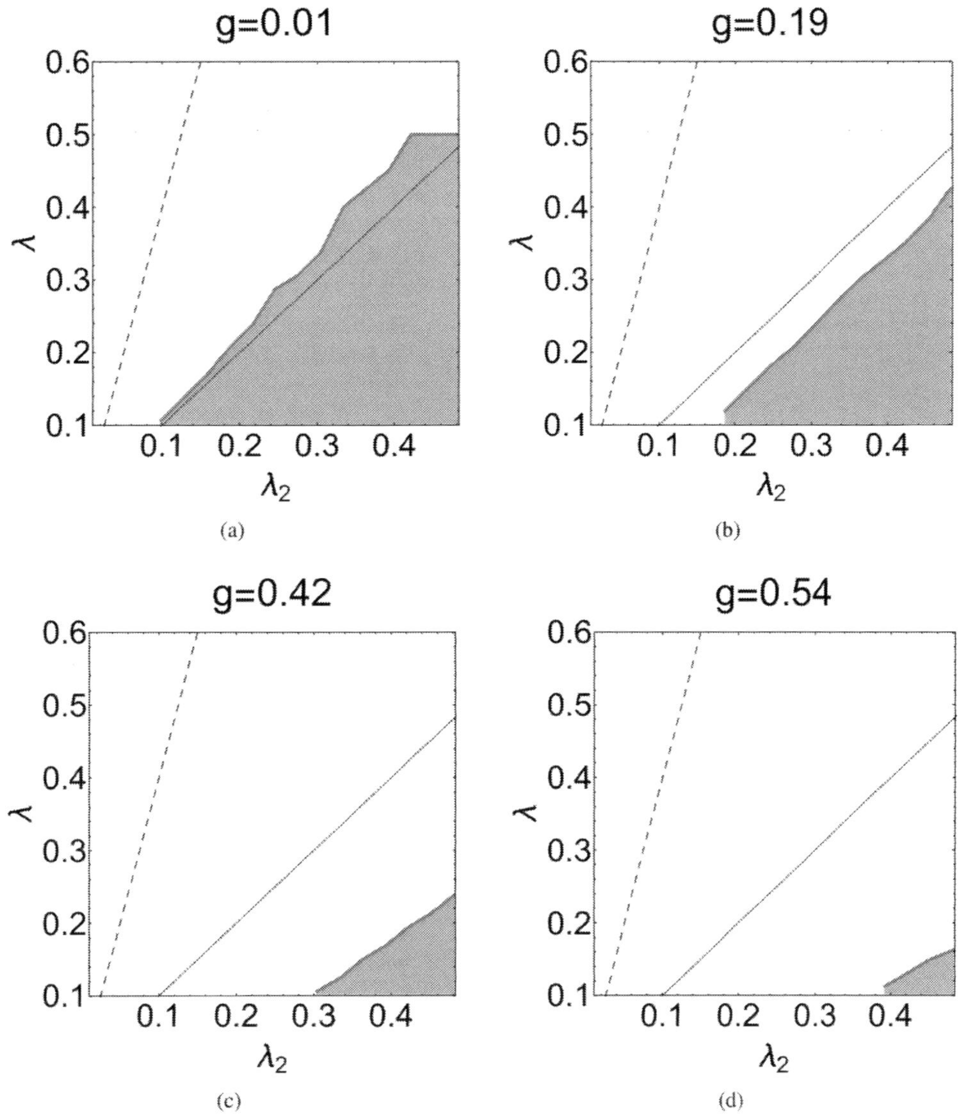

Figure 2. Slices through the parameter space of the couplings for different values of g. The shaded area is the region of meta-stability, its complement constitutes regions of instability. The dashed line is the projection of the plane of weak-coupling results, see text. The solid line represents $\lambda = \lambda_2$, which, in the case of vanishing gauge coupling, should yield the boundary of stability/meta-stability. Note that the metastability region can be characterized by $\lambda_1 \lesssim -0.16g$. For fixed values of g and λ_1, a change in λ_2 has almost no effect on the stability boundary.

vortices according to

$$\Phi = \left(\prod_{i=1}^{N} \beta(r_i)\right) \exp\left\{ i \sum_{i=1}^{N} n_i \theta_i \right\} \bar{\phi} \mathbf{1}_{3\times 3}, \tag{13}$$

which is just a superposition of the N vortices. At a given point (r, θ), the variable r_i indicates the distance to a vortex at position i, and θ_i is the angle with respect to the vortex at position i. Even though we plan to simulate systems of multiple vortices at non-zero gauge coupling in the unstable regime, we start with the most simple case of zero gauge coupling and in the meta-stable region of couplings. In this case, the Ansatz (13) is all we need. Since the superfluid vortices feature a repulsive behavior, they separate quickly once we start our simulation, and, if we use the boundary conditions of our previous study, approach the boundary and pin there. As far as semi-superfluid strings are concerned, the fixed boundary case of our initial work should serve our purpose, since the boundary is repulsive to the individual flux tubes. Together with the repulsion of the semi-superfluid vortices among themselves, this will allow us to find the lattice structure of the multi-fluxtube arrangement. However, at this point we have not yet observed the decay into semi-superfluid fluxtubes in the multi-vortex system, and in this first step we intend to study the numerically much cheaper approach of global vortices only. We thus have to use a modification of the first boundary condition. Using fixed boundary conditions without any modifications also introduces the problem that, far away from the center, many vortices constitute a large overall winding number, which in turn causes the energy density to increase with decreasing distance to the boundary. In order to cancel some of the winding, we introduced a grid of vortices which resides *beyond* the boundary, that is, outside of the simulation region. The presence of those vortices, however, is taken into account when the initial condition is computed, and, in particular, is frozen into the fixed boundary, see Figure 3.

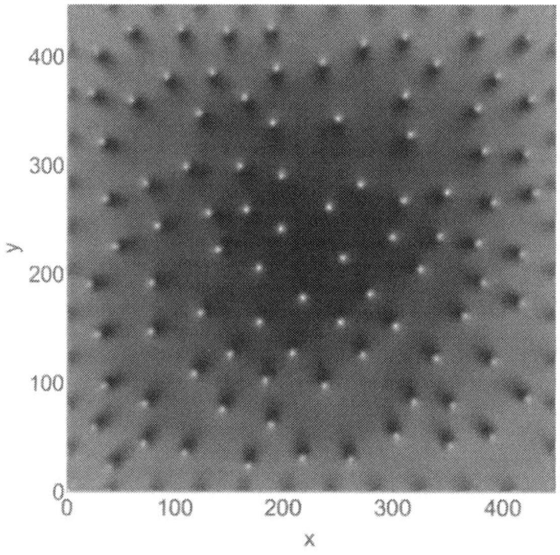

Figure 3. Example of the energy density of an initial configuration of 90 vortices. The vortex positions have been chosen randomly, with the additional constraint of a minimal vortex separation of 35 lattice units. In addition to the 90 vortices, one can see the vortices sitting *beyond* the simulation region. They appear as 'shadows' along the edges of the lattice. Those vortices have been frozen into the fixed boundary condition at initial time, and help in providing a better initial setting for large vortex number arrangements, see text.

3 Time evolution of a multi-superfluid-vortex arrangement

In this section we show the time evolution of a random setting of 90 vortices on a 448×448 lattice at vanishing gauge coupling in the meta-stable regime of the theory. In this case, the superfluid vortices won't undergo the decay to the semi-superfluid fluxtubes, but are expected to form a hexagonal lattice

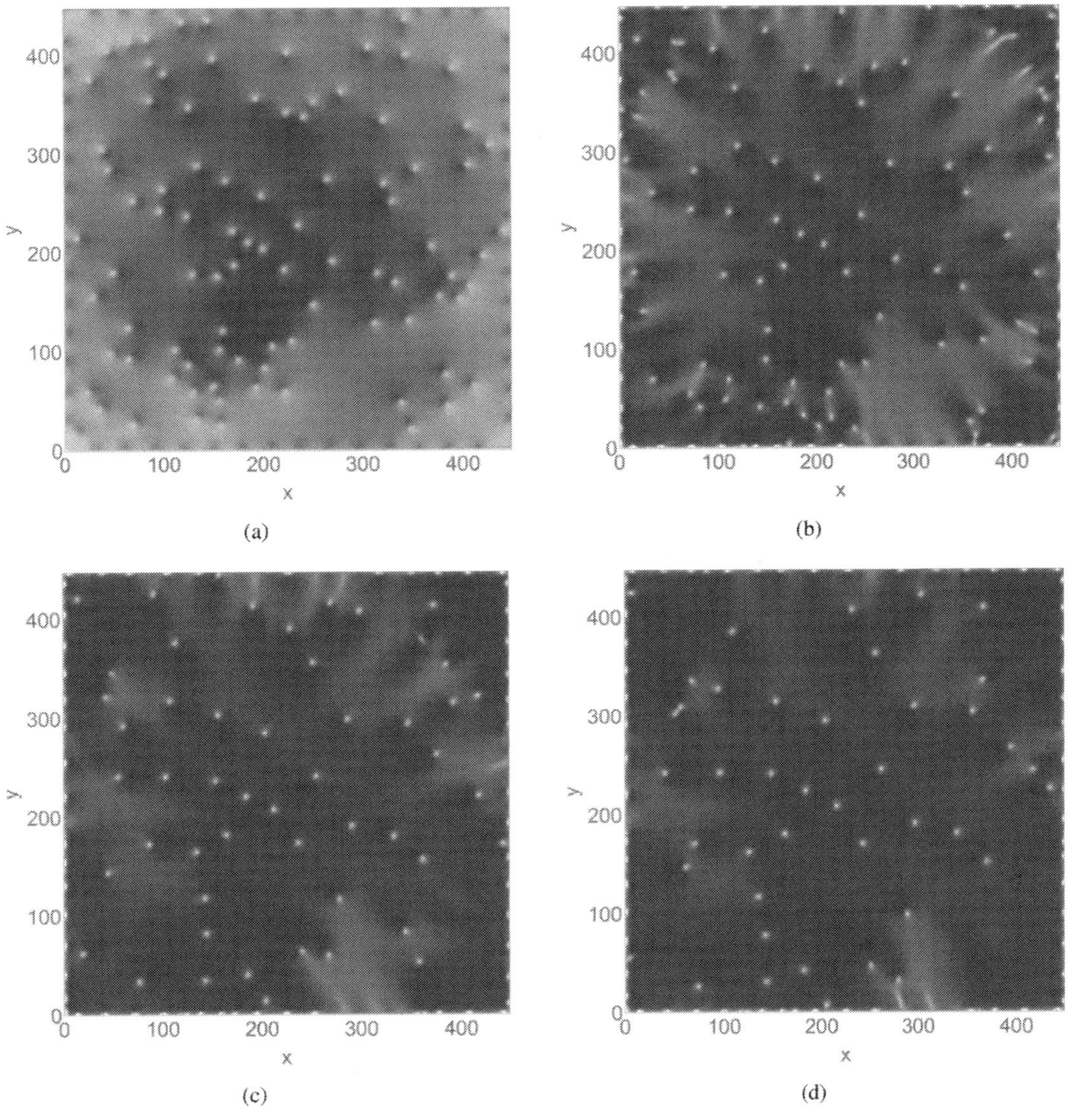

Figure 4. Energy density snapshots of the time evolution of an initial arrangement of 90 vortices at random positions with minimal initial vortex-vortex separation of 15 lattice units, using the improved fixed boundary condition discussed in Section 3. The panels 4(a),4(b),4(c) and 4(d) show the state of the system at initial time, and at t = 1000, t = 2000 and t = 3000 respectively. In order to prevent a violent separation of the vortices, a high damping factor has been used in this evaluation. The lighter color shading of the initial state in panel 4(a) comes from the fact that the initial condition has a lot of excess energy, which is then dissipated throughout the time-evolution.

structure. Snapshots of the time evolution of a first simulation run are shown in Figure 4. In this run we have used a very high damping factor, in order to relax the system towards its lowest energy state as gentle as possible. This is necessary, since the initial state has a high vortex (winding) number, and since our Ansatz is not the lowest energy state, it has a lot of excess energy. The downside of a high damping factor is a very slow time evolution, which is evident from the comparably small changes in the configurations shown in Figure 4(a), despite the large time separation of the configurations. In this first simulation run we successfully managed to simulate the time evolution of a state with very high vorticity in a controlled and stable way. The vortices separate, and seem to prefer to occupy the boundary, which requires some improvement of the boundary condition. So far we did not observe the expected hexagonal lattice structure, however, this should be easily achievable by an appropriate choice of initial condition and simulation parameters.

4 Conclusions and outlook

In this progress report we have reviewed our findings from our previous study [1], where we have found numerical proof of the instability of superfluid vortices in CFL quark matter. We identified regions of stability and metastability in the parameter space of the couplings, and connected our result to the physically relevant regime by extrapolating weak coupling results. This indicates that superfluid vortices are unstable in CFL quark matter. An analysis of the coupling dependence of the stability/metastability boundary revealed that it seems to be independent of the coupling λ_2, which raises the question of the role of the condensate self-couplings in the decay process. We also constructed a mode that is sufficient to trigger the decay of a superfluid vortex into the triplet of semi-superfluid flux-tubes. The phenomenological consequences of the instability, for example on neutron star glitches, remain elusive for now, since that requires a thorough understanding of a pinning mechanism. As a first step towards a better understanding of the dynamics of a system with high vorticity we extended our framework and presented first results for the time evolution of a multi-superfluid-vortex system. We modified the boundary conditions slightly, which allows for a better initial setting of a multi-vortex state, and simulated a system of 90 superfluid vortices in a controlled and stable way. Because it is numerically less expensive, we started with the zero-gauge coupling case, and focused on the stable regime of condensate self-couplings. As a next step, we intend to study the non-zero gauge coupling case. This is numerically more expensive, since, on the lattice, gauge fields are represented as link variables corresponding to group elements, and their time evolution involves a matrix exponential. In order to speed up the code, we thus plan to use Graphics Processing Units (GPUs), which allows for much faster evolution on larger lattices.

Acknowledgments

We thank Andreas Schmitt for helpful discussions. This study has been supported by the U.S. Department of Energy, Office of Science, Office of Nuclear Physics under Award number #DE-FG02-05ER41375. AW acknowledges support through the Austrian Science Fund (FWF) Schrödinger Fellowship J 3800-N27. TV acknowledges support by the U.S. Department of Energy, Office of Science, Office of High Energy Physics under Award number #DE-SC0013605.

References

[1] M.G. Alford, K. Rajagopal, F. Wilczek, Phys. Lett. **B422**, 247 (1998), hep-ph/9711395.

[2] M.G. Alford, S.K. Mallavarapu, T. Vachaspati, A. Windisch, Phys. Rev. **C93**, 045801 (2016), 1601.04656.
[3] A.P. Balachandran, S. Digal, T. Matsuura, Phys. Rev. **D73**, 074009 (2006), hep-ph/0509276.
[4] E. Nakano, M. Nitta, T. Matsuura, Phys. Lett. **B672**, 61 (2009), 0708.4092.
[5] K. Iida, G. Baym, Phys. Rev. **D63**, 074018 (2001), [Erratum: Phys. Rev.D66,059903(2002)], hep-ph/0011229.
[6] I. Giannakis, H.c. Ren, Phys. Rev. **D65**, 054017 (2002), hep-ph/0108256.
[7] M. Eto, M. Nitta, N. Yamamoto, Phys. Rev. Lett. **104**, 161601 (2010), 0912.1352.

Chiral spiral induced by a strong magnetic field

Hiroaki Abuki[1,⋆]

[1] *Department of Education, Aichi University of Education, 1 Hirosawa, Igaya-cho, Kariya 448-8542, Japan*

Abstract. We study the modification of the chiral phase structure of QCD due to an external magnetic field. We first demonstrate how the effect of magnetic field can systematically be incorporated into a generalized Ginzburg-Landau framework. We then analyze the phase structure in the vicinity of the chiral critical point. In the chiral limit, the effect is found to be so drastic that it brings a "continent" of chiral spiral in the phase diagram, by which the chiral tricritical point is totally washed out. This is the case no matter how small the intensity of magnetic field is. On the other hand, the current quark mass protects the chiral critical point from a weak magnetic field. However, the critical point will eventually be covered by the chiral spiral phase as the magnetic field grows.

1 Introduction

There has recently been a growing interest on possible crystal structures formed by the chiral condensates in QCD at finite density [1–3]. On the other hand, the effect of magnetic field on QCD has also been the subject of intensive studies. Phenomenologically, exploring possible forms of strongly interacting matter under the magnetic field is relevant to the physics of magnetars; the compact stellar objects known to have a strong magnetic field, $B \sim 10^{10}$T [4]. It also brings some impacts on the physics of heavy ion collisions which could produce a huge magnetic field $B \sim 10^{14}$T in the very early stage of noncentral collisions [5]. There have been made a lot of theoretical approaches to the effects of magnetic field on QCD phase diagram. These include computations based on phenomenological models [6–8] as well as the lattice QCD simulations [9, 10]. The former approaches predict the "*magnetic catalysis*", while the latter gives the opposite effect known as the "*inverse magnetic catalysis*". The mechanism for the magnetic catalysis is rather transparent, but that for the inverse one remains still a matter of active debates [11–14].

In this article, we report our recent study on the effect of strong magnetic fields on the chiral phases with a particular focus put on how it modifies the phase structure in the vicinity of the critical point. Several studies are already devoted on how the magnetic field affects the critical points [15, 16]. For example, a new critical point is suggested to appear in the presence of a strong magnetic field [17]. There are also some work related to inhomogeneous phases under a strong magnetic field; these include the widening of the phase for solitonic modulation [18], the hybrid chiral condensate where the space varying phase is attached to the Real-Kink Crystal (RKC) [19]. The effect of current quark mass is also studied in [20]; it was shown that the chiral spiral aka the dual chiral density wave

⋆e-mail: abuki@auecc.aichi-edu.ac.jp

(DCDW) survives as the "massive dual chiral density wave" where the complex phase of condensate gets skewed from a linear function of space coordinate, say, z.

We here concentrate on the neighborhood of the critical point. We first show that in this case it is possible to derive systematically the generalized Ginzburg-Landau (gGL) action without specifying any details about the spatial form of the chiral condensate. We derive this functional up to the first nontrivial order in the current quark mass h and the magnetic field B. Based on the derived functional, we analyze the phases near the critical point. It turns out that these two ingredients have competing effects on inhomogeneous phases. In particular, the condensate accompanied by the complex phase, the chiral spiral, is found to be favored by the magnetic field [19], and accordingly the phase diagram gets drastically changed once the effect of magnetic field prevails.

2 Deriving generalized Ginzburg-Landau action

The generalized Ginzburg-Landau (gGL) action density in the absence of the external magnetic field can be derived in the same way as described in [21, 22]. The quark loop contribution to the effective action can be expanded in the power of the quark self-energy $\Sigma(\mathbf{x}) = m_q + \sigma(\mathbf{x}) + i\gamma_5 \boldsymbol{\tau} \cdot \boldsymbol{\pi}(\mathbf{x})$ as

$$\delta S_{\text{eff}} = \frac{T}{2} \sum_n \int d\mathbf{x} \int d\mathbf{y} \text{tr} \left[S(i\omega_n, \mathbf{x} - \mathbf{y}) \Sigma(\mathbf{y}) S(i\omega_n, \mathbf{y} - \mathbf{x}) \Sigma(\mathbf{x}) \right] + O(\Sigma^4). \tag{1}$$

Here $S(i\omega_n, \mathbf{x}) = -\int d\mathbf{p} e^{i\mathbf{p}\cdot\mathbf{x}} \frac{i\omega_n \gamma_0 - \mathbf{p}\cdot\boldsymbol{\gamma}}{\omega_n^2 + \mathbf{p}^2}$ is the quark propagator with $\omega_n = \pi T(2n-1)$ being the Matsubara frequency. Expressing $\Sigma(\mathbf{y}) = \Sigma(\mathbf{x}) + \sum_{i=1}^{\infty} \frac{1}{i!} [(\mathbf{y} - \mathbf{x}) \cdot \nabla \Sigma(\mathbf{x})]^i$, we can perform a systematic derivative expansion of the effective action. Writing the action with the gGL action density ω as $S_{\text{eff}} = \int d\mathbf{x} \omega(\mathbf{x})$, the result is found up to the sixth order in σ, π_a ($a = 1, 2, 3$) and $\nabla \equiv \partial_{\mathbf{x}}$ as

$$\begin{aligned} \omega(\mathbf{x}) &= \delta_m \omega(\mathbf{x}) + \frac{\alpha_2}{2} \phi^2 + \frac{\alpha_4}{4} \left(\phi^4 + (\nabla \phi)^2 \right) \\ &+ \frac{\alpha_6}{6} \left(\phi^6 + 3[\phi^2 (\nabla \phi)^2 - (\phi \cdot \nabla \phi)^2] + 5(\phi \cdot \nabla \phi)^2 + \frac{1}{2} (\Delta \phi)^2 \right), \end{aligned} \tag{2}$$

where we have switched to the chiral four-vector notation $\phi = (\sigma, \boldsymbol{\pi})$. $\delta_m \omega(\mathbf{x}) = -h\sigma$, which is the explicit symmetry breaking term associated with the current quark mass m_q, hereafter, called "h-term". h and α_n ($n = 2, 4, 6$), are the Ginzburg-Landau (GL) couplings which depend on temperature T and chemical potential μ. h is proportional to m_q, and its explicit form is

$$h = m_q \left(4 N_c N_f T \sum_n \int \frac{d\mathbf{p}}{(2\pi)^3} \frac{1}{(\omega_n - i\mu)^2 + \mathbf{p}^2} \right).$$

$N_{c(f)}$ is the number of color (flavor). The integral is divergent in ultra-violet and needs some regularization scheme to be evaluated. In the spirit of the GL approach, we simply take h as a parameter characterizing the explicit symmetry breaking. Similarly the expressions for α_n can be found. There is an extra tree-level counter-contribution to α_2 for the case of the standard NJL model [2]:

$$\alpha_{2i} = \frac{\delta_{i,1}}{2G} + (-1)^i 4 N_c N_f T \sum_n \int \frac{d\mathbf{p}}{(2\pi)^3} \frac{1}{((\omega_n - i\mu)^2 + \mathbf{p}^2)^i},$$

where G is the NJL coupling constant for a four-quark (chiral-invariant) interaction. The integral is divergent for α_2 and α_4. These parameters are zero at the tricritical point ($\mu_{\text{TCP}}, T_{\text{TCP}}$) which is expected to show up in the phase diagram in the chiral limit $m_q = 0$ ($h = 0$).

Now we come to consider the effect of an external magnetic field. There is a direct effect on the quark propagator, whereas that on gluon sector is somehow indirect. It is easy to expand quark propagator in the power of magnetic field along with the line described in [23]:

$$S(i\omega_n, \mathbf{p}) \to S(i\omega_n, \mathbf{p}) + (QB_i)\frac{\not{p}_{\parallel} + \not{\mu}}{[(i\omega_n + \mu)^2 - \mathbf{p}^2]^2}\frac{i\epsilon_{ijk}\gamma^j\gamma^k}{2} + O(Q\mathbf{B})^2, \tag{3}$$

where we have used the four vector notation $p_{\parallel}^{\mu} = (i\omega_n + \mu, \mathbf{p}_{\parallel})$ with $\mathbf{p}_{\parallel} = (\mathbf{p}\cdot\boldsymbol{B})\boldsymbol{B}/|\boldsymbol{B}|^2$ being the parallel component of momentum. $Q = \text{diag.}(2e/3, -e/3)$ is the electric charge matrix in the flavor space. The first nontrivial term depending on $\boldsymbol{B}$ comes from the second order term in Eq. (1). Plugging Eq. (3) into the integrand of Eq. (1), and extracting the term linear in $\boldsymbol{B}$, we have

$$\delta\omega_B(\mathbf{x}) = \frac{T}{2}\sum_n\int\frac{d\mathbf{p}}{(2\pi)^3}\frac{1}{[(i\omega_n + \mu)^2 - \mathbf{p}^2]^8}\text{tr}\left[\not{p}_{\parallel}(B_i\epsilon_{ijk}\gamma^j\gamma^k)Q\Sigma(\mathbf{x})\not{p}\gamma^l\not{p}(\partial_l\Sigma(\mathbf{x}))\right].$$

Performing the traces over the Dirac, color and flavor spaces, we have

$$\delta\omega_B(\mathbf{x}) = e\boldsymbol{B}\cdot(\pi_3\boldsymbol{\nabla}\sigma - \sigma\boldsymbol{\nabla}\pi_3)N_cT\sum_n\int\frac{d\mathbf{p}}{(2\pi)^3}\frac{4(i\omega_n + \mu)}{[(i\omega_n + \mu)^2 - \mathbf{p}^2]^3}.$$

The Matsubara sum and integral over $\mathbf{p}$ can be done analytically, and the result will be expressed by the generalized zeta function. However, its explicit form is of no importance here. Instead of writing the result, we only note that the result can be written with the derivative of α_4 with respect to μ.

$$\delta\omega_B(\mathbf{x}) = -\frac{1}{4N_f}\frac{\partial\alpha_4}{\partial\mu}e\boldsymbol{B}\cdot(\pi_3\boldsymbol{\nabla}\sigma - \sigma\boldsymbol{\nabla}\pi_3) \equiv -\boldsymbol{b}\cdot(\sigma\boldsymbol{\nabla}\pi_3 - \pi_3\boldsymbol{\nabla}\sigma).$$

We introduced a new GL coupling $\boldsymbol{b}$ whose magnitude serves as a measure of the intensity of the external magnetic field. The above extra term adds to the gGL potential (2) when the magnetic field is on. While h-term only breaks the chiral symmetry to the isospin SU(2), the $\boldsymbol{b}$-term explicitly breaks several symmetries: the time reversal symmetry, the rotational symmetry, in addition to the isospin SU(2) symmetry which is broken down to $\text{U}_\text{Q}(1)$.

Once we assume $\pi_1 = \pi_2 = 0$, and take the complex notation for the condensate $\Delta = \sigma + i\pi_3$, the gGL potential density can be cast into the more intuitive form

$$\begin{aligned}\omega(\mathbf{x}) \;&=\; -\boldsymbol{b}\cdot\text{Im}\left[\Delta^*\boldsymbol{\nabla}\Delta\right] - h\text{Re}\left[\Delta\right]\\ &\quad+\frac{\alpha_2}{2}|\Delta|^2 + \frac{\alpha_4}{4}\left(|\Delta|^4 + |\boldsymbol{\nabla}\Delta|^2\right) + \frac{\alpha_6}{6}\left(|\Delta|^6 + 3|\Delta|^2|\boldsymbol{\nabla}\Delta|^2 + 2\left(\text{Re}[\Delta^*\boldsymbol{\nabla}\Delta]\right)^2 + \frac{1}{2}|\boldsymbol{\nabla}^2\Delta|^2\right).\end{aligned} \tag{4}$$

The first two terms are the symmetry breaking sources, responsible for the current quark mass and the magnetic field, respectively. It can be easily guessed that the h-term favors the RKC, while the $\boldsymbol{b}$-term stabilizes the complex condensate such as the chiral spiral. We note that our $\boldsymbol{b}$-term is exactly in the same form as the one obtained in one-dimensional Gross-Neveau model [24], where it was shown that the spiral phase dominates the phase diagram. This term is forbidden in the three dimensional NJL model because it breaks the rotational symmetry. The magnetic field induces this term so that it opens the possibility that the complex condensate comes into play in the QCD phase diagram.

3 How do magnetic fields modify the phase diagram?

Let us first begin with the case of the chiral limit. This corresponds to ignoring the h-term in the gGL energy density (2). We measure every dimensionful quantity with the proper power of $(\alpha_6)^{-1/2}$.

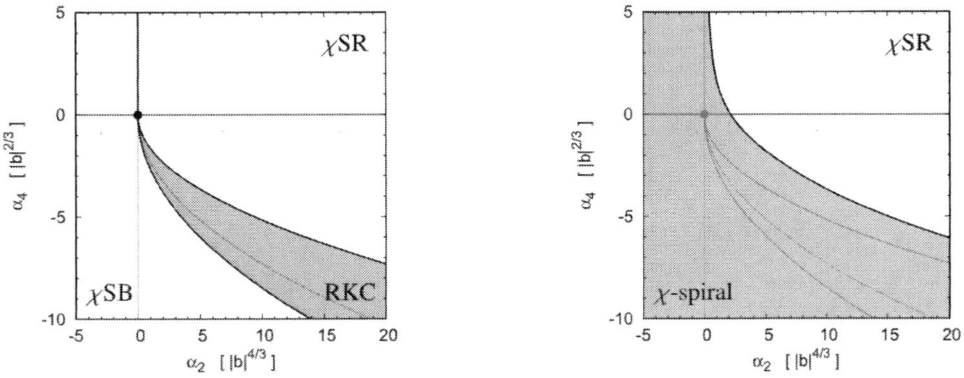

Figure 1. The phase diagrams in the chiral limit, $h = 0$. Left panel: The phase diagram for zero magnetic field. Right panel: The phase diagram for nonvanishing magnetic field $\boldsymbol{b}$.

Then, we can scale out the effect of $\boldsymbol{b}$, by taking $|b|^{4/3}$ ($|b|^{2/3}$) for the unit of α_2 (α_4). The phase diagram for $|b| = 0$ is depicted in the left panel of Fig. 1. First, note that the Lifshitz tricritical point (LTCP) is located at the origin which, in principle, has a unique map onto the ($\mu_{\rm TCP}, T_{\rm TCP}$) in QCD phase diagram. Second, the RKC enters in between the chiral symmetric phase (χSR) and the chiral symmetry broken phase (χSB). One might wonder why $|b|$ comes in the units of α_2 and α_4 in spite of zero magnetic field $b = 0$. This is just for a convenience, and in this case $|b|$ is arbitrary. In fact, the phase boundaries are independent of $|b|$, since any critical lines are expressed by $\alpha_4^2 \propto \alpha_2$. In the right panel, the phase diagram for nonvanishing $|b|$ is displayed. The phase structure is completely changed by the emergence of a complex chiral spiral, $\Delta(\mathbf{x}) = \Delta_0 e^{i\boldsymbol{q}\cdot\mathbf{x}}$, denoted by "$\chi$-spiral" in the figure. In this phase the direction of $\boldsymbol{q}$ is locked to the direction of the magnetic field. The LTCP is killed by the stabilization of the χ-spiral phase, and there is only a second order phase transition line between χSR and χ-spiral phases. We stress that this drastic change happens for an arbitrary intensity of magnetic field. It means that the standard χSB phase becomes unstable against the formation of density wave, and the LTCP will never be realized in the presence of an external magnetic field.

Next we consider the effect of current quark mass h together with the magnetic field $\boldsymbol{b}$. We show in Fig. 2 the phase diagrams for four different values of magnetic fields. The phase diagram displayed in Fig. 2(a) is for $8b = 0.2 \times h^{3/5}$, which is the case where the effect of $\boldsymbol{b}$ is relatively weaker than the current quark mass (h-term) effect. Note, however, even in this case the magnetic energy is roughly estimated as $\sqrt{eB} \sim 20$MeV corresponding to a quite large intensity of magnetic field, $B \sim 7 \times 10^{12}$T. We see that the phase diagram is not much modified at this magnetic intensity. The magnetic field replaces only a tiny thin region near the phase boundary between the χSR and RKC phases with a *modified* χ-spiral defined by $\Delta = M_0 + \Delta_0 e^{i\boldsymbol{q}\cdot\mathbf{x}}$ with M_0, $\boldsymbol{q}$ and Δ_0 the variational parameters. However, a major part of the RKC and the Lifshitz critical point (LCP) itself remain intact. We conclude that the current quark mass protects the LCP and the RKC phase from a weak magnetic field. Fig. 2(b) presents the phase diagram for $8b = 1.0 \times h^{3/5}$, that roughly corresponds to $\sqrt{eB} \sim 40$MeV ($B \sim 3 \times 10^{13}$T). At this magnetic intensity, we see a sizable region for the χ-spiral. Accordingly the LCP is killed and replaced by a new critical point, where the second order transition from the χ-spiral to the χSR turns into a first order one from the χ-spiral to the RKC (or χSB). Depicted in Fig. 2(c) is the phase diagram for a stronger magnetic field $8b = 5.0 \times h^{3/5}$, roughly, $\sqrt{eB} \sim 90$MeV ($B \sim 10^{14}$T). The region for the χ-spiral gets significantly magnified, and the original LCP is now completely covered by the spiral phase. There is a new critical point, denoted by a black square, where the second order

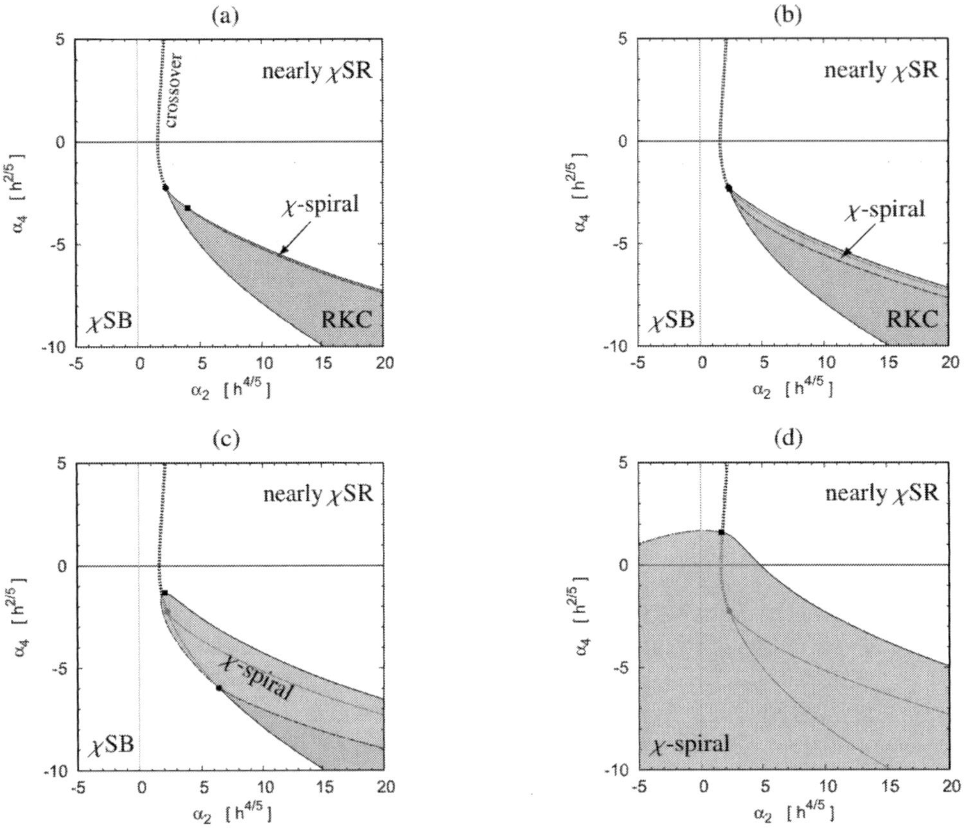

Figure 2. The phase diagrams off the chiral limit. (a): $8b = 0.2 \times h^{3/5}$. (b): $8b = 1.0 \times h^{3/5}$. (c): $8b = 5.0 \times h^{3/5}$. (d): $8b = 15 \times h^{3/5}$.

phase transition at which the χ-spiral ends at high density (large α_2) side, changes into a first order one at low density side (small α_2). Fig. 2(d) represents the phase diagram at an even stronger magnetic field $8b = 15 \times h^{3/5}$, that is estimated roughly $\sqrt{eB} \sim 150$MeV ($B \sim 4 \times 10^{14}$T). In this extreme case, the effect of magnetic field completely dominates over that from h-term. The RKC phase is replaced by the χ-spiral, which now spreads over a wide region. We see that the critical point still exists on the phase boundary, where the second order phase transition turns into a first order one.

4 Conclusions

We studied the effects of an external magnetic field on the chiral phase structure of QCD within the generalized Ginzburg-Landau (gGL) effective action. We first derived the gGL action performing the derivative expansion up to the sixth order in condensates and spatial derivatives. Expanding the action also up to the lowest nontrivial order in a current quark mass and a magnetic field, we obtained the explicit symmetry breaking sources, h-term and $\boldsymbol{b}$-term, respectively. The h-term explicitly breaks the chiral symmetry to the diagonal isospin SU(2), while the $\boldsymbol{b}$-term violates the time reversal symmetry, and reduces the isospin SU(2) down to $U_Q(1)$, the spatial rotation symmetry SO(3) down to O(2), the rotation about the magnetic axis. It is clearly seen in the obtained gGL action that these two symmetry breaking terms have competing effects on the condensate; the former prefers the real condensate, while the latter favors the complex condensate spatially modulated in the direction of magnetic field. We have computed the phase diagrams for nonvanishing magnetic fields. In the chiral limit, the effect of an external magnetic field is such drastic that it completely washes out the tricritical point as well as the Real-Kink Crystal (RKC) phase. There is only a second order phase transition at which the spiral terminates. On the other hand, the effect of current quark mass was found to protect the RKC phase and the Lifshitz critical point from the erosion by a weak magnetic field. However, as the intensity of magnetic field increases, the χ-spiral phase gradually invades the coast region of the high density boundary between the RKC and nearly symmetric phases. When the magnetic field strength is large enough, the effect of magnetic field prevails over that of current quark mass, and the RKC phase gets completely beaten by the chiral spiral phase. We confirmed that, in the regime of strong magnetic fields, the shape of the phase structure approaches the extreme one obtained in the chiral limit.

Acknowledgements

I thank R. Yoshiike, K. Nishiyama, and T. Tatsumi for useful discussions. I would like to express my sincere gratitude to the organizers of QCD@work 2016 at Martina Franca, especially to Pietro Colangelo and Fulvia De Fazio for their kind hospitality.

References

[1] E. Nakano and T. Tatsumi, Phys. Rev. D **71**, 114006 (2005).
[2] D. Nickel, Phys. Rev. Lett. **103**, 072301 (2009).
[3] For a recent review; see; M. Buballa and S. Carignano, Prog. Part. Nucl. Phys. **81**, 39 (2015).
[4] R. C. Duncan and C. Thompson, Astrophys. J. **392**, L9 (1992).
[5] D. E. Kharzeev, L. D. McLerran and H. J. Warringa, Nucl. Phys. A **803**, 227 (2008).
[6] H. Suganuma and T. Tatsumi, Annals Phys. **208**, 470 (1991).
[7] S. P. Klevansky and R. H. Lemmer, Phys. Rev. D **39**, 3478 (1989).
[8] V. P. Gusynin, V. A. Miransky and I. A. Shovkovy, Phys. Rev. Lett. **73**, 3499 (1994).
[9] G. S. Bali, et al., Phys. Rev. D **86**, 071502 (2012).
[10] G. Endrodi, JHEP **1507**, 173 (2015).
[11] F. Bruckmann, G. Endrodi and T. G. Kovacs, JHEP **1304**, 112 (2013).
[12] K. Fukushima and Y. Hidaka, Phys. Rev. Lett. **110**, no. 3, 031601 (2013).
[13] J. Chao, P. Chu and M. Huang, Phys. Rev. D **88**, 054009 (2013).
[14] F. Preis, A. Rebhan and A. Schmitt, JHEP **1103**, 033 (2011).
[15] M. Ruggieri, L. Oliva, P. Castorina, R. Gatto and V. Greco, Phys. Lett. B **734**, 255 (2014).
[16] P. Costa, et al., Phys. Rev. D **92**, no. 3, 036012 (2015).
[17] T. Tatsumi, K. Nishiyama and S. Karasawa, Phys. Lett. B **743**, 66 (2015).
[18] G. Cao and A. Huang, Phys. Rev. D **93**, no. 7, 076007 (2016).
[19] K. Nishiyama, S. Karasawa and T. Tatsumi, Phys. Rev. D **92**, 036008 (2015).
[20] R. Yoshiike and T. Tatsumi, Phys. Rev. D **92**, no. 11, 116009 (2015).
[21] H. Abuki, Phys. Lett. B **728**, 427 (2014); Phys. Rev. D **87**, no. 9, 094006 (2013).
[22] H. Abuki, D. Ishibashi and K. Suzuki, Phys. Rev. D **85**, 074002 (2012).
[23] E. V. Gorbar, V. A. Miransky, I. A. Shovkovy and X. Wang, Phys. Rev. D **88**, no. 2, 025043 (2013); Phys. Rev. D **88**, no. 2, 025025 (2013).
[24] C. Boehmer, M. Thies and K. Urlichs, Phys. Rev. D **75**, 105017 (2007).

Chiral medium produced by parallel electric and magnetic fields

Marco Ruggieri[1,⋆], *Guang Xiong* Peng[1,2,⋆⋆], and *Maxim* Chernodub[3,4,⋆⋆⋆]

[1] *College of Physics, University of Chinese Academy of Sciences, Yuquanlu 19A, Beijing 100049, China*
[2] *Theoretical Physics Center for Science Facilities, Institute of High Energy Physics, Beijing 100049, China*
[3] *Laboratoire de Mathématiques et Physique Théorique UMR 7350, Université de Tours, 37200, France*
[4] *Laboratory of Physics of Living Matter, Far Eastern Federal University, Sukhanova 8, Vladivostok, Russia*

Abstract. We compute (pseudo)critical temperature, T_c, of chiral symmetry restoration for quark matter in the background of parallel electric and magnetic fields. This field configuration leads to the production of a chiral medium on a time scale τ, characterized by a nonvanishing value of the chiral density that equilibrates due to microscopic processes in the thermal bath. We estimate the relaxation time τ to be about $\approx 0.1 - 1$ fm/c around the chiral crossover; then we compute the effect of the fields and of the chiral medium on T_c. We find T_c to be lowered by the external fields in the chiral medium.

1 Introduction

In recent years it has been realized that in ultrarelativistic heavy ion collisions very strong electric and magnetic fields can be produced, at least in the very early stages of the collision [1, 2]. In particular the strong magnetic field and the interplay with chiral (ABJ) anomaly in Quantum Chromodynamics (QCD) [3, 4] has suggested the possibility of observation of chiral magnetic effect (CME) in relativistic heavy ion collisions [5, 6]; a recent review and an exhaustive list of references can be found in [7]. Naming $n_{R/L}$ the number densities of right- and left-handed species, and $\mathcal{E}$, $\mathcal{B}$ a color-electric and color-magnetic field respectively, CME is related to the generation of chiral density, $n_5 = n_R - n_L$ induced by interaction of quarks with gluon fields characterized by $\mathcal{E}\cdot\mathcal{B} \neq 0$ named the QCD sphalerons.

Because of chiral anomaly it is possible to induce a chiral density also by the interaction of quark matter with external electric, E, and magnetic, B, fields with $E \cdot B \neq 0$. The chiral anomaly will then inject n_5 into the system, and this chiral density can equilibrate thanks to microscopic processes that change quark chirality. The purpose of the study presented here is to study this equilibration, which leads to the formation of a medium that we name chiral medium, and to compute its effect on the critical temperature of QCD. The work presented here is mainly based on [8–12].

⋆*Speaker.* e-mail: marco.ruggieri@ucas.ac.cn
⋆⋆e-mail: gxpeng@ucas.ac.cn
⋆⋆⋆e-mail: maxim.chernodub@lmpt.univ-tours.fr

2 Chiral medium produced by E and B

We consider quark matter in the background of parallel Abelian (electromagnetic) electric, E, and magnetic, B, fields [8–10] (see also [13–15]). In this case the evolution of n_5 with time is given by

$$\frac{dn_5}{dt} = -n_5\Gamma + N_c\frac{e^2 E\cdot B}{2\pi^2}\sum_f q_f^2 e^{-\frac{\pi M_q^2}{|q_f eE|}}, \tag{1}$$

where the first term on the right hand side describes relaxation of chiral density due to chirality-changing processes in the thermal medium, while the second term comes from the ABJ chiral anomaly [3, 4, 13]. M_q denotes the quark mass and q_f the electric charge of the flavor f in units of the proton charge. The quantity Γ in Eq. (1) corresponds to the rate of the chirality flips while its inverse defines the chiral relaxation time,

$$\tau = 1/\Gamma. \tag{2}$$

The physical interpretation of Eq. (1) is that in the background of $E \parallel B$ fields the ABJ anomaly creates a chiral imbalance n_5; the growth of n_5 competes however with chiral relaxation due to microscopic processes that flip the chirality of quarks. For time $t \approx \tau$ the value of chiral density exponentially equilibrates at:

$$n_5^{\rm eq} = N_c\frac{e^2 E\cdot B}{2\pi^2}\tau\sum_f q_f^2 e^{-\frac{\pi M_q^2}{|q_f eE|}}. \tag{3}$$

The main message encoded in Eq. (3) is that the combined effect of the external fields and of the chirality changing processes in the thermal bath is the formation of chiral medium, characterized by a finite value of chiral density and to which a chiral chemical potential μ_5 can be associated, namely

$$n_5^{\rm eq} = -\frac{\partial\Omega}{\partial\mu_5}, \tag{4}$$

where Ω corresponds to the thermodynamic potential.

The knowledge of the relaxation time τ is crucial since it allows to determine the equilibrium value of chiral density, $n_5^{\rm eq}$. In the context of QCD, τ been computed in [9] for temperatures around T_c within an NJL model. The microscopic processes considered in [9] are quark-quark scattering mediated by collective modes with the quantum numbers of pions and σ−mesons, hence we refer to these processes as one-pion or one-σ exchange for simplicity. While referring the interested reader to [9] for all the technical details, we remind here the main quantities that have to be computed to estimate τ. The starting point is the Boltzmann collision integral,

$$\frac{df_R(p)}{dt} = \int d\Pi\frac{(2\pi)^4\delta^4(p+k-p'-k')}{2E_p}|\mathcal{M}|^2 F, \tag{5}$$

where f_R denotes the distribution function for the right-handed quarks (a similar equation holds for left-handed quarks) and $d\Pi$ corresponds to the momentum space measure,

$$d\Pi = \frac{d^3k}{(2\pi)^3 2E_k}\frac{d^3k'}{(2\pi)^3 2E'_k}\frac{d^3p'}{(2\pi)^3 2E'_p}; \tag{6}$$

the kernel F takes into account the population of the incoming and outgoing particles in the process,

$$F(p,k,p',k') = f_L(p')f_L(k')[1-f_R(p)][1-f_R(k)] - f_R(p)f_R(k)[1-f_L(p')][1-f_L(k')], \tag{7}$$

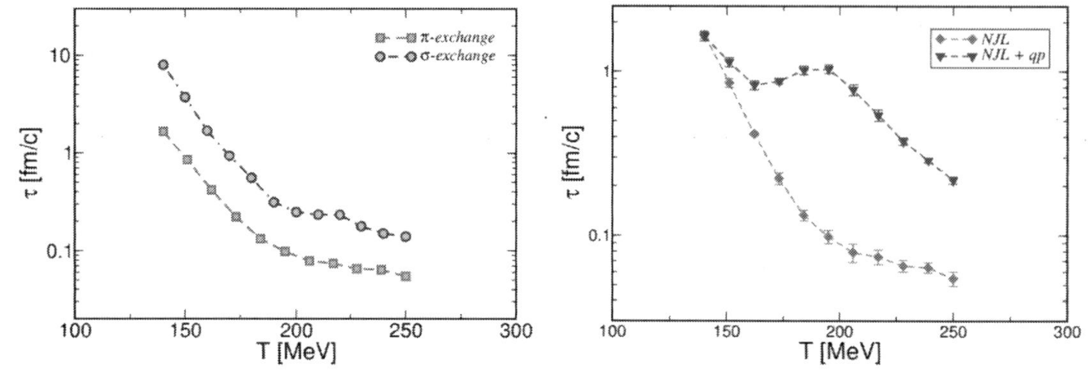

Figure 1. Left panel: relaxation time versus temperature in the crossover region for π−exchange (squares) and σ−exchange (circles). Right panel: relaxation time for π−exchange for the cases of the pure NJL model (green diamonds) and of the NJL+quasiparticle model (blue triangles). Adapted from [9].

for the scattering of two incoming R quarks giving two outgoing L quarks, with $f_{R/L}(p) = (1 + e^{\beta\omega_\pm})^{-1}$ and $\omega_s = \sqrt{(p + s\mu_5)^2 + M_q^2}$ with $s = \pm 1$. In [9] any dependence of τ on E and B has been neglected for simplicity, and an expansion to the lowest nontrivial order in μ_5/T has been used to further simplify the numerical evaluation of the collision integral.

In Eq.(5) the squared transition amplitude $|\mathcal{M}|^2$ can be computed once a microscopic process for the chirality change has been chosen, namely the π−exchange and σ−exchange in [9]. From the definition of n_5 we get the equation governing its time evolution, namely

$$\frac{dn_5}{dt} = N_c N_f \int \frac{d^3p}{(2\pi)^3} \left(\frac{df_R}{dt} - \frac{df_L}{dt} \right). \tag{8}$$

Then from Eq.(1) with $E = B = 0$ we get the collision rate $n_5\Gamma = -dn_5/dt$; the relation between n_5 and μ_5 has been computed in [9] within the NJL model. Once the reaction rate Γ is known the relaxation time can be computed by Eq.(2).

In the left panel of Fig. 1 we plot the relaxation time versus temperature in the chiral crossover region for π−exchange (squares) and σ−exchange (circles). We notice that the σ−exchange is less efficient than π−exchange in changing chirality of quarks in the thermal bath, resulting in a larger relaxation time. This can be understood because of the larger thermal mass of the σ−meson with respect to the pion mass.

An interesting feature of the results shown in Fig. 1 is that the relaxation time decreases with temperature. This feature might sound counterintuitive: naively a chirality change in the system should be related to the quark mass operator, that in the NJL model is related to M_q which becomes smaller with temperature, implying Γ becomes also smaller. However, while this naive argument works well for the transition matrix element, it does not necessarily work for Γ: in fact the latter is determined also by the phase space available for collisions, which instead becomes larger with temperature because the quark mass becomes smaller and the distribution functions broader. As a result, the collision rate increases with temperature and the relaxation time becomes smaller. For a more detailed discussion see [9].

In [9] the NJL model calculation of τ has been compared with a model calculation where the quark mass function at large temperature is taken to be that of a quasiparticle model [16]. The latter is

DOI: 10.1051/epjconf/201612900037

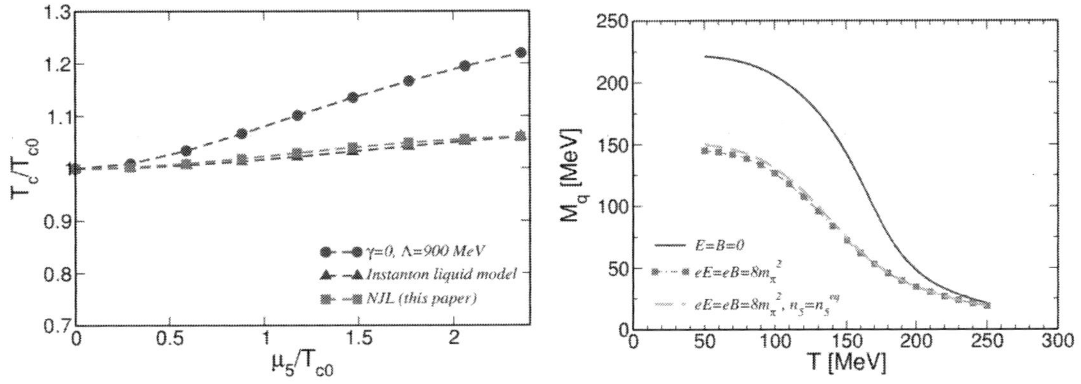

Figure 2. Left panel: Critical temperature versus μ_5. Data taken from [11]. Right panel: M_q versus temperature for the case $eE = eB = 8m_\pi^2$ and $n_5 = 0$ (green dots), same values of E and B but with n_5 given by its equilibrium value Eq. (3) (orange dashed line). For comparison we plot by indigo solid line the data corresponding to $E = B = 0$. Adapted from [10].

assumed to take into account many-body effects that are ignored within the NJL model: in particular, the quasiparticle mass is larger than temperature, at least for $T \approx T_c$, while the NJL mass becomes much smaller than the temperature itself because of chiral symmetry restoration. The smooth transition from the NJL mass at low temperature to the quasiparticle mass at large temperature introduced in [9] models a phase space volume reduction at $T \approx T_c$, hence potentially lowering Γ and increasing τ at the crossover. In the right panel of Fig. 1 we plot the relaxation time for π−exchange computed within the NJL model (green diamonds) and the NJL+quasiparticle model (blue triangles). The main difference of the relaxation time computed within the NJL+quasiparticle model in [9], with respect to the one obtained within the pure NJL model, is that in the former case τ stays almost constant at the crossover instead of decreasing as in the pure NJL model calculation; moreover, its order of magnitude is $\tau \approx 1$ fm/c while for the NJL calculation it drops down to $\tau \approx 0.1$ fm/c for $T > T_c$. Clearly the merging of NJL and quasiparticle models of [9] is too naive: nevertheless the results about τ show that the many body interactions, usually neglected in NJL model calculations, can have important effects on the computation of physical quantities around and above T_c, and more serious studies about these effects should be performed.

3 Critical temperature for chiral symmetry restoration

A natural question that arises is: can the formation of the chiral medium lead to any observable effect on equilibrium properties of the bulk? In particular, in our study we are interested to the QCD critical temperature in presence of the electric and magnetic fields, therefore we can ask whether the equilibrated n_5 can affect T_c. The answer is that in principle this is possible. To elaborate more on this point we plot in the left panel of Fig. 2 the critical temperature versus μ_5 computed in [11] within NJL models with several interaction kernels for $E = B = 0$. The results in Fig. 2 show that T_c increases with μ_5: the chiral chemical potential acts as a catalyzer of chiral symmetry breaking, see also [17–20]. Since in the context of our study the chemical potential is produced by the external fields, the net effect of the field on T_c has to be considered taking into account the equilibrated n_5.

The effect of μ_5 on T_c is opposite to the one induced by the fields. In fact on the right panel of Fig. 2 we plot the constituent quark mass versus temperature for $eE = eB = 8m_\pi^2$, for the cases in

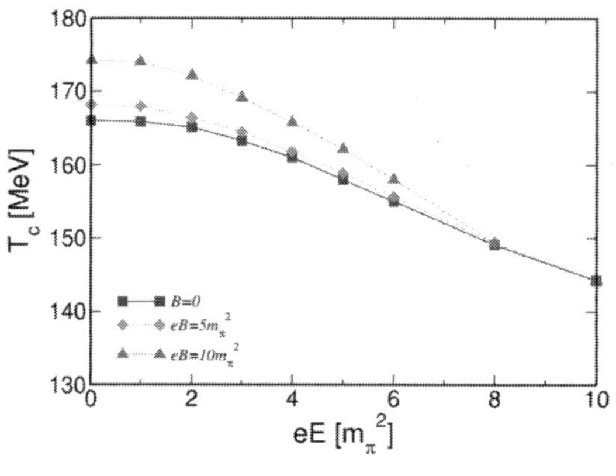

Figure 3. Critical temperature versus electric field strength for several values of B. Adapted from [10].

which we ignore the chiral density at equilibrium (green squares). We also plot the data corresponding to $E = B = 0$ by the solid indigo line. Data have been obtained from [10]. The results collected in the right panel of Fig. 2 show that the fields lower both M_q and T_c in comparison with the $E = B = 0$ case: the fields thus act as inhibitors of spontaneous chiral symmetry breaking, see also [14].

Given the different response of T_c to the fields on the one hand, and to μ_5 on the other hand, it is important to consider both of the effects to make a firm computation of T_c in presence of $E \parallel B$. This program has been accomplished in [10] where the QCD critical temperature in the background of $E \parallel B$ has been computed taking into account the equilibration of chiral density. In the right panel of Fig. 2 we plot by orange dashed line M_q versus temperature for $eE = eB = 8m_\pi^2$ for the case in which the equilibrated chiral density has been taken into account, by solving simultaneously the gap and the number equations within the NJL model. As expected, the effect of the chiral medium is to increase the value of M_q with respect to the $n_5 = 0$ case, as shown by the comparison of the orange and the green lines. On the other hand, the net effect of the chiral medium is not so strong, since the increase of M_q due to $n_5 \neq 0$ is always within the 10% of the $n_5 = 0$ result; as a consequence we conclude that the effect of $E \parallel B$ is much stronger than the one induced by n_5 and the net effect of the fields will be a lowering of T_c. Similar results have been recently obtained in [8].

In [10] the picture drawn here has been completed by the computation of the bulk structure of the chiral medium at equilibrium: in particular, it has been found that the chiral chemical potentials for u and d quarks are different, a result that can be easily understood because n_5 is produced by the electromagnetic coupling of quarks to E and B so a dependence on the quark electric charge is expected. Hence the external fields induce an imbalanced population of u and d quarks, which might have some effect on the charged pion condensation.

In Fig. 3, adapted from [10], we plot the critical temperature as a function of E for several values of B. The general trend of the results shown in the figure is that the electric field lowers T_c. An interesting feature of data is that even in the case of the largest magnetic field considered in the figure, which increases increase T_c with respect to the case $B = 0$, a modest electric field $eE \approx 4m_\pi^2$ is enough to bring T_c down to the zero field case. Hence the conclusion is that $E \parallel B$ acts as an inhibitor of chiral symmetry breaking, as long as E is not much smaller than B.

4 Conclusions

In this talk we have discussed our results on chiral symmetry restoration at finite temperature, for quark matter in the presence of parallel electric, E, and magnetic, B, fields. This background leads to the formation of a chiral medium, namely a medium with a finite chiral density at equilibrium, $n_5 = n_R - n_L$, because of the combined effect of the chiral anomaly which pumps n_5 into the system, and chirality changing microscopic processes in the thermal bath that instead suppress the formation of n_5. The characteristic time scale necessary for the formation of this medium corresponds approximately to the relaxation time of chiral density, τ. We have estimated τ around the chiral crossover of QCD, finding $\tau \approx 0.1 - 1$ fm/c.

Then we have used this value of τ to compute n_5 at equilibrium simultaneously with the in-medium chiral condensate at finite temperature, eventually computing the critical temperature, T_c, as a function of the external field magnitude. Our result is summarized in Fig. 3 where we plot T_c versus the electric field strength for several values of the magnetic field; the critical lines have been computed by taking into account the value of n_5 at equilibrium computed self-consistently by virtue of Eq. (3). We find that the $E \parallel B$ background acts as an inhibitor of chiral symmetry breaking, at least for values of E not too smaller than B.

An interesting follow up of the work presented here might be the inclusion of neutral pion condensation which is known to be induced by the chiral anomaly [15]; also it might be worth studying the charged pion condensation induced by the mismatch of u and d chemical potentials in the chiral medium. We plan to report on these subjects in the near future.

Acknowledgements. M. R and G. X. P. would like to thank the CAS President's International Fellowship Initiative (Grant No. 2015PM008), and the NSFC projects (11135011 and 11575190). M. R. acknowledges discussions with V. Greco, Z. Y. Lu and F. Scardina.

References

[1] V. Skokov, A. Y. Illarionov and V. Toneev, Int. J. Mod. Phys. A **24**, 5925 (2009).
[2] A. Bzdak and V. Skokov, Phys. Lett. B **710**, 171 (2012).
[3] S. L. Adler, Phys. Rev. **177**, 2426 (1969).
[4] J. S. Bell and R. Jackiw, Nuovo Cim. A **60**, 47 (1969).
[5] D. E. Kharzeev, L. D. McLerran and H. J. Warringa, Nucl. Phys. A **803**, 227 (2008).
[6] K. Fukushima, D. E. Kharzeev and H. J. Warringa, Phys. Rev. D **78**, 074033 (2008).
[7] K. Hattori and X. G. Huang, arXiv:1609.00747 [nucl-th].
[8] M. Ruggieri, Z. Y. Lu and G. X. Peng, arXiv:1608.08310 [hep-ph].
[9] M. Ruggieri, G. X. Peng and M. Chernodub, Phys. Rev. D **94**, no. 5, 054011 (2016).
[10] M. Ruggieri and G. X. Peng, Phys. Rev. D **93**, no. 9, 094021 (2016).
[11] M. Ruggieri and G. X. Peng, arXiv:1602.05250 [hep-ph].
[12] M. Ruggieri and G. X. Peng, arXiv:1602.03651 [hep-ph].
[13] H. J. Warringa, Phys. Rev. D **86**, 085029 (2012).
[14] A. Y. Babansky, E. V. Gorbar and G. V. Shchepanyuk, Phys. Lett. B **419**, 272 (1998).
[15] G. Cao and X. G. Huang, Phys. Lett. B **757**, 1 (2016).
[16] S. Plumari, W. M. Alberico, V. Greco and C. Ratti, Phys. Rev. D **84**, 094004 (2011).
[17] L. Yu, H. Liu and M. Huang, Phys. Rev. D **94**, no. 1, 014026 (2016).
[18] V. V. Braguta *et al.*, Phys. Rev. D **93**, no. 3, 034509 (2016).
[19] V. V. Braguta *et al.*, JHEP **1506**, 094 (2015).
[20] V. V. Braguta and A. Y. Kotov, Phys. Rev. D **93**, no. 10, 105025 (2016).

Theta Dependence in Yang-Mills from Holography

Francesco Bigazzi[1,⋆], *Aldo L.* Cotrone[1,2], and *Roberto* Sisca[3]

[1] *INFN, Sezione di Firenze; Via G. Sansone 1, I-50019 Sesto Fiorentino (Firenze), Italy*
[2] *Dipartimento di Fisica e Astronomia, Università di Firenze; Via G. Sansone 1, I-50019 Sesto Fiorentino (Firenze), Italy*
[3] *Department of Mathematics, University of Surrey; Guildford, GU2 7XH, UK*

Abstract. In this contribution we review how holographic tools can be used to study the dependence on the CP-breaking θ parameter of various observables in a large N_c Yang-Mills model. Relevant examples are the ground-state energy density, the string tension, the glueball mass spectrum and the critical temperature for deconfinement. These observables have been also studied, in the small θ regime and up to very few powers of θ^2, in pure Yang-Mills on the Lattice. The holographic results qualitatively agree with available Lattice data and, being exact in θ, provide benchmarks for higher order corrections.

1 Introduction

Instanton configurations naturally induce a topological term in the Lagrangian for a $SU(N_c)$ Yang-Mills gauge theory

$$\mathcal{L}_E = \frac{N_c}{2\lambda}\left[\mathrm{Tr}F_{\mu\nu}F^{\mu\nu} - i\frac{\lambda}{16\pi^2}\frac{\theta}{N_c}\epsilon^{\mu\nu\rho\sigma}\mathrm{Tr}F_{\mu\nu}F_{\rho\sigma}\right]. \tag{1}$$

Here $\lambda = g^2 N_c$ is the 't Hooft coupling and Euclidean signature is used. The θ parameter multiplies the so called topological charge density, a term which breaks parity and time reversal (hence CP) symmetries. Since the 4d integral of the topological charge density is an integer, the θ parameter behaves like an angle: the physics has to be periodic under $\theta \to \theta + 2\pi$ shifts.

Studying how observables depend on θ is both interesting and challenging. It is interesting because the parameter affects the vacuum structure of the theory. It is challenging because of the non-perturbative nature of the θ term. The use of numerical simulations on Euclidean Lattices is limited since the θ term is imaginary and Monte Carlo methods fail to converge. This is the reason why on the Lattice one can extract results only up to few terms in the expansion around $\theta = 0$, either analytically continuing from imaginary θ values or computing certain correlators at $\theta = 0$. See [1] for an excellent review on the subject.

Some hints on the physics beyond the small θ regime can be obtained studying the large N_c limit [2] where θ/N_c is held fixed. Periodicity under 2π shifts and θ/N_c scaling can be reconciled if there are infinite possible vacua, labeled by an integer k, which become stable but non-degenerate in the $N_c = \infty$ limit. The ground-state energy density, for a given value of θ, is obtained by minimizing,

⋆e-mail: bigazzi@fi.infn.it

with respect to k, a function of $(\theta + 2\pi k)/N_c$. The resulting expression, periodic in θ, shows cusps at particular values of θ, accounting for jumps between different branches.

Remarkably, these features can be explicitly realized in a large N_c Yang-Mills model introduced by Witten in [3]. The reason is that the model, under certain assumptions, can be explored at the non-perturbative level and for any value of θ using the holographic gauge/gravity correspondence [4–6]. The latter is a proposed equivalence map between classes of quantum field theories (QFT) and theories of gravity in at least one extra dimension, which finds its concrete realization in the framework of string theory. Open strings contain gauge degrees of freedom, while closed strings contain gravitons. The open/closed string duality, which naturally emerges in the theory practically implements the gauge/gravity correspondence.

The correspondence is relevant since it works as a duality: the large N_c, strong 't Hooft coupling regime of, say, a $SU(N_c)$ gauge theory, can be mapped into a classical theory of gravity. Gauge invariant operators correspond to gravity fields, QFT local symmetries are mapped into isometries on the gravity side. Finally, the QFT partition function on a given vacuum, is computed by the on-shell gravity action on the dual background corresponding to that vacuum.

The map between a QFT and a theory of gravity is one-to-one, when the dual pairs are embedded into a consistent ten dimensional string theory model. This feature is crucially based on the properties of the so called Dp-branes, $p + 1$ dimensional hyperplanes over which the open strings can end (here "D" stands for Dirichlet boundary conditions). These D-branes are actually physical objects, with their own tension and charge (they are electrically charged under C_{p+1} potentials): as such they deform the ambient spacetime. The resulting curved background and the low energy effective QFT on such D-branes are the dual pairs in the game.

When a stack of N_c $D3$-branes in flat space is considered, the low energy QFT is the maximally supersymmetric Yang-Mills theory in $3 + 1$ dimensions. This is also a conformal field theory (CFT). The dual description is provided by the near-horizon geometry sourced by the $D3$-branes: the direct product of a five dimensional Anti de Sitter (AdS) spacetime and a five dimensional sphere [4].

In order to describe a non supersymmetric and non conformal Yang-Mills theory one has thus to go beyond this AdS/CFT correspondence. This is the case for the model [3] we will focus on. The starting point is a 5d supersymmetric $SU(N_c)$ gauge theory describing the low energy dynamics of N_c $D4$-branes. One of the spatial directions, call it x_4, is compactified on a circle of radius $1/M_{KK}$, where antiperiodic boundary conditions on the fermions are imposed. The resulting low energy model is a 4d non supersymmetric $SU(N_c)$ Yang Mills coupled with a tower of adjoint massive Kaluza-Klein (KK) modes, with mass scale set by M_{KK}. A θ term can be easily added to the model and the holographic correspondence allows to explore the related physics without encountering any sign problem. To leading order in θ/N_c this has been already done by Witten in [7]. It is not difficult, however, to go beyond this limit and so to consider finite generic values of θ. The corresponding gravity solution has been found in [8, 9] and the related physics has been further explored in [10]. The present contribution contains a brief review of the latter work.

We will first provide some details about the gauge theory model and its dual gravity description. Then, we will present a selected class of results, ranging from the θ-dependence of the ground state energy density, the lightest glueball mass and the string tension to that of the finite temperature phase diagram. As we will see, the holographic model qualitatively agrees with available Lattice data in the small θ regime, confirming that mass scales get reduced by θ to leading order and providing possible benchmarks for higher order corrections. Moreover, as it is shown in [10], holographic tools allow to extract the θ-dependence of observables like the 't Hooft loop, the baryon mass (vertex) and the entanglement entropy, which are not under control using other techniques. It is also possible to introduce massive quarks and study the physics of the associated QCD-like holographic model, obtaining interesting results concerning for example the vacuum structure of the theory or the calculation of the neutron electric dipole moment [11].

2 The θ term in Witten's Holographic Yang Mills

The gauge/gravity dual pair we want to focus on is described by the following actions. On the gravity side, the relevant one is (in string frame) the following consistent truncation of the 10d type IIA supergravity action

$$S_{gravity} = \frac{1}{2k_0^2}\int d^{10}x\sqrt{-g}\left[e^{-2\phi}\left(\mathcal{R}+4(\partial\phi)^2\right)-\frac{1}{2}|F_4|^2-\frac{1}{2}|F_2|^2\right]. \tag{2}$$

Here $2k_0^2 = (2\pi)^7 l_s^8$ where $l_s \equiv \sqrt{\alpha'}$ is the string length, $F_4 = dC_3$ is a four-form which is magnetically sourced by the N_c $D4$-branes, $F_2 = dC_1$, and ϕ is the dilaton.

The gauge theory action (omitting the contribution of the KK fields) is obtained as the low energy limit of the action for N_c $D4$-branes wrapping a circle S_{x^4}

$$S_{gauge} = \frac{1}{8\pi g_s l_s M_{KK}}\int d^4x TrF^2 - \frac{i}{32\pi^2 l_s}\int_{S_{x^4}} C_1 \int d^4x\epsilon^{\mu\nu\rho\sigma} TrF_{\mu\nu}F_{\rho\sigma}. \tag{3}$$

Comparing this with (1), one gets the following identifications

$$\lambda = g^2 N_c \equiv 2\lambda_4 \equiv 2g_{YM}^2 N_c = 4\pi g_s N_c l_s M_{KK}, \qquad \theta + 2\pi k = \frac{1}{l_s}\int_{S_{x_4}} C_1, \tag{4}$$

where k is an integer (accounting for the fact that the integral of C_1 is gauge invariant only modulo $2\pi\mathbb{Z}$), g_s is the string coupling and λ_4, g_{YM} are introduced to match with a standard notation adopted for Witten's model.

Holography allows to study the gauge theory above by means of a dual classical gravity solution, provided we take the limits $\lambda_4 \gg 1$ and $N_c \gg 1$. Beyond these limits one should add α' and string loop corrections to the gravity action.

The family (parameterized by the integer k) of gravity backgrounds corresponding to the k-branches of field theory vacua has been found in [8, 9]. The metric reads

$$ds_{10}^2 = \left(\frac{u}{R}\right)^{3/2}\left[\sqrt{H_0}\,dx_\mu dx^\mu + \frac{f}{\sqrt{H_0}}dx_4^2\right] + \left(\frac{R}{u}\right)^{3/2}\sqrt{H_0}\left[\frac{du^2}{f} + u^2 d\Omega_4^2\right], \tag{5}$$

where

$$f = 1 - \frac{u_0^3}{u^3}, \qquad H_0 = 1 - \frac{u_0^3}{u^3}\frac{\Theta^2}{1+\Theta^2}. \tag{6}$$

The background also includes a running dilaton, a one-form and a four-form field strength given by

$$e^\phi = g_s\left(\frac{u}{R}\right)^{3/4} H_0^{3/4}, \qquad C_1 = \frac{\Theta}{g_s}\frac{f}{H_0}dx^4, \qquad F_4 = 3R^3\omega_4, \tag{7}$$

with $R = (\pi g_s N_c)^{1/3} l_s$ and

$$\Theta \equiv \frac{\lambda_4}{4\pi^2}\left(\frac{\theta + 2k\pi}{N_c}\right). \tag{8}$$

In the expressions above $\mu = 0, 1, 2, 3$ are the $1+3$ Minkowski directions where the Yang-Mills theory is defined, $d\Omega_4^2$ is the metric of a four-sphere S^4 of unit radius, u is a radial coordinate $u \in [u_0, \infty)$ -

holographically mapped to the renormalization group scale of the dual field theory - x_4 is the compact coordinate of length $2\pi M_{KK}^{-1}$ and R is a curvature radius. The isometry group of S^4 corresponds to a global $SO(5)$ symmetry group in the dual field theory; this acts non-trivially on the Kaluza-Klein massive modes signaling that these are not decoupled. Finally ω_4 is the volume form of the transverse S^4, of volume $V_{S^4} = 8\pi^2/3$.

The S_{x_4} circle shrinks to zero size at $u = u_0$ and absence of conical singularities is guaranteed if

$$u_0 = \frac{4R^3}{9} M_{KK}^2 \frac{1}{1+\Theta^2}, \tag{9}$$

which implies that the (x_4, u) subspace has the topology of a disk.

The above gravity background is regular and in particular, since $g_{00}(u_0) \neq 0$, it describes a confining gauge theory. This will be made explicit in a moment when the string tension of the model will be computed.

2.1 The ground-state energy density

Holography relates the field theory partition function with the (renormalized) on-shell gravity action on the θ-dependent background reviewed above. In particular

$$Z = e^{-V_4 f(\theta)} \approx e^{-S^{\rm ren}_{E\,\rm on-shell}}, \tag{10}$$

where $f(\theta)$ is the Yang Mills ground state energy density, V_4 is the (infinite) 4d Euclidean spacetime volume and

$$S^{\rm ren}_{E\,\rm on-shell} = S_E + S_{GH} + S_{c.t.}\,. \tag{11}$$

Here S_E is the Euclidean version of the action (2). Moreover

$$S_{GH} = -\frac{1}{k_0^2}\int d^9x\,\sqrt{h}e^{-2\phi}K, \tag{12}$$

is the Gibbons-Hawking term, where h is the determinant of the boundary metric and K is the trace of the extrinsic curvature of the boundary. Finally, the counter-term action is given by [12]

$$S^{bulk}_{c.t.} = \frac{1}{k_0^2}\left(\frac{g_s^{1/3}}{R}\right)\int d^9x\,\sqrt{h}\,\frac{5}{2}e^{-7\phi/3}. \tag{13}$$

Using the holographic map (10) it follows that

$$f(\theta) = \min_k f(\Theta), \quad \text{where} \quad f(\Theta) = -\frac{2N_c^2\lambda_4}{3^7\pi^2}\frac{M_{KK}^4}{(1+\Theta^2)^3}. \tag{14}$$

The ground state energy density (see figure 1) turns out to be a periodic function of θ, as expected. To any given interval, of length 2π, of possible values of θ, it corresponds a value of k. For example, $k = 0$ when $\theta \in (-\pi, \pi)$. The cusps at $\theta = \pm\pi$ and odd multiples of these values, signal first order CP-breaking phase transitions. In the small θ limit one gets

$$f(\theta) - f(0) = \frac{1}{2}\chi_g\theta^2\left[1 + \bar{b}_2\frac{\theta^2}{N_c^2} + \bar{b}_4\frac{\theta^4}{N_c^4} + O(\theta^6)\right], \tag{15}$$

with the topological susceptibility given by $\chi_g = \lambda_4^3 M_{KK}^4/(4(3\pi)^6)$ and the expansion coefficients given by $\bar{b}_2 = -\lambda_4^2/(8\pi^4)$, $\bar{b}_4 = 5\lambda_4^4/(384\pi^8)$.

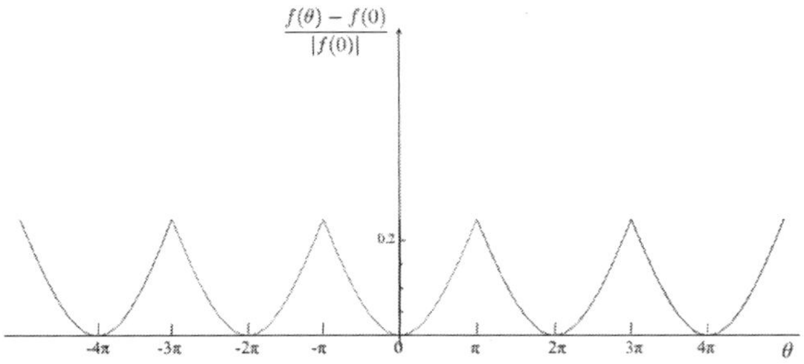

Figure 1. The (normalized) ground state energy density at finite θ. Different colors correspond to different branches of the vacuum energy. The plot has been obtained setting $\lambda_4/(4\pi^2 N_c) = 0.1$.

These results can be compared with those obtained for pure Yang-Mills on the Lattice (see [1]), where extrapolating from the $N_c = 3, 4, 6$ results one gets $\bar{b}_2 \approx -0.2$. The sign of this coefficient matches with that obtained in the holographic model.

Large errors do not allow to fix the sign of b_4 on the Lattice: recent results [13, 14] just provide a bound on its absolute value $|b_4| < 0.001$. Our results suggest that b_4 should have a positive sign.

2.2 The string tension

Another standard holographic map [15] relates the VEV of a Wilson loop on a contour C with the (renormalized) Nambu-Goto on-shell action for a fundamental open string whose end-points span that contour

$$\langle W[C]\rangle \sim e^{-S^r_{NG}}. \tag{16}$$

For a rectangular contour with sides of length t along the time direction and length L along one spatial direction, the Wilson loop allows to obtain the static quark-antiquark potential $V(L)$. In the large L limit in fact $\langle W[C]\rangle \approx e^{-tV(L)}$. Linear confinement of the chromoelectric flux tube implies that $V(L) = T_s L$ at large L where T_s is the string tension.

The holographic computation of the rectangular Wilson loop just requires minimizing the Nambu-Goto action of an open string, localized in the transverse S^4, with worldvolume coordinates $\tau = x^0 \in [0, t]$, $\sigma = x \in [-L/2, L/2]$, and embedded in such a way that $u = u(x)$

$$S_{NG} = -\frac{1}{2\pi\alpha'}\int d\tau d\sigma\sqrt{-g_{\tau\tau}g_{\sigma\sigma}} = -\frac{1}{2\pi\alpha'}t\int dx\sqrt{-g_{00}(g_{xx} + g_{uu}u'(x)^2)}, \tag{17}$$

where g is the (induced) string frame metric. In the large L limit the string profile which minimize this action is well approximated by a bath-tube-shaped configuration: a string stretching for most of its length at $u = u_0$, turning vertically, around the extrema $x = -L/2$ and $x = L/2$, up to a UV cutoff $u = u_\Lambda \to \infty$ where it is attached to a probe brane. The two vertical parts of the string give divergent contributions to the action when the cutoff is sent to infinity. They are interpreted as due to the (infinite) masses of the static quark-antiquark pair and must be subtracted in order to get the renormalized Nambu-Goto action. All in all, in the large L limit one gets

$$S^r_{NG} \approx -\frac{1}{2\pi\alpha'} t\sqrt{-g_{00}g_{xx}}|_{u=u_0} L\,, \tag{18}$$

which in turn, on the θ-deformed background (5), gives the string tension

$$T_s = \frac{2\lambda_4}{27\pi} M^2_{KK} \frac{1}{(1+\Theta^2)^2}\,, \tag{19}$$

which becomes a periodic function after the minimization over k is performed. Around $\theta \to 0$

$$T_s = \frac{2\lambda_4}{27\pi} M^2_{KK} \left(1 - \frac{\lambda_4^2}{8\pi^4}\frac{\theta^2}{N_c^2} + \frac{3\lambda_4^4}{256\pi^8}\frac{\theta^4}{N_c^4} + O(\theta^6)\right). \tag{20}$$

Notice that the $O(\theta^2)$ correction is negative. This, together with the expected scaling with θ/N_c, is precisely what has been found on the Lattice in [1]

$$T_{s\,lat} = T_{s\,lat}(0)\left[1 + \bar{s}_2 \frac{\theta^2}{N_c^2} + \dots\right], \tag{21}$$

where, using the large N_c extrapolation of the $N_c = 3, ..., 6$ results one gets $\bar{s}_2 \approx -0.9$. The holographic result (20) also predicts that the $O(\theta^4)$ correction to the string tension in pure Yang-Mills should be positive.

2.3 The scalar glueball mass

Just as in pure Yang-Mills, Witten's model has a mass gap in the glueball spectrum [3]. Each glueball state corresponds to a mode in the dual string picture. The 0^{++} glueball closely related to that of pure Yang-Mills can be identified with a scalar fluctuation involving the dilaton. The equation of motion for the fluctuation (see [10] for details)

$$H''(u) + \frac{4u^3 - u_0^3}{u(u^3 - u_0^3)} H'(u) - \frac{M^2 R^3}{u^3 - u_0^3} H(u) = 0\,, \tag{22}$$

has to be solved by imposing regularity at $u = u_0$ and normalizability at $u \to \infty$. The possible solution exist only if $M^2 > 0$ and the spectrum is discrete. From equation (9) it follows that the mass of the 0^{++} glueball is formally given by $M(\Theta) = M(\Theta = 0)(1 + \Theta^2)^{-1/2}$. Around $\theta = 0$

$$M(\theta) = M(\theta = 0)\left(1 - \frac{\lambda_4^2}{32\pi^4}\frac{\theta^2}{N_c^2} + \frac{3\lambda_4^4}{2048\pi^8}\frac{\theta^4}{N_c^4} + O(\theta^6)\right), \tag{23}$$

where the leading correction in θ^2 has a negative sign, as in Lattice Yang-Mills [16]. The holographic model predicts that the $O(\theta^4)$ correction has a positive sign.

2.4 Finite temperature: θ-dependence of the phase diagram

Going to finite temperature in Witten's holographic Yang-Mills model, amounts to compactifying the Euclidean time on a circle of length $\beta = 1/T$. There are two possible solutions of the gravity equations of motion for which this condition can be satisfied (see e.g. [17]). A first solution is just the Euclidean continuation (with a compact time circle) of the one presented above. This solution corresponds to the confined phase of the gauge theory. The free energy density in this phase has the same expression as in the $T = 0$ case (14).

The second solution has a black hole event horizon and corresponds to a deconfined phase. The S_{x^4} circle does not shrink anymore and the equations of motion are solved by $C_1 = \theta dx^4$ which implies that $F_2 = 0$. As a result the background metric looses any θ-dependence

$$ds^2 = \left(\frac{u}{R}\right)^{3/2}\left[\tilde{f}(u)dx_0^2 + dx_a dx^a + dx_4^2\right] + \left(\frac{u}{R}\right)^{-3/2}\left[\frac{du^2}{\tilde{f}(u)} + u^2 d\Omega_4^2\right],$$

$$\tilde{f}(u) = 1 - \frac{u_T^3}{u^3}\,. \tag{24}$$

The dilaton and the F_4 form do not change w.r.t. the $T = 0$ solution. In the equation above, $a = 1, 2, 3$. Regularity is now guaranteed by the relation $9\beta^2 u_T = 16\pi^2 R^3$, where u_T is the radial position of the horizon.

The gauge theory free energy density (hence minus the pressure) in the deconfined phase is holographically mapped into that of the black hole background and reads

$$f_{dec} = -p_{dec} = -\frac{1}{6}\frac{256 N_c^2 \pi^4 \lambda_4}{729 M_{KK}^2} T^6\,, \tag{25}$$

where the scaling with T follows from the higher dimensional UV completion of the model. The free energy is independent from θ and thus the topological susceptibility is zero in the deconfined phase. This is actually what one expects in the large N_c limit of Yang-Mills, where the susceptibility is exponentially suppressed.

Comparing the free energies (or pressures) in the two phases one gets that the dual field theory experiences a first order confinement-deconfinement phase transition at a critical temperature T_c formally given by

$$T_c(\Theta) = \frac{M_{KK}}{2\pi}\frac{1}{\sqrt{1+\Theta^2}}\,, \tag{26}$$

which corresponds to a periodic function $T_c(\theta)$ as shown in figure 2.

Remarkably the holographic model explicitly realizes the expected but still conjectural (see e.g. [18]) structure of the phase diagram, with triple critical points where the first order deconfinement transition line meets the first order lines related to the CP breaking transitions in the confined phase. One can conjecture that the same cusped structure should appear in pure Yang-Mills at large N_c.

Around $\theta = 0$ we find (at $O(\theta^2)$ the result was already found in [19])

$$T_c(\theta) = \frac{M_{KK}}{2\pi}\left[1 - \frac{\lambda_4^2}{32\pi^4}\frac{\theta^2}{N_c^2} + \frac{3\lambda_4^4}{2048\pi^8}\frac{\theta^4}{N_c^4} + O(\theta^6)\right], \tag{27}$$

where, again, both the quadratic dependence and the sign of the leading θ-dependent correction, agree with what has been found (for $N_c = 3$) on the Lattice [18]

$$T_c(\theta)_{lat} = T_c(0)_{lat}\left[1 - R_\theta \theta^2 + O(\theta^4)\right], \quad R_\theta = 0.0175(7)\,. \tag{28}$$

Our model, then, predicts that the $O(\theta^4)$ correction has a positive coefficient. It would be nice to check this and the previous predictions on the Lattice.

Acknowledgments

FB is grateful to the organizers of the Workshop "QCD at Work 2016" for their kind invitation.

DOI: 10.1051/epjconf/201612900038

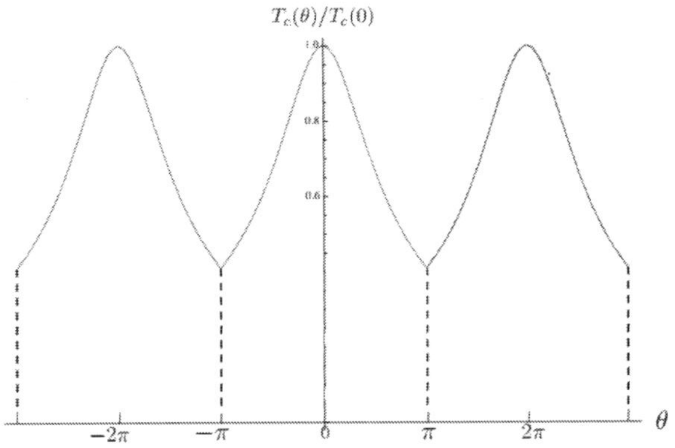

Figure 2. The (T, θ) phase diagram. Different colors correspond to different branches of the vacuum energy.

References

[1] E. Vicari and H. Panagopoulos, Phys. Rept. **470**, 93 (2009).
[2] E. Witten, Annals Phys. **128**, 363 (1980).
[3] E. Witten, Adv. Theor. Math. Phys. **2**, 505 (1998).
[4] J. M. Maldacena, Int. J. Theor. Phys. **38**, 1113 (1999) [Adv. Theor. Math. Phys. **2**, 231 (1998)].
[5] E. Witten, Adv. Theor. Math. Phys. **2**, 253 (1998).
[6] S. S. Gubser, I. R. Klebanov and A. M. Polyakov, Phys. Lett. B **428**, 105 (1998).
[7] E. Witten, Phys. Rev. Lett. **81**, 2862 (1998).
[8] J. L. F. Barbon and A. Pasquinucci, Phys. Lett. B **458**, 288 (1999).
[9] S. Dubovsky, A. Lawrence and M. M. Roberts, JHEP **1202**, 053 (2012).
[10] F. Bigazzi, A. L. Cotrone and R. Sisca, JHEP **1508**, 090 (2015).
[11] L. Bartolini, F. Bigazzi, S. Bolognesi, A. L. Cotrone and A. Manenti, to appear.
[12] D. Mateos, R. C. Myers and R. M. Thomson, JHEP **0705**, 067 (2007).
[13] H. Panagopoulos and E. Vicari, JHEP **1111**, 119 (2011).
[14] C. Bonati, M. D'Elia, P. Rossi and E. Vicari, "θ dependence of 4D $SU(N)$ gauge theories in the large-N limit," arXiv:1607.06360 [hep-lat].
[15] J. M. Maldacena, Phys. Rev. Lett. **80**, 4859 (1998). S. J. Rey and J. T. Yee, Eur. Phys. J. C **22**, 379 (2001).
[16] L. Del Debbio, G. M. Manca, H. Panagopoulos, A. Skouroupathis and E. Vicari, JHEP **0606**, 005 (2006).
[17] O. Aharony, J. Sonnenschein and S. Yankielowicz, Annals Phys. **322**, 1420 (2007).
[18] M. D'Elia and F. Negro, Phys. Rev. Lett. **109**, 072001 (2012); Phys. Rev. D **88**, no. 3, 034503 (2013).
[19] F. Bigazzi and A. L. Cotrone, JHEP **1501**, 104 (2015).

Strongly interacting matter from holographic QCD model

Yidian Chen[1,⋆] , Danning Li[2,⋆⋆], and Mei Huang[1,3,⋆⋆⋆]

[1] *Institute of High Energy Physics, Chinese Academy of Sciences, Beijing 100049, China*
[2] *Physics Department, Jinan University, Guangzhou 510632, China*
[3] *Theoretical Physics Center for Science Facilities, Chinese Academy of Sciences, Beijing 100049, China*

Abstract. We introduce the 5-dimension dynamical holographic QCD model, which is constructed in the graviton-dilaton-scalar framework with the dilaton background field Φ and the scalar field X responsible for the gluodynamics and chiral dynamics, respectively. We review our results on the hadron spectra including the glueball and light meson spectra, QCD phase transitions and transport properties in the framework of the dynamical holographic QCD model.

1 Introduction

Quantum chromodynamics (QCD) is the fundamental theory of strong interaction. In the infrared (IR) regime, it remains as an outstanding challenge of solving nonperturbative QCD physics, e.g. the chiral symmetry breaking and color confinement. The anti-de Sitter/conformal field theory (AdS/CFT) correspondence or gauge/gravity duality [1–3] provides a novel method to tackle the problem of strongly coupled gauge theories, and has been widely used in investigating hadron physics, strongly coupled quark gluon plasma and condensed matter physics. In general, holography maps a D-dimensional quantum field theory (QFT) to a quantum gravity in (D + 1)-dimensions, and the gravitational description becomes classical when the QFT is strongly-coupled. The extra dimension can be regarded as an energy scale or renormalization group (RG) flow in the QFT [4].

In this talk, I give a brief review on the dynamical holographic QCD (DhQCD) model recently developed in our group. The DhQCD model is constructed in the graviton-dilaton-scalar framework, and it resembles the renormalization group from ultraviolet (UV) to infrared (IR). The dilaton background field $\Phi(z)$ and the scalar field $X(z)$ are dual to the gluon operator and the quark operator at the UV boundary, and in the IR region, the dilaton background field and scalar field are responsible for nonperturbative gluodynamics and chiral dynamics, respectively. The metric structure at IR is automatically deformed by the nonperturbative gluon condensation and chiral condensation in the vacuum. I also give a summary of our results on hadron spectra, QCD phase transitions, equation of state and transport properties in the framework of DhQCD model.

⋆e-mail: chenyd@mail.ihep.ac.cn
⋆⋆e-mail: lidanning@jnu.edu.cn
⋆⋆⋆e-mail: huangm@mail.ihep.ac.cn

2 The glueball spectra and meson spectra

The detailed description of DhQCD model can be found in Ref.[5]. The pure gluon part of QCD is modelled by the 5D graviton-dilaton coupled action:

$$S_G = \frac{1}{16\pi G_5}\int d^5x\,\sqrt{g_s}e^{-2\Phi}\left(R_s + 4\partial_M\Phi\partial^M\Phi - V_G^s(\Phi)\right), \tag{1}$$

with G_5 the 5D Newton constant, g_s, Φ and V_G^s the 5D metric, and the dilaton field and dilaton potential in the string frame, respectively. The metric takes the following form

$$g_{MN}^s = b_s^2(z)(dz^2 + \eta_{\mu\nu}dx^\mu dx^\nu),\;\; b_s(z) \equiv e^{A_s(z)}, \tag{2}$$

and is automatically deformed by the dilaton background $\Phi(z) = \mu_G^2 z^2$ as in [6].

The glueball spectra are excitations from the pure gluon background, i.e. the deformed AdS_5 background solved from the graviton-dilaton coupled system. In order to distinguish even and odd parity, we introduce the positive and negative coupling between the dilaton field and glueballs, respectively. With this setup, we have the 5D actions for the scalar, vector and tensor glueballs $\mathcal{G}(x,z)$ taking the following form:

$$S_{\mathcal{G}} = -\frac{1}{2}\int d^5x\,\sqrt{g_s}e^{-p\Phi}(\partial_M\mathcal{G}\partial^M\mathcal{G} + M_{\mathcal{G},5}^2(z)\mathcal{G}^2) \tag{3}$$

$$S_V = -\frac{1}{2}\int d^5x\,\sqrt{g_s}e^{-p\Phi}(\frac{1}{2}F^{MN}F_{MN} + M_{\mathcal{V},5}^2(z)\mathcal{V}^2), \tag{4}$$

$$\begin{aligned} S_T = &-\frac{1}{2}\int d^5x\,\sqrt{g_s}e^{-p\Phi}(\nabla_L h_{MN}\nabla^L h^{MN} - 2\nabla_L h^{LM}\nabla^N h_{NM} \\ &+2\nabla_M h^{MN}\nabla_N h - \nabla_M h\nabla^M h + M_{h,5}^2(z)(h^{MN}h_{MN} - h^2)), \end{aligned} \tag{5}$$

where $M_5^2(z) = M_5^2 e^{-2\Phi/3}$, $p = 1$ for even parity and $p = -1$ for odd parity. The equation of motion for any glueball $\mathcal{A}$ can be brought into Schroedinger-like equation

$$-\mathcal{A}_n'' + V_{\mathcal{A}}\mathcal{A}_n = m_{\mathcal{A},n}^2\mathcal{A}_n, \tag{6}$$

with the 5D effective Schroedinger potential

$$V_{\mathcal{A}} = \frac{cA_s'' - p\Phi''}{2} + \frac{(cA_s' - p\Phi')^2}{4} + e^{2A_s - \frac{2}{3}\Phi}M_{\mathcal{A},5}^2, \tag{7}$$

where $c = 1$ for 1-form and $c = 3$ for 0-form and 2-form, and $M_{\mathcal{A},5}^2$ the 5-dimension mass.

The final results for two-gluon and three-gluon glueball spectra in the DhQCD model is shown in Fig. 1(a). With the only one parameter $\mu_G = 1$GeV, which is fixed by the Regge slope of the scalar glueball spectra, we can produce almost all glueballs spectra agree well with lattice data [7], except three trigluon glueball states 0^{--}, 0^{+-} and 2^{+-}, whose masses are 1.5 GeV lighter than lattice results.

We add light flavors in terms of meson fields on the gluodynamical background. The total 5D action for the graviton-dilaton-scalar system takes the form of

$$S = S_G + \frac{N_f}{N_c}S_{KKSS}, \tag{8}$$

with

$$S_G = \frac{1}{16\pi G_5}\int d^5x\sqrt{g_s}e^{-2\Phi}(R+4\partial_M\Phi\partial^M\Phi - V_G(\Phi)), \tag{9}$$

$$S_{KKSS} = -\int d^5x\sqrt{g_s}e^{-\Phi}Tr(|DX|^2 + V_X(X^+X,\Phi) + \frac{1}{4g_5^2}(F_L^2+F_R^2)). \tag{10}$$

The meson spectra produced from this model is summarized in Fig. 1(b). It is observed that in the graviton-dilaton-scalar system, the generated meson spectra agree well with experimental data.

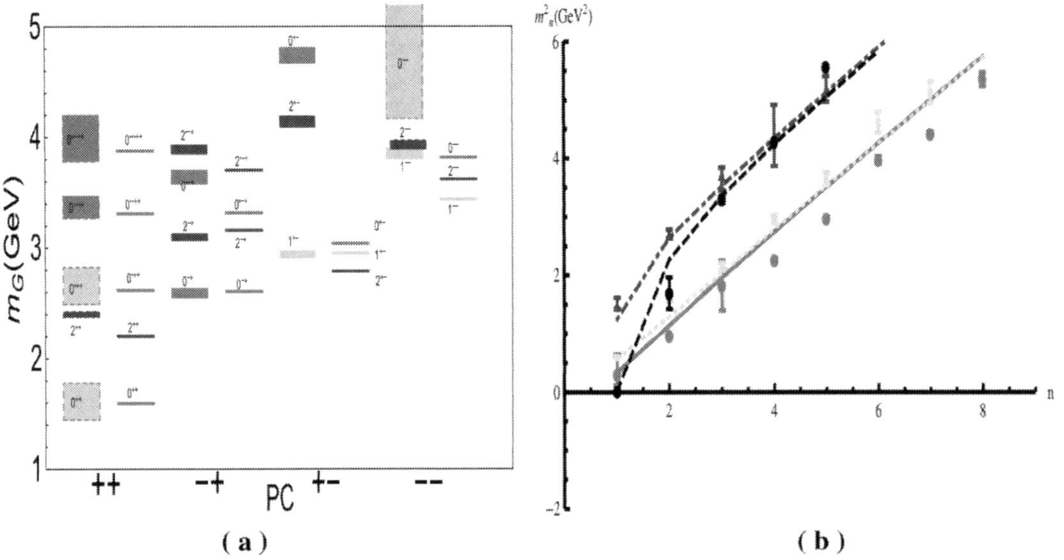

Figure 1. The glueball spectra (a) and light meson spectra (b) in the DhQCD model.

3 Thermodynamical and transport properties

In order to study strongly interacting matter at finite temperature, we introduce a black hole in the 5D gravity side and study the black hole thermodynamics. The metric ansatz at finite temperature in string frame takes the form of

$$ds_S^2 = e^{2A_s}\left(-f(z)dt^2 + \frac{dz^2}{f(z)} + dx^i dx^i\right). \tag{11}$$

With this metric ansatz, from the Einstein equations we derive the equations of motion:

$$-A_s'' + A_s'^2 + \frac{2}{3}\Phi'' - \frac{4}{3}A_s'\Phi' = 0, \tag{12}$$

$$f''(z) + (3A_s'(z) - 2\Phi'(z))f'(z) = 0, \tag{13}$$

$$\frac{8}{3}\partial_z\left(e^{3A_s(z)-2\Phi}f(z)\partial_z\Phi\right) - e^{5A_s(z)-\frac{10}{3}\Phi}\partial_\Phi V_G^E = 0, \tag{14}$$

with $V_G^E = e^{4\Phi/3} V_G^s$. The temperature of the solution is identified with the Hawking temperature

$$T = \frac{e^{-3A_s(z_h)+2\Phi(z_h)}}{4\pi \int_0^{z_h} e^{-3A_s(z')+2\Phi(z')} dz'}. \tag{15}$$

For the pure gluon system, when the dilaton profile takes the form of $\Phi = \mu_G z^2$, we can get the analytic solution of the metric prefactor

$$A_s(z) = \log(\frac{L}{z}) - \log({}_0F_1(5/4, \frac{\mu_G^4 z^4}{9})) + \frac{2}{3}\mu_G^2 z^2, \tag{16}$$

where L is the AdS radius.

We obtain the equation of state of the hot gluonic matter, and compare the trace anomaly $\epsilon - 3p$ with the pure SU(3) lattice data [8] in Fig.2 (a). It is observed that the thermodynamical quantities in the quenched dynamical holographic QCD model can describe the pure gluon system quite well, especially, the near T_c sharp peak of the trace anomaly shows that we have encoded the correct IR physics in the 5D model.

We also investigated the transport properties of the gluonic matter, e.g. the jet quenching parameter, the shear viscosity and bulk viscosity in the DhQCD model.

Jet quenching measures the energy loss rate of an energetic parton passing thorough the created hot dense medium, and the jet quenching parameter is related to the adjoint light like Wilson loop by the following AdS/CFT dictionary [10]

$$W^{Adj}[C] \approx exp(-\frac{1}{4\sqrt{2}}\hat{q}L^- L^2). \tag{17}$$

The expectation value of Wilson loop is dual to the on-shell value of the string Nambu-Goto action with proper string configuration, and the $\hat{q}$ can be obtained [9]

$$\hat{q} = \frac{\sqrt{2}\sqrt{\lambda}}{\pi z_h^3 \int_0^1 dv \sqrt{\frac{e^{-4A_s(vz_h)}}{z_h^4}\frac{1-f(vz_h)}{2} f(vz_h)}}. \tag{18}$$

In pure AdS background, $\hat{q}/T^3$ is a constant for all temperatures. In Fig.2 (b), we show $\hat{q}/T^3$ as a function of the temperature in the DhQCD model. It is seen that there is a peak with the height around 40 at around $T = 1.1T_c$, which coincides with the peak of the trace anomaly. This indicates that the DhQCD model has encoded novel property of the deconfinement phase transition.

For a conformal system with AdS_5 background, the shear viscosity over entropy density is $1/4\pi$ an the bulk viscosity is zero. However, for a system with phase transitions, it has been observed that the shear viscosity over entropy density ratio η/s has a minimum in the phase transition region in systems of water, helium, nitrogen [11]. It is expected that the same feature also shows up for nonconformal QCD system. The lattice QCD shows that the temperature dependence of bulk viscosity over entropy density [12] exhibits a peak around the phase transition. We derive the shear/bulk viscosity in the DhQCD model, and the details can be found in [13].

The bulk viscosity can be extracted from the Kubo formula

$$\zeta = \frac{1}{9}\lim_{\omega\to 0}\frac{1}{\omega}\mathrm{Im}\langle T_{xx}(\omega)T_{xx}(0)\rangle, \tag{19}$$

with T_{xx} is the $x-x$ component of the stress tensor. We have obtained the bulk viscosity from the DhQCD model in [13], which is shown in Fig.2(c). It is found that the bulk viscosity over entropy

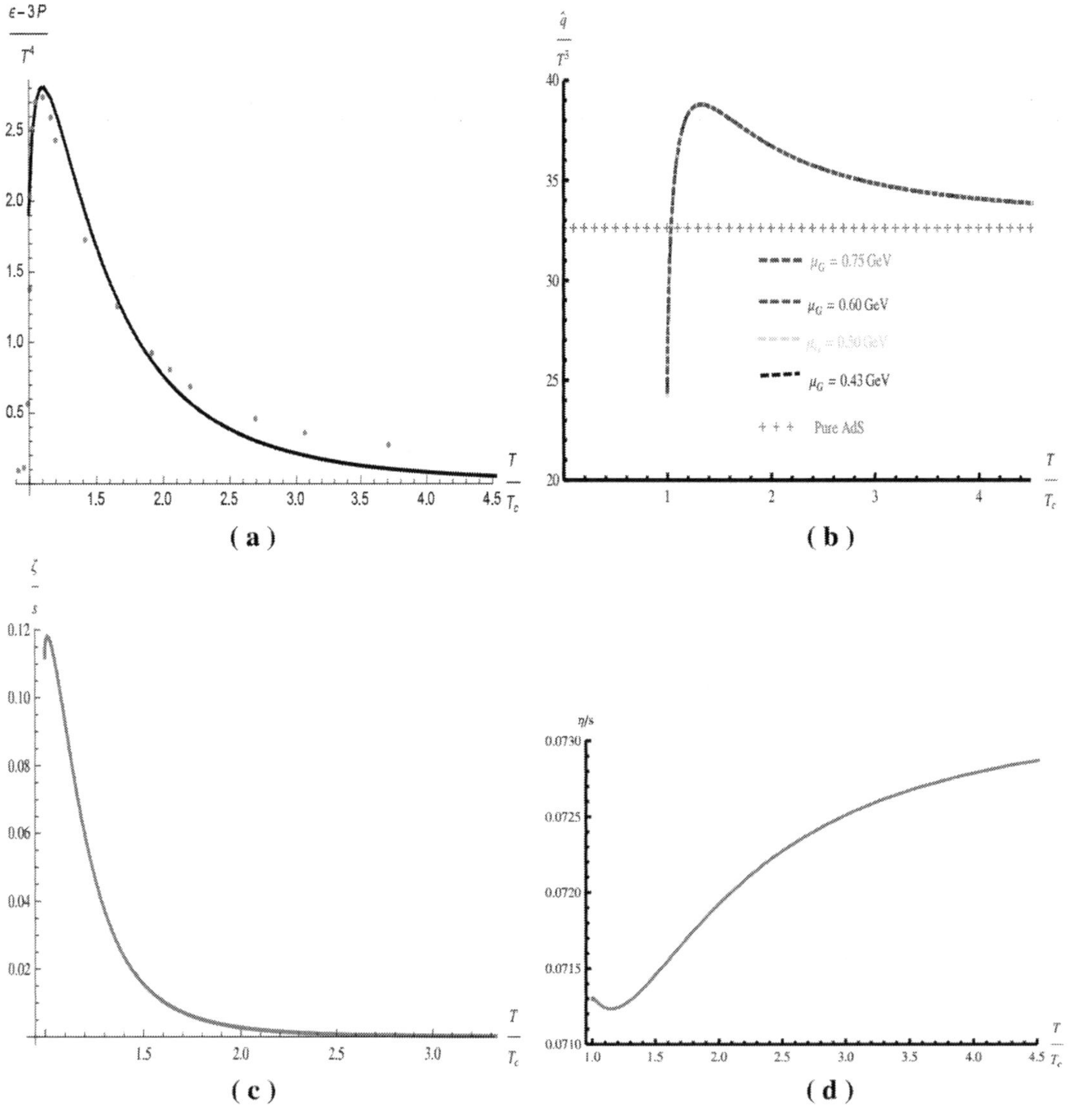

Figure 2. (a) The trace anomaly $\frac{\epsilon-3p}{T^4}$ as a function of T_c scaled temperature T/T_c in the DhQCD model (Solid black line) comparing with the pure SU(3) lattice data [8]; (b) The jet quenching parameter $\widehat{q/T}^3$ as a function of temperature T; (c) Bulk viscosity over entropy density and (d) shear viscosity over entropy density as functions of the temperature.

density shows a sharp peak near the transition temperature, and this feature is in agreement with lattice results in [12].

For shear viscosity, any isotropic Einstein gravity system generates the universal result of $\eta/s = 1/4\pi$. In order to introduce temperature dependence of the shear viscosity, we have to introduce higher derivative corrections of the following form [14]

$$S = \frac{1}{16\pi G_5}\int d^5x\sqrt{-g}(R - \frac{4}{3}\partial_\mu\Phi\partial^\mu\Phi - V(\Phi) + \beta e^{\sqrt{2/3}\gamma\Phi}R_{\mu\nu\lambda\rho}R^{\mu\nu\lambda\rho}).$$

The shear viscosity can be extracted through the Kubo formula

$$\eta = \lim_{\omega\to 0}\frac{1}{\omega}\mathrm{Im}\langle T_{xy}(\omega)T_{xy}(0)\rangle. \tag{20}$$

Up to the order of $O(\beta)$, the shear viscosity over entropy density ratio results read

$$\eta/s = \frac{1}{4\pi}\left(1 - \frac{\beta}{c_0}e^{\sqrt{2/3}\Phi_h}(1 - \sqrt{2/3}\gamma z_h\Phi'(z_h))\right), \tag{21}$$

with $c_0 = -z_h^5\partial_z\left((1 - z^2/z_h^2)^2 e^{2A}/(8f(z)z^2)\right)|_{z=z_h}$. Fig.2(d) shows the numerical result of η/s when $\beta = 0.01, \gamma = -\sqrt{8/3}$, it is observed that there is a valley at around $T = 1.1T_c$, which is almost the same location of the peak position of $\hat{q}/T^3$ and ζ/s.

4 Chiral phase transition

For QCD phase transitions, most studies focus on confinement/deconfinement phase transition. However, it is difficult to realize the chiral phase transition of QCD in the framework of AdS/CFT from both top-down and bottom-up approaches. We show how to correctly realize chiral symmetry breaking in the vacuum and restoration at finite temperature in the soft-wall holographic QCD models [15, 16].

In this part, we focus on chiral symmetry breaking and restoration, therefore, we only take the scalar part of the $SU(N_f)_L\times SU(N_f)_R$ 5D action, which takes the form of

$$S = -\int d^5x\sqrt{-g}e^{-\Phi}Tr(D_mX^+D^mX + V_X(|X|)). \tag{22}$$

Where Φ is the dilaton field, and the 5D mass of the complex scalar field X can be determined as $M_5^2 = -3$ from the AdS/CFT dictionary. For simplicity, we do not consider the back-reaction of the dilaton field and scalar field to the background geometry, and g is the AdS_5 metric background.

If the scalar field X gets a non-vanishing vacuum expectation value X_0, then $SU(N_f)_L\times SU(N_f)_R$ is spontaneously broken. We consider the 2-flavor case with $m_u = m_d$ and 3-flavor case with $m_u = m_d = m_s$, so we expect that the symmetry is broken to $SU(N_f)$ and $X_0 = \frac{\chi(z)}{\sqrt{2N_f}}I_{N_f}$. Here I_{N_f} is the $N_f\times N_f$ identity matrix and $\chi(z)$ is only dependent on the fifth coordinate z. Inserting the expectation value of X, we obtain the the effective action in terms of χ of the following form

$$S_\chi = -\int d^5x\sqrt{-g}e^{-\Phi}(\frac{1}{2}g^{zz}\chi'^2 + V(\chi)), \tag{23}$$

with

$$V(\chi) \equiv Tr(V_X(|X|)) = -\frac{3}{2}\chi^2 + v_3\chi^3 + v_4\chi^4. \tag{24}$$

The leading term of $V(\chi)$ is the mass term and it is fixed to be $-\frac{3}{2}\chi^2$, the quartic term $v_4\chi^4$ keeps $\chi\leftrightarrow-\chi$ symmetry, and the cubic term $v_3\chi^3$ is for the three-flavor mixing term and vanishes for the two-flavor case. The equation of motion for χ can be derived as

$$\chi'' + (3A_s' - \Phi' + \frac{f'}{f})\chi' - \frac{e^{2A_s}}{f}\partial_\chi V(\chi) = 0. \tag{25}$$

 DOI: 10.1051/epjconf/201612900039

In addition, we take the following form for the dilaton field

$$\Phi(z) = -\mu_1 z^2 + (\mu_1 + \mu_0) z^2 \tanh(\mu_2 z^2), \tag{26}$$

which is negative in the UV and positive at IR.

The chiral phase transition is shown in Fig. 3 (a) and (b) for 2-flavor and 3-flavor cases, respectively. We have realized the spontaneous chiral symmetry breaking in the vacuum and its restoration at finite temperature in the holographic QCD framework. The results are in good agreement with lattice result: In the chiral limit, the phase transition is of second order for two-flavor case and of 1st-order for three-flavor case, and in the case of finite current quark mass, the phase transition turns to crossover in both two-flavor and three-flavor cases.

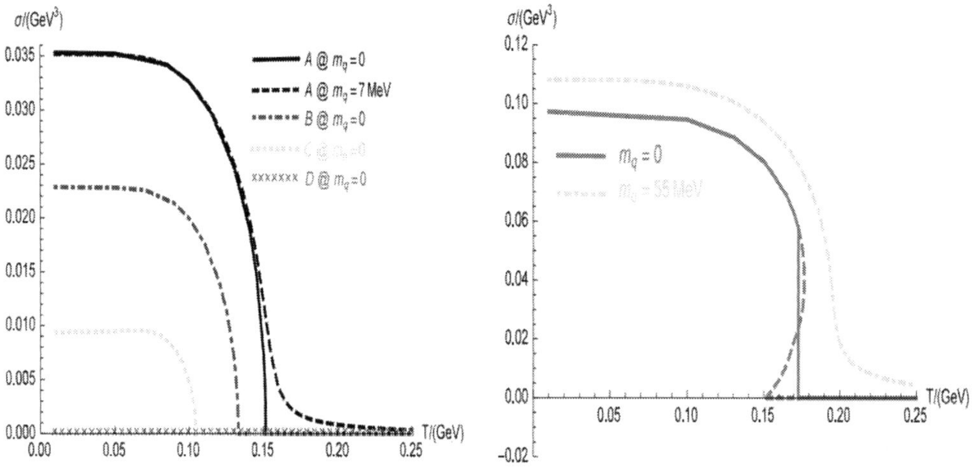

(a) (b)

Figure 3. The chiral condensate σ as a function of the temperature T in 2-flavor case (a) and 3-flavor case (b), respectively.

5 Discussion and summary

In summary, we report our DhQCD model which is constructed in the graviton-dilaton-scalar framework, where the dilaton background field and scalar field are responsible for the gluodynamics and chiral dynamics, respectively. The dynamical holographic model can resemble the renormalization group from ultraviolet (UV) to infrared (IR), and the metric structure at IR is automatically deformed by the nonperturbative gluon condensation and chiral condensation in the vacuum.

We review our results on hadron spectra, and the produced glueball spectra and the light-flavor meson spectra in the DhQCD model agree well with lattice data and experimental data. We also show that for pure gluonic matter, the equation of state is in agreement with lattice result, and the transport properties show temperature dependent behavior. Especially, we observe that the transport properties can reflect the phase transition information, e.g. the ratio of the jet quenching parameter over cubic temperature $\hat{q}/T^3$ and the bulk viscosity over entropy density show a peak around the critical temperature T_c, and the shear viscosity over entropy density shows a valley around phase transition, at which the trace anomaly $(\epsilon - 3p)/T^4$ also shows a peak.

We also show that chiral phase transition can be successfully realized in the holographic QCD framework. In summary, our dynamical holographic QCD model can describe hadron physics, QCD phase transition, thermodynamical properties and transport properties quite successfully.

Acknowledgements

This work is supported by the NSFC under Grant Nos. 11275213, 11261130311(CRC 110 by DFG and NSFC), CAS key project KJCX2-EW-N01, and Youth Innovation Promotion Association of CAS.

References

[1] Maldacena J. M., Adv. Theor. Math. Phys. **2**, 231 (1998).
[2] Gubser S. S., Klebanov I. R. and Polyakov A. M., Phys. Lett. B **428**, 105 (1998).
[3] Witten E., Adv. Theor. Math. Phys. **2**, 253 (1998).
[4] Adams A., Carr L. D., Schaefer T., Steinberg P. and Thomas J. E., New J. Phys. **14**, 115009 (2012).
[5] Li D. N., Huang M. and Yan Q. -S. , Eur.Phys.J.C(2013) 73:2615; D. Li and M. Huang, JHEP **1311**, 088 (2013).
[6] A. Karch, E. Katz, D. T. Son and M. A. Stephanov, Phys. Rev. D **74** (2006) 015005.
[7] Meyer H. B., hep-lat/0508002; Lucini B. and Teper M. , JHEP **0106** (2001) 050; Morningstar C. J. and Peardon M. J., Phys. Rev. D **60** (1999) 034509; Chen Y., Alexandru A., Dong S. J., Draper T., *et al.*, Phys. Rev. D **73** (2006) 014516.
[8] G. Boyd, J. Engels, F. Karsch, E. Laermann, C. Legeland, M. Lutgemeier and B. Petersson, Nucl. Phys. B **469**, 419 (1996).
[9] D. Li, J. Liao and M. Huang, arXiv:1401.2035 [hep-ph].
[10] H. Liu, K. Rajagopal and U. A. Wiedemann, Phys. Rev. Lett. **97**, 182301 (2006) [hep-ph/0605178].
[11] L. P. Csernai, J. I. Kapusta and L. D. McLerran, Phys. Rev. Lett. **97**, 152303 (2006).
[12] D. Kharzeev and K. Tuchin, arXiv:0705.4280 [hep-ph]. F. Karsch, D. Kharzeev and K. Tuchin, Phys. Lett. B **663**, 217 (2008). H. B. Meyer, Phys. Rev. Lett. **100**, 162001 (2008).
[13] D. Li, S. He and M. Huang, JHEP **1506**, 046 (2015) doi:10.1007/JHEP06(2015)046 [arXiv:1411.5332 [hep-ph]].
[14] S. Cremonini, U. Gursoy and P. Szepietowski, JHEP **1208** (2012) 167 [arXiv:1206.3581 [hep-th]].
[15] K. Chelabi, Z. Fang, M. Huang, D. Li and Y. L. Wu, Phys. Rev. D **93**, no. 10, 101901 (2016) doi:10.1103/PhysRevD.93.101901 [arXiv:1511.02721 [hep-ph]].
[16] K. Chelabi, Z. Fang, M. Huang, D. Li and Y. L. Wu, JHEP **1604**, 036 (2016) doi:10.1007/JHEP04(2016)036 [arXiv:1512.06493 [hep-ph]].

EPJ Web of Conferences **129**, 00040 (2016) DOI: 10.1051/epjconf/201612900040
QCD@Work 2016

Thermodynamics of Resonant Scalars in AdS/CFT and implications for QCD*

Eugenio Megías[1,2,**] and *Manuel* Valle[2,***]

[1] *Max-Planck-Institut für Physik (Werner-Heisenberg-Institut), Föhringer Ring 6, D-80805 Munich, Germany*
[2] *Departamento de Física Teórica, Universidad del País Vasco UPV/EHU, Apartado 644, 48080 Bilbao, Spain*

Abstract. We explore the thermodynamics of a simple 5D Einstein-dilaton gravity model with a massive scalar field, with asymptotically AdS behavior in the UV. The holographic renormalization is addressed in details, and analytical results are obtained at high temperatures. We study the power corrections predicted by the model, and compare with lattice data in the deconfined phase of gluodynamics. Finally, it is discussed the role played by the conformal anomaly for integer values of the dimension of the condensate dual to the scalar field.

1 Introduction

The physics of the quark-gluon plasma has turned out to be more difficult than initially expected. Quantum Chromodynamics (QCD) is asymptotically free at high energies and temperatures, and this allows us to use perturbative methods to study its thermodynamic properties in this regime. Nevertheless close, and above, the phase transition non perturbative effects become very important, and usual perturbative approaches are not applicable. Lattice calculations at finite temperature show that there is a first order phase transition in gluodynamics for $N_c \geq 3$ [1, 2] and a crossover in QCD [3]. In addition, they seem to indicate the existence of power corrections in temperature in the deconfined phase that cannot be explained within a naive perturbative approach, see e.g. [4–6].

New techniques have appeared recently that can help in a true understanding of the underlying mechanism of strongly coupled gauge theories. The postulated gauge/gravity duality between a strongly coupled super Yang-Mills theory and a weakly coupled supergravity theory in $\mathrm{AdS}_5\times\mathrm{S}^5$ space [7], became a powerful tool in recent years to understand the non perturbative dynamics of QCD close to the phase transition. While a rigorous top-down approach is still far from giving any experimental prediction, several bottom-up models that can quantitatively reproduce many results of QCD in the low energy limit have been proposed in the literature. The core of the bottom-up scenario is to find a reasonable non-conformal metric of the AdS_5 space which mimics the physical properties of QCD, see e.g. [8–10]. In this work we use this approach, and explore the thermodynamics of a simple holographic model based on 5D Einstein-dilaton gravity with a massive scalar field. Some related works in the literature are e.g. [11–15].

*Talk given by E. Megías at the QCD@Work: International Workshop on QCD, 27-30 June 2016, Martina Franca, Italy.
**e-mail: emegias@mppmu.mpg.de
***e-mail: manuel.valle@ehu.es

2 Power corrections as a signal of non perturbative effects in the thermodynamics of QCD

Gluodynamics is invariant under scale and conformal transformations at the classical level. This classical invariance is, however, broken by quantum corrections due to the necessary regularization of the UV divergences. The divergence of the dilatation current equals the trace of the energy-momentum tensor T^{μ}_{μ} [16] yielding the *trace anomaly* [17]. At finite temperature, the energy density ε and the pressure P enter as [18]

$$\langle T^{\mu}_{\mu} \rangle = -\varepsilon + 3P = -\frac{\beta(g)}{2g}\langle (F^{a}_{\mu\nu})^2 \rangle , \tag{1}$$

where $F_{\mu\nu}$ is the field strength tensor and $\beta = \mu \partial g / \partial \mu$ is the beta function. While in the conformal limit, i.e. at very high temperatures, one has $\varepsilon = 3P$, a non-vanishing value of the trace anomaly provides a measure of the interaction. This means that $\varepsilon - 3P$ is crucial to understand the deconfinement process, as the non perturbative nature of QCD plays a prominent role in this regime.

The equation of state of QCD has been studied for a long time by using different methods. A naive weak coupling expansion turns out to be poorly convergent in the regime of temperatures close to the phase transition, see e.g. [19]. It has been proposed in the literature several methods to resum the perturbative expansion, one of the most popular being the Hard Thermal loop (HTL), currently computed up to 3-loops order [20]. However, all these methods fail to reproduce the lattice data for the trace anomaly in the regime $T_c \lesssim T \lesssim (2.5-3)T_c$, which corresponds to a strongly interacting quark-gluon plasma picture, see Fig. 1 (left).

It was studied in [2, 5, 6, 21] the possible existence and characterization of power corrections in the equation of state of gluodynamics. We show in Fig. 1 (right) a plot of the lattice data of the trace anomaly of gluodynamics $(\varepsilon - 3P)/T^4$ as a function of $(T_c/T)^2$. A clear linear behavior appears for temperatures $T \gtrsim 1.13T_c$. This behavior contradicts perturbation theory which contains no powers but only logarithms in the temperature, a feature shared by HTL and other resummation techniques. Basically, this result can be summarized in the formula

$$\Delta(T) \equiv \frac{\varepsilon - 3P}{T^4} = a_{\Delta}(\mu_T) + b_{\Delta}\left(\frac{T_c}{T}\right)^2 , \tag{2}$$

where $a_{\Delta}(\mu_T) \sim 1/\log T$ is the perturbative contribution. A fit of the lattice data with this formula, using $a_{\Delta}(\mu_T) = \Delta_{\mathrm{HTL}}(\mu_T)$, leads to $b_{\Delta} = (3.57 \pm 0.28)$, see [21]. This motivates the need of new methods to account for non perturbative effects.

3 Thermodynamics of AdS/QCD

The gauge/gravity duality is nowadays a powerful tool to study the strong coupling properties of gauge theories, and in particular of QCD, either at zero or finite temperature. One of the most important applications of this duality is the physics of strongly coupled plasmas. In particular, we can study the thermodynamics of a field theory (or QCD) from the classical computation of the thermodynamics of black holes in the gravity dual. This duality can be expressed in the form

$$S_{\text{Black Hole}}(T) = \frac{A(r_{\text{horizon}})}{4G_D} \longleftrightarrow S_{\text{QCD}}(T) . \tag{3}$$

The entropy of a black hole can be obtained classically from the famous Bekenstein-Hawking entropy formula, where $A(r_{\text{horizon}})$ is the area of the black hole horizon, see e.g. [12, 13, 22, 23]. In conformal AdS_5 the metric has a horizon in the bulk space at $r_T = \pi \ell^2 T$ where ℓ is the size of the AdS space, and the entropy scales like $S_{\text{Black Hole}} \propto r_T^3 \propto T^3$. However, in order to have a reliable extension of this duality to SU(N_c) Yang-Mills theory, the first task is to control the breaking of conformal invariance.

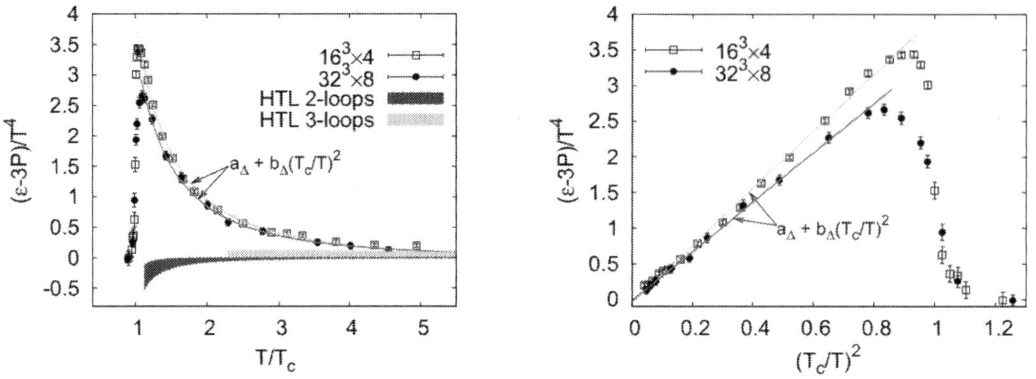

Figure 1. Trace anomaly density $(\varepsilon - 3P)/T^4$ as a function of T (left) and $1/T^2$ (right) (in units of T_c). Lattice data are from [1] for gluodynamics with $N_c = 3$. The fits use Eq. (2) with $\mu_T \sim T$ and b_Δ as adjustable parameters. See [6, 21] for further details.

3.1 The 5D Einstein-dilaton model at finite temperature

The bottom-up approach turns out to be quite useful to study the thermodynamics of QCD in the strongly coupling regime. It is based on the building of a gravity dual of QCD, including the main properties of QCD. One of the most successful models within this approach is the 5D Einstein-dilaton model, with the Euclidean action [12]

$$S = \frac{1}{2\kappa^2}\int d\rho d^4x\sqrt{G}\left(-R + G^{\mu\nu}\partial_\mu\Phi\partial_\nu\Phi + 2V(\Phi)\right) - \int_{\rho=\upsilon} d^4x\sqrt{\gamma}2K + S_{\text{boundary}}\,, \tag{4}$$

where κ^2 is the 5D Newton constant related to the 5D Planck mass by $\kappa^2 = 1/(2M^3)$, Φ is a scalar field, and υ is a cut-off surface near the boundary. Taking the limit $\upsilon \to 0$, one takes the surface to the AdS_5 boundary. The first boundary term is the usual Gibbons-Hawking term, while the second one S_{boundary} might depend on $\Phi'(\upsilon)$. The introduction of a scalar breaks conformal invariance, and the form of the scalar potential $V(\Phi)$ is phenomenologically adjusted to describe some observables of QCD, like for instance the trace anomaly, see Fig. 1. There are in the literature many different proposals for the dilaton potential, but in this work we will restrict to the much simpler form

$$V(\Phi) = -\frac{6}{\ell^2} + \frac{1}{2}m^2\Phi^2\,, \qquad m^2\ell^2 = \Delta(\Delta-4)\,, \qquad 0 \le \Delta \le 4\,, \tag{5}$$

where ℓ is the radius of AdS_5, and $m^2 < 0$ is a tachyonic mass. A black hole solution of the form [1]

$$ds^2_{\text{Black Hole}} = G_{\mu\nu}dx^\mu dx^\nu = \frac{\ell^2}{4\rho^2}d\rho^2 + \frac{\ell^2}{\rho}g_{\tau\tau}(\rho)d\tau^2 + \frac{\ell^2}{\rho}g_{xx}(\rho)d\vec{x}^2\,, \tag{6}$$

with an horizon in the extra coordinate $g_{\tau\tau}(\rho_h) = 0$, allows to describe the deconfined phase of the field theory dual. From the *AdS/CFT dictionary*, the model is the holographic dual of a Conformal Field Theory (CFT) with a deformation which breaks conformal invariance, i.e.

$$\mathscr{L} = \mathscr{L}^{CFT} + \lambda\mathscr{O}\,, \tag{7}$$

where $\mathscr{O}$ is an operator dual of the scalar field with $\dim\mathscr{O} = \Delta$, and λ is the source of the operator with $\dim\lambda = 4 - \Delta$. In the following we will focus on the so-called *Resonant Scalars*, i.e. the cases $\Delta \in \mathbb{Z}$, as they have some special properties.

[1] We work in Fefferman-Graham coordinates, as this allows for a clean treatment of the renormalization. See e.g. [24].

3.2 Holographic Renormalization

In order to study the holographic renormalization of the model, one should perform a near boundary expansion of the fields. If one considers the case $\Delta \notin \mathbb{Z}$, then the expansion of the scalar field reads

$$\Phi(\rho) = \phi_0 \cdot \rho^{(4-\Delta)/2} + \phi_{(2\Delta-4)} \cdot \rho^{\Delta/2} + \cdots \qquad \rho \to 0\,. \tag{8}$$

When $\Delta > 2$ then ϕ_0 is interpreted as the source λ, and $\phi_{(2\Delta-4)}$ as the condensate $\langle \mathscr{O} \rangle$ in Eq. (7). [2] From now on we will focus on the cases $\Delta = 1$ and 3, as their physical implications for QCD turn out to be highly relevant. In this case the expansion of the scalar field and the metric read

$$\Phi(\rho) = \chi \cdot \rho^{1/2} + \frac{\chi^3}{6}\rho^{3/2}\log\rho \; + \psi \cdot \rho^{3/2} + \cdots, \tag{9}$$

$$g_{xx}(\rho) = 1 - \frac{\chi^2}{6}\rho + g_{(4)xx}\rho^2 - \frac{\chi^4}{24}\rho^2\log\rho + \cdots, \tag{10}$$

$$g_{\tau\tau}(\rho) = 1 - \frac{\chi^2}{6}\rho - 3g_{(4)xx}\rho^2 - \chi\psi\rho^2 + \frac{\chi^4}{9}\rho^2 - \frac{\chi^4}{24}\rho^2\log\rho + \cdots, \tag{11}$$

where, roughly speaking, χ is interpreted as the source and ψ as the condensate when $\Delta = 3$ (the opposite when $\Delta = 1$). The $\log\rho$ contributions appear as a consequence of a resummation of higher orders in Eq. (8) that become dominant in the limit $\Delta \to 1$ or 3. Regularity at the horizon $\rho = \rho_h$ allows to fix the coefficients ψ and $g_{(4)xx}$. Without loss of generality we can choose $S_{\text{boundary}} = 0$, and then the counterterms needed to regularize the action read

$$S_{\text{ct}} = \frac{1}{\kappa^2\ell}\int_{\rho=\upsilon} d^4x\,\sqrt{\gamma(\upsilon)}\left(3 + \frac{\Phi^2(\upsilon)}{2} + \frac{\Phi^4(\upsilon)}{12}\log\left(\upsilon\mu_0^2\right)\right), \tag{12}$$

where μ_0 is an arbitrary renormalization scale. The variation of the regularized action $S_{\text{reg}} = S_{\text{on-shell}}[\upsilon]$ reads

$$\delta S_{\text{reg}} = -\frac{1}{\kappa^2\ell}\int_{\rho=\upsilon} d^4x\,\sqrt{\gamma}\,2\upsilon\Phi'(\upsilon)\delta\Phi(\upsilon) + \cdots, \tag{13}$$

and from here one can compute the condensate as the variation of the action with respect to the source,

$$\langle \mathscr{O}_3 \rangle = \frac{\delta(S_{\text{reg}} + S_{\text{ct}})}{\delta\chi} = \lim_{\upsilon\to 0}\left(\frac{1}{\upsilon^{-1/2}}\frac{\delta(S_{\text{reg}} + S_{\text{ct}})}{\delta\Phi(\upsilon)}\right) = \frac{\ell^3}{\kappa^2}\left(-2\psi - \frac{\chi^3}{3} + \frac{\chi^3}{3}\log\mu_0^2\right). \tag{14}$$

For $\Delta = 1$ the computation is the same but the interpretation is different. In this case $\dim\chi = 1$ so that χ should be the condensate, i.e. $\langle \mathscr{O}_1 \rangle = \chi$. This means that $\frac{\delta(S_{\text{reg}}+S_{\text{ct}})}{\delta\chi}$ is interpreted as the renormalized source of the operator $\mathscr{O}_1$.

3.3 Conformal Anomaly

By using the previous ingredients, one can obtain general formulas for the thermodynamics of a CFT deformed by $\int d^4x\chi\mathscr{O}_3$. The energy density and pressure are obtained from the $T^{\tau\tau}$ and T^{xx} components of the energy-momentum tensor. A straightforward computation from the regularized action leads to [25]

$$\varepsilon \;=\; \langle T^{\tau\tau} \rangle = -2\frac{\delta S_{\text{ren}}}{\delta g_{\tau\tau}(\upsilon)} = \frac{\ell^3}{\kappa^2}\left(6g_{(4)xx} + \chi\psi - \frac{5}{24}\chi^4 + \frac{\chi^4}{12}\log\mu_0^2\right), \tag{15}$$

$$P \;=\; \langle T^{xx} \rangle = -2\frac{\delta S_{\text{ren}}}{\delta g_{xx}(\upsilon)} = \frac{\ell^3}{\kappa^2}\left(2g_{(4)xx} + \chi\psi - \frac{\chi^4}{72} - \frac{\chi^4}{12}\log\mu_0^2\right). \tag{16}$$

[2] In the case $\Delta < 2$ the interpretation of these quantities is the inverse: ϕ_0 is the condensate and $\phi_{(2\Delta-4)}$ is the source.

From these expressions one can easily check that the following Ward identity for the trace anomaly

$$-\varepsilon + 3P = \langle T^{\mu}_{\mu} \rangle = (\Delta - 4)\chi \langle \mathscr{O}_\Delta \rangle + \mathscr{A}, \tag{17}$$

is fulfilled when $\Delta = 3$, with $\mathscr{A} = -\frac{\ell^3}{6\kappa^2}\chi^4$. We have checked that Eq. (17) is also valid when $\Delta = 1$ and 2. The term $\mathscr{A}$ is the holographic conformal anomaly, it is a zero temperature contribution and it appears only when $\Delta \in \mathbb{Z}$. [3] This anomaly has been studied also in [26] in a resonant theory in AdS_4.

3.4 Solution for the Equation of State

It is possible to obtain some analytical results for the equation of state by making a high temperature expansion. The temperature and entropy density are obtained from the metric of Eq. (6) as

$$T = \frac{1}{2\pi}\sqrt{2\rho_h g''_{\tau\tau}(\rho_h)}, \qquad s = \frac{2\pi}{\kappa^2}\left(\frac{\ell^2}{\rho_h} g_{xx}(\rho_h)\right)^{3/2}, \tag{18}$$

respectively. We will focus on the case $\Delta = 3$. From the equations of motion one can easily observe that the general dependence of the entropy density is $s = T^3\sigma(\chi/T)$, where σ is some specific function. By applying the standard thermodynamic relations one gets that the pressure can be written as $P = T^4 H_1(\chi/T) + \chi^4 H_2(\chi/\mu_0)$, where H_2 is a zero temperature contribution. From a computation of $-\varepsilon + 3P$ with this formula and its comparison with Eq. (17), one can make the identification

$$\frac{\partial P}{\partial \chi} = -\langle \mathscr{O}_3 \rangle, \quad -\frac{\chi^5}{\mu_0} H_2'\left(\frac{\chi}{\mu_0}\right) = \mathscr{A}. \tag{19}$$

The dependence in μ_0 arises only from the conformal anomaly. It is possible to make a computation of $H_1(\chi/T)$ at high temperature as a power series expansion in χ/T. The result for the trace anomaly is [4]

$$\varepsilon - 3P = \frac{\pi^4\ell^3 T^4}{2\kappa^2}\left[-4c_3\left(\frac{\chi}{\pi T}\right)^2 + \mathscr{O}(\bar{\chi}^4 \log \bar{\chi})\right] - \frac{2\ell^3}{3\kappa^2}\chi^4 \log\frac{\chi}{\mu_0}, \qquad \bar{\chi} \equiv \chi/T, \tag{20}$$

where $c_3 = -\Gamma\left[\frac{3}{4}\right]^2/(\Gamma\left[\frac{1}{4}\right]\Gamma\left[\frac{5}{4}\right]) \approx -0.457$ with $\Gamma[n]$ the Euler gamma function. We find that the thermodynamics of a deformation $\int d^4x\, \chi \mathscr{O}_3$ leads to a power series expansion in T^2 which can explain very precisely the lattice data of the trace anomaly in the deconfined phase of gluodynamics, as shown in Fig. 1. From the fit of Eq. (2) one gets $\chi = \frac{\kappa}{\ell^{3/2}}(0.629 \pm 0.025)T_c$. [5]

4 Conclusions and outlook

The non perturbative behavior of gluodynamics/QCD near and above T_c is characterized by power corrections in T^2. While perturbative methods fail to reproduce the non perturbative regime of QCD at finite temperature, AdS/CFT serves as a powerful tool to study this regime by using much simpler classical gravity techniques. In this work we have studied a simple holographic model of conformal

[3] The expressions of $\mathscr{A}$ for other values of Δ will be presented in [25].

[4] If one chooses the Newton constant to reproduce the Stefan-Boltzmann limit of gluodynamics at high temperatures, $P_{\text{gluons}} \to (N_c^2 - 1)\frac{\pi^2}{45}T^4$, then one has $\frac{\ell^3}{\kappa^2} = (N_c^2 - 1)\frac{2}{45\pi^2}$.

[5] While these power corrections in T^2 were studied in a similar model in [14], we have performed here an analysis that shows the consistency of the result with general formulas like Eq. (17). Our analysis conveniently accounts for the holographic conformal anomaly, and this is only possible with a careful holographic renormalization when $\Delta \in \mathbb{Z}$.

symmetry breaking in 5D based on dilatons that allows to account for power corrections in the deconfined phase of QCD in an elegant way. The model is dual to a CFT deformed by an operator $\mathcal{O}$. We find that the power corrections in the equation of state can be conveniently described with a deformation of $\dim \mathcal{O} = 3$. Finally, we have studied in details the holographic renormalization of the theory, and the role of the holographic conformal anomaly when $\dim \mathcal{O} \in \mathbb{Z}$.

We plan to extend the computation of the equation of state of the model for any value of Δ in the interval $0 \leq \Delta \leq 4$. This will be performed numerically in the whole regime of temperatures, as well as analytically at high and low temperatures. There remain also some open questions, like the physical interpretation of the condensate of dimension 3. These and other issues will be addressed in [25].

Acknowledgement

This work has been supported by Plan Nacional de Altas Energías grant FPA2012-34456, Spanish Consolider-Ingenio 2010 Programme CPAN (CSD2007-00042) and by the Basque Government under grant IT559-10. The research of E.M. is supported by the European Union under a Marie Curie Intra-European Fellowship (FP7-PEOPLE-2013-IEF) with project number PIEF-GA-2013-623006.

References

[1] G. Boyd, J. Engels, F. Karsch, E. Laermann, C. Legeland, M. Lutgemeier, B. Petersson, Nucl. Phys. **B469**, 419 (1996).
[2] M. Panero, Phys. Rev. Lett. **103**, 232001 (2009).
[3] Y. Aoki, G. Endrodi, Z. Fodor, S.D. Katz, K.K. Szabo, Nature **443**, 675 (2006).
[4] E. Megias, E. Ruiz Arriola, L.L. Salcedo, JHEP **01**, 073 (2006).
[5] R.D. Pisarski, Prog. Theor. Phys. Suppl. **168**, 276 (2007).
[6] E. Megias, E. Ruiz Arriola, L.L. Salcedo, Phys. Rev. **D80**, 056005 (2009).
[7] J.M. Maldacena, Adv. Theor. Math. Phys. **2**, 231 (1998).
[8] O. Andreev, V.I. Zakharov, JHEP **04**, 100 (2007).
[9] U. Gursoy, E. Kiritsis, F. Nitti, JHEP **02**, 019 (2008).
[10] B. Galow, E. Megias, J. Nian, H.J. Pirner, Nucl. Phys. **B834**, 330 (2010).
[11] T. Hertog, G.T. Horowitz, Phys. Rev. Lett. **94**, 221301 (2005).
[12] U. Gursoy, E. Kiritsis, L. Mazzanti, F. Nitti, JHEP **05**, 033 (2009).
[13] E. Megias, H.J. Pirner, K. Veschgini, Phys. Rev. **D83**, 056003 (2011).
[14] F. Zuo, JHEP **06**, 143 (2014).
[15] H. Lu, C.N. Pope, Q. Wen, JHEP **03**, 165 (2015).
[16] C.G. Callan, Jr., S.R. Coleman, R. Jackiw, Annals Phys. **59**, 42 (1970).
[17] J.C. Collins, A. Duncan, S.D. Joglekar, Phys. Rev. **D16**, 438 (1977).
[18] N.P. Landsman, C.G. van Weert, Phys. Rept. **145**, 141 (1987).
[19] E. Braaten, A. Nieto, Phys. Rev. **D53**, 3421 (1996).
[20] J.O. Andersen, M. Strickland, N. Su, Phys. Rev. Lett. **104**, 122003 (2010).
[21] E. Megias, E. Ruiz Arriola, L.L. Salcedo, Phys. Rev. **D81**, 096009 (2010).
[22] J. Alanen, K. Kajantie, V. Suur-Uski, Phys. Rev. **D80**, 126008 (2009).
[23] D. Li, S. He, M. Huang, Q.S. Yan, JHEP **09**, 041 (2011).
[24] S. de Haro, S.N. Solodukhin, K. Skenderis, Commun. Math. Phys. **217**, 595 (2001).
[25] E. Megias, M. Valle, (in progress) (2016).
[26] A. Anabalon, D. Astefanesei, D. Choque, C. Martinez, JHEP **03**, 117 (2016).

EPJ Web of Conferences **129**, 00041 (2016) DOI: 10.1051/epjconf/201612900041
QCD@Work 2016

Equilibration of a strongly interacting plasma: holographic analysis of local and nonlocal probes

Loredana Bellantuono[1,2,⋆]

[1] *Dipartimento di Fisica, Università di Bari,via Orabona 4, I-70126 Bari, Italy*
[2] *INFN, Sezione di Bari, via Orabona 4, I-70126 Bari, Italy*

Abstract. The relaxation of a strongly coupled plasma towards the hydrodynamic regime is studied by analyzing the evolution of local and nonlocal observables in the holographic approach. The system is driven in an initial anisotropic and far-from equilibrium state through an impulsive time-dependent deformation (quench) of the boundary spacetime geometry. Effective temperature and entropy density are related to the position and area of a black hole horizon, which has formed as a consequence of the distortion. The behavior of stress-energy tensor, equal-time correlation functions and Wilson loops of different shapes is examined, and a hierarchy among their thermalization times emerges: probes involving shorter length scales thermalize faster.

1 Boundary sourcing

The deconfined and strongly-coupled plasma produced in relativistic heavy ion collisions experiments is characterized by an initial configuration which is dense, hot and highly anisotropic. Such system expands and cools down, reaching a hydrodinamic regime $\sim$ 1 fm/c after the collision. A possible framework for studying the equilibration process is holographic QCD. This approach is based on Maldacena's conjecture, according to which a strongly coupled Super-Yang Mills gauge theory on a 4-dimensional Minkowski space (boundary) is dual to a classical gravity theory in the space $AdS_5 \times S^5$, with AdS_5 the 5-dimensional anti-de Sitter manifold (bulk) and S^5 a 5-dimensional sphere [1]. In order to mimic the effects of heavy ion collisions and reproduce the initial far-from-equilibrium state, a time-dependent deformation (quench) is introduced to the metric on the boundary geometry (boundary sourcing) [2]. When the quench becomes static, the system starts to evolve towards the hydrodynamic regime. The duality correspondence is employed to transfer such a description in the bulk geometry. The evolution of the system is obtained by solving the Einstein equations in the curved 5-dimensional spacetime. Let us set the 4-dimensional coordinates $x^\mu = \left(x^0, x^1, x^2, x^3\right)$, with $x^3 = x_\parallel$ the direction in which ion collisions and plasma expansion occurs. Boost-invariance along that axis is imposed, as well as translational and rotational $O(2)$ invariance in the transverse plane $\mathbf{x}_\perp = (x_1, x_2)$. The above symmetries suggest us to introduce the proper time τ and the spacetime rapidity y, defined through the relations $x^0 = \tau \cosh y$ and $x_\parallel = \tau \sinh y$. The 4-dimensional Minkowski line element $ds_4^2 = -d\tau^2 + dx_\perp^2 + \tau^2 dy^2$ is distorted impulsively:

$$ds_4^2 = -d\tau^2 + e^{\gamma(\tau)} dx_\perp^2 + \tau^2 e^{-2\gamma(\tau)} dy^2 \, ; \tag{1}$$

⋆e-mail: loredana.bellantuono@ba.infn.it

DOI: 10.1051/epjconf/201612900041

note that this deformation leaves the spatial three-volume invariant and respects the aforementioned symmetries. The 5-dimensional dual metric can be parametrized in terms of Eddington-Finkelstein coordinates as

$$ds_5^2 = 2drd\tau - Ad\tau^2 + \Sigma^2 e^B dx_\perp^2 + \Sigma^2 e^{-2B} dy^2 \,, \tag{2}$$

with r the fifth holographic coordinate. The boundary of this geometry is obtained in the limit $r \to \infty$ and corresponds to the 4-dimensional spacetime (1). The metric functions A, B and Σ depend only on r and τ because of the imposed symmetries. They are solutions of Einstein's equations with negative cosmological constant, that can be rephrased as [2]:

$$\Sigma(\dot{\Sigma})' + 2\Sigma'\dot{\Sigma} - 2\Sigma^2 = 0 \tag{3}$$

$$\Sigma(\dot{B})' + \frac{3}{2}\left(\Sigma'\dot{B} + B'\dot{\Sigma}\right) = 0 \tag{4}$$

$$A'' + 3B'\dot{B} - 12\frac{\Sigma'\dot{\Sigma}}{\Sigma^2} + 4 = 0 \tag{5}$$

$$\ddot{\Sigma} + \frac{1}{2}\left(\dot{B}^2\Sigma - A'\dot{\Sigma}\right) = 0 \tag{6}$$

$$\Sigma'' + \frac{1}{2}B'^2\Sigma = 0 \,. \tag{7}$$

The notations $\xi' \equiv \partial_r\xi$ and $\dot{\xi} \equiv \partial_\tau\xi + (A/2)\partial_r\xi$ for a generic function $\xi(r,\tau)$ indicate directional derivatives along the infalling radial null geodesics and the outgoing radial null geodesics, respectively. The system (3)–(7) involves partial differential equations that should be solved recursively: an efficient procedure for the resolution is discussed in detail in [3]. A, B and Σ are determined by imposing two boundary conditions:

- As $r \to \infty$ the metric (2) reduces to (1);
- At the initial time $\tau = \tau_i$, when the quench is turned on, the bulk metric coincides with the AdS_5 form

$$ds^2 = r^2\left[-d\tau^2 + dx_\perp^2 + \left(\tau + \frac{1}{r}\right)^2 dy^2\right] + 2drd\tau \,. \tag{8}$$

A strategy to monitor the plasma during thermalization is to compute characteristic probes in the boundary sourcing approach and to compare them with the corresponding viscous hydrodynamic expressions. A quench profile $\gamma(\tau)$ chosen for the analysis is shown in the top left panel of figure 1; it can be viewed as the overlap of two signals having different time scales: a short pulse ending at $\tau_f = 5$ and a step function approaching a constant value.

2 Viscous hydrodynamic regime

Let us present an overview on viscous hydrodynamics in a homogeneous 4-dimensional spacetime, whose metric is invariant under boosts along $x_\parallel = x_3$, and under translations and rotations in the $\mathbf{x}_\perp = (x_1, x_2)$ plane. We focus on the stress-energy tensor T^μ_ν, whose components are the energy density ε and the pressures transverse and longitudinal with respect to the collisional axis, $p_\perp$ and $p_\parallel$. As a consequence of the imposed symmetries, these quantities depend only on the proper time τ [4]. Moreover, a conserved and traceless stress-energy tensor can be expressed in terms of a single function $f(\tau)$ as

$$T^\nu_\mu(\tau) = diag\left(-f(\tau),\, f(\tau) + \frac{1}{2}\tau f'(\tau),\, f(\tau) + \frac{1}{2}\tau f'(\tau),\, -f(\tau) - \tau f'(\tau)\right) . \tag{9}$$

For a perfect fluid, the equation of state $\varepsilon = 3p$ and the isotropy condition $p = p_\perp = p_\parallel$ fix the τ-dependence: $\varepsilon(\tau) \propto \tau^{-4/3}$, which is modified if viscous effects are included [5]. The effective temperature $T_{eff}(\tau)$, defined through the relation $\varepsilon(\tau) = 3\pi^4 T_{eff}(\tau)^4/4$, has been computed in a late-time expansion of $\mathcal{N} = 4$ SYM [6], with the result

$$T_{eff}(\tau) = \frac{\Lambda}{(\Lambda\tau)^{1/3}}\left[1 - \frac{1}{6\pi(\Lambda\tau)^{2/3}} + \frac{-1+\log 2}{36\pi^2(\Lambda\tau)^{4/3}} + \frac{-21+2\pi^2+51\log 2 - 24(\log 2)^2}{1944\pi^3(\Lambda\tau)^2} + O\left(\frac{1}{(\Lambda\tau)^{8/3}}\right)\right], \tag{10}$$

The corresponding viscous hydrodynamic expressions for the components of T^μ_ν at large τ are:

$$\varepsilon(\tau) = \frac{3\pi^4\Lambda^4}{4(\Lambda\tau)^{4/3}}\left[1 - \frac{2c_1}{(\Lambda\tau)^{2/3}} + \frac{c_2}{(\Lambda\tau)^{4/3}} + O\left(\frac{1}{(\Lambda\tau)^2}\right)\right], \tag{11}$$

$$p_\perp(\tau) = \frac{\pi^4\Lambda^4}{4(\Lambda\tau)^{4/3}}\left[1 - \frac{c_2}{(\Lambda\tau)^{4/3}} + O\left(\frac{1}{(\Lambda\tau)^2}\right)\right], \tag{12}$$

$$p_\parallel(\tau) = \frac{\pi^4\Lambda^4}{4(\Lambda\tau)^{4/3}}\left[1 - \frac{6c_1}{(\Lambda\tau)^{2/3}} + \frac{5c_2}{(\Lambda\tau)^{4/3}} + O\left(\frac{1}{(\Lambda\tau)^2}\right)\right], \tag{13}$$

with $c_1 = 1/(3\pi)$ and $c_2 = (1+2\log 2)/(18\pi^2)$. The parameter Λ appearing in Eqs. (10)–(13) will be fixed in Sect. 3. A bulk metric reproducing via holographic renormalization the viscous hydrodynamic stress-energy tensor has been identified [7, 8].

3 Thermalization probes

As it can be deduced by solving Einstein's equations, the boundary distorsion $\gamma(\tau)$ determines a time-dependent horizon $r = r_h(\tau)$ in the bulk geometry (2). The gravitational radiation produced by the quench propagates in the fifth dimension and induces the creation of a black hole, whose horizon is $r_h(\tau)$ [2]. We can thus define, following the prescriptions of the holographic principle, the effective temperature and entropy of the system, which are proportional to the horizon position and area, respectively. Note that the quantity $\Sigma^3(r_h, \tau)$ coincides with the horizon area per unit rapidity, and thus it can be used to characterize the entropy density. As shown in the left column of figure 1, T_{eff} and Σ^3 manifest, respectively, a decreasing and a constant behavior in the time interval before the impulsive part of the quench γ. The stress-energy tensor on the boundary, $T^\mu_\nu(\tau)$, has been computed by the holographic renormalization procedure from the explicit metric functions $A(r,\tau)$, $B(r,\tau)$ and $\Sigma(r,\tau)$ in (2). As a result of comparing its components $\varepsilon(\tau)$, $p_\perp(\tau)$ and $p_\parallel(\tau)$ with the corresponding hydrodynamic expressions (11)–(13), the parameter $\Lambda = 1.12$ has been fitted. The results for such observables are displayed in figure 1.

Nonlocal observables allowing to examine the evolution of a strongly-coupled plasma are the two-point correlation function of boundary theory operators, and expectation values of Wilson loops defined on the boundary [9]. Let us consider a boundary scalar operator O having conformal dimension $\Delta \gg 1$. In AdS/CFT, its correlation function between two equal-time points with a spatial separation ℓ along one of the transverse directions, $P = (t_0, -\ell/2, x_2, y)$ and $Q = (t_0, \ell/2, x_2, y)$, can be written as

$$\langle O(t,\mathbf{x})O(t,\mathbf{x}')\rangle \simeq \sum_{\text{geodesics}} e^{-\mathcal{L}\Delta}. \tag{14}$$

DOI: 10.1051/epjconf/201612900041

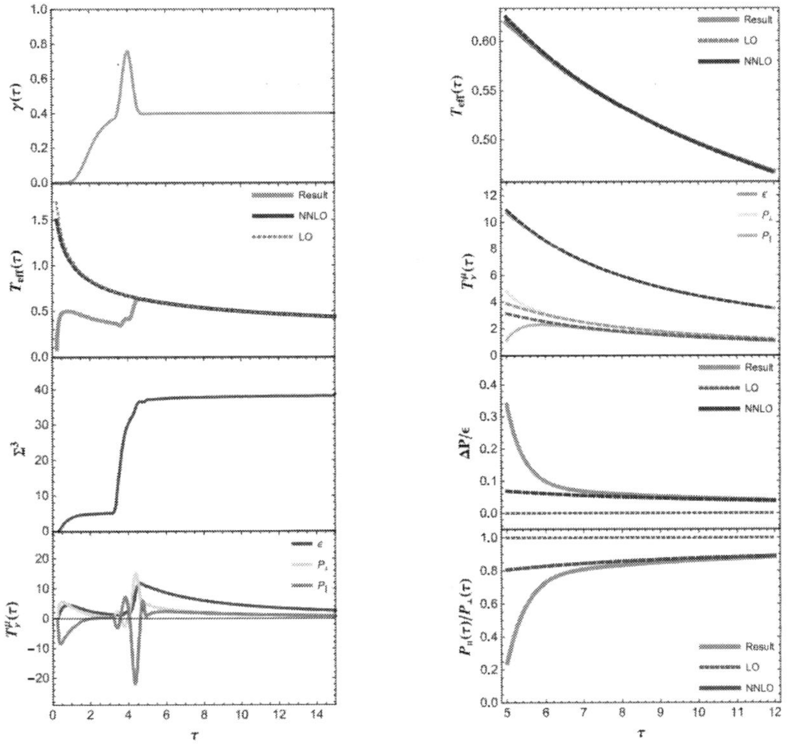

Figure 1. Left, from top to bottom: (i) quench profile $\gamma(\tau)$, (ii) temperature $T_{eff}(\tau)$, (iii) horizon area per unit rapidity $\Sigma^3\,(r_h(\tau),\tau)$ and (iv) the three components $\varepsilon(\tau)$, $p_\perp(\tau)$ and $p_\parallel(\tau)$ of the stress-energy tensor. Right, from top to bottom: (i) $T_{eff}(\tau)$, (ii) $\varepsilon(\tau)$, $p_\perp(\tau)$ and $p_\parallel(\tau)$, (iii) the pressure anisotropy $\Delta p/\varepsilon = (p_\perp - p_\parallel)/\varepsilon$ and (iv) the ratio $p_\parallel/p_\perp$ as compared to the corresponding hydrodynamic forms (LO and NNLO results in the $1/\tau$ expansion), at times after the end of the quench pulse [3].

$\mathcal{L}$ is the geodesic length, i.e. the length of the parametrized extremal trajectory $x^M(\lambda)$ (x_2 and y fixed) connecting P and Q:

$$\mathcal{L} = \int_P^Q d\lambda\,\sqrt{\pm g_{MN}\bar{x}^M\bar{x}^N}\,, \tag{15}$$

with $\bar{x}^M \equiv dx^M/d\lambda$, g_{MN} the bulk metric, and $\pm$ signs in the square root for a space-like or time-like curve, respectively. The approximation (14) is reliable only for boundary operators with $\Delta \gg 1$. The top panel of figure 2 displays the regularized length $\mathcal{L} - \mathcal{L}_{AdS}$, with $\mathcal{L}_{AdS}$ the geodesics length in the AdS_5 geometry (8), as a function of the boundary time t_0. This quantity follows the quench profile $\gamma(t_0)$ with a delay that increases with the distance ℓ between the two points in the correlation function. A nonlocal probe of thermalization is the difference $\Delta\mathcal{L}$ between the geodesics length in the distorted geometry and the one computed in a bulk metric reproducing the viscous hydrodynamic behavior (11)–(13) of T^μ_ν [7, 8]. The top right panel of figure 2 shows the relaxation of $\Delta\mathcal{L}$ for several sizes ℓ, starting from the value of the physical time $\tilde{t}_0(\ell)$ at which it is no longer affected by the quench. The expectation value of the Wilson loop along a closed path C of linear length scale ℓ, fixed time t_0 and

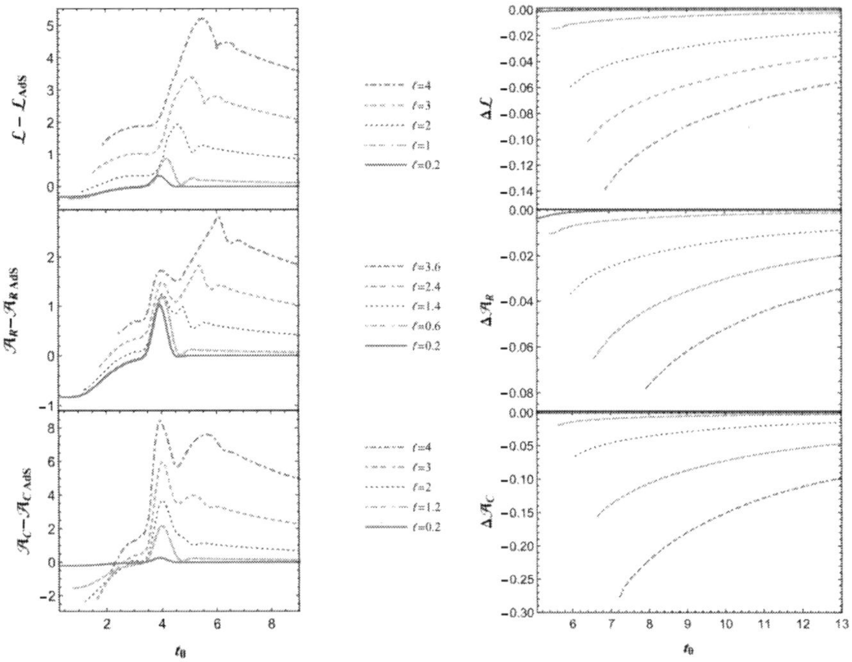

Figure 2. Left, from top to bottom: regularized lengths of the geodesics, regularized areas of the minimal surfaces for rectangular and circular Wilson loops. Regularization has been obtained subtracting from $\mathcal{L}, \mathcal{A}_R, \mathcal{A}_C$ the corresponding AdS_5 expressions computed from Eq. (8). Right, from top to bottom: differences between the observables $\mathcal{L}, \mathcal{A}_R, \mathcal{A}_C$ in the quenched geometry and the ones obtained using the hydrodynamic metric [8].

living on the boundary, can be computed through a similar approximation:

$$\langle W_C \rangle \simeq e^{-\mathcal{A}(t_0,\ell)}, \quad \text{with} \quad \mathcal{A}(t_0,\ell) = \frac{1}{2\pi\alpha'}\int d^2\xi \sqrt{det\left[g_{MN}\partial_\alpha x^M \partial_\beta x^N\right]}. \tag{16}$$

$\mathcal{A}(t_0,\ell)$ is the Nambu-Goto action, representing the area of the extremal surface bounded by C on the boundary and plunging in the bulk at fixed y, while ξ^α $(\alpha,\beta = 1,2)$ and $x^M(\xi^\alpha)$ are respectively the worldsheet coordinates and the embedding of the surface into the bulk. Let us consider two different contours C of linear size ℓ: a rectangle of finite width ℓ along the x_1 axis and infinite length along x_2, and a circumference of diameter ℓ in the transverse plane $\mathbf{x}_\perp = (x_1, x_2)$. The regularized areas of the extremal surfaces bounded by these paths have been computed in the boundary sourced geometry; the hydrodynamization of such quantities can be monitored through the nonlocal probes $\Delta\mathcal{A}_R$ and $\Delta\mathcal{A}_C$. The results, resumed in figure 2, are similar to the ones discussed for the geodesics.

4 Hierarchy among thermalization times

The right columns in figures 1 and 2 display the physical observables of the distorted geometry in comparison with their hydrodynamic forms. The results can be summarized as follows.

The effective temperature $T_{eff}(\tau)$ reaches the hydrodynamic regime as soon as the boundary deformation becomes stationary. Since the energy density is related to the effective temperature, it also acquires the hydrodynamical form at $\tau_f = 5$, when the deformation pulse is turned off and $\gamma(\tau)$ approaches a constant value. On the other hand, pressure isotropy is restored with a time delay

$\tau_p - \tau_f = 1.74$, where the isotropization time $\tau_p = 6.74$ has been identified as the value at which the computed ratio $p_\parallel/p_\perp$ differs from its hydrodynamical analogue by less than 5%. These results can be expressed in physical units introducing an energy scale in the system. In particular, if the effective temperature at the end of the impulsive quench is set to $T_{eff} = 500$ MeV, the pressures reach a common value after a time interval of 0.42 fm/c. Such a delay is comparable to the values inferred from phenomenological analyses in heavy ion collisions experiments.

The time a nonlocal probe needs for reaching the viscous hydrodynamic behavior depends on its size: for larger boundary separation between the geodesics extremes or among the contour points, the geodesics length and the areas bounded by Wilson loops take longer to equilibrate. It is interesting to evaluate the half-life $t_{1/2}(\ell)$, defined as the value of t_0 at which the Δ-probes are reduced by a half with respect to their values at time $\tilde{t}_0(\ell)$, when the effect of the quench pulse on them vanishes. As shown in figure 3, the half-lives of the considered nonlocal probes exceed the pressure isotropization time τ_p for boundary separations $\ell \simeq 1$, and increase linearly with ℓ for larger sizes. The hierarchy among the hydrodinamization times of the energy density, pressures and large probes has been identified, and shows that thermalization proceeds top-down, from short to large distances: UV thermalizes first. Such analysis can be extended to further nonlocal observables, e.g. the entanglement entropy [10, 11].

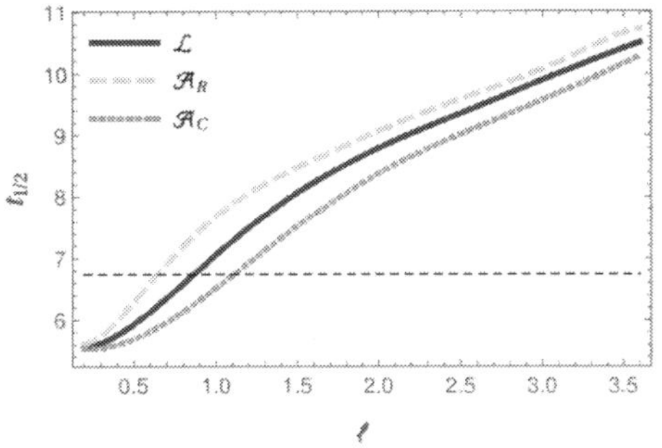

Figure 3. Half-life $t_{1/2}(\ell)$ as a function of the size ℓ for the three nonlocal probes. The horizontal dashed line indicates the pressure isotropization time $\tau_p = 6.74$ [8].

Acknowledgments
I thank P. Colangelo, F. De Fazio, F. Giannuzzi and S. Nicotri for collaboration.

References

[1] J. M. Maldacena, Int.J.Theor.Phys. **38**, 1113–1133 (1999).
[2] P. M. Chesler and L. G. Yaffe, Phys.Rev.Lett. **102**, 211601 (2009); Phys.Rev. **D82**, 026006 (2010).
[3] L. Bellantuono, P. Colangelo, F. De Fazio and F. Giannuzzi, JHEP **07**, 053 (2015).
[4] J. Bjorken, Phys.Rev. **D27**, 140 (1983).
[5] R. A. Janik and R. B. Peschanski, Phys.Rev. **D73**, 045013 (2006).
[6] M. P. Heller and R. A. Janik, Phys.Rev. **D76**, 025027 (2007).
[7] W. van der Schee, Phys. Rev. **D87**, 061901 (2013).
[8] L. Bellantuono *et al.*, Phys. Rev. **D94**, 025005 (2016).
[9] V. Balasubramanian *et al.*, Phys. Rev. **D84**, 026010 (2011).
[10] C. Ecker, D. Grumiller and S. A. Stricker, JHEP **1507**, 146 (2015).
[11] C. Ecker *et al.*, arXiv:1609.03676 [hep-th].

Hadronic structure of the photon at small x in holographic QCD

Akira Watanabe[1,2,⋆] and *Hsiang-nan* Li[2,⋆⋆]

[1] *Department of Physics and Center for High Energy Physics, Chung-Yuan Christian University, Chung-Li 32023, Taiwan, Republic of China*
[2] *Institute of Physics, Academia Sinica, Taipei 11529, Taiwan, Republic of China*

Abstract. We present our analysis on the photon structure functions at small Bjorken variable x in the framework of the holographic QCD. In the kinematic region, a photon can fluctuate into vector mesons and behaves like a hadron rather than a pointlike particle. Assuming the Pomeron exchange dominance, the dominant hadronic contribution to the structure functions is computed by convoluting the probe and target photon density distributions obtained from the wave functions of the U(1) vector field in the five-dimensional AdS space and the Brower-Polchinski-Strassler-Tan Pomeron exchange kernel. Our calculations are in agreement with both the experimental data from OPAL collaboration at LEP and those calculated from the parton distribution functions of the photon proposed by Glück, Reya, and Schienbein. The predictions presented here will be tested at future linear colliders, such as the planned International Linear Collider.

1 Introduction

A solid understanding of the nature of a photon has been one of the most longstanding problems in high energy physics for several decades. A photon can fluctuate into quark-antiquark pairs or vector mesons in high energy scattering processes, and this feature provides us with an opportunity to investigate its partonic structure. Experimentally, it can be studied via electron-photon deep inelastic scattering (DIS), and historically the data of photon structure functions measured at LEP have significantly deepened our knowledge of the energetic photon. If the scattering is hard, perturbative QCD approaches are basically applicable because the pointlike component of the photon gives a dominant contribution, and a lot of studies have successfully been done so far, which strongly support the predictive power of them in high energy QCD processes. However, in the small Bjorken x region, the situation becomes quite different, because the photon behaves like a hadron rather than a pointlike particle. When $x < 0.1$, we cannot neglect the hadronic contribution, and it becomes dominant at $x < 0.01$. Since the photon consists of so many gluons with tiny momenta in this kinematic region, approaches motivated by effective models are practically the only way to investigate its nature. Here we present our analysis on the photon structure functions at small x in the framework of the holographic QCD, one of the effective models of QCD, based on our recent paper [1].

The AdS/CFT correspondence (more generally, gauge/string correspondence) has broadly gathered theoretical interests due to its strong predictive power for physical quantities in strongly coupled

⋆e-mail: watanabe@cycu.edu.tw
⋆⋆e-mail: hnli@phys.sinica.edu.tw

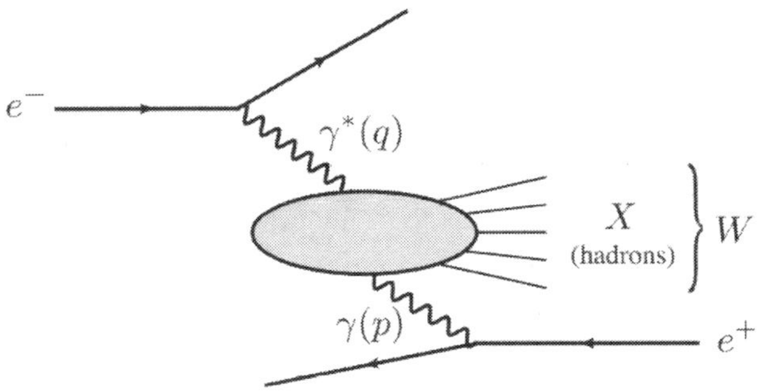

Figure 1. Deeply inelastic electron-photon scattering process.

systems, while the nontrivial correspondence itself has not yet been mathematically proved. In particular, its application to QCD processes is called holographic QCD, and it has been enthusiastically studied for over a decade. As for DIS, after Polchinski and Strassler first performed the string calculations and discussed the Callan-Gross relation for the structure functions [2], a lot of elaborated studies have been done so far. Brower, Polchinski, Strassler, and Tan (BPST) focused on the small x region, and proposed the Pomeron exchange kernel [3], which gives a contribution of the Pomeron exchange to the total cross section. Utilizing the BPST kernel, various phenomenological investigations have been done [1, 4–6], and their successes support further applications for high energy scattering processes in the kinematic region, where the Pomeron exchange gives a dominant contribution.

In this study, assuming the Pomeron dominance, we express the quasi-real photon structure functions by convoluting the BPST Pomeron exchange kernel and density distributions of the probe and target photons in the five-dimensional AdS space. In contrast to the case of the nucleon structure functions, both the probe and target are pointlike particles, whose density distributions are expressed by the wave functions of the 5D U(1) vector field. Also, two of three adjustable parameters of the model have been fixed in previous studies of the nucleon structure functions, and the remaining one is an overall factor which controls the magnitude of the structure functions. Therefore we can test the predictive power of the model involving the BPST kernel through this study.

It will be demonstrated that our predictions for the photon structure functions are in agreement with the experimental data measured by OPAL collaboration at LEP [7], although available data are limited in the small x region. We will show that our calculations are also consistent with those calculated from the parton distribution functions (PDFs) of the photon proposed by Glück, Reya, and Schienbein [8], and discuss differences between the results of this work and previous studies of the nucleon structure functions. The predictions particularly for the very low x region will be tested at future linear colliders, e.g., the planned International Linear Collider.

2 Model setup

The electron-photon DIS is described in Fig. 1, where q and p are the four-momenta of the probe and target photons, respectively, and W denotes the invariant mass of the hadronic final state X. The differential cross section of this scattering process is given by

$$\frac{d^2\sigma_{e\gamma\to eX}}{dxdQ^2} = \frac{2\pi\alpha^2}{xQ^4}\left[\left\{1+(1-y)^2\right\}F_2^\gamma\left(x,Q^2\right) - y^2F_L^\gamma\left(x,Q^2\right)\right], \quad (1)$$

where α is the fine structure constant, and y is the inelasticity. The Bjorken scaling variable x is written by

$$x = \frac{Q^2}{Q^2 + W^2 + P^2}, \tag{2}$$

with $Q^2 = -q^2$ and $P^2 = -p^2$. In this study, we focus on the kinematic region, where $W^2 \gg Q^2 \gg P^2$, so that the Bjorken variable can be approximated as $x \approx Q^2/W^2$.

With the BPST Pomeron exchange kernel, the two structure functions, $F_2^\gamma(x, Q^2)$ and $F_L^\gamma(x, Q^2)$, are expressed as [1, 9, 10]

$$F_i^\gamma(x, Q^2) = \frac{\alpha g_0^2 \rho^{3/2} Q^2}{32\pi^{5/2}} \int dz dz' P_{13}^{(i)}(z, Q^2) P_{24}(z', P^2)(zz') \mathrm{Im}[\chi_c(W^2, z, z')], \tag{3}$$

where $i = 2, L$, and g_0^2 and ρ are adjustable parameters which govern the magnitude and the energy dependence of the structure functions, respectively. P_{13} and P_{24} are density distributions of the incident and target particles, respectively, in the five-dimensional AdS space. To specify them, we consider the wave functions of the massless 5D U(1) vector field, which satisfies the Maxwell equation in the background spacetime and couple to leptons at the ultraviolet (UV) boundary [2]. We utilize it for the probe photon density distribution, then $P_{13}^{(i)}(z, Q^2)$ can be expressed as

$$P_{13}^{(2)}(z, Q^2) = Q^2 z \left[K_0^2(Qz) + K_1^2(Qz)\right], \tag{4}$$

$$P_{13}^{(L)}(z, Q^2) = Q^2 z K_0^2(Qz). \tag{5}$$

The first (second) term on the right-hand side of Eq. (4) represents the longitudinal (transverse) component of the wave function, and $K_0(Qz)$ and $K_1(Qz)$ are the modified Bessel functions of the second kind. In this study, we concentrate on the unpolarized DIS process, so $P_{24}(z', P^2)$ takes the same functional form as $P_{13}^{(2)}(z, Q^2)$. As to the four-momentum squared of the target photon, we focus on the real photon case, but P^2 cannot be exactly zero in practice. Events with $P^2 \sim O(0.01)$ GeV^2 have been measured [11, 12], so we assume a quasi-real photon with $P^2 = 0.01$ GeV^2 as the target particle here [8].

The imaginary part of the conformal kernel was derived within the conformal field theory, and can be expressed as [3, 10]

$$\mathrm{Im}[\chi_c(W^2, z, z')] = e^{(1-\rho)\tau} e^{-[(\log^2 z/z')/\rho\tau]} / \sqrt{\tau}, \tag{6}$$

where τ is the conformal invariant defined by $\tau = \log(\rho z z' W^2/2)$. In the considered kinematic region, where $x \le 10^{-2}$ and $Q^2 \le 10$ GeV^2, nonperturbative hadronic contribution becomes important, so we also employ the modified kernel, whose imaginary part is given by [3, 4]

$$\mathrm{Im}[\chi_{mod}(W^2, z, z')] \equiv \mathrm{Im}[\chi_c(W^2, z, z')] + \mathcal{F}(z, z', \tau)\mathrm{Im}[\chi_c(W^2, z, z_0^2/z')], \tag{7}$$

$$\mathcal{F}(z, z', \tau) = 1 - 2\sqrt{\rho\pi\tau} e^{\eta^2} \mathrm{erfc}(\eta), \tag{8}$$

$$\eta = \left(-\log\frac{zz'}{z_0^2} + \rho\tau\right) / \sqrt{\rho\tau}. \tag{9}$$

The second term on the right-hand side of Eq. (7) mimics the confinement effect in QCD, and suppresses the structure functions in the small x region at fixed Q^2. The strength of this effect is controlled by the additional parameter z_0, whose inverse is of order of the QCD scale Λ_{QCD}, and its smaller value gives a stronger suppression. It should be also noted here that we treat the conformal and modified kernels as different models, and fix the parameter sets for them separately.

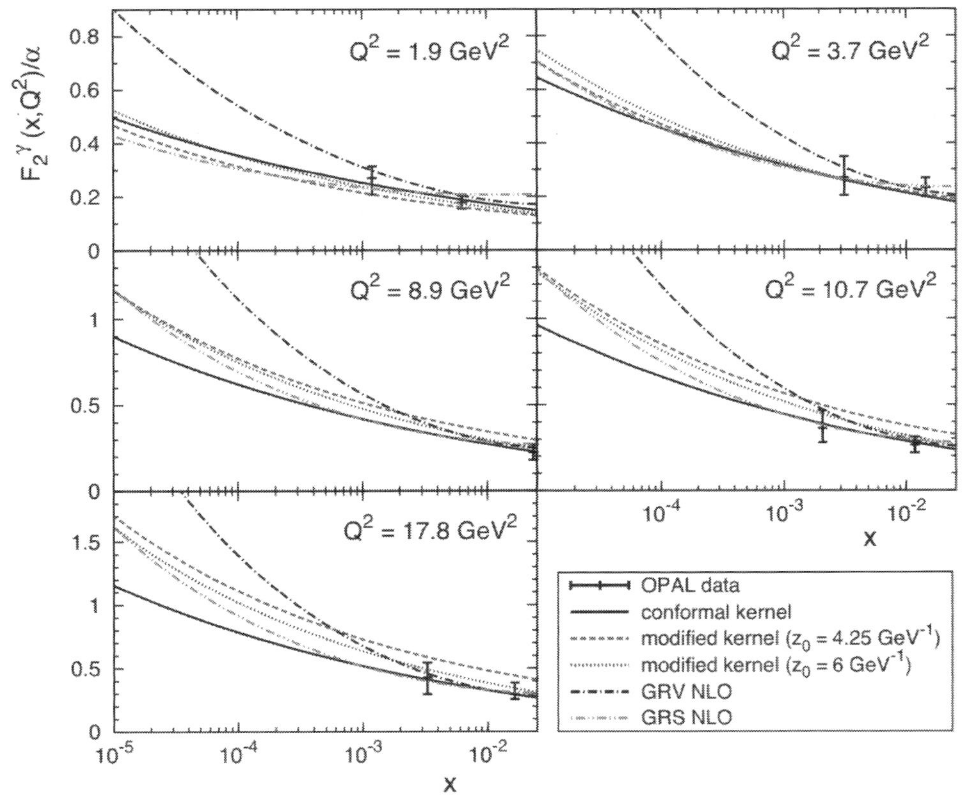

Figure 2. $F_2^\gamma(x, Q^2)$ as a function of x for various Q^2. In each panel, experimental data from the OPAL collaboration at LEP [7] are denoted with error bars. The solid, dashed, and dotted curves represent our calculations. The dash-dotted and dashed double-dotted curves are from PDF sets, GRV [13] and GRS [8], respectively.

3 Numerical results

To evaluate the structure functions, we take the parameter sets obtained in Ref. [6]. In the previous studies on nucleon structure functions, since the target nucleon is an on-shell particle, the normalizable condition, $\int dz' P_{24}(z') = 1$, is satisfied. However, the target quasi-real photon is a non-normalizable mode, so we have to newly fix the overall factor g_0^2 with experimental data. We use 9 data points with $x \leq 0.025$ from the OPAL collaboration at LEP [7], and obtain $g_0^2 = 17.51$ and 49.01 for the conformal and modified kernel cases, respectively. When we perform the integration in Eq. (3), we need to avoid collecting unphysical contributions from the region, where the conformal invariant τ takes negative values. To overcome this difficulty, we replace z and z' in the definition of τ with their average values defined by

$$\bar{z} \equiv \frac{\int dz z^2 P_{13}^{(i)}\left(z, Q^2\right)}{\int dz z P_{13}^{(i)}\left(z, Q^2\right)}, \qquad \bar{z}' \equiv \frac{\int dz' z'^2 P_{24}\left(z', P^2\right)}{\int dz' z' P_{24}\left(z', P^2\right)}, \tag{10}$$

respectively.

We display in Fig. 2 our predictions of the photon structure function F_2 with the conformal and modified Pomeron exchange kernels, experimental data from OPAL collaboration at LEP [7], and those calculated from the well-known PDF sets, GRV [13] and GRS [8], at next-to-leading-order accuracy. Both GRV and GRS results include the charm-quark contribution in the kinematic region, where $W \geq 2m_c$. The hadronic contributions in the PDFs by GRV and GRS are parameterized by using the pion PDFs based on the vector meson dominance and the assumed similarity between the vector meson and the pion. For the modified kernel results, we show two versions with different values of z_0, $z_0 = 4.25$ GeV^{-1} and 6 GeV^{-1}, and the former value was obtained in Ref. [6]. One can see in Fig. 2 that all the curves are in agreement with the OPAL data. In the smaller $x < 10^{-3}$ region, the x dependence of the GRV results is obviously stronger than that of others. The difference between GRV and GRS results was discussed in Ref. [14]. Also in the region, x dependence of the conformal kernel results is weakest for $Q^2 \geq 3.7$ GeV^2. This can be understood via the results of Q^2 dependencies of the Pomeron intercept with the two kernels, which were shown in Refs. [4–6]. In those results, it is seen that the Q^2 dependence of the conformal kernel results is obviously weaker than that of the modified kernel results. At fixed x, the suppression effect by the second term on the right-hand side of Eq. (7) becomes weaker with Q^2.

At $Q^2 \geq 8.9$ GeV^2, the conformal kernel results are more consistent with the experimental data than those of the modified kernel with $z_0 = 4.25$ GeV^{-1}, although it is known that the modified kernel can reproduce the nucleon structure function F_2 better compared to the conformal kernel [4–6]. We find that the modified kernel results can be improved by considering the relationship between the model parameter z_0 and the number of active quark flavors f. In the nucleon case, its density distribution is localized in the infrared (IR) region, and there is almost no component near the UV boundary. Due to this, heavy quark contributions are suppressed, and we get $f = 4$. On the other hand, the density distribution of the quasi-real photon spreads broadly from the UV to IR region, which leads to the participation of all flavors, and we get $f = 6$. Since the QCD scale $\Lambda_{\rm QCD}$ decreases with f, the parameter $z_0 \sim 1/\Lambda_{\rm QCD}$ for the quasi-real photon DIS should be larger than that for the nucleon DIS. We choose $z_0 = 6$ GeV^{-1} and $g_0^2 = 21.60$, and show the resulting $F_2^\gamma(x, Q^2)$ in Fig. 2. The results are perfectly consistent with the data within errors, implying that one of the fundamental QCD features is realized in the present model setup.

Next we display in Fig. 3 (a) the ratio $R_{\gamma/N} = F_2^\gamma(x, Q^2)/[\alpha F_2^N(x, Q^2)]$ to see the details of defferences between our and GRS results. For $F_2^N(x, Q^2)$, we take the results in Ref. [6], which are in good agreement with the HERA data. Both curves from our predictions are nearly linear and their slopes are small, which means the x dependencies of the nucleon and photon structure functions F_2 are similar. This may imply the universal feature of the BPST Pomeron exchange kernel. Our and GRS results are close to each other at $Q^2 = 10$ GeV^2, because both results are in agreement with the data at this scale as shown in Fig. 2. However, at $Q^2 = 1$ GeV^2, the difference between two curves is much larger, and this difference can be seen in Fig. 2 also. Particularly around $x \sim 10^{-2}$, the GRS results are nearly constant, while ours show the obvious x dependence.

Finally, we investigate the longitudinal-to-transverse ratio of the structure functions defined as $R_{L/T} = F_L(x, Q^2)/F_T(x, Q^2)$, where $F_T = F_2 - F_L$, in Fig. 3 (b). The photon results are compared with the nucleon results, which are taken from Ref. [6]. It is seen that the photon results are almost independent of x, but the Q^2 dependence is substantial. Similar behaviors are observed for the nucleon results, which may imply universal features of the BPST Pomeron.

4 Conclusion

In this study, we have investigated the electron-photon DIS, and calculated the structure functions F_2 and F_L of the quasi-real photon in the kinematic region with the small Bjorken variable x by the holographic QCD approach. The photon structure functions are expressed by combining the BPST

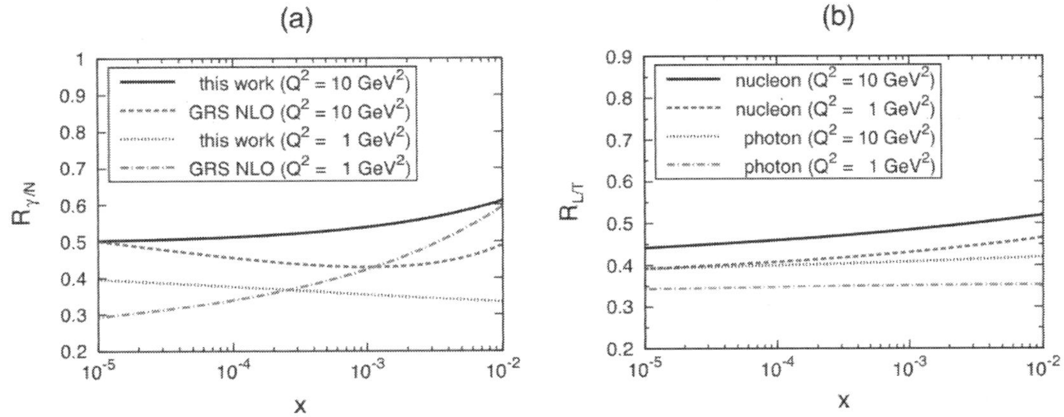

Figure 3. Ratios (a) $R_{\gamma/N} = F_2^\gamma(x, Q^2)/[\alpha F_2^N(x, Q^2)]$ and (b) $R_{L/T} = F_L(x, Q^2)/F_T(x, Q^2)$ as functions of x for $Q^2 = 1$ and 10 GeV^2. The GRS results are calculated from the PDF set [8]. $F_i^\gamma(x, Q^2)$ are obtained by employing the modified kernel with $z_0 = 4.25$ GeV^{-1}. The nucleon results are taken from Ref. [6], and common to all the curves in the panel (a).

Pomeron exchange kernel and the density distributions of the probe and target photons in the five-dimensional AdS space, which can be described with the wave functions of the 5D U(1) vector field. Our predictions are in agreement with the OPAL data, which supports the predictive power of this model. We have also compared our results with those calculated from the well-known PDF sets, GRV and GRS, and found that ours are consistent with those from the latter one. Since the hadronic component of the GRS PDF set is expressed with the pion PDFs, the consistency may imply the realization of the vector meson dominance in the present model setup. Currently available experimental data at small x are quite limited, but future linear colliders will enable us to further investigate the nature of a photon in the nonperturbative region.

References

[1] A. Watanabe, H.n. Li, Phys. Lett. **B751**, 321 (2015).
[2] J. Polchinski, M.J. Strassler, JHEP **0305**, 012 (2003).
[3] R.C. Brower, J. Polchinski, M.J. Strassler, C.I. Tan, JHEP **0712**, 005 (2007).
[4] R.C. Brower, M. Djuric, I. Sarcevic, C.I. Tan, JHEP **1011**, 051 (2010).
[5] A. Watanabe, K. Suzuki, Phys.Rev. **D86**, 035011 (2012).
[6] A. Watanabe, K. Suzuki, Phys. Rev. **D89**, 115015 (2014).
[7] G. Abbiendi et al., Eur. Phys. J. **C18**, 15 (2000).
[8] M. Gluck, E. Reya, I. Schienbein, Phys. Rev. **D60**, 054019 (1999).
[9] R.C. Brower, M.J. Strassler, C.I. Tan, JHEP **0903**, 050 (2009).
[10] R.C. Brower, M.J. Strassler, C.I. Tan, JHEP **0903**, 092 (2009).
[11] R. Nisius, Phys. Rept. **332**, 165 (2000).
[12] M. Krawczyk, A. Zembrzuski, M. Staszel, Phys. Rept. **345**, 265 (2001).
[13] M. Gluck, E. Reya, A. Vogt, Phys. Rev. **D46**, 1973 (1992).
[14] M. Gluck, E. Reya, I. Schienbein, Phys. Rev. **D64**, 017501 (2001).

EPJ Web of Conferences **129**, 00043 (2016) DOI: 10.1051/epjconf/201612900043
QCD@Work 2016

From OPE to chiral perturbation theory in holographic QCD

Luigi Cappiello[1,2,⋆], *Giancarlo* D'Ambrosio[1,⋆⋆], and *David* Greynat[1,2,⋆⋆⋆]

[1] *INFN-Sezione di Napoli, Via Cintia, 80126 Napoli, Italia*
[2] *Dip. di Scienze Fisiche, Universitá di Napoli "Federico II", Via Cintia, 80126 Napoli, Italia*

Abstract. The soft wall model in holographic QCD has Regge trajectories but wrong operator product expansion (OPE) for the two-point vectorial QCD Green function. We modify the dilaton potential to comply with the OPE. We study also the axial two-point function using the same modified dilaton field and an additional scalar field to address chiral symmetry breaking. OPE is recovered adding a boundary term and low energy chiral parameters, F_π and L_{10}, are well described analytically by the model in terms of Regge spacing and QCD condensates. The model nicely supports and extends previous theoretical analyses advocating Digamma function to study QCD two-point functions in different momentum regions.

1 Introduction

This proceeding is based on our previous work [1]. As matter of fact, one knows partially QCD Green functions on three ranges of energies *(i)* deep Euclidean, where perturbative QCD and OPE expansion methods can be applied, *(ii)* an intermediate Minkowski region, where resonances are described by Regge trajectories and *(iii)* the strong interacting, low-energy region, described by chiral perturbation theory (χPT) [2]. Unfortunately, global analytic expressions for each of them are unknown due to the complexity of the non-perturbative properties of QCD. It is natural to ask ourself if it is possible to reconstruct an approximate analytic expression satisfying all QCD properties valid in any region. In this approach, we think that Regge trajectories, *i.e.* a linear growth of the square of the resonance masses like $M_V^2(n) = \sigma n$ GeV^2 for the vectors, were conjectured long ago, before QCD, and very well verified phenomenologically [3] play a crucial role [4–7].

In the three energy sectors, the AdS/CFT correspondence [8–10] has been useful to describe the QCD properties. At low energy, QCD properties like the chiral symmetry breaking (χSB) parameters, F_π and the Gasser-Leutwyler coefficients L_i's of the $O(p^4)$ chiral Lagrangian have been studied in several five dimensional set-ups. The basic feature of Hard-Wall (HW) models is to simulate confinement by cutting drastically the extradimension of the AdS_5 in the IR, producing an infinite spectrum of Kaluza-Klein (KK) states to be identified with vector resonances of increasing masses. χSB in the axial sector is triggered either by a scalar field in the bulk [11, 12] or by appropriate boundary conditions [13]. A recent review of all these models and their relation with the light-front holographic QCD approach is Ref [15].

⋆e-mail: luigi.cappiello@na.infn.it
⋆⋆e-mail: gdambros@na.infn.it
⋆⋆⋆Speaker, e-mail: david.greynat@gmail.com

All these models have a good description of the deep Euclidean region; for instance the correct two point function recover both the partonic log and a pretty well description of low energy QCD, obtaining values of the chiral parameters F_π and the Gasser-Leutwyler coefficients L_i's which are close to the physical ones. In fact, it was observed that these HW models [11–13] have the same two point vectorial Green function as the one obtained by Migdal long ago who proposed an *ad hoc* prescription to perform the analytic continuation of the perturbative deep Euclidean QCD result to the Minkowski region [16, 17].

In order to reproduce resonances masses with a Regge spacing, one can consider a 5D model with AdS metric and an additional field, the dilaton [18]. However, in this model it can be shown that the partonic log of the two point vectorial Green function receives $^1/_{Q^2}$-corrections differently from what holds in perturbative QCD Operator Product Expansion (OPE) [19–22]. Nevertheless, we find extremely interesting to have a model where the Regge region would be analytically related to the deep Euclidean region even though with the wrong OPE. This will be exactly our starting point: OPE tells us the correct Green functions in the deep Euclidean in terms of gluon and quarks condensates. Is it possible to modify the dilaton profile, $\phi(z) \mapsto \phi(z) + \delta\phi(z)$, such to comply with QCD requirements in the intermediate (Regge) region and UV (OPE) region? Once this question has been answered in the affirmative, we are naturally led to address another important problem, that does not have a satisfying answer in this "linear confinement" holographic approach [18]: how to implement the effects of chiral symmetry breaking into our model. In fact, implementing the correct OPE and mass spectrum for the axial sector, we will propose our post-diction of F_π and L_{10} (one of the the chiral $O(p^4)$ coefficients) in terms of our input parameters, *i.e.* the Regge spacing and the QCD condensates. Our model, beside being a novel proposal for holographic QCD, links naturally to previous 4D QCD work where a phenomenological matching between low energy and OPE was realised through Regge theory (or Veneziano model) [2, 5, 23–27].

2 Construction of the model

The Soft-Wall model is a five dimensional model where the additional coordinate, z, has the range $0 < z < \infty$ and background fields consist in a five-dimensional AdS metric and a dilaton field $\Phi(z)$. The AdS metric is written as

$$g_{MN}\, \mathrm{d}x^M \mathrm{d}x^N = \frac{1}{z^2}\left(\eta_{\mu\nu}\mathrm{d}x^\mu \mathrm{d}x^\nu - \mathrm{d}z^2\right), \tag{1}$$

where $\eta_{\mu\nu} = \mathrm{Diag}\,(1,-1,-1,-1)$, the Greek indices $\mu,\nu = (0,1,2,3)$ referring to the usual 4-dimensions, and the capital Latin ones $M,N = (0,1,2,3,z)$ to the 5 dimensions. It was shown, in [18], that with the choice of a quadratic profile of the dilaton field

$$\Phi(z) \doteq \kappa^2 z^2, \tag{2}$$

the spectrum of vector resonances followed a Regge trajectory.

Our model is based on the following five dimensional action,

$$S_5 = \int \mathrm{d}^4x \int_0^\infty \mathrm{d}z\; w_0(z)\, \mathrm{Tr}\Big\{ g^{MN} \left(D_M \mathbb{X}\right)^\dagger \left(D_N \mathbb{X}\right) - m^2 \mathbb{X}^2 \\ - \frac{1}{4g_5^2} g_{MN} g_{RS} \left(\mathbb{F}_V^{MR}\mathbb{F}_V^{NS} + \mathbb{F}_A^{MR}\mathbb{F}_A^{NS}\right)\Big\}, \tag{3}$$

since the generic field strength (for vector and identical for axial vector) is $\mathbb{F}_{V\,MN} = \partial_M \mathbb{V}_M - \partial_N \mathbb{V}_M - i\,[\mathbb{V}_M, \mathbb{V}_N]$ and $g_5^2 = {}^{12\pi^2}/_{N_c}$ is the 5D coupling constant where N_c is the number of colors of QCD, $D^M \mathbb{X} = \partial^M \mathbb{X} - i[\mathbb{V}^M, \mathbb{X}] - i\{\mathbb{A}^M, \mathbb{X}\}$, where the 5D fields $\mathbb{V}_M$, $\mathbb{A}_M$ and $\mathbb{X}$ are respectively the vectorial, axial and scalar fields. We defined also,

$$w_0(z) \doteq \frac{\mathrm{e}^{-\Phi(z)}}{z} = \frac{\mathrm{e}^{-\kappa^2 z^2}}{z} \,. \tag{4}$$

2.1 The vectorial sector

Let's consider first the vectorial sector. We shall work in the axial gauge $\mathbb{V}_z = 0$. The AdS/CFT correspondence prescribes that the boundary value of the 5D gauge field $\mathbb{V}_\mu$ has to be identified with the classical source v_μ coupled to the the 4-dimensional vectorial current $J^a_{V\,\mu} =: \bar{q}\,\gamma_\mu\, t^a q :$,

$$\lim_{z\to 0} \mathbb{V}^a_{\mu,z}(x,z) = \mathrm{v}^a_\mu(x) \,. \tag{5}$$

The corresponding equation of motion for the gauge field, derived from the Lagrangian (3), is more easily written in terms of the 4-dimensional Fourier transform $f_V(-q^2, z)$ of the field $\mathbb{V}_M$,

$$f_V(-q^2, z)\, \hat{\mathrm{v}}_\mu(q) \doteq \int \mathrm{d}^4 x\, \mathrm{e}^{-iq\cdot x}\, \mathbb{V}_{\mu,0}(x,z) \,, \tag{6}$$

where $\hat{\mathrm{v}}_\mu$ is the Fourier transform of the source v_μ. Using gauge invariance, one may assume the $\hat{\mathrm{v}}_\mu$ to be transverse and one obtains

$$\partial_z^2 f_V + \partial_z [\ln w_0(z)]\, \partial_z f_V + q^2 f_V = 0 \,, \tag{7}$$

with the boundary conditions (5)

$$f_V(-q^2, 0) = 1 \qquad \text{and} \qquad f_V(-q^2, \infty) = 0 \,. \tag{8}$$

In this context, the vector-vector correlator Π_V,

$$i \int \mathrm{d}^4 x\, \mathrm{e}^{-iq\cdot x} \left\langle J^a_{V\,\mu}(x)\, J^b_{V\,\nu}(0) \right\rangle = \delta^{ab} \left(q_\mu q_\nu - q^2 \eta_{\mu\nu} \right) \Pi_V(-q^2) \,. \tag{9}$$

where q is the 4D momentum and the current $J^a_{V\,\mu}$ is the one defined before equation (5) is extracted from the 5D expressions by taking the limit (with the Euclidean momentum $Q^2 \doteq -q^2$)

$$Q^2 \Pi_V(Q^2) = \frac{1}{g_5^2} \lim_{z\to 0} w_0(z) f_V(Q^2, z)\, \partial_z f_V(Q^2, z) \,. \tag{10}$$

Our main concern is that the two-point function Π_V coming from the model built in 5D has the following properties:

(i) **The progression of its poles in the Minkowski region (the resonances) follows a Regge trajectory.**

In the Large-N$_c$ limit of QCD, the two-point function Π_V could be written as a sum over an infinite set of stable vector resonances [29, 30]

$$\Pi_V(Q^2) = \sum_{n=0}^{\infty} \frac{F_V(n)^2}{Q^2 + M_V(n)^2} \,, \tag{11}$$

where $F_V(n)$ are named decay constants and the $M_V(n)$ are the masses associated to the resonances of the vectorial channel: $\rho, \rho', \rho'', \ldots$. We assume that these resonances follow, in a first approximation, a Regge progression [3],

$$M_V(n)^2 \underset{n\to\infty}{\sim} \sigma n \,, \tag{12}$$

where the integer n is the radial excitation number and σ is related to the confining string tension as explained in [18] and we can evaluate $\sigma \approx 0.90$ GeV^2 from [31, 32].

(ii) **Π_V has the correct OPE.** One of the very well-known properties of this two point function is its OPE [33],

$$\Pi_V(Q^2) \underset{Q^2\to\infty}{\sim} \frac{1}{2}\frac{N_c}{12\pi^2}\ln\left(\frac{\Lambda_V^2}{Q^2}\right) + \langle O_2\rangle \frac{1}{Q^2} + \langle O_4\rangle \frac{1}{Q^4} + \langle O_6\rangle_V \frac{1}{Q^6} \tag{13}$$

where in the large-N$_c$ limit the coefficients of the OPE are given by

$$\left\{ \begin{array}{l} \langle O_2\rangle = 0 \\ \langle O_4\rangle = \frac{1}{2}\frac{1}{12\pi}\,\alpha_s\,\left\langle G^2\right\rangle \\ \langle O_6\rangle_V = \frac{1}{2}\left(-\frac{28\pi}{9}\right)\,\alpha_s\,\langle\bar\psi\psi\rangle^2 \end{array}\right. , \tag{14}$$

here α_s stands for the strong coupling constant, $\left\langle G^2\right\rangle$ for the gluon condensate, and $\langle\bar\psi\psi\rangle$ for the quark condensate.

If one solves the differential equation (7) with the boundary conditions (8), one obtains

$$Q^2\Pi_V^{(0)}(Q^2) = -\frac{2\kappa^2}{g_5^2}\left(\frac{Q^2}{4\kappa^2}\right)\left[\gamma_E + \psi\left(\frac{Q^2}{4\kappa^2}+1\right)\right] , \tag{15}$$

which gives the following OPE,

$$\Pi_V^{(0)}(Q^2) \underset{Q^2\to\infty}{\sim} \frac{1}{2g_5^2}\ln\left(\frac{4\kappa^2 e^{-\gamma_E}}{Q^2}\right) + \frac{\kappa^2}{g_5^2}\frac{1}{Q^2} . \tag{16}$$

Unfortunately, this solution (15) does not satisfy the full OPE (13) of the vectorial two-point function. In order to cure this problem, we shall assume that the effects of the OPE on the vector current two-point functions in QCD can be encoded in a new profile for the dilaton field of the SW model. More explicitly, we assume that OPE is related to the behaviour of the dilaton profile around the UV boundary, $z = 0$ of the 5D, while keeping the metric to have the AdS form (1). Moreover, as we want to keep the Regge behaviour induced by the dilaton profile $\Phi(z)$, we are led to modify the original quadratic profile of the SW dilaton by adding new terms which we collect in a function B,

$$\Phi(z) \longmapsto \Phi(z) + B(z) . \tag{17}$$

We assume that the function B can be represented *for all* z by a polynomial of degree $2K$ (with no constant term),

$$B(z) = \sum_{k=1}^{K} \frac{b_{2k}}{2k} z^{2k} . \tag{18}$$

We prove in [1] then that the equation of motion (7) becomes

$$\partial_z^2 f_V + \partial_z [\ln w(z)]\, \partial_z f_V - Q^2 f_V = 0 \,, \tag{19}$$

with $w(z) \doteq w_0(z)\, e^{-B(z)}$ and can be decomposed into a solvable hierarchical system of differential equations order by order.

This general solution drives to an expression for the vectorial correlator,

$$Q^2 \Pi_V(Q^2) = \sum_k \mathcal{P}_k \left(\frac{Q^2}{4\kappa^2}\right) \psi^{(k)} \left(\frac{Q^2}{4\kappa^2}\right) , \tag{20}$$

where, respectively, $\mathcal{P}_k$ are polynomials and $\psi^{(k)}$ is the kth derivative of the Digamma ψ function defined as the logarithmic derivative of the Euler's Γ function. The coefficients of $\mathcal{P}_k$ depend only on κ^2 and the coefficients b_k of the dilaton (18).

This solution fulfils perfectly our two requirements *(i)* and *(ii)*. Indeed, the Digamma function and its derivatives in (20) have poles only at all negative integers $(-n)$,

$$\frac{-q^2}{4\kappa^2} = -n \,, \tag{21}$$

so the usual Regge spectrum is verified,

$$M(n)^2 = 4\kappa^2 n = \sigma n \,, \tag{22}$$

fixing $\kappa^2 = {}^{\sigma}\!/_4$.

On the other hand, performing the expansion at $Q^2 \to \infty$ of (20), we prove in [1] that we identify

$$b_{2k} = \begin{cases} -2\kappa^2 & k = 1 \\ g_5^2 \frac{2}{\sqrt{\pi}} \frac{\Gamma(\frac{1}{2}+k)}{\Gamma(k)\Gamma^2(k+1)} \langle O_{2k} \rangle_V & k > 1 \end{cases} , \tag{23}$$

in order to reproduce the OPE of the vectorial two-point function (13).

2.2 Axial sector

In the previous section, we have presented a SW model with a deformed dilaton profile which describes Regge theory and allows us to obtain the correct OPE of the vectorial two-point function. The natural question is now if it is possible to do a similar construction for the axial correlator. The problem to recover the right axial spectrum and the right axial OPE is more complicated if we assume, as usual, that the coupling of the 5D vector and axial vector gauge field to the metric and the dilaton is the same. Actually, under these assumptions, the form of the dilaton profile is already fixed by the requirements of a correct OPE of the vector two-point function, with the coefficients b_k identified to the vectorial OPE coefficients (23). So we introduce a 5D scalar field $\mathbb{X}(x,z)$ in (3) to reproduce the QCD patterns of chiral symmetry breaking as in [11, 12]. They showed that by taking

$$\mathbb{X}^a \doteq \frac{v(z)}{2} \mathbb{I}^a , \tag{24}$$

the equation of motion for the Fourier transform over the 4D space of the axial field $\mathbb{A}$, $f_A(-q^2, z)$, becomes

$$\partial_z^2 f_A + \partial_z [\ln w(z)]\, \partial_z f_A - Q^2 f_A = g_5^2 \left(\frac{v(z)}{z}\right)^2 f_A \,, \tag{25}$$

while the equation of motion for the vector field remains unchanged.

In this approach, chiral symmetry is broken by the 5D scalar field, and in particular by a non vanishing profile $v(z)$. The form of $v(z)$ near the origin $z \sim 0$ is dictated in the AdS/CFT correspondence by asking the field $\mathbb{X}$ to be dual to the bilinear quark $q\bar{q}$ operator, whose non vanishing VEV is responsible for spontaneous χSB in QCD. In the following, we shall assume for the dilaton profile $B(z)$ the one obtained from the OPE coefficients of the vectorial fields and then use a suitable form of the profile $v(z)$ to encode the properties of the axial sector:

- The axial spectrum contains a pion pole at $q^2 = 0$ and a Regge spectrum starting at $q^2 = M_{a_1}^2$ with the same spacing than the vectorial spectrum.
- The axial OPE has the following expression

$$\Pi_A(Q^2) \underset{Q^2\to\infty}{\sim} \frac{N_c}{24\pi^2} \ln\left(\frac{\Lambda_A^2}{Q^2}\right) + \langle O_2\rangle \frac{1}{Q^2} + \langle O_4\rangle \frac{1}{Q^4} + \langle O_6\rangle_A \frac{1}{Q^6} \tag{26}$$

and

$$\langle O_6\rangle_A = -\frac{11}{7}\langle O_6\rangle_V \ . \tag{27}$$

with the same definitions than in (14).

In order to satisfy the axial spectral properties, we make the following first ansatz for the contribution due to the scalar profile in (25):

$$\left(\frac{v(z)}{z}\right)^2 = \beta_0 + \beta^* z\,\delta(z) \ . \tag{28}$$

Let focus first on the Regge progression of the axial spectrum. We notice that, phenomenologically, taking the first axial pole at $q^2 = M_{a_1}^2 \simeq 2M_\rho^2 \simeq 2\sigma$ is quite an acceptable approximation for the axial spectrum. Then our prescription to obtain the axial spectrum from the vectorial one is to perform a "shift" over the vectorial spectrum like $Q^2 \to Q^2 + 4\kappa^2$. This fixes directly in (25)

$$g_5^2\beta_0 = 4\kappa^2 = \sigma \ . \tag{29}$$

The full axial spectrum is not yet realised, indeed, since the axial spectrum is a shifted version of the vectorial spectrum, it does not contain yet any pion pole and moreover on the Euclidean side, this shift implies a modifcation of the OPE as

$$\Pi_A(Q^2) \underset{Q^2\to\infty}{\sim} \frac{1}{2g_5^2} \ln\left(\frac{4\kappa^2 e^{-\gamma_E}}{Q^2}\right) + \frac{2\kappa^2}{g_5^2}\frac{1}{Q^2} \ , \tag{30}$$

where absence of dimension two operator in the axial OPE is violated.

The introduction of the Dirac delta function in (28) has the nice properties to cure this two problems at the same time. This term in (25) generates only one exact contribution $-\beta^*/Q^2$ (for any Q^2) such that the pion pole appears naturally and modifying the OPE as

$$\Pi_A(Q^2) \underset{Q^2\to\infty}{\sim} \frac{1}{2g_5^2} \ln\left(\frac{4\kappa^2 e^{-\gamma_E}}{Q^2}\right) + \left(\frac{2\kappa^2}{g_5^2} - \beta^*\right)\frac{1}{Q^2} \ , \tag{31}$$

then by taking

$$\beta^* = \frac{2\kappa^2}{g_5^2} \ , \tag{32}$$

the axial OPE properties is satisfied again.

The generation of the rest of the axial OPE terms derives exactly from the same procedure used in the vectorial section provided that we add two other terms in the expression (28) such as

$$\left(\frac{v(z)}{z}\right)^2 = \beta_0 + \beta^* z\,\delta(z) + \beta_2 z^2 + \beta_4 z^4 \,. \tag{33}$$

With slight modifications, the iterative method we used to obtain the corrections to the vector two-point function, can be applied to the axial case too. All the details can be found in [1]. Then one has by identification: $\Lambda_A^2 = 4\kappa^2 e^{-\gamma_E} = \Lambda_V^2$, the coefficients β_2 and β_4 are fixed by matching with QCD axial OPE (26), (27):

$$\begin{cases} \beta_2 &= -\frac{6\kappa^4}{g_5^2} \\ \beta_4 &= -\frac{10\kappa^2}{3g_5^2} - 5\kappa^2 \langle O_4 \rangle + \frac{45}{28} \langle O_6 \rangle_V \,. \end{cases} \tag{34}$$

The polynomial part of the scalar profile $v(z)$ is

$$\begin{aligned} v(z) &= z\sqrt{\beta_0 + \beta_2 z^2 + \beta_4 z^4} \\ &\underset{z\to 0}{\sim} \frac{2\kappa}{g_5} z - \frac{3\kappa^3}{2g_5} z^3 + \left(-\frac{67\kappa^5}{48 g_5} - \frac{5 g_5 \kappa}{4} \langle O_4 \rangle + \frac{45 g_5}{112\kappa} \langle O_6 \rangle_V \right) z^5 \,, \end{aligned} \tag{35}$$

where the first two terms, which are the leading terms near $z = 0$, are exactly the ones of a scalar field dual to the bilinear quark $q\bar{q}$ operator, *i.e.* the one required in Ref. [28] and compatible with the approach in [34].

3 Analytic continuation in the chiral sector: the left-right correlator

We have explicitly shown how to implement the constraints on an axial and vectorial two-point functions coming from two different regions in the q^2–plane: the deep Euclidean region where we reproduce the matching with the OPE of QCD, and the Minkowski region where the two-point function have poles following Regge trajectories. Having built explicit expressions for Π_V and Π_A valid on the whole complex plane we now turn to the analysis of their prediction for chiral quantities defined at low Q^2.

Since we are now interested to the chiral sector, *i.e.* the low Q^2 expansions, it is more pertinent to consider the Π_{LR} correlator,

$$\Pi_{LR}(Q^2) = \frac{1}{2}\left(\Pi_V(Q^2) - \Pi_A(Q^2)\right) \,. \tag{36}$$

which is an order parameter of the chiral symmetry breaking mechanism in QCD.

The low Q^2 limit can be obtained by the analytic continuation of our expressions for the axial and vectorial two-point function, this allows us to extract from Π_{LR} for instance the following chiral constants

$$F_\pi^2 = 2\,\mathrm{Res}\left[\Pi_{LR}\left(Q^2\right), 0\right] \tag{37}$$

$$L_{10} = \frac{1}{2}\frac{\mathrm{d}}{\mathrm{d}Q^2}\left[Q^2 \Pi_{LR}(Q^2)\right]\Big|_{Q^2=0} \,. \tag{38}$$

From our analytic expressions for the vectorial and axial correlator, one obtains the following analytic expressions for F_π^2 and L_{10},

$$F_\pi^2 = \frac{N_c\kappa^2\left(180\zeta(3)+191-41\pi^2\right)}{72\pi^2} + \frac{5\left(\pi^2-10\right)}{2}\frac{\langle O_4\rangle}{\kappa^2} - \frac{45\left(\pi^2-10\right)}{56}\frac{\langle O_6\rangle_V}{\kappa^4}\,, \tag{39}$$

using $N_c = 3$, $\kappa = \sqrt{1.43/4}\,\text{GeV} \simeq 0.6\,\text{GeV}$ and the values of the condensates $\langle O_4\rangle = (-0.635 \pm 0.04)\cdot 10^{-3}\,\text{GeV}^4$ and $\langle O_6\rangle_V = (14\pm 3)\cdot 10^{-4}\,\text{GeV}^6$ from [5], we obtain

$$F_\pi \simeq \sqrt{4099.9+579+1147.8}\ \text{MeV} \simeq 76\ (\pm 3)_{\text{ext.}}\ \text{MeV}\,, \tag{40}$$

the error in (40) are coming from the errors quoted for σ and the condensates.

The expression for L_{10} is

$$L_{10} = \frac{N_c(8010\zeta(3)+495-585\pi^2-46\pi^4)}{8640\pi^2} + \frac{-72\zeta(3)-12+11\pi^2}{64}\frac{\langle O_4\rangle}{\kappa^4} + \frac{5[5216\zeta(3)+67-33\pi^2]}{1792}\frac{\langle O_6\rangle_V}{\kappa^6}\,, \tag{41}$$

then with the same numerical values used for the evaluation of F_π,

$$10^3 L_{10} \simeq -4.6-0.8+0.1 \simeq -5.3\,(\pm 1)_{\text{ext.}}\,, \tag{42}$$

the error in (42) are coming from the errors quoted for σ and the condensates [1].

The values of F_π and L_{10} are quite reasonable face with our model hypothesis, if one admits the usual 30% error coming from Large-N$_c$ QCD limit, compared to the range of variation of F_π in the chiral limit, $66 < F_\pi < 84$ MeV, and compared to the value $10^3 L_{10} = -5.3 \pm 0.13$ according to [35] and references therein.

The relative contributions to F_π and L_{10} from the gluon and the quark condensates are consistent with previous evaluation [36, 37].

4 Conclusions

There have been several models describing QCD properties of the two point vectorial and axial Green function. Already an interpolation among low energy chiral properties and perturbative QCD is good [2]. The intermediate region could be phenomenologically matched with a tower of resonance states equally spaced (à la Regge) [5, 7, 24, 27]. Indeed, there are excellent dynamical reasons that Regge trajectories are dynamically generated. Holographic QCD gives us a fundamental theoretical tool with the SW model to start directly from a theory where Regge trajectories are analytically implemented. The original model [18] had several difficulties: the absence of a satisfying description of chiral symmetry breaking and of the axial sector, wrong OPE. For some further attempts to cure these problems one can refer to [39, 40].

It is natural to modify opportunely the SW model to comply with the OPE. While it was done already for the HW model [41], it is a novelty for the vector correlator in the SW model: we obtain the solution for the vector field f_V profile and the analytic expression $\Pi_V(Q^2)$ in eq. (20) in terms of the Digamma function ψ and its derivatives. As a result, we support previous works [24, 27, 38].

[1] Let us notice that if we had made the choice for $\sigma \simeq 0.9\,\text{GeV}^2$ as in [18] the values obtained would have been $F_\pi \simeq 80$ MeV and $10^3 L_{10} \simeq -6.2$ which remain quite acceptable too.

The coefficients of the polynomial of the dilaton profile (b_k's), fixed by the requirement of a correct vectorial OPE, appear also in the differential equation (25) of the axial vector field f_A, whose expression extends the analogous ones in Ref. [11, 12, 14, 28] with the presence of a scalar field in the bulk with a non-trivial vacuum profile, $v(z)$. A coherent, complete and phenomenologically consistent solution emerges: due the phenomenologically observed relation $8\kappa^2 = M_{a1}^2$ the solution for the axial in eq. (25) can be obtained from the vectorial one as $\Pi_A(Q^2) = \Pi_V(Q^2 + 4\kappa^2)$ + "corrections". This solution generates the pion pole, the correct axial spectrum and axial OPE, if the profile of the vacuum, $v(z)$, is not only a polynomial but contains also a Dirac δ-function term, *i.e.* a boundary term for the axial field, which is needed in order to comply with the axial OPE.

Our analytic solution for F_π and L_{10} respectively in eq. (39) and (41) are very successfully phenomenologically and show how these parameter are linked to Regge spacing and QCD condensates; compared to previous literature [5, 7, 24, 27] our results are fully analytical.

Acknowledgements

D.G. wants to thank the organizers to have given the nice opportunity to present this work. G. D'A. acknowledges partial support by MIUR under project 2010YJ2NYW. L.C. and D.G. were supported in part by MIUR under project 2010YJ2NYW and by INFN research initiative PhenoLNF.

References

[1] L. Cappiello, G. D'Ambrosio and D. Greynat, "Assuming Regge trajectories in holographic QCD: from OPE to Chiral Perturbation Theory", Eur. Phys. J. C **75**, no. 10, 465 (2015).
[2] S. Peris, M. Perrottet and E. de Rafael, "Matching long and short distances in large N(c) QCD", JHEP **9805**, 011 (1998).
[3] P.D.B. Collins, "Regge theory and particle physics", Phys. Reports **1** (1971) 103.
[4] G. 't Hooft, Nucl. Phys. B75 (1974)461.
C.G. Callan et al., Phys. Rev. D13 (1976) 1649.
M.B. Einhorn, Phys. Rev. D14 (1976) 3451.
[5] M. Golterman and S. Peris, "Large N(c) QCD meets Regge theory: The Example of spin one two point functions" JHEP **0101**, 028.
[6] A. B. Kaidalov, "Regge poles in QCD", In Shifman, M. (ed.): At the frontier of particle physics, vol. 1 603-636.
[7] M. Shifman and A. Vainshtein, "Highly Excited Mesons, Linear Regge Trajectories and the Pattern of the Chiral Symmetry Realization", Phys. Rev. D **77**, 034002 (2008).
[8] J. M. Maldacena, "The large N limit of superconformal field theories and supergravity", Adv. Theor. Math. Phys. **2**, 231 (1998).
[9] S. S. Gubser, I. R. Klebanov and A. M. Polyakov, "Gauge theory correlators from noncritical string theory", Phys. Lett. B **428**, 105 (1998).
[10] E. Witten, "Anti-de Sitter space and holography", Adv. Theor. Math. Phys. **2**, 253 (1998).
[11] J. Erlich, E. Katz, D. T. Son and M. A. Stephanov, "QCD and a holographic model of hadrons", Phys. Rev. Lett. **95**, 261602 (2005).
[12] L. Da Rold and A. Pomarol, "Chiral symmetry breaking from five dimensional spaces", Nucl. Phys. B **721**, 79 (2005).
[13] J. Hirn and V. Sanz, "Interpolating between low and high energy QCD via a 5-D Yang-Mills model", JHEP **0512**, 030 (2005).

[14] D. T. Son and M. A. Stephanov, "QCD and dimensional deconstruction", Phys. Rev. D **69**, 065020 (2004).

[15] S. J. Brodsky, G. F. de Teramond, H. G. Dosch and J. Erlich, "Light-Front Holographic QCD and Emerging Confinement", arXiv:1407.8131 [hep-ph].

[16] A. A. Migdal, "Series Expansion for Mesonic Masses in Multicolor QCD", Annals Phys. **110**, 46 (1978).

[17] M. Shifman, "Highly excited hadrons in QCD and beyond", hep-ph/0507246.

[18] A. Karch, E. Katz, D. T. Son and M. A. Stephanov, "Linear confinement and AdS/QCD", Phys. Rev. D **74**, 015005 (2006).

[19] O. Andreev, "1/q**2 corrections and gauge/string duality", Phys. Rev. D **73**, 107901 (2006).

[20] O. Cata, "Towards understanding Regge trajectories in holographic QCD", Phys. Rev. D **75**, 106004 (2007).

[21] F. Zuo and T. Huang, "Comments on the two-dimensional power correction in the soft wall model", Chin. Phys. Lett. **25**, 3601 (2008).

[22] C. Csaki and M. Reece, "Toward a systematic holographic QCD: A Braneless approach", JHEP **0705**, 062 (2007).

[23] G. Veneziano, "Construction of a crossing - symmetric, Regge behaved amplitude for linearly rising trajectories", Nuovo Cim. A **57**, 190 (1968).

[24] M. Golterman, S. Peris, B. Phily and E. De Rafael, "Testing an approximation to large N(c) QCD with a toy model", JHEP **0201**, 024 (2002).

[25] C. A. Dominguez, "Pion form-factor in large N(c) QCD", Phys. Lett. B **512**, 331 (2001).

[26] S. S. Afonin, A. A. Andrianov, V. A. Andrianov and D. Espriu, "Matching Regge theory to the OPE", JHEP **0404**, 039 (2004).

[27] O. Cata, M. Golterman and S. Peris, "Duality violations and spectral sum rules", JHEP **0508**, 076 (2005).

[28] H. J. Kwee and R. F. Lebed, "Pion Form Factor in Improved Holographic QCD Backgrounds", Phys. Rev. D **77**, 115007 (2008).

[29] G. 't Hooft, "A Planar Diagram Theory for Strong Interactions", Nucl. Phys. **B72** (1974) 461.

[30] E. Witten, "Baryons in the 1/N Expansion", Nucl. Phys. **B160** (1979) 57.

[31] J. Beringer *et al.* [Particle Data Group Collaboration], "Review of Particle Physics (RPP)", Phys. Rev. D **86**, 010001 (2012).

[32] P. Masjuan, E. Ruiz Arriola and W. Broniowski, "Systematics of radial and angular-momentum Regge trajectories of light non-strange $q\bar{q}$-states", Phys. Rev. D **85**, 094006 (2012).

[33] M. Shifman, A. Vainshtein and V. Zakharov, "QCD and Resonance Physics. The rho-omega Mixing", Nucl. Phys. **B147** (1979) 385; 447.

[34] P. Cox and T. Gherghetta, "A Soft-Wall Dilaton", JHEP **1502** (2015) 006.

[35] G. Ecker, "Facets of chiral perturbation theory", Nucl. Phys. Proc. Suppl. **245** (2013) 1.

[36] E. de Rafael, "Chiral Lagrangians and kaon CP violation", In *Boulder 1994, Proceedings, CP violation and the limits of the standard model* 15-84, and Marseille Cent. Theor. Phys. - 95-P-3161 (95/01,rec.Feb.) 71 p. (507515).

[37] J. Hirn, N. Rius and V. Sanz, "Geometric approach to condensates in holographic QCD", Phys. Rev. D **73**, 085005 (2006).

[38] M. A. Shifman, *Quark-hadron duality*. Published in the Boris Ioffe Festschrift 'At the Frontier of Particle Physics Handbook of QCD', ed. M. Shifman (World Scientific, Singapore, 2001).

[39] B. Batell and T. Gherghetta, "Dynamical Soft-Wall AdS/QCD", Phys. Rev. D **78**, 026002 (2008).

[40] G. F. de Teramond and S. J. Brodsky, "Light-Front Holography and Gauge/Gravity Duality: The Light Meson and Baryon Spectra", Nucl. Phys. Proc. Suppl. **199**, 89 (2010).
[41] L. Cappiello and G. D'Ambrosio, "On the Evaluation of Gluon Condensate Effects in the Holographic Approach to QCD", Eur. Phys. J. C **69**, 315 (2010).

EPJ Web of Conferences **129**, 00044 (2016) DOI: 10.1051/epjconf/201612900044
QCD@Work 2016

The multidimensional nucleon structure

Barbara Pasquini[1,2,⋆] and *Cédric* Lorcé[3,⋆⋆]

[1] *Dipartimento di Fisica, Università degli Studi di Pavia, Pavia, Italy*
[2] *Istituto Nazionale di Fisica Nucleare, Sezione di Pavia, Pavia, Italy*
[3] *Centre de Physique Théorique, École polytechnique, CNRS, Université Paris-Saclay, Palaiseau, France*

Abstract. We discuss different kinds of parton distributions, which allow one to obtain a multidimensional picture of the internal structure of the nucleon. We use the concept of generalized transverse momentum dependent parton distributions and Wigner distributions, which combine the features of transverse-momentum dependent parton distributions and generalized parton distributions. We show examples of these functions within a phenomenological quark model, with focus on the role of the spin-spin and spin-orbit correlations of quarks.

1 Introduction

During the past two decades, a lot of attention has been paid to generalized parton distributions (GPDs) [1–7] and to transverse-momentum dependent parton distributions (TMDs) [8–12]. Those objects are of particular interest because they describe the three-dimensional parton structure of hadrons — the distribution of the parton's longitudinal momentum and transverse position in the case of GPDs, and the distribution of the parton's longitudinal momentum and transverse momentum in the case of TMDs. Even though GPDs and TMDs already are quite general entities, the maximum possible information about the (two-) parton structure of strongly interacting systems is encoded in GTMDs [13–15]. GTMDs can reduce to GPDs and to TMDs in certain kinematical limits, and therefore they are often denoted as *mother distributions*. Further limits/integrations reduce them to collinear parton distribution functions and form factors. The Fourier transform of GTMDs can be interpreted as Wigner distributions [16–18], the quantum-mechanical analogue of classical phase-space distributions. A classification of GTMDs for quarks (through twist-4) for a spin-0 target was given in Ref. [13], followed by a corresponding work for a spin-$\frac{1}{2}$ target [14]. In Ref. [15], the counting of quark GTMDs was independently confirmed, and a complete classification of gluon GTMDs was provided as well. Gluon GTMDs appear when describing high-energy diffractive processes like vector meson production or Higgs production [19–21] in the so-called k_T-factorization in Quantum Chromodynamics. Recently, it was also suggested that the Wigner distributions of small-x gluons are closely related to the color dipole scattering amplitude [22].
For a comprehensive review of both the actual status and future developments in the theoretical and experimental investigation of GPDs, TMDs and Wigner functions we refer to a series of papers collected in a "Topical issue on the 3-D structure of the nucleon" [23]. Here, we review some aspects

⋆e-mail: barbara.pasquini@unipv.it
⋆⋆e-mail: cedric.lorce@polytechnique.edu

related to modelling and imaging of the nucleon structure [24]. In Sect. 2, we revisit the definition of the Wigner distributions obtained by Fourier transform of the GTMDs to the impact-parameter space. We identify all the possible correlations between target polarization, quark polarization and quark orbital angular momentum (OAM) encoded in these phase-space distributions, and we show results for a few selected examples. In Sect. 3 and 4, we discuss some of the complementary information encoded in GPDs and TMDs, in particular with regards to the information on the quark OAM. Concluding remarks are given in the final section.

2 Wigner distributions

The quark GTMD correlator is defined as [14, 15]

$$W^{ab}_{\Lambda'\Lambda} \equiv \int \mathrm{d}k^- \int \frac{\mathrm{d}^4 z}{(2\pi)^4}\, e^{ik\cdot z}\, \langle P+\tfrac{\Delta}{2},\Lambda'|\overline{\psi}_b(-\tfrac{z}{2})\,\mathcal{W}\,\psi_a(\tfrac{z}{2})|P-\tfrac{\Delta}{2},\Lambda\rangle, \tag{1}$$

where $\mathcal{W}$ is an appropriate Wilson line ensuring color gauge invariance, k is the quark average four-momentum conjugate to the quark field separation z, and $|p,\Lambda\rangle$ is the spin-1/2 target state with four-momentum p and light-front helicity Λ. A proper definition of the GTMDs should include also a soft-factor contribution [25]. However, it is not relevant for the following multipole analysis. At leading twist, one can interpret

$$W_{\vec{S}\vec{S}^q} = \tfrac{1}{8}\sum_{\Lambda',\Lambda}(1+\vec{S}\cdot\vec{\sigma})_{\Lambda\Lambda'}\,\mathrm{Tr}[W_{\Lambda'\Lambda}\Gamma_{\vec{S}^q}] \tag{2}$$

with $\Gamma_{\vec{S}^q} = \gamma^+ + S^q_L\gamma^+\gamma_5 + S^{qj}_T\, i\sigma^{j+}_T\gamma_5$, as the GTMD correlator describing the distribution of quarks with polarization $\vec{S}^q$ inside a target with polarization $\vec{S}$ [26].

The corresponding phase-space distribution is obtained by an appropriate Fourier transform [18]

$$\rho_{\vec{S}\vec{S}^q}(x,\boldsymbol{k}_T,\boldsymbol{b}_T;\hat{P},\eta) = \int \frac{\mathrm{d}^2\Delta_T}{(2\pi)^2}\, e^{-i\Delta_T\cdot\boldsymbol{b}_T}\, W_{\vec{S}\vec{S}^q}(P,k,\Delta)\big|_{\xi=0}, \tag{3}$$

where $x = k^+/P^+$ and $\boldsymbol{k}_T$ are, respectively, the longitudinal fraction and transverse component of the quark average momentum, $\boldsymbol{b}_T$ is the quark average impact parameter conjugate to the transverse-momentum transfer $\boldsymbol{\Delta}_T$, $\xi = -\Delta^+/2P^+$ is the fraction of longitudinal momentum transfer, and $\eta = +1$ ($\eta = -1$) indicates a future- (past-) pointing Wilson line. This phase-space distribution can be interpreted semi-classically as giving the quasi-probability of finding a quark with polarization $\vec{S}^q$, transverse position $\boldsymbol{b}_T$ and light-front momentum $(xP^+,\boldsymbol{k}_T)$ inside a spin-1/2 target with polarization $\vec{S}$ [18]. Thanks to the hermiticity properties of the GTMD correlator (2), these phase-space distributions are always real-valued [27], consistently with their quasi-probabilistic interpretation. Considering the various polarization configurations, one finds that there are 16 Wigner distributions just like there are 16 possible GTMDs [14, 15]. By construction, the real and imaginary parts of these GTMDs have opposite behavior under naive time-reversal transformation. Similarly, each Wigner distribution can be separated into naive T-even and T-odd contributions, $\rho_{\vec{S}\vec{S}^q} = \rho^e_{\vec{S}\vec{S}^q} + \rho^o_{\vec{S}\vec{S}^q}$, with $\rho^{e,o}_{\vec{S}\vec{S}^q}(x,\boldsymbol{k}_T,\boldsymbol{b}_T;\hat{P},\eta) = \pm\rho^{e,o}_{\vec{S}\vec{S}^q}(x,\boldsymbol{k}_T,\boldsymbol{b}_T;\hat{P},-\eta) = \pm\rho^{e,o}_{\vec{S}\vec{S}^q}(x,-\boldsymbol{k}_T,\boldsymbol{b}_T;-\hat{P},\eta)$. We can interpret the T-even contributions as describing the intrinsic distribution of quarks inside the target, whereas the naive T-odd contributions describe how initial- and final-state interactions modify these distributions. The relativistic phase-space distribution is linear in $\vec{S}$ and $\vec{S}^q$

$$\rho_{\vec{S}\vec{S}^q} = \rho_{UU} + S_L\rho_{LU} + S^q_L\rho_{UL} + S_L S^q_L\rho_{LL} \;+ S^i_T(\rho_{T^iU} + S^q_L\rho_{T^iL}) + S^{qi}_T(\rho_{UT^i} + S_L\rho_{LT^i}) + S^i_T S^{qj}_T\rho_{T^iT^j}, \tag{4}$$

Table 1. Correlations between target polarization (S_L, S_T), quark polarization (S^q_L, S^q_T) and quark OAM (ℓ^q_L, ℓ^q_T) encoded in the various phase-space distributions ρ_X, with $X = UU, LU, \ldots$.

ρ_X	U	L	T_x	T_y
U	$\langle 1\rangle$	$\langle S^q_L \ell^q_L\rangle$	$\langle S^q_x \ell^q_x\rangle$	$\langle S^q_y \ell^q_y\rangle$
L	$\langle S_L \ell^q_L\rangle$	$\langle S_L S^q_L\rangle$	$\langle S_L \ell^q_L S^q_x \ell^q_x\rangle$	$\langle S_L \ell^q_L S^q_y \ell^q_y\rangle$
T_x	$\langle S_x \ell^q_x\rangle$	$\langle S_x \ell^q_x S^q_L \ell^q_L\rangle$	$\langle S_x S^q_x\rangle$	$\langle S_x \ell^q_x S^q_y \ell^q_y\rangle$
T_y	$\langle S_y \ell^q_y\rangle$	$\langle S_y \ell^q_y S^q_L \ell^q_L\rangle$	$\langle S_y \ell^q_y S^q_x \ell^q_x\rangle$	$\langle S_y S^q_y\rangle$

and can further be decomposed into two-dimensional multipoles in both $\boldsymbol{k}_T$ and $\boldsymbol{b}_T$ spaces [28]. While there is no limit in the multipole order, parity and time-reversal give constraints on the allowed multipoles. It is therefore more appropriate to decompose the Wigner distributions ρ_X as follows

$$\rho_X(x, \boldsymbol{k}_T, \boldsymbol{b}_T; \hat{P}, \eta) = \sum_{m_k, m_b} \rho_X^{(m_k,m_b)}(x, \boldsymbol{k}_T, \boldsymbol{b}_T; \hat{P}, \eta), \qquad (X = UU, LU, \ldots), \tag{5}$$

$$\rho_X^{(m_k,m_b)}(x, \boldsymbol{k}_T, \boldsymbol{b}_T; \hat{P}, \eta) = B_X^{(m_k,m_b)}(\hat{k}_T, \hat{b}_T; \hat{P}, \eta)\, C_X^{(m_k,m_b)}[x, \boldsymbol{k}_T^2, (\boldsymbol{k}_T \cdot \boldsymbol{b}_T)^2, \boldsymbol{b}_T^2], \tag{6}$$

where $B_X^{(m_k,m_b)}$ are the basic (or simplest) multipoles allowed by parity and time-reversal symmetries, multiplied by the coefficient functions $C_X^{(m_k,m_b)}$ which depend on P and T-invariant variables only. The couple of integers (m_k, m_b) gives the basic multipole order in both $\boldsymbol{k}_T$ and $\boldsymbol{b}_T$ spaces. All the contributions ρ_X can be understood as encoding all the possible correlations between target and quark angular momenta, see Table 1.
In order to obtain a two-dimensional representation of the Wigner functions, we integrate these phase-space distributions over x and discretize the polar coordinates of $\boldsymbol{b}_T$. We also set $\eta = +1$ and choose $\hat{P} = \vec{e}_z = (0, 0, 1)$ so that $\hat{b}_T = (\cos\phi_b, \sin\phi_b, 0)$ and $\hat{k}_T = (\cos\phi_k, \sin\phi_k, 0)$. The resulting transverse phase-space distributions are then represented as sets of distributions in $\boldsymbol{k}_T$-space

$$\rho_X(\boldsymbol{k}_T|\,\boldsymbol{b}_T) = \int \mathrm{d}x\, \rho_X(x, \boldsymbol{k}_T, \boldsymbol{b}_T; \hat{P} = \vec{e}_z, \eta = +1)\big|_{\boldsymbol{b}_T \text{ fixed}} \tag{7}$$

with the origin of axes lying on circles of radius $|\boldsymbol{b}_T|$ at polar angle ϕ_b in impact-parameter space. In particular, we choose to represent only eight points in impact-parameter space lying on a circle with radius $|\boldsymbol{b}_T| = 0.4$ fm and $\phi_b = k\pi/4$ with $k \in \mathbb{Z}$. Also, for a better legibility, the $\boldsymbol{k}_T$-distributions are normalized to the absolute maximal value over the whole circle in impact-parameter space.

In the following, we will show results from the light-front constituent quark model (LFCQM) [29], which has been extensively applied to calculate form factors [30], TMDs [31–34] and GPDs [35–38], showing a typical accuracy of 30%. Since our purpose is simply to illustrate the multipole structure, we computed only the naive T-even contributions in this model. The naive T-odd contributions are obtained by extracting the coefficient functions from the naive T-even part and multiplying them by the appropriate basic multipoles. Therefore, the global sign of these naive T-odd contributions has been chosen arbitrarily. Only a proper calculation including initial- and/or final-state interactions can determine these global signs.

As an example, we discuss here the Wigner distribution for unpolarized quark in a longitudinal polarized target. We find only two phase-space distributions

$$\rho^e_{LU} = \rho^{(1,1)}_{LU}, \qquad \rho^o_{LU} = \rho^{(2,2)}_{LU}, \tag{8}$$

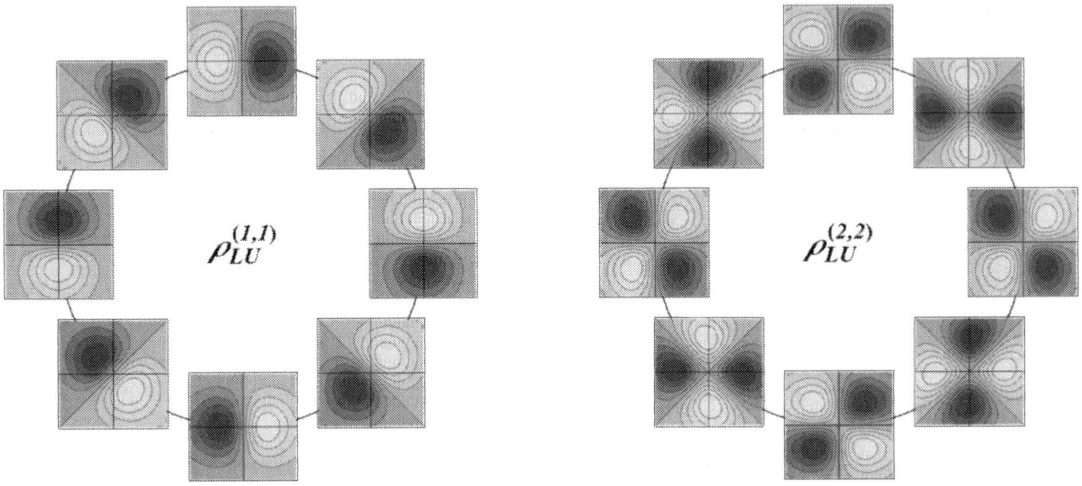

Figure 1. Naive T-even (left) and T-odd (right) contributions to the transverse phase-space distribution ρ_{LU}. Light and dark regions represent, respectively, positive and negative domains of the phase-space distributions.

which are represented in Fig. 1. The corresponding basic multipoles are

$$S^q_L B^{(1,1)}_{UL}(\hat{k}_T, \hat{b}_T; \hat{P}, \eta) = -S^q_L \epsilon^{ij}_T D^i_k D^j_b = S^q_L (\hat{b}_T \times \hat{k}_T)_L, \tag{9}$$

$$S^q_L B^{(2,2)}_{UL}(\hat{k}_T, \hat{b}_T; \hat{P}, \eta) = -\eta S^q_L \epsilon^{ij}_T Q^{il}_k Q^{jl}_b = \eta S^q_L (\hat{b}_T \times \hat{k}_T)_L (\hat{k}_T \cdot \hat{b}_T). \tag{10}$$

None of these survive integration over $\boldsymbol{k}_T$ or $\boldsymbol{b}_T$. Both therefore represent completely new information which is not accessible *via* GPDs or TMDs at leading twist. The $\boldsymbol{k}_T$-dipole in $\rho^{(1,1)}_{LU}$ signals the presence of a net longitudinal component of quark OAM correlated with the target longitudinal polarization S_L. By reversing the target longitudinal polarization S_L, one reverses also the orbital flow. The coefficient function $C^{(1,1)}_{LU}$ then gives the amount of longitudinal quark OAM in a longitudinally polarized target $\langle S_L \ell^q_L \rangle$ [18]. Similarly, the contribution $\rho^{(2,2)}_{LU}$ gives the difference of radial flows between quarks with opposite OAM $\langle S_L \ell^q_L \rangle$, with the coefficient function $C^{(2,2)}_{LU}$ representing in some sense the strength of the $\langle S_L \ell^q_L \rangle$-dependent part of the force felt by the quark due to initial- and final-state interactions. As a matter of fact, the quark OAM can be obtained from the ρ_{LU} distribution as [18, 27, 39]

$$\ell^q_z = \int \mathrm{d}x\, \mathrm{d}^2\boldsymbol{k}_T\, \mathrm{d}^2\boldsymbol{b}_T\, (\boldsymbol{b}_T \times \boldsymbol{k}_T)_z\, \rho_{UL}(x, \boldsymbol{k}_T, \boldsymbol{b}_T; \hat{P} = \vec{e}_z, \eta). \tag{11}$$

Since we are working here with a staple-like gauge link, ℓ^q_z is the canonical version of quark OAM [40]. On the other side, using a straight gauge link, Eq. (11) gives the kinetic OAM appearing in the Ji decomposition, as discussed in Sect. 4. From Fig. 1, we also note the ρ^o_{LU} cannot contribute to the quark OAM, and hence the quark OAM is η–independent [41–43].

3 Transverse Momentum Dependent Parton distributions

The forward limit ($\Delta = 0$) of the correlator (1) corresponds to a transverse momentum dependent quark-quark correlator that can be parametrized in terms of 8 TMDs, depending on the possible polarizations for the nucleon and quark (see Table 2). They are a natural extension of collinear parton

 DOI: 10.1051/epjconf/201612900044

Table 2. Twist-2 TMDs, for different quark (columns) and nucleon (rows) polarizations.

		quark pol.			
		U	T_x	T_y	L
nucleon pol.	U	f_1	$\frac{k_y}{M} h_1^\perp$	$-\frac{k_x}{M} h_1^\perp$	0
	T_x	$\frac{k_y}{M} f_{1T}^\perp$	$h_1 + \frac{k_x^2-k_y^2}{2M^2} h_{1T}^\perp$	$\frac{k_x k_y}{M^2} h_{1T}^\perp$	$\frac{k_x}{M} g_{1T}$
	T_y	$-\frac{k_x}{M} f_{1T}^\perp$	$\frac{k_x k_y}{M^2} h_{1T}^\perp$	$h_1 - \frac{k_x^2-k_y^2}{2M^2} h_{1T}^\perp$	$\frac{k_y}{M} g_{1T}$
	L	0	$\frac{k_x}{M} h_{1L}^\perp$	$\frac{k_y}{M} h_{1L}^\perp$	g_{1L}

distributions from one to three dimensions in momentum space, being functions of both the longitudinal quark momentum fraction x and the square of the transverse momentum k_T. The knowledge of TMDs allows us to build tomographic images of the inner structure of the nucleon in momentum space. The multipole pattern in $\boldsymbol{k}_T$ is clearly visible in Table 2. The TMDs f_1, g_{1L} and h_1 give the strength of monopole contributions and correspond to matrix elements without a net change of helicity between the initial and final states. The TMDs $f_{1T}^\perp$, g_{1T}, $h_1^\perp$, and $h_{1L}^\perp$ give the strength of dipole contributions and correspond to matrix elements involving one unit of helicity flip, either on the nucleon side ($f_{1T}^\perp$ and g_{1T}) or on the quark side ($h_1^\perp$ and $h_{1L}^\perp$). Finally, the TMD $h_{1T}^\perp$ corresponds to matrix elements where both the nucleon and quark helicities flip, but in opposite directions. Conservation of total angular momentum tells us that helicity flip is compensated by a change of OAM [31], which manifests itself by powers of $\boldsymbol{k}_T/M$. The corresponding quadrupole structure in $\boldsymbol{k}_T$ is shown in Fig. 2, as obtained from a LFCQM [31]. The function $h_{1T}^\perp$ has attracted a particular interest, since it has been suggested, on the basis of some quark-model calculations [44, 45], that it may be related to the quark OAM:

$$\mathcal{L}_z^q = -\int \mathrm{d}x\mathrm{d}^2\boldsymbol{k}_T \frac{k_T^2}{2M^2} h_{1T}^{\perp q}(x, k_T^2). \tag{12}$$

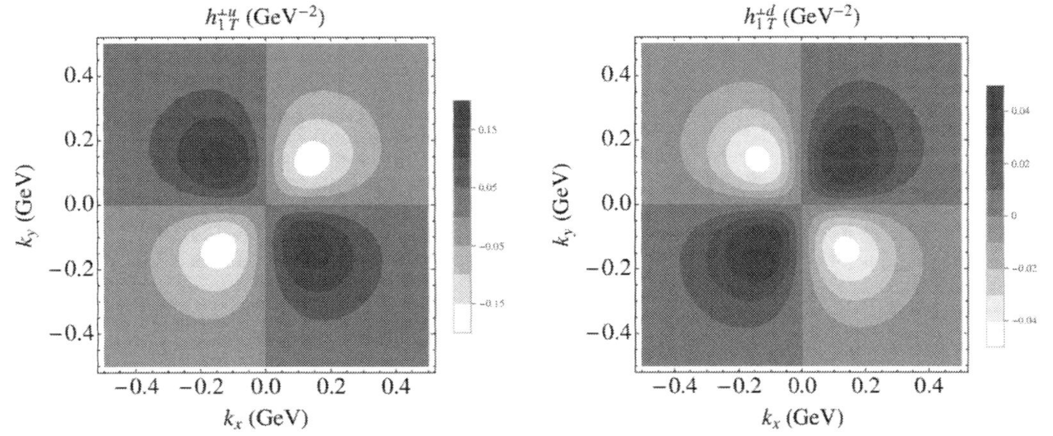

Figure 2. Quark density in the transverse-momentum plane for transverse polarization of quarks and nucleon in perpendicular directions. Left (right) panel is for up (down) quarks.

However, Eq. (12) is not a rigorous expression, and holds only in a restricted class of quark models [26, 46]. In general, no direct quantitative relations between OAM and TMDs should actually be expected as the former represents a correlation between parton position and momentum, whereas the latter only provide information about the momentum distribution. Nevertheless, TMDs do provide some indirect information about the OAM content of the nucleon, for example, by constraining the nucleon light-front wave functions that are eigenstates of the OAM [27].

4 Generalized Parton Distributions

When integrating the correlator (1) over $\boldsymbol{k}_T$, one obtains an off-diagonal quark-quark correlator that can be parametrized in terms of combinations of 8 GPDs (see Table 3). Comparing the entries in Tables 3 and 2, one notices that the same multipole pattern appears, where the role of $\boldsymbol{k}_T$ in Table (2) is played by $\boldsymbol{\Delta}_T$ in Table (3). Although direct links between GPDs and TMDs cannot exist [14, 47], this correspondence leads us to expect correlations between signs or similar orders of magnitude [48, 49]. The GPDs are functions of x, ξ, and $t = -\Delta^2$. Just like GTMDs and Wigner distributions are connected by Fourier transform at $\xi = 0$, one also finds a direct connection between GPDs at $\xi = 0$ and parton distributions in impact-parameter space. The multipole structure in $\boldsymbol{b}_T$ is the same as in Table 3, with the only difference that there are no polarization effects in the impact-parameter distributions for longitudinally polarized quarks in a transversely polarized proton and vice versa, since they are forbidden by time reversal. For example, the distribution of unpolarized quarks in a transversely polarized target reads as

$$q_X(x, \boldsymbol{b}_T) = \int \frac{d^2\boldsymbol{\Delta}_T}{(2\pi)^2} e^{-i\boldsymbol{\Delta}_T\cdot\mathbf{b}_T} H(x, 0, -\boldsymbol{\Delta}_T^2) - \frac{1}{2M}\frac{\partial}{\partial b_y}\int \frac{d^2\boldsymbol{\Delta}_T}{(2\pi)^2} e^{-i\boldsymbol{\Delta}_T\cdot\mathbf{b}_T} E(x, 0, -\boldsymbol{\Delta}_T^2).$$

This quark density is given by the sum of a nucleon spin-independent contribution related to the GPD H and a nucleon spin-dependent contribution from the GPD E, corresponding to monopole and dipole distributions in impact-parameter space, respectively. The dipole contribution introduces a large distortion perpendicular to both the nucleon spin and the momentum of the proton, with opposite sign for u and d quarks, as shown in Fig. 3. While details of the GPD E are not known, its x-integral is equal to the Pauli form factor F_2, which allows one to constrain the average deformation model-independently to the contribution from each quark flavor q to the nucleon anomalous magnetic moment.

Even before 3D imaging was introduced, there was great interest in GPDs due to their connection with the form factors of the energy-momentum tensor and thus to the angular momentum (spin plus

Table 3. Combinations of twist-2 GPDs, as defined in Ref. [26], for different quark (columns) and nucleon (rows) polarizations.

		quark pol.			
		U	T_x	T_y	L
nucleon pol.	U	$\mathcal{H}$	$i\frac{\Delta_y}{2M}\mathcal{E}_T$	$-i\frac{\Delta_x}{2M}\mathcal{E}_T$	0
	T_x	$i\frac{\Delta_y}{2M}\mathcal{E}$	$\mathcal{H}_T + \frac{\Delta_x^2-\Delta_y^2}{2M^2}\tilde{\mathcal{H}}_T$	$\frac{\Delta_x\Delta_y}{M^2}\tilde{\mathcal{H}}_T$	$\frac{\Delta_x}{2M}\tilde{\mathcal{E}}$
	T_y	$-i\frac{\Delta_x}{2M}\mathcal{E}$	$\frac{\Delta_x\Delta_y}{M^2}\tilde{\mathcal{H}}_T$	$\mathcal{H}_T - \frac{\Delta_x^2-\Delta_y^2}{2M^2}\tilde{\mathcal{H}}_T$	$\frac{\Delta_y}{2M}\tilde{\mathcal{E}}$
	L	0	$\frac{\Delta_x}{2M}\tilde{\mathcal{E}}_T$	$\frac{\Delta_y}{2M}\tilde{\mathcal{E}}_T$	$\tilde{\mathcal{H}}$

DOI: 10.1051/epjconf/201612900044

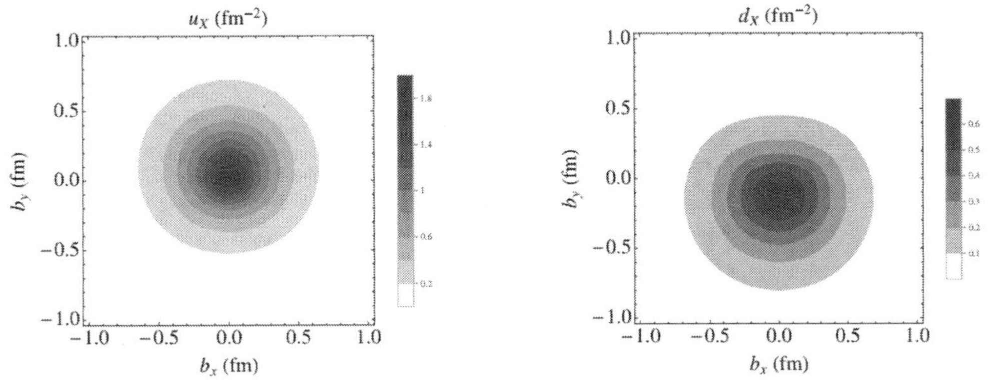

Figure 3. Quark density in the impact parameter space $\boldsymbol{b}_T$ for unpolarized quarks in a transversely polarized nucleon along the $\hat{x}$ direction, as obtained within a LFCQM [37]. The left (right) panel is for up (down) quark.

orbital) carried by quarks of flavor q as described by the Ji relation [2]

$$J_q = \frac{1}{2}\int_0^1 dx\, x\left[H_q(x,\xi,0) + E_q(x,\xi,0)\right], \tag{13}$$

where the ξ-dependence on the r.h.s. disappears upon x-integration. The quark kinetic OAM is then obtained by subtracting the quark spin contribution, which is given by the first Mellin moment of the quark helicity distribution.

5 Conclusions

We discussed the leading-twist quark Wigner distributions in the nucleon, introducing a multipole analysis in the four-dimensional transverse phase space. In this approach, the multipole structures are constrained by parity and time-reversal symmetries and are multiplied by coefficient functions which depend on P and T-invariant variables only. This representation provides a clear interpretation of all the amplitudes in terms of the possible correlations between target and quark angular momenta in the transverse phase space, and it gives a convenient basis to make a direct connection with TMDs in transverse-momentum space and GPDs in impact-parameter space. In particular, we discussed results for the Wigner distributions of unpolarized quarks in a longitudinally polarized nucleon that allow us to calculate the phase-space average of the quark OAM. Other ways to access information about the quark OAM from TMDs and GPDs have been also discussed, showing results within a light-front constituent quark model.

References

[1] D. Müller, D. Robaschik, B. Geyer, F. -M. Dittes and J. Hořejši, Fortsch. Phys. **42**, 101 (1994).
[2] X.-D. Ji, Phys. Rev. Lett. **78**, 610 (1997).
[3] A. V. Radyushkin, Phys. Lett. B **380**, 417 (1996).
[4] K. Goeke, M. V. Polyakov and M. Vanderhaeghen, Prog. Part. Nucl. Phys. **47**, 401 (2001).
[5] M. Diehl, Phys. Rept. **388**, 41 (2003).
[6] A. V. Belitsky and A. V. Radyushkin, Phys. Rept. **418**, 1 (2005).
[7] S. Boffi and B. Pasquini, Riv. Nuovo Cim. **30**, 387 (2007).

[8] A. Bacchetta, M. Diehl, K. Goeke, A. Metz, P. J. Mulders and M. Schlegel, JHEP **0702**, 093 (2007).
[9] U. D'Alesio and F. Murgia, Prog. Part. Nucl. Phys. **61**, 394 (2008).
[10] M. Burkardt, C. A. Miller and W. D. Nowak, Rept. Prog. Phys. **73**, 016201 (2010).
[11] V. Barone, F. Bradamante and A. Martin, Prog. Part. Nucl. Phys. **65**, 267 (2010).
[12] C. A. Aidala, S. D. Bass, D. Hasch and G. K. Mallot, Rev. Mod. Phys. **85**, 655 (2013).
[13] S. Meissner, A. Metz, M. Schlegel and K. Goeke, JHEP **0808**, 038 (2008).
[14] S. Meissner, A. Metz and M. Schlegel, JHEP **0908**, 056 (2009).
[15] C. Lorcé and B. Pasquini, JHEP **1309**, 138 (2013).
[16] X. d. Ji, Phys. Rev. Lett. **91**, 062001 (2003).
[17] A. V. Belitsky, X. d. Ji and F. Yuan, Phys. Rev. D **69**, 074014 (2004).
[18] C. Lorcé and B. Pasquini, Phys. Rev. D **84**, 014015 (2011).
[19] A. D. Martin, M. G. Ryskin and T. Teubner, Phys. Rev. D **62**, 014022 (2000).
[20] V. A. Khoze, A. D. Martin and M. G. Ryskin, Eur. Phys. J. C **14**, 525 (2000).
[21] A. D. Martin and M. G. Ryskin, Phys. Rev. D **64**, 094017 (2001).
[22] Y. Hatta, B. W. Xiao and F. Yuan, Phys. Rev. Lett. **116**, no. 20, 202301 (2016).
[23] M. Anselmino, M. Guidal, and P. Rossi, Eur. Phys. J. A**52**, 164 (2016).
[24] M. Burkardt and B. Pasquini, Eur. Phys. J. A **52**, no. 6, 161 (2016).
[25] M. G. Echevarria, A. Idilbi, K. Kanazawa, C. Lorcé, A. Metz, B. Pasquini, and M. Schlegel, Phys. Lett. B **759**, 336 (2016).
[26] C. Lorcé and B. Pasquini, Phys. Rev. D **84**, 034039 (2011).
[27] C. Lorcé, B. Pasquini, X. Xiong and F. Yuan, Phys. Rev. D **85**, 114006 (2012).
[28] C. Lorcé and B. Pasquini, Phys. Rev. D **93**, no. 3, 034040 (2016).
[29] C. Lorcé, B. Pasquini and M. Vanderhaeghen, JHEP **1105**, 041 (2011).
[30] B. Pasquini and S. Boffi, Phys. Rev. D **76**, 074011 (2007).
[31] B. Pasquini, S. Cazzaniga and S. Boffi, Phys. Rev. D **78**, 034025 (2008).
[32] B. Pasquini and F. Yuan, Phys. Rev. D **81**, 114013 (2010).
[33] S. Boffi, A. V. Efremov, B. Pasquini and P. Schweitzer, Phys. Rev. D **79**, 094012 (2009).
[34] B. Pasquini and P. Schweitzer, Phys. Rev. D **83**, 114044 (2011).
[35] S. Boffi, B. Pasquini and M. Traini, Nucl. Phys. B **649**, 243 (2003).
[36] S. Boffi, B. Pasquini and M. Traini, Nucl. Phys. B **680**, 147 (2004).
[37] B. Pasquini and S. Boffi, Phys. Lett. B **653**, 23 (2007).
[38] B. Pasquini, M. Pincetti and S. Boffi, Phys. Rev. D **72**, 094029 (2005).
[39] K. Kanazawa, C. Lorcé, A. Metz, B. Pasquini and M. Schlegel, Phys. Rev. D **90**, 014028 (2014).
[40] K. F. Liu and C. Lorcé, Eur. Phys. J. A **52**, no. 6, 160 (2016).
[41] C. Lorcé, Phys. Lett. B **719**, 185 (2013).
[42] Y. Hatta, Phys. Lett. B **708**, 186 (2012).
[43] X. Ji, X. Xiong and F. Yuan, Phys. Rev. D **88**, no. 1, 014041 (2013).
[44] J. She, J. Zhu and B. Q. Ma, Phys. Rev. D **79**, 054008 (2009).
[45] H. Avakian, A. V. Efremov, P. Schweitzer and F. Yuan, Phys. Rev. D **81**, 074035 (2010).
[46] C. Lorcé and B. Pasquini, Phys. Lett. B **710**, 486 (2012).
[47] S. Meissner, A. Metz and K. Goeke, Phys. Rev. D **76**, 034002 (2007).
[48] M. Burkardt, Nucl. Phys. A **735**, 185 (2004).
[49] M. Burkardt and D. S. Hwang, Phys. Rev. D **69**, 074032 (2004).

EPJ Web of Conferences **129**, 00045 (2016) DOI: 10.1051/epjconf/201612900045
QCD@Work 2016

Pion polarizabilities: ChPT vs Experiment

Mikhail A. Ivanov[1,*]

[1] *Bogoliubov Laboratory of Theoretical Physics, Joint Institute for Nuclear Research, 141980 Dubna, Russia*

Abstract. The values of charged pion polarizabilities obtained in the framework of chiral perturbation theory at the level of two-loop accuracy are compared with the experimental result recently reported by COMPASS Collaboration. It is found that the calculated value for the dipole polarizabilities $(\alpha-\beta)_{\pi^\pm} = (5.7 \pm 1.0) \times 10^{-4}\,\mathrm{fm}^3$ fits quite well the experimental result $(\alpha-\beta)_{\pi^\pm} = (4.0 \pm 1.2_{\rm stat} \pm 1.4_{\rm syst}) \times 10^{-4}\,\mathrm{fm}^3$.

1 Introduction

Almost ten years ago the amplitude $\gamma\gamma \to \pi\pi$ have been evaluated in the framework of chiral perturbation theory (ChPT) [1–3] at two-loop order. It was done for both neutral pions [4] and charged ones [5]. The obtained results were found to be in agreement with the only previous calculation performed at this accuracy [6, 7], provided that the same set of low-energy constants (LECs) is used. With updated LECs at order p^4 [8, 9], the values of dipole polarizabilities of charged pions were found equal to [5]

$$(\alpha-\beta)_{\pi^\pm} = (5.7 \pm 1.0) \times 10^{-4}\,\mathrm{fm}^3, \qquad (\alpha+\beta)_{\pi^\pm} = 0.16 \times 10^{-4}\,\mathrm{fm}^3. \tag{1}$$

At that time the MAMI Collaboration [10] has reported the experimental result

$$(\alpha-\beta)_{\pi^\pm} = (11.6 \pm 1.5_{\rm stat} \pm 3.0_{\rm syst} \pm 0.5_{\rm mod}) \times 10^{-4}\,\mathrm{fm}^3\,. \tag{2}$$

The index “mod” denotes the uncertainty generated by the theoretical models used to analyze the data. The ChPT calculation was clearly in conflict with the MAMI result, see also [11] for a discussion.

The COMPASS collaboration at CERN has recently investigated pion Compton scattering by using the Primakoff effect. The pion polarizability has been determined to be [12]

$$(\alpha-\beta)_{\pi^\pm} = (4.0 \pm 1.2_{\rm stat} \pm 1.4_{\rm syst}) \times 10^{-4}\,\mathrm{fm}^3 \tag{3}$$

under the assumption $(\alpha+\beta)_{\pi^\pm} = 0$. This result is in agreement with the expectation from chiral perturbation theory.

The concept of the polarizability of molecules, atoms and nuclei was applied for the first time to hadrons in Refs. [13–15]. By using the general properties of quantum field theory it was shown that an expansion of the Compton scattering amplitude for hadrons with spin one half in small photon energy up to the second order contains two structure parameters called the electric and magnetic hadron polarizabilities. The classical sum rule for these quantities has been derived in Ref. [16]. Further

*e-mail: ivanovm@theor.jinr.ru

theoretical investigation of the pion polarizabilities has been pursued since the early 1970s. In the current algebra + PCAC approach of Terent'ev [17], the fundamental low-energy theorem has been proven which allows one to relate the pion polarizability to the ratio $\gamma = h_A(0)/h_V(0)$ of the vector and axial form factors in radiative pion decay $\pi \to e\nu\gamma$. By using recent precise measurements of the pion weak form factors by the PIBETA collaboration [18] one finds $\alpha_{\pi^\pm} = -\beta_{\pi^\pm} = 2.78(10) \times 10^{-4}\,\mathrm{fm}^3$.

There were many calculations of the pion polarizabilities by employing various models: the linear σ-model with quarks [19], the chiral quark model [20], the superconductor quark model [21], some chiral models [22] and so on.

Almost all of them except Terent'ev approach predicted a value of the electric polarizability within the range $4.0 \times 10^{-4}\,\mathrm{fm}^3 \leq \alpha_{\pi^\pm} \leq 6.0 \times 10^{-4}\,\mathrm{fm}^3$ which we call large-valued results. We note that models not based on a chiral Lagrangian, i.e., dispersion relations and finite-energy sum rules, also obtained the polarizability within this range of values [23, 24]. The pion and kaon polarizabilities have been calculated in the quark confinement model [25] in which the emphasis is placed on quark confinement and the composite nature of hadrons. It was found for charged pions $\alpha_{\pi^\pm} \sim 3.6\times10^{-4}\,\mathrm{fm}^3$ which is smaller than the large-valued results but slightly larger than Terent'ev's prediction.

The first correct calculation of the cross section $\gamma\gamma \to \pi\pi$ within chiral perturbation theory to next-to-leading order (one-loop accuracy) was performed in [26]. It was shown in [27] that chiral symmetry relates the low-energy constants (LECs) appearing in the $\gamma\gamma \to \pi\pi$-amplitude with the axial form factor $h_A(0)$. Thus it was shown explicitly that Terent'ev's low-energy theorem follows from one-loop calculation of the $\gamma\gamma \to \pi\pi$ process within chiral perturbation theory.

Note that the axial form factor $h_A(0)$ can be expressed through the dispersion integral of the difference of the vector and axial spectral densities [28]. By using this sum rule the pion polarizability was estimated in [29] and found to be in perfect agreement with chiral perturbation theory.

An actual two-loop ChPT calculation of the $\gamma\gamma \to \pi\pi$ amplitude was done in [6] (neutral pions) and [7] (charged pions). Because the effective Lagrangian at order p^6 was not available at that time, the ultraviolet divergences were evaluated in the $\overline{\mathrm{MS}}$ scheme, then dropped and replaced with a corresponding polynomial in the external momenta. The three new counterterms which enter at this order in the low-energy expansion were estimated with resonance saturation. Whereas such a procedure is legitimate from a technical point of view, it does not make use of the full information provided by chiral symmetry.

Later on, considerable progress has been made in this field, both in theory and experiment. As for theory, the Lagrangian at order p^6 has been constructed [30, 31], and its divergence structure has been determined [32]. This provides an important check on the above calculations: adding the counterterm contributions from the p^6 Lagrangian to the $\overline{\mathrm{MS}}$ amplitude evaluated in [6] and in [7] must provide a scale independent result. Also in the theory, improved techniques to evaluate the two-loop diagrams that occur in these amplitudes have been developed [33]. The updated calculation of the $\gamma\gamma \to \pi\pi$ amplitude to two loops was then performed in [4] (neutral pions) and [5] (charged pions). The final results for the pion polarizabities were presented in a rather compact algebraic form. By using updated values for the LECs one obtains the values of the pion polarizabilities given in Eq. (1).

A comprehensive review of the modern status of this field maybe found in Refs. [34, 35].

Finally, one has to mention that research on pion polarizabities using lattice simulation is currently conducted by several groups, see, for instance, Refs. [36–39].

2 Definition of pion polarizabilities

The electric (α_H) and magnetic (β_H) polarizabilities characterize the response of hadron to two-photon interactions. These quantities must be considered as fundamental as the electromagnetic mean square radii, static magnetic moments, etc. They are defined by the expansion of the Compton scattering amplitude in small photon momenta and energies. Since our interest here is the pion polarizabilities, we plot in Fig. 1 the diagram describing the Compton scattering by charged pion. Expanding the

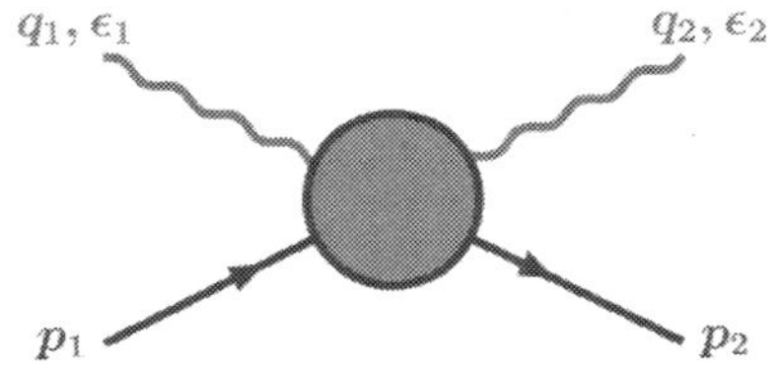

Figure 1. A schematic illustration of Compton scattering by a pion.

Compton scattering amplitude in small photon momenta and energies, one finds

$$T_{\gamma\pi^+\to\gamma\pi^+} = \underbrace{-2\,e^2\,\vec{\epsilon}_1\cdot\vec{\epsilon}_2}_{\textbf{Born term}} + \underbrace{8\,\pi\,M_\pi\left\{\alpha_\pi\,\omega_1\omega_2\,\vec{\epsilon}_1\cdot\vec{\epsilon}_2+\beta_\pi\,(\vec{\epsilon}_1\times\vec{q}_1)\cdot(\vec{\epsilon}_2\times\vec{q}_2)\right\}}_{\textbf{el-mag polarizabilities}}+\ldots \tag{4}$$

It is convenient to use the linear combinations of the electric and magnetic polarizabilities: $(\alpha-\beta)_\pi$ and $(\alpha+\beta)_\pi$ which are obtained from the helicity flip and helicity non-flip amplitudes, respectively.

As follows from the definition, the dipole pion polarizabilities are proportional to $\alpha_\pi(\beta_\pi) \sim \frac{\alpha}{M_\pi}\frac{1}{\Lambda^2} \approx 4\times 10^{-4}\,\mathrm{fm}^3$ where the hadronic scale $\Lambda \sim 4\pi F_\pi \sim 1\,\mathrm{GeV}$ was used. Then a natural choice of units for the polarizabilities is $10^{-4}\,\mathrm{fm}^3$.

3 Effective Lagrangian

An effective Lagrangian of QCD with two flavors in the isospin symmetry limit $m_u = m_d = \hat{m}$ at next-to-next-to-leading order (NNLO) is written as [2]

$$\mathcal{L}_{\rm eff} = \mathcal{L}_2 + \mathcal{L}_4 + \mathcal{L}_6\,. \tag{5}$$

The subscripts refer to the chiral order. The expression for $\mathcal{L}_2$ is

$$\begin{aligned} \mathcal{L}_2 &= \frac{F^2}{4}\langle D_\mu U\,D^\mu U^\dagger + M^2(U+U^\dagger)\rangle\,,\\ D_\mu U &= \partial_\mu U - i(QU-UQ)A_\mu\,,\ Q=\frac{e}{2}\mathrm{diag}(1,-1)\,, \end{aligned} \tag{6}$$

where e is the electric charge, and A_μ denotes the electromagnetic field. The quantity F denotes the pion decay constant in the chiral limit, and M^2 is the leading term in the quark mass expansion of the pion (mass)^2, $M_\pi^2 = M^2(1+O(\hat{m}))$. Further, the brackets $\langle\ldots\rangle$ denote a trace in flavor space. In Eq. (6), we have retained only the terms relevant for the present application, i.e., we have dropped additional external fields. We choose the unitary 2×2 matrix U in the form

$$U = \sigma + i\pi/F\,,\quad \sigma^2+\frac{\pi^2}{F^2} = \mathbf{1}_{2\times2}\,,\quad \pi = \begin{pmatrix} \pi^0 & \sqrt{2}\pi^+ \\ \sqrt{2}\pi^- & -\pi^0 \end{pmatrix}. \tag{7}$$

The Lagrangian at NLO has the structure [2]

$$\mathcal{L}_4 = \sum_{i=1}^{7} l_i K_i + \sum_{i=1}^{3} h_i \bar{K}_i = \frac{l_1}{4}\langle D_\mu U\,D^\mu U^\dagger\rangle^2 + \cdots\,, \tag{8}$$

where l_i, h_i denote low-energy couplings, not fixed by chiral symmetry. At NNLO, one has [30–32] $\mathcal{L}_6 = \sum_{i=1}^{57} c_i P_i$. As was shown in Ref. [40] the number of operators P_i can be reduced by at least one from 57 to 56. For the explicit expressions of the polynomials K_i, $\bar{K}_i$ and P_i, we refer the reader to Refs. [2, 30–32]. The vertices relevant for $\gamma\gamma \to \pi^+\pi^-$ involve $l_1, \ldots, l_6$ from $\mathcal{L}_4$ and several c_i's from $\mathcal{L}_6$, see below.

The couplings l_i and c_i absorb the divergences at order p^4 and p^6, respectively,

$$\begin{aligned} l_i &= (\mu c)^{d-4} \{ l_i^r(\mu, d) + \gamma_i \Lambda \} , \qquad \Lambda = \frac{1}{16\pi^2(d-4)} , \ \ln c = -\frac{1}{2} \{ \ln 4\pi + \Gamma'(1) + 1 \} , \\ c_i &= \frac{(\mu c)^{2(d-4)}}{F^2} \left\{ c_i^r(\mu, d) - \gamma_i^{(2)} \Lambda^2 - (\gamma_i^{(1)} + \gamma_i^{(L)}(\mu, d)) \Lambda \right\} . \end{aligned} \tag{9}$$

The physical couplings are $l_i^r(\mu, 4)$ and $c_i^r(\mu, 4)$, denoted by l_i^r, c_i^r in the following. The coefficients γ_i are given in [2], and $\gamma_i^{(1,2,L)}$ are tabulated in [32]. We shall use the scale independent quantities $\bar{l}_i$ introduced in [2], $l_i^r = \frac{\gamma_i}{32\pi^2} (\bar{l}_i + l)$, where the *chiral logarithm* is $l = \ln(M_\pi^2/\mu^2)$.

4 Evaluation of the diagrams

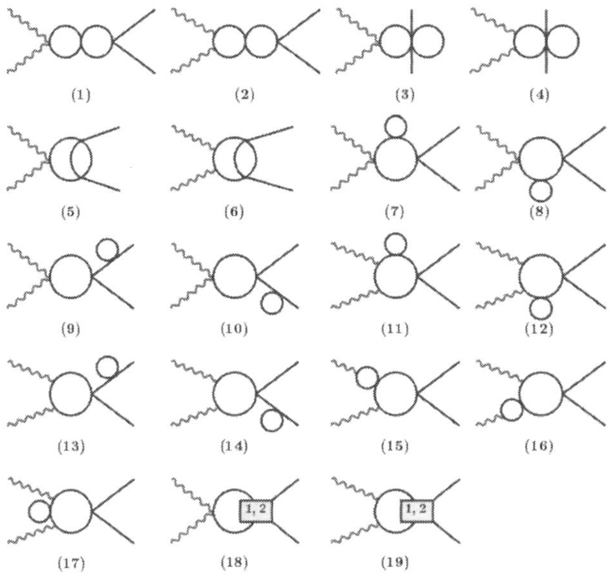

Figure 2. A set of two-loop diagrams generated by $\mathcal{L}_2$ and one-loop diagrams generated by $\mathcal{L}_4$.

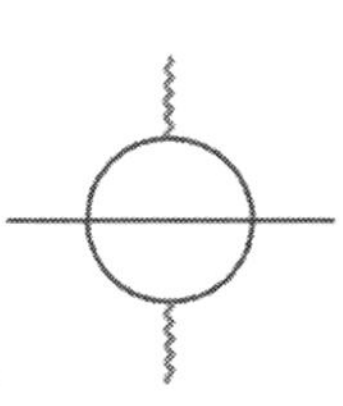

Figure 3. Acnode diagram.

The lowest-order contributions to the scattering amplitude are described by tree- and one-loop diagrams. These contributions were calculated in [26]. The two-loop reducible diagrams may be reduced to tree-diagrams by using Ward identities. They sum up to the expression $2 Z_\pi g^{\mu\nu} - \left\{(2p_1 - q_1)^\mu (2p_2 - q_2)^\nu \left[(M_\pi^2 - t)^{-1} - Z_\pi R(t)\right] + \text{crossed}\right\}$, where Z_π is the pion renormalization constant. The function $R(t)$ starts at order $1/F_\pi^4$ and can be obtained from the full pion propagator. The two-loop diagrams displayed in Figs. 2 are evaluated by using contraction of the $\gamma\gamma \to \pi\pi$ Born amplitude with the d-dimensional one-loop off-shell $\pi\pi$-scattering amplitude. The acnode diagram displayed in Fig. 3 is the most complicated for calculation. This diagram was evaluated in [4] by using a dispersion relation for the relevant master integral. The remaining diagrams at order p^6, e.g. one-loop graphs generated by $\mathcal{L}_4$, and counterterm contributions from $\mathcal{L}_6$, are easy to evaluate.

The evaluation of the diagrams was done in the manner described in [4, 33] by invoking FORM [43]. In particular, we have verified that the counterterms from the Lagrangian $\mathcal{L}_6$ [32] remove all ultraviolet divergences, which is a very non-trivial check on our calculation. Furthermore, we have checked that the (ultra-violet finite) amplitude so obtained is scale independent.

5 Chiral expansion for pion polarizabilities

Using the same notation as in [7], we find for the *dipole* polarizabilities

$$(\alpha \pm \beta)_{\pi^+} = \frac{\alpha}{16\pi^2 F_\pi^2 M_\pi} \left\{ c_{1\pm} + \frac{M_\pi^2 d_{1\pm}}{16\pi^2 F_\pi^2} + O(M_\pi^4) \right\}, \tag{10}$$

where $c_{1+} = 0$, $c_{1-} = \frac{2}{3}\bar{l}_\Delta$,

$$\begin{aligned} d_{1+} &= 8 b^r - \frac{4}{9}\left\{ l\left(l + \frac{1}{2}\bar{l}_1 + \frac{3}{2}\bar{l}_2\right) - \frac{53}{24} l + \frac{1}{2}\bar{l}_1 + \frac{3}{2}\bar{l}_2 + \frac{91}{72} + \Delta_+ \right\}, \\ d_{1-} &= a_1^r + 8 b^r - \frac{4}{3}\left\{ l\left(\bar{l}_1 - \bar{l}_2 + \bar{l}_\Delta - \frac{65}{12}\right) - \frac{1}{3}\bar{l}_1 - \frac{1}{3}\bar{l}_2 + \frac{1}{4}\bar{l}_3 - \bar{l}_\Delta \bar{l}_4 + \frac{187}{108} + \Delta_- \right\}, \end{aligned} \tag{11}$$

with

$$\Delta_+ = \frac{8105}{576} - \frac{135}{64}\pi^2, \qquad \Delta_- = \frac{41}{432} - \frac{53}{64}\pi^2. \tag{12}$$

It would be interesting to numerically compare the values of $\Delta_\pm$ given by Eq. (12) with those obtained in Refs. [7]. One has

$$\Delta_+ = \begin{cases} -6.75 & \text{our} \\ -8.69 & \text{Burgi} \end{cases} \qquad \Delta_- = \begin{cases} -8.08 & \text{our} \\ -8.73 & \text{Burgi} \end{cases}$$

We shall use the following values for the p^4 LECs [8]

$$\bar{l}_1 = -0.4 \pm 0.6, \quad \bar{l}_2 = 4.3 \pm 0.1, \quad \bar{l}_3 = 2.9 \pm 2.4, \quad \bar{l}_4 = 4.4 \pm 0.2, \tag{13}$$

and $\bar{l}_\Delta \doteq \bar{l}_6 - \bar{l}_5 = 3.0 \pm 0.3$ obtained from radiative pion decay to two loop accuracy [9, 41].

The constants c_i^r occur in the combinations

$$\begin{aligned} a_1^r &= -4096\pi^4 \left(6 c_6^r + c_{29}^r - c_{30}^r - 3 c_{34}^r + c_{35}^r + 2 c_{46}^r - 4 c_{47}^r + c_{50}^r\right), \\ b^r &= -128\pi^4 \left(c_{31}^r + c_{32}^r - 2 c_{33}^r - 4 c_{44}^r\right). \end{aligned}$$

As follows from the resonance exchange model [7] $\left(a_1^r, b^r\right) = \left(-3.2, 0.4\right)$. The values of these constants were obtained in the ENJL model [42] $(a_1^r, b^r) = (-8.7, 0.38)$. One can see that only b^r

DOI: 10.1051/epjconf/201612900045

agrees in the two approaches. We shall use $b^r = 0.4 \pm 0.4$. The combinations $(\alpha \pm \beta)_{\pi^\pm}$ are determined precisely by the chiral expansion to two loops, once a_1^r is fixed. We will then simply display this quantity as a function of a_1^r - the result turns out to be rather independent of its exact value.

The uncertainty in the prediction for the polarizability has two sources. First, the low-energy constants are not known precisely. Second, we are dealing here with an expansion in powers of the momenta and of the quark masses up to and including terms of order p^6. The discussion of estimating uncertainties may be found in our paper [5]. It was shown that the value for the dipole polarizability $(\alpha - \beta)_{\pi^\pm}$ is rather reliable - there is no sign of any large, uncontrolled correction to the two-loop result. The maximum deviation 1.0 from the central value 5.7 has been used as the final theoretical uncertainty for the dipole polarizability $(\alpha - \beta)_{\pi^\pm} = (5.7 \pm 1.0) \times 10^{-4}\,\mathrm{fm}^3$. The chiral expansion for the combination $(\alpha + \beta)_{\pi^\pm} = 0.16 \times 10^{-4}\,\mathrm{fm}^3$ starts out at order p^6 so we have determined only its leading order term.

6 Experimental information

There are three types of experiments aiming to measure the pion polarizabilities:

- scattering of pions off the Coulomb field of heavy nucleus using the Primakoff effect,
- radiative pion photoproduction from the proton,
- pion pair production in photon-photon collisions.

Schematically, they are shown in Fig. 4. The possibility to measure the pion polarizability via the Primakoff reaction was proposed in the early 1980s in [44]. The measurement of the pion-photon Compton scattering amplitude by using the Primakoff effect was performed in an experiment at Serpukhov [45] , but the small data sample led to only an imprecise value for the polarizability of $\alpha_\pi = (6.8 \pm 1.4_{\mathrm{stat}} \pm 1.2_{\mathrm{syst}}) \times 10^{-4}\,\mathrm{fm}^3$. Low statistics made it difficult to evaluate the systematic uncertainty.

COMPASS has now achieved a modern Primakoff experiment, using a 190 GeV pion beam from the Super Proton Synchrotron at CERN directed at a Nickel target. It is important that COMPASS was also able to use a muon, which is a point-like particle, to calibrate the experiment. The Compton $\pi^-\gamma \to \pi^-\gamma$ scattering is extracted by selecting events from the Coulomb peak at small momentum transfer $Q^2 < 0.0015\ \mathrm{GeV}^2$. From the analysis of a sample of 63,000 events, the collaboration [12] obtained a value of the pion polarizability given by Eq. (3).

The cross section for the radiative pion photoproduction $\gamma p \to \gamma\pi^+ n$ has been measured at the Lebedev Institute [46]. By using an extrapolation to the pion pole in the unphysical region the value of the electric polarizability was obtained $\alpha_{\pi^+} = -\beta_{\pi^+} = (20 \pm 12) \times 10^{-4}\,\mathrm{fm}^3$. Similar experiment

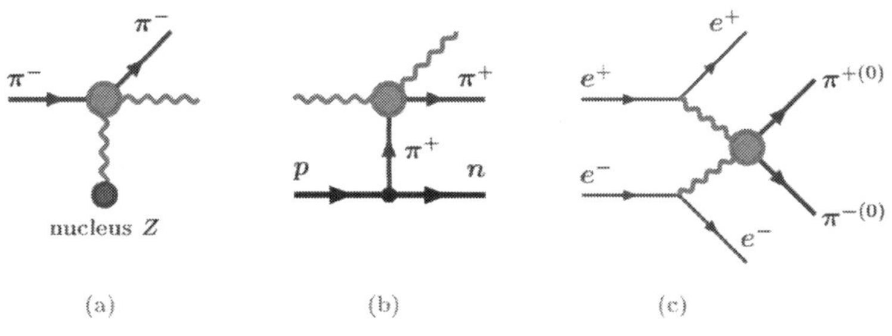

Figure 4. Schematic images of the available experiments.

Table 1. Experimental information on $(\alpha-\beta)_{\pi^\pm}$, in units of 10^{-4} fm^3. We indicate the reaction and data used. In [52], [47] and [12] α_π was determined, using as a constraint $\alpha_\pi = -\beta_\pi$. To obtain $(\alpha-\beta)_{\pi^\pm}$, we multiplied the results by a factor of 2.

Experiments		$(\alpha-\beta)_{\pi^\pm}$
$\pi^- Z \to \gamma\pi^- Z$	Serpukhov (1983) [45]	$15.6 \pm 6.4_{\text{stat}} \pm 4.4_{\text{syst}}$
$\gamma p \to \gamma\pi^+ n$	Lebedev Inst. (1984) [46]	40 ± 24
D. Babusci *et al.* (1992) [47]		
$\gamma\gamma \to \pi^+\pi^-$	PLUTO [48]	$38.2 \pm 9.6 \pm 11.4$
	DM 1 [49]	34.4 ± 9.2
	DM 2 [50]	52.6 ± 14.8
	MARK II [51]	4.4 ± 3.2
J.F. Donoghue, B. Holstein (1993) [52]		5.4
$\gamma\gamma \to \pi^+\pi^-$	MARK II [51]	
A. Kaloshin, V. Serebryakov (1994) [53]		5.25 ± 0.95
$\gamma\gamma \to \pi^+\pi^-$	MARK II [51]	
	Crystal Ball Coll. [54]	
$\gamma p \to \gamma\pi^+ n$	Mainz (2005) [10]	$11.6 \pm 1.5_{\text{stat}} \pm 3.0_{\text{syst}} \pm 0.5_{\text{mod}}$
L. Fil'kov, V. Kashevarov (2005) [24]		$13.0^{+2.6}_{-1.9}$
$\gamma\gamma \to \pi^+\pi^-$	MARK II [51],	
TPC/2γ [55], CELLO [56],		
VENUS [57], ALEPH [58], BELLE [59]		
$\pi^- Z \to \gamma\pi^- Z$	COMPASS (2015) [12]	$4.0 \pm 1.2_{\text{stat}} \pm 1.4_{\text{syst}}$

was performed at the Mainz Microtron MAMI [10], but the pion polarizability has been extracted by a comparison of the data with the predictions of two different models yielding the value given by Eq.(2).

Another possibility to obtain the value for the pion polarizability is to extrapolate the data from the pion pair production in photon-photon collisions $\gamma\gamma \to \pi\pi$ to the region of the Compton scattering threshold by using crossing symmetry and analyticity. Normally, the procedure involves the construction of the dispersion relations with one or two subtractions. The most recent analysis performed in [24] produced the value $(\alpha-\beta)_{\pi^\pm} = 13.0^{+2.6}_{-1.9} \times 10^{-4}$ fm^3, which is close to the MAMI data. There are plenty of previous studies in this direction which give quite a broad region for the value of the pion polarizability. The available experimental information is shown in Table 1.

Acknowledgments

I would like to thank Juerg Gasser and Mikko Sainio for their collaboration in which the results for pion polarizabilities at two-loop level accuracy were obtained. I am grateful to Pietro Colangelo for inviting me to give a talk at the workshop QCD@Work.

References

[1] S. Weinberg, Physica A, **96**, 327 (1979).

[2] J. Gasser and H. Leutwyler, Ann. Phys., **158**, 142 (1984).
[3] J. Gasser and H. Leutwyler, Nucl. Phys. B, **250**, 465 (1985).
[4] J. Gasser, M. A. Ivanov and M. E. Sainio, Nucl. Phys. B, **728**, 31 (2005).
[5] J. Gasser, M. A. Ivanov and M. E. Sainio, Nucl. Phys. B, **745**, 84 (2006).
[6] S. Bellucci, J. Gasser and M. E. Sainio, Nucl. Phys. B, **423**, 80 (1994), [Erratum-ibid. B **431**, 413 (1994)].
[7] U. Burgi, Nucl. Phys. B, **479**, 392 (1996), Phys. Lett. B, **377**, 147 (1996).
[8] G. Colangelo, J. Gasser and H. Leutwyler, Nucl. Phys. B, **603**, 125 (2001).
[9] J. Bijnens and P. Talavera, Nucl. Phys. B, **489**, 387 (1997).
[10] J. Ahrens et al., Eur. Phys. J. A, **23**, 113 (2005).
[11] S. Scherer, Eur. Phys. J. A, **28**, 59 (2006).
[12] COMPASS Collab. (C. Adolph et al.), Phys. Rev. Lett., **114**, 062002 (2015).
[13] A. Klein, Phys. Rev., **99**, 998 (1955).
[14] A.M. Baldin, Nucl. Phys., **18**, 310 (1960).
[15] V.A. Petrunkin, JETP, **13**, 804 (1961).
[16] V. A. Petrunkin, Nucl. Phys., **55**, 197 (1964).
[17] M. V. Terentev, Sov. J. Nucl. Phys., **16**, 87 (1973); **19**, 664 (1974).
[18] M. Bychkov et al., Phys. Rev. Lett., **103**, 051802 (2009). [arXiv:0804.1815 [hep-ex]].
[19] A. I. L'vov, Sov. J. Nucl. Phys., **34**, 289 (1981).
[20] M. K. Volkov and D. Ebert, Sov. J. Nucl. Phys., **34**, 104 (1981); Phys. Lett. B, **101**, 252 (1981).
[21] M. K. Volkov and A. A. Osipov, Sov. J. Nucl. Phys., **41**, 659 (1985).
[22] V. Bernard, B. Hiller and W. Weise, Phys. Lett. B, **205**, 16 (1988).
[23] L. V. Filkov, I. Guiasu and E. E. Radescu, Phys. Rev. D, **26**, 3146 (1982).
[24] L. V. Fil'kov and V. L. Kashevarov, Phys. Rev. C, **73**, 035210 (2006).
[25] M. A. Ivanov and T. Mizutani, Phys. Rev. D, **45**, 1580 (1992).
[26] J. Bijnens and F. Cornet, Nucl.Phys. B, **296**, 557 (1988).
[27] J. F. Donoghue and B. R. Holstein, Phys. Rev. D, **40**, 2378 (1989).
[28] T. Das, V. S. Mathur and S. Okubo, Phys. Rev. Lett., **19**, 859 (1967).
[29] A. E. Dorokhov and W. Broniowski, Eur. Phys. J. C, **32**, 79 (2003).
[30] H. W. Fearing and S. Scherer, Phys. Rev. D, **53**, 315 (1996).
[31] J. Bijnens, G. Colangelo and G. Ecker, JHEP, **9902**, 020 (1999).
[32] J. Bijnens, G. Colangelo and G. Ecker, Ann. Phys., **280**, 100 (2000).
[33] J. Gasser and M. E. Sainio, Eur. Phys. J. C, **6**, 297 (1999).
[34] V. A. Petrunkin, Sov. J. Part. Nucl., **12**, 278 (1981).
[35] B. R. Holstein and S. Scherer, Ann. Rev. Nucl. Part. Sci., **64**, 51 (2014).
[36] E. V. Luschevskaya, O. E. Solovjeva, O. A. Kochetkov and O. V. Teryaev, Nucl. Phys. B, **898**, 627 (2015).
[37] M. Lujan, A. Alexandru, W. Freeman and F. Lee, Phys. Rev. D, **89**, 074506 (2014).
[38] W. Detmold, B. C. Tiburzi and A. Walker-Loud, Phys. Rev. D, **79**, 094505 (2009).
[39] J. Hu, F. J. Jiang and B. C. Tiburzi, Phys. Rev. D, **77**, 014502 (2008).
[40] C. Haefeli, M. A. Ivanov, M. Schmid and G. Ecker, arXiv:0705.0576 [hep-ph].
[41] C. Q. Geng, I. L. Ho and T. H. Wu, Nucl. Phys. B, **684**, 281 (2004).
[42] J. Bijnens and J. Prades, Nucl. Phys. B, **490**, 239 (1997).
[43] J. A. M. Vermaseren, arXiv:math-ph/0010025.

[44] A. S. Galperin, G. Mitselmakher, A. G. Olszewski and V. N. Pervushin, Yad. Fiz., **32**, 1053 (1980).
[45] Y. M. Antipov et al., Z. Phys. C, **26** (1985) 495; Phys. Lett. B, **121** (1983) 445.
[46] T. A. Aibergenov et al., Czech. J. Phys. B, **36**, 948 (1986).
[47] D. Babusci, S. Bellucci, G. Giordano, G. Matone, A. M. Sandorfi and M. A. Moinester, Phys. Lett. B, **277**, 158 (1992).
[48] PLUTO Collab. (C. Berger et al.), Z. Phys. C. **26**, 199 (1984).
[49] DM1-DM2 Collab. (A. Courau et al.), Nucl. Phys. B. **271**, 1 (1986).
[50] DM1-DM2 Collab. (Z. Ajaltouni et al.), Phys. Lett. B, **194**, 573 (1987); [Erratum-ibid. Phys. Lett. B, **197**, 565 (1987)].
[51] MARK-II Collab. (J. Boyer et al.), Phys. Rev. D, **42**, 1350 (1990).
[52] J. F. Donoghue and B. R. Holstein, Phys. Rev. D, **48**, 137 (1993).
[53] A. E. Kaloshin and V. V. Serebryakov, Z. Phys. C, **64**, 689 (1994).
[54] Crystal Ball Collab. (H. Marsiske et al.), Phys. Rev. D, **41**, 3324 (1990).
[55] TPC/Two-Gamma Collab. (H. Aihara et al.) , Phys. Rev. Lett., **57**, 404 (1986).
[56] CELLO Collab. (H. J. Behrend et al.), Z. Phys. C, **56**, 381 (1992).
[57] VENUS Collab. (F. Yabuki et al.), J. Phys. Soc. Jap., **64**, 435 (1995).
[58] ALEPH Collab. (A. Heister et al.), Phys. Lett. B, **569**, 140 (2003).
[59] BELLE Collab. (H. Nakazawa et al.), Phys. Lett. B, **615**, 39 (2005).

Recent results and prospects from NA62

Andrea Bizzeti[1,2,★]

[1] *Department of Physics, Informatics and Mathematics, University of Modena and Reggio Emilia, Italy*
[2] *Istituto Nazionale di Fisica Nucleare - Sezione di Firenze, Italy*

Abstract. A large sample of charged kaon decays in 2007 has been collected by the NA62 experiment at CERN SPS using the experimental setup of the former NA48 experiment. Its intense kaon beam provides an abundant source of tagged neutral pions in vacuum. A measurement of the electromagnetic transition form factor slope of the neutral pion from 1.05×10^6 fully reconstructed π^0 Dalitz decays is presented. The obtained preliminary value $a = (3.70 \pm 0.53_{\rm stat} \pm 0.36_{\rm syst}) \times 10^{-2}$ is the first 5.8σ observation of a non-zero slope in the time-like region of momentum transfer.
$K^+ \to \pi^+ \nu\bar{\nu}$ is a theoretically very clean decay where indirect effects of new physics may be detectable. The NA62 apparatus has been significantly upgraded between 2008 and 2014 in order to measure the branching ratio of this decay with 10% precision. The NA62 experiment took data with the new setup in pilot runs in 2014 and 2015, reaching the design beam intensity. Results of first data quality studies in view of the 2016-2017 physics runs are presented.

★for the NA62 Collaboration: G. Aglieri Rinella, R. Aliberti, F. Ambrosino, R. Ammendola, B. Angelucci, A. Antonelli, G. Anzivino, R. Arcidiacono, I. Azhinenko, S. Balev, M. Barbanera, J. Bendotti, A. Biagioni, L. Bician, C. Biino, A. Bizzeti, T. Blazek, A. Blik, B. Bloch-Devaux, V. Bolotov, V. Bonaiuto, M. Boretto, M. Bragadireanu, D. Britton, G. Britvich, M.B. Brunetti, D. Bryman, F. Bucci, F. Butin, E. Capitolo, C. Capoccia, T. Capussela, A. Cassese, A. Catinaccio, A. Cecchetti, A. Ceccucci, P. Cenci, V. Cerny, C. Cerri, B. Checcucci, O. Chikilev, S. Chiozzi, R. Ciaranfi, G. Collazuol, A. Conovaloff, P. Cooke, P. Cooper, G. Corradi, E. Cortina Gil, F. Costantini, F. Cotorobai, A. Cotta Ramusino, D. Coward, G. D'Agostini, J. Dainton, P. Dalpiaz, H. Danielsson, J. Degrange, N. De Simone, D. Di Filippo, L. Di Lella, S. Di Lorenzo, N. Dixon, N. Doble, B. Dobrich, V. Duk, V. Elsha, J. Engelfried, T. Enik, N. Estrada, V. Falaleev, R. Fantechi, V. Fascianelli, L. Federici, S. Fedotov, M. Fiorini, J. Fry, J. Fu, A. Fucci, L. Fulton, S. Gallorini, S. Galeotti, E. Gamberini, L. Gatignon, G. Georgiev, A. Gianoli, M. Giorgi, S. Giudici, L. Glonti, A. Goncalves Martins, F. Gonnella, E. Goudzovski, R. Guida, E. Gushchin, F. Hahn, B. Hallgren, H. Heath, F. Herman, T. Husek, O. Hutanu, D. Hutchcroft, L. Iacobuzio, E. Iacopini, E. Imbergamo, O. Jamet, P. Jarron, E. Jones, T. Jones K. Kampf, J. Kaplon, V. Kekelidze, S. Kholodenko, G. Khoriauli, A. Khotyantsev, A. Khudyakov, Yu. Kiryushin, A. Kleimenova, K. Kleinknecht, A. Kluge, M. Koval, V. Kozhuharov, M. Krivda, Z. Kucerova, Yu. Kudenko, J. Kunze, G. Lamanna, G. Latino, C. Lazzeroni, G. Lehmann-Miotto, R. Lenci, M. Lenti, E. Leonardi, P. Lichard, R. Lietava, L. Litov, R. Lollini, D. Lomidze, A. Lonardo, M. Lupi, N. Lurkin, K. McCormick, D. Madigozhin, G. Maire, C. Mandeiro, I. Mannelli, G. Mannocchi, A. Mapelli, F. Marchetto, R. Marchevski, S. Martellotti, P. Massarotti, K. Massri, P. Matak, E. Maurice, A. Mefodev, E. Menichetti, E. Minucci, M. Mirra, M. Misheva, N. Molokanova, J. Morant, M. Morel, M. Moulson, S. Movchan, D. Munday, M. Napolitano, I. Neri, F. Newson, A. Norton, M. Noy, G. Nuessle, T. Numao, V. Obraztsov, A. Ostankov, S. Padolski, R. Page, V. Palladino, G. Paoluzzi, C. Parkinson, E. Pedreschi, M. Pepe, F. Perez Gomez, M. Perrin-Terrin, L. Peruzzo, P. Petrov, F. Petrucci, R. Piandani, M. Piccini, D. Pietreanu, J. Pinzino, I. Polenkevich, L. Pontisso, Yu. Potrebenikov, D. Protopopescu, F. Raffaelli, M. Raggi, P. Riedler, A. Romano, P. Rubin, G. Ruggiero, V. Russo, V. Ryjov, A. Salamon, G. Salina, V. Samsonov, C. Santoni, G. Saracino, F. Sargeni, V. Semenov, A. Sergi, M. Serra, A. Shaikhiev, S. Shkarovskiy, I. Skillicorn, D. Soldi, A. Sotnikov, V. Sugonyaev, M. Sozzi, T. Spadaro, F. Spinella, R. Staley, A. Sturgess, P. Sutcliffe, N. Szilasi, D. Tagnani, S. Trilov, M. Valdata-Nappi, P. Valente, M. Vasile, T. Vassilieva, B. Velghe, M. Veltri, S. Venditti, P. Vicini, R. Volpe, M. Vormstein, H. Wahl, R. Wanke, P. Wertelaers, A. Winhart, R. Winston, B. Wrona, O. Yushchenko, M. Zamkovsky, A. Zinchenko.

1 The NA62 experiment

NA62 is a fixed target experiment using a secondary hadron beam from SPS accelerator to perform flavor physics studies mainly in the charged kaon sector. NA62 collected data in 2007 using the same beam line and detecting apparatus of the earlier NA48 experiment[1], to study lepton universality in kaon decays through the measurement of the ratio between K_{e2} and $K_{\mu 2}$ leptonic decays[2] $R_K = \Gamma(K^\pm \to e^\pm \nu)/\Gamma(K^\pm \to \mu^\pm \nu)$. After an R&D and construction phase started in 2008, the experimental apparatus has been replaced by a new one designed to measure precisely the branching ratio of the $K^+ \to \pi^+ \nu\bar{\nu}$ decay. First data with the new NA62 setup have been collected in 2014/2015, mainly to commission detectors and perform data quality studies.

2 Measurement of the π^0 transition form factor

The neutral pion is the lightest meson and plays an important role in the study of low-energy properties of the strong interaction. The differential decay width of the subleading Dalitz decay $\pi^0 \to \gamma e^+ e^-$ (π^0_D) normalized to the leading decay $\pi^0 \to \gamma\gamma$ ($\pi^0_{\gamma\gamma}$) as a function of the particle 4-momenta is given by:

$$\frac{1}{\Gamma(\pi^0_{\gamma\gamma})}\frac{\mathrm{d}^2\Gamma(\pi^0_D)}{\mathrm{d}x\,\mathrm{d}y} = \frac{\alpha}{4\pi}\frac{(1-x)^3}{x}\left(1 + y^2 + \frac{r^2}{x}\right)|\mathcal{F}(x)|^2\,[1 + \delta(x,y)]\,, \tag{1}$$

where $x = (P_{e^+} + P_{e^-})^2/m^2_{\pi^0}$, $y = 2P_{\pi^0}\cdot(P_{e^+} - P_{e^-})/[m^2_{\pi^0}(1-x)]$, $r^2 = 4m_e^2/m^2_{\pi^0} \le x \le 1$, $\mathcal{F}(x)$ is the transition form factor (TFF) describing the hadronic physics in the $\gamma\gamma^*$ vertex and $\delta(x,y)$ contains the radiative corrections to the π^0 Dalitz decay. The TFF is usually parametrized as $\mathcal{F}(x) = 1 + ax$, where a is called the TFF slope parameter. The TFF enters in the predictions of observable quantities like the rate of $\pi^0 \to e^+e^-$ decay and the anomalous magnetic moment of the muon. It has been extensively studied theoretically[3–8] and the slope parameter has been measured experimentally in the time-like[9–12] as well as in the space-like domain[13–16].

The charged kaons decaying in flight in the NA62 beam provide an abundant source of π^0s, mainly from the $K^\pm \to \pi^\pm\pi^0$ decay. NA62 collected about 2×10^{10} $K^\pm$ decays in 2007, allowing the extraction of the slope parameter a through the study of the x spectrum of the π^0_D decay, according to eq.1 integrated over y. The radiative correction $\delta(x,y)$ has been studied extensively[17–20] and the most up-to-date calculation[20] has been implemented by NA62 in the Monte Carlo simulation of the π^0_D decay.

2.1 The NA62 setup in 2007 data taking

The NA62 experiment detects in-flight decays of charged kaons at CERN SPS. The NA48 beam line and apparatus were used in NA62 2007 run, with different beam parameteres. The primary 400 GeV/c proton beam from CERN SPS impinged a beryllium target, producing charged particles, 6% of which were $K^\pm$. A 100 m long beam line selected, focused and transported two simultaneous oppositely charged secondary beams of (74.0 ± 1.4) GeV/c momentum down to a 100 m long evacuated region. Particles produced by kaon decays were detected by the experimental apparatus located downstream. Charged particles were traced by a magnetic spectrometer consisting of a dipole magnet and four drift chamber stations. An array of horizontal and vertical plastic scintillator slabs provided fast signals for triggering and sub-ns timestamping of charged particles. The energy and position of photons and electrons were precisely measured by a quasi-homogeneous liquid krypton electromagnetic calorimeter (LKr). A muon detector (MUV) composed of iron absorbers and three planes of scintillator counters was used to identify muons. An iron-scintillator hadron calorimeter and several photon veto counters completed the experimental apparatus, a detailed description of which can be found in Ref. [1]. In 2007 run the trigger was optimized to select electron events by measuring the energy deposit in LKr.

2.2 Data analysis and preliminary results

Events containing a $K^{\pm} \to \pi^{+}\pi^{0}$ decays followed by the $\pi^{0} \to e^{+}e^{-}\gamma$ decay are selected by performing a full kinematic reconstruction. Three tracks in the magnetic spectrometer are required, coming from a common vertex inside the fiducial decay region. The photon is identified by a single energy deposit in LKr, separated in space from the track impact points at the calorimeter surface. The total reconstructed momentum from photon and tracks has to be consistent with the nominal beam momentum magnitude and direction within the experimental resolution. Assuming the track with charge opposite to $K^{\pm}$ to be the $e^{\mp}$, events are selected if one and only one of the two possible $(e^{\pm}, \pi^{\pm})$ mass hypotheses for the two remaining tracks is compatible with the kinematics of the $K^{+} \to \pi^{0}_{D}$ decay within the resolution of the reconstructed π^{0} and kaon mass. In order to be consistent with the simulation of the trigger conditions, at least one of the $e^{\pm}$ is required to have a momentum larger than 5.5 GeV/c and to deposit more than 80% of its energy in the LKr; moreover, the total energy detected in the LKr must not exceed 14 GeV. Due to the acceptance not being well reproduced in the simulation for events with low x, the signal region is defined as $x > 0.01$, equivalent to $m_{ee} > 13.5\,\mathrm{MeV/c^2}$.

About 1.05×10^{6} π^{0}_{D} decays have been fully reconstructed. The TFF slope parameter a is extracted by fitting the x distribution of data and simulation, using equipopolous bins. The different hypotheses are tested by reweighting the MC events simulated with a known slope $a_{\mathrm{sim}} = 0.032$. The main contributions to the systematic uncertainty come from the simulation of the beam spectrum and the calibration of the spectrometer global momentum scale. The NA62 preliminary measurement of the π^{0} TFF slope parameter is:

$$a = (3.70 \pm 0.53_{\mathrm{stat}} \pm 0.36_{\mathrm{syst}}) \times 10^{-2}$$

This measurement represents the first $5.8\,\sigma$ observation of a positive π^{0} electromagnetic TFF slope in the time-like region of the momentum transfer. The best fit result is shown in figure 1 together with the comparison with previous measurements from π^{0} Dalitz decays. The NA62 result improves the precision of previous measurement by an order of magnitude, and is consistent with theoretical predictions.

3 Measurement of the $K^{+} \to \pi^{+}\nu\bar{\nu}$ branching ratio

The Flavour Changing Neutral Current $K \to \pi\nu\bar{\nu}$ decays proceed through box and electroweak penguin diagrams and are heavily suppressed in the Standard Model (SM). Using the values of the Cabibbo-Kobayashi-Maskawa (CKM) matrix elements as external inputs, the Standard Model predicts[21, 22]:

$$\mathrm{BR_{SM}}(K^{+} \to \pi^{+}\nu\bar{\nu}) = (8.4 \pm 1.0) \cdot 10^{-11} \tag{2}$$

$$\mathrm{BR_{SM}}(K_{L} \to \pi^{0}\nu\bar{\nu}) = (3.4 \pm 0.6) \cdot 10^{-11} \tag{3}$$

where the uncertainties are dominated by the experimental knowledge of the external inputs.

The $K \to \pi\nu\bar{\nu}$ decays are very sensitive to physics beyond the SM. New particles at mass scales up to 100 TeV are expected to produce significant variations of the $K \to \pi\nu\bar{\nu}$ branching ratios from SM predictions, which could be detected with a 10% precise measurement.

Only the charged mode has been observed so far. The present status of BR measurements is[23–25]:

$$\mathrm{BR}(K^{+} \to \pi^{+}\nu\bar{\nu}) = (17.3^{+11.5}_{-10.5}) \times 10^{-11} \tag{4}$$

$$\mathrm{BR}(K_{L} \to \pi^{0}\nu\bar{\nu}) < 2.6 \times 10^{-8} \quad (\,90\%\ \mathrm{C.L.}\,) \tag{5}$$

EPJ Web of Conferences **129**, 00046 (2016) DOI: 10.1051/epjconf/201612900046
QCD@Work 2016

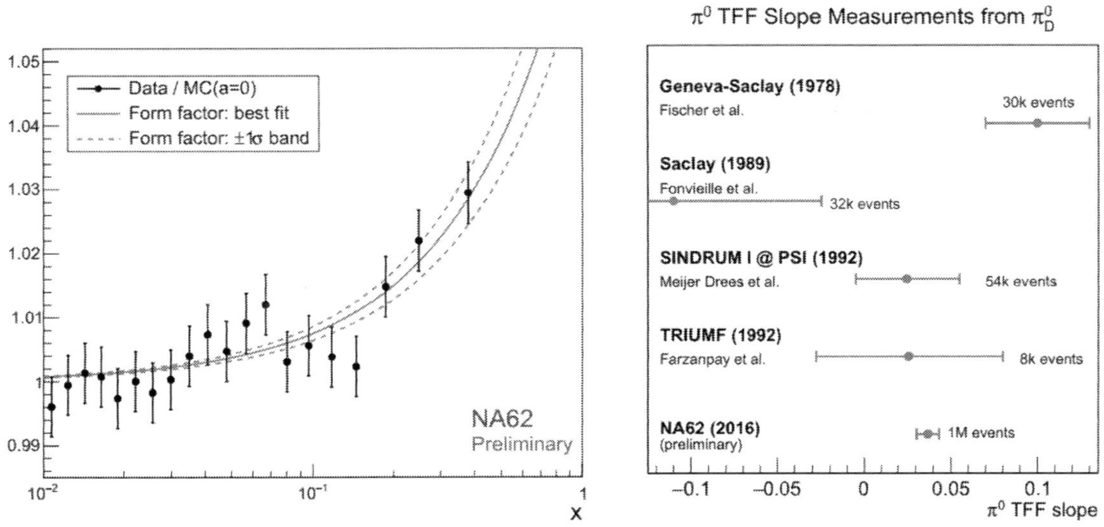

Figure 1. Measurement of the π^0 TFF slope parameter a. (left) Fit result showing the ratio data/MC as a function of the x variable, with the MC sample weighted to obtain a flat TFF (a = 0). The events are divided into 20 equipopulous bins. The solid line represents the squared TFF function $|\mathcal{F}(x)|^2$ with a value of the slope parameter a equal to the fit central value. Dashed lines indicate the 1σ band. (right) Comparison of this result with previous experiments measuring the TFF slope from π^0_D decays.

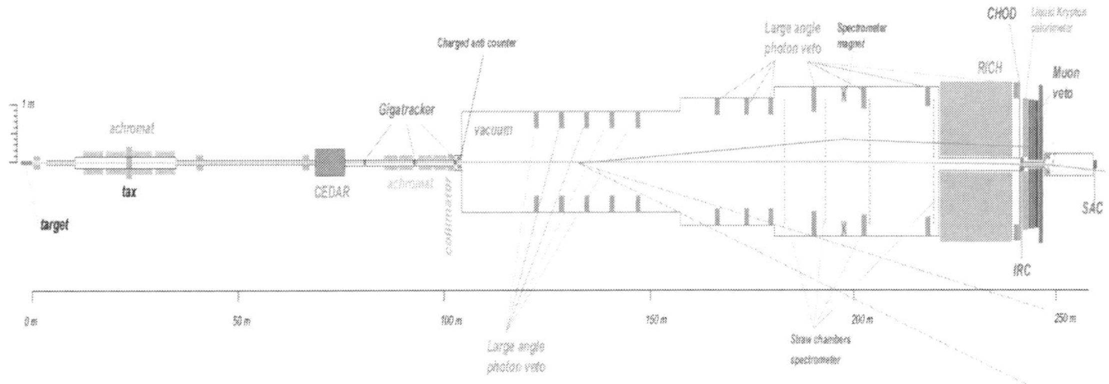

Figure 2. The NA62 layout for $K^+ \to \pi^+ \nu\bar{\nu}$ measurement.

3.1 The NA62 experimental apparatus for $K^+ \to \pi^+ \nu\bar{\nu}$

The main goal of the NA62 experiment at CERN SPS[26, 27] is to measure the $K^+ \to \pi^+ \nu\bar{\nu}$ branching ratio with a precision of 10% or better. The experiment plans to collect about 10^{13} kaon in-flight decays in a few years, with a 10% acceptance. The extremely small branching ratio of the BR($K^+ \to \pi^+ \nu\bar{\nu}$) decay combined with the weak signature of the signal events requires a very challenging $O(10^{12})$ suppression of the main K^+ decay modes. Several independent experimental techniques are required to reach this level of background suppression.

The NA62 apparatus for the $K^+ \to \pi^+\nu\bar{\nu}$ measurement, developed and built between 2008 and 2014 and replacing the NA48 one, is sketched in figure 2. The primary 400 GeV/c proton beam from CERN SPS impinges on a Be target producing secondary hadrons. A 100 m long beam line selects, collimates, focuses and transports positively charged particles of (75 ± 0.8) GeV/c momentum down to a 100 m long evacuated decay region. Kaons provide about 6% of the 750 MHz particle rate in the charged beam. They are identified and timestamped by a nitrogen-filled Čerenkov counter (KTAG) located on the beam line. A beam spectrometer with three silicon pixel stations (GTK) of 6×3 cm^2 surface traces and timestamps all the beam particles upstream of the evacuated decay region. A guard ring detector (CHANTI) at the entrance of the decay volume tags upstream hadronic interactions. About 10% of beam kaons decay in the vacuum region between the beam line and the downstream detectors. Large angle annular lead glass electromagnetic calorimeters (LAV) surround the decay and downstream volumes to detect γ up to 50 mrad. Charged particles are traced by a magnetic spectrometer with four straw tubes stations in vacuum, located at the downstream end of the decay volume. Holes around the beam axis in the spectrometer chambers allow undecayed particles to pass through without interacting. The large vacuum volume ends at the last spectrometer station and downstream the beam passes in vacuum inside a beam pipe.

A 17 m long neon-filled RICH counter provides particle identification for π^+, μ^+ and e^+ up to 40 GeV/c. The time of charged particles is measured both with the RICH and with two scintillator hodoscope planes (CHOD). The NA48 liquid kripton eletromagnetic calorimeter (LKr) detects forward γ and contributes to charged particle identification. A shashlik small-angle annular calorimeter (IRC) in front of LKr efficiently detects γ which would otherwise interact with the LKr inner edge outside the LKr acceptance. An hadronic calorimeter made by two modules of iron-scintillator sandwiches (MUV1 and MUV2) provides further π^+/μ^+ separation and hadronic energy based triggers. A fast scintillator array (MUV3) downstream of the calorimeters identifies and triggers muons with sub-nanosecond time resolution. In order to detect γ down to zero angle a shashlik calorimeter (SAC) is located further downstream on the beam line after a dipole magnet deflecting undecayed charged particles outside its acceptance. All the downstream detectors see less than 1% of the beam rate.

A multi-level trigger architecture is used. The level zero trigger is based on timing information from CHOD, RICH and MUV3 and on calorimetric variables from electromagnetic and hadronic calorimeters, determined using FPGA mounted on the TEL62 readout boards [28]. Data from KTAG, LAV and magnetic spectrometer are used in the software-based higher level trigger.

NA62 collected data in 2014 and 2015. The hardware and readout of the detectors have been commissioned at 10% of the nominal beam intensity, the beam line up to the nominal intensity. The GTK ran with a partial hardware configuration. Data samples with beam intensity varying from percent of the nominal up to the nominal one have been recorded in 2015. Data at low intensity have been triggered with CHOD only. At higher intensities $K^+ \to \pi^+\nu\bar{\nu}$-like events have been triggered at level zero by calorimetric information. A preliminary analysis of a subset of the low intensity sample, aiming to exploit the data quality in view of the $K^+ \to \pi^+\nu\bar{\nu}$ measurement, is described in the following sections. The analysis of the full 2015 data set is in progress.

3.2 Principles of the measurement

The $K^+ \to \pi^+\nu\bar{\nu}$ signature in NA62 consists of a track in the GTK compatible with a kaon hypothesis, a charged particle in the detectors downstream and nothing else. Other (more abundant) kaon decays and beam related activity are the main sources of background.

For each event we define the squared missing mass as $m_{\text{miss}} = (P_K - P_\pi)^2$, where P_K is the 4-momentum of the beam kaon and P_π that of the downstream track under pion mass hypothesis. As can be seen in figure 3, the expected squared missing mass distribution of the signal has a broad shape

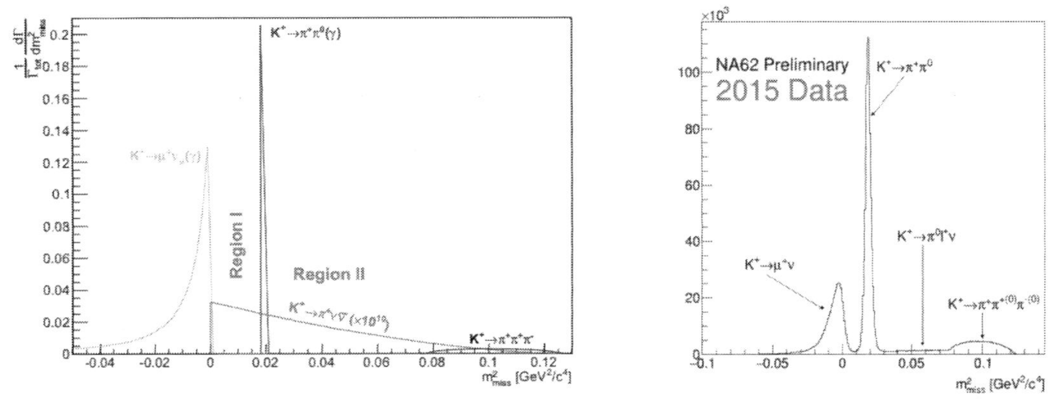

Figure 3. (left) Squared missing mass distribution expected for $K^+ \to \pi^+ \nu\bar{\nu}$ signal (multiplied by 10^{10}) and for main kaon decays $K^+ \to \mu^+ \nu_\mu$ ($K_{\mu 2}$), $K^+ \to \pi^+ \pi^0$ ($K_{2\pi}$) and $K^+ \to \pi^+ \pi^+ \pi^-$($K_{3\pi}$). (right) Squared missing mass distribution measured on single-track selected events collected in 2015.

typical of three-body decays, while more than 90% of the charged kaon decays are peaking. Two signal regions are defined outside the $K^+ \to \pi^+ \pi^0$ and $K^+ \to \mu^+ \nu_\mu$ peaks. Semileptonic decays, radiative processes, beam induced tracks and reconstruction tails of main K^+ decays are expected sources of background in these regions. Precise kinematic reconstruction, efficient photon detection, particle identification and sub-nanosecond timing coincidences between subdetectors must be employed to reach the required background rejection.

3.3 Data quality analysis for the $K^+ \to \pi^+ \nu\bar{\nu}$ measurement

The quality of data for $K^+ \to \pi^+ \nu\bar{\nu}$ measurement is studied after applying the following selection. Tracks reconstructed in the straw spectrometer are selected, matching in space CHOD signals and energy depositions in calorimeters. The event is required to contain a single track, i.e. a track not forming a vertex with any other in-time track in the decay region between the last GTK station and the first plane of the straw spectrometer. A vertex is defined as the average position of two tracks projected back in the decay region at their closest approach, provided their distance (CDA) is less than 1.5 cm. In order to select a single track originating from a kaon decay, a GTK track is required to match the downstream track both in time and space, forming a vertex in the decay region with it, and also to be in time with a kaon-like signal in KTAG. The time resolution of RICH, KTAG, GTK track, and CHOD have been measured in the range between 70 and 200 ps, matching the design values. Figure 4(left) shows the distribution of events in the track momentum-squared missing mass plane for 2015 data recorded at low intensity. Regions populated by main K^+ decays are clearly visible. The same distribution for events not related to kaons, selected with KTAG in anti-coincidence, is plotted in figure 4(right). The analysis of these events show that beam π^+ decays, elastic scattering of beam particles in the material along the beam line and inelastic scatterings in the last GTK station are the main sources of downstream tracks from beam-related activity. Kinematic resolution, particle identification and γ rejection are studied using the sample of single track events from kaon decays selected above.

The resolution on m^2_{miss} measured from the width of the $K^+ \to \pi^+\pi^0$ peak is about 1.2×10^{-3} GeV^2/c^4, close to the 10^{-3} GeV^2/c^4 design value. Taking the average beam momentum in

DOI: 10.1051/epjconf/201612900046

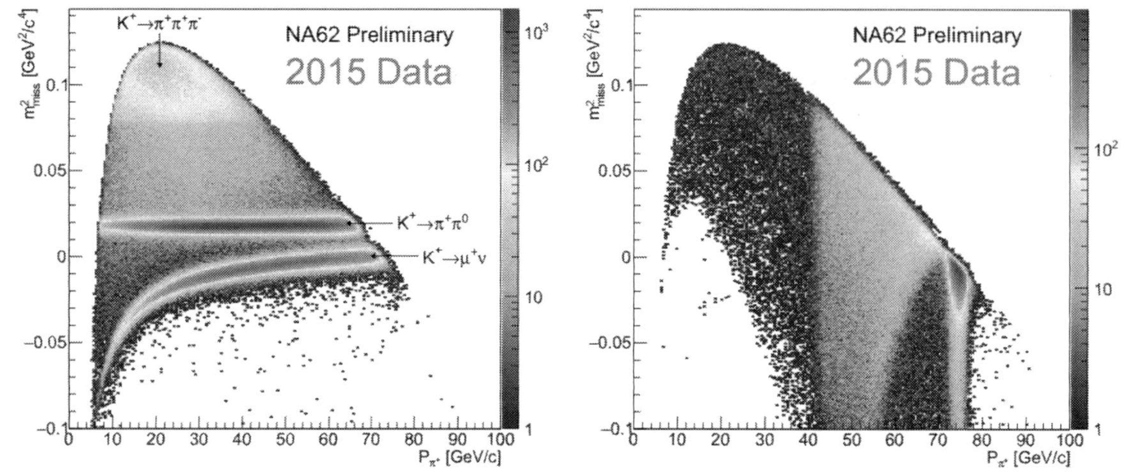

Figure 4. Squared missing mass ditribution vs. pion momentum, for: kaon-related events selected with KTAG in coincidence (left), non-kaon-related events with KTAG in anti-coincidence (right).

place of the event-by-event kaon momentum measured with the Gigatracker increases the resolution by a factor 3. The m^2_{miss} variable obtained from NA62 spectrometers is used to reject $K^+ \to \pi^+\pi^0$ and $K^+ \to \mu^+\nu$ events. The measured suppression for $K^+ \to \pi^+\pi^0$ is of the order of 10^3, still below the design value ($10^4 - 10^5$). This is due to m^2_{miss} tails caused by beam track mis-reconstruction in the still partially equipped GTK.

The NA62 particle identification, based on a combined usage of RICH and calorimeters, is designed to separate π^+ from μ^+ and e^+ and reach a 10^7 suppression of $K^+ \to \mu^+\nu$ in addition to the kinematic rejection. Pure samples of $K^+ \to \pi^+\pi^0$ and $K^+ \to \mu^+\nu$ events are used to study the π^+/μ^+ separation in RICH and calorimeters. A 10^2 muon suppression with 80% pion efficiency is obtained by the RICH in the required track momentum region ($15 \div 35$) GeV/c, as shown in figure 5. A simple cut-based analysis of calorimeters data provide an additional muon suppression factor between 10^4 and 10^6 for a pion efficiency ranging between 90% and 50%. More efficient analysis techniques are under study.

The NA62 layout is designed to provide a 10^8 suppression for $K^+ \to \pi^+\pi^0$ events in addition to the kinematic rejection by detecting at least one photon from π^0 decay in one of the electromagnetic calorimeters LAV, LKr, IRC and SAC, covering an overall angular region between 0 and 50 mrad. The suppression of π^0s from $K^+ \to \pi^+\pi^0$ decays is measured on kinematically selected $K^+ \to \pi^+\pi^0$ events. The measurement of π^0 suppression is statistically limited at a lower limit of $\approx 10^6$.

Commissioning of NA62 for $K^+ \to \pi^+\nu\bar{\nu}$ is almost completed. The preliminary data quality analysis of low intensity 2015 data shows that NA62 is approaching the $K^+ \to \pi^+\nu\bar{\nu}$ design sensitivity.

References

[1] V. Fanti et al. (NA48 Collaboration), Nucl. Instrum. Methods A **574**, 433 (2007).

[2] C. Lazzeroni et al. (NA62 Collaboration), Phys. Lett. B **719**, 326 (2013).

[3] M. Gell-Mann and F. Zachariasen, Phys. Rev. **124**, 953 (1961).

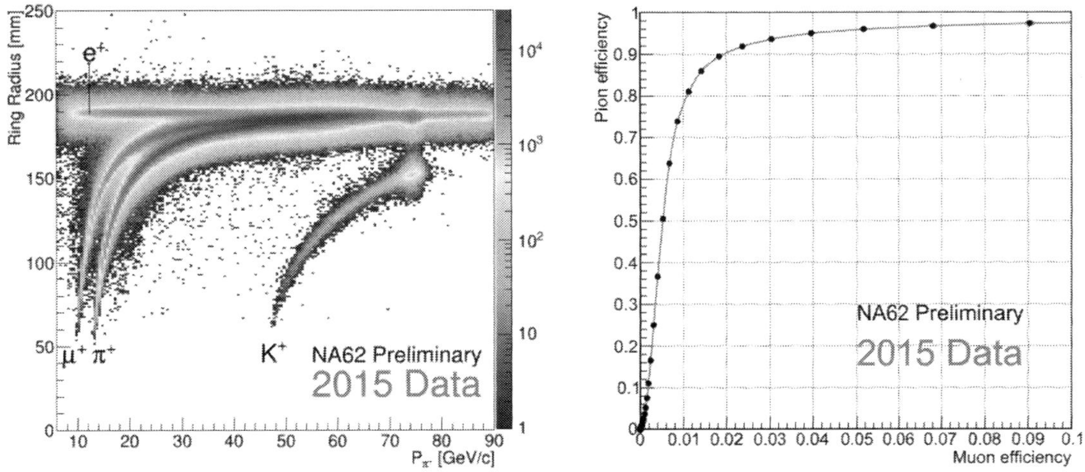

Figure 5. Particle identification with NA62 RICH: (left) RICH ring radius vs. momentum; (right) Pion efficiency vs. muon suppression (efficiency), for particle momentum between 15 and 35 GeV/c.

[4] P. Lichard, Phys. Rev. D **83**, 037503 (2011).
[5] K. Kampf et al., Eur. Phys. J. C **46**, 191 (2006).
[6] T. Husek and S. Leupold, Eur. Phys. J. C **75**, 586 (2015).
[7] P. Masjuan, Phys. Rev. D **86**, 094021 (2012).
[8] M. Hoferichter et al., Eur. Pjys. J. C **74**, 3180 (2014).
[9] J. Fisher et al.(Geneva-Saclay Collaboration), Phys. Lett. B **73**, 359 (1978).
[10] H. Fonvieille et al., Phys. Lett. B **233**, 65 (1989).
[11] H. Forzanpay et al., Phys. Lett. B **278**, 413 (1992).
[12] R. Meijer Drees et al.(SINDRUM Collaboration), Phys. Rev. D **45**, 1439 (1992).
[13] H.J. Berend et al.(CELLO Collaboration), Zeit. Phys. C **49**, 401 (1991).
[14] et al.(CLEO Collaboration), Phys. Rev. D **57**, 33 (1998).
[15] et al.(BaBar Collaboration), Phys. Rev. D **80**, 052002 (2009).
[16] et al.(Belle Collaboration), Phys. Rev. D **86**, 092007 (2012).
[17] D.W. Joseh, Il Nuovo Cimento **16**, 997 (1960).
[18] B.E. Lautrup and J. Smith, Phys. Rev. D **3**, 1122 (1971).
[19] M. Mikaelian and J. Smith, Phys. Rev. D **5**, 1763 (1972).
[20] T. Husek et al., Phys. Rev. D **92**, 054027 (2015).
[21] J. Brod et al., Phys. Rev. D **83**, 034030 (2011).
[22] A.J. Buras et al., JHEP **11**, 033 (2015).
[23] S. Adler et al. (E787 Collaborations), Phys. Rev. D **77**, 052003 (2008).
[24] A.V Artamonov et al. (E949 Collaborations), Phys. Rev. D **79**, 092004 (2009).
[25] J.K. Ahn et al. (E381a Collaboration), Phys. Rev. D **81**, 072004 (2010).
[26] G. Anelli et al., CERN-SPSC-2005-013; SPSC-P-326.
[27] NA62 Technical Design Document, NA62-10-07.
[28] B. Angelucci et al., JINST **7**, C02046 (2012).

EPJ Web of Conferences **129**, 00047 (2016) DOI: 10.1051/epjconf/201612900047
QCD@Work 2016

Janus-Facedness of the Pion: Analytic Instantaneous Bethe–Salpeter Models

Wolfgang Lucha[1,⋆]

[1] *Institute for High Energy Physics, Austrian Academy of Sciences, Nikolsdorfergasse 18, A-1050 Vienna, Austria*

Abstract. Inversion enables the construction of interaction potentials underlying — under fortunate circumstances even analytic — instantaneous Bethe–Salpeter descriptions of all lightest pseudoscalar mesons as quark–antiquark bound states of Goldstone-boson nature.

1 Introduction: quark–antiquark bound states of Goldstone-boson identity

Within quantum chromodynamics, the pions or, as a matter of fact, all *light pseudoscalar mesons* must be interpretable as both *quark–antiquark bound states* and almost massless (pseudo) *Goldstone bosons* related to the spontaneously (and, to a minor extent, also explicitly) broken chiral symmetries of QCD.

Relativistic quantum field theory describes bound states by their Bethe–Salpeter amplitudes, $\Phi(p)$, controlled by the *homogeneous Bethe–Salpeter equation* defined (for two bound particles of individual and relative momenta $p_{1,2}$ and p) by their full propagators $S_{1,2}(p_{1,2})$ and the integral kernel $K(p,q)$ that encompasses their interactions (notationally suppressing dependences on the total momentum p_1+p_2):

$$\Phi(p) = \frac{\mathrm{i}}{(2\pi)^4}\, S_1(p_1) \int \mathrm{d}^4q\, K(p,q)\, \Phi(q)\, S_2(-p_2)\,.$$

The application of suitably adapted *inversion* techniques [1] allows us to retrieve all the underlying interactions — rooted, of course, in QCD — analytically in the form of a (configuration-space) *central potential* $V(r)$, $r \equiv |\boldsymbol{x}|$, from presumed solutions to the Bethe–Salpeter equation [2]. By that, we are put in a position to construct *exact analytic* Bethe–Salpeter *solutions* for all massless pseudoscalar mesons [3] in the sense of establishing in a rigorous manner the analytic relationships between interactions and resulting solutions: all analytic findings [4] can be confronted with associated numerical outcomes [5].

2 Sequence of simplifying assumptions crucial for the inversion formalism

By a few steps, we cast the Bethe–Salpeter equation into a shape that allows us to talk about potentials.

1. Assuming, for each involved quark, both *instantaneous* interactions and *free* propagation, with a mass dubbed as *constituent*, simplifies the Bethe–Salpeter equation to a bound-state equation for the *Salpeter amplitude* $\phi(\boldsymbol{p})$, obtained from the Bethe–Salpeter amplitude by integration over p_0:

$$\phi(\boldsymbol{p}) \propto \int \mathrm{d}p_0\, \Phi(p)\,.$$

⋆e-mail: Wolfgang.Lucha@oeaw.ac.at

Generically, for a spin-$\frac{1}{2}$ fermion and a spin-$\frac{1}{2}$ antifermion of equal constituent masses m, bound to a spin-singlet state (which, for instance, clearly is the case for any such pseudoscalar state), its three-dimensional wave function involves just two independent components, here called $\varphi_{1,2}(\boldsymbol{p})$:

$$\phi(\boldsymbol{p}) = \left[\varphi_1(\boldsymbol{p})\,\frac{\gamma_0\,(\boldsymbol{\gamma}\cdot\boldsymbol{p}+m)}{E(p)} + \varphi_2(\boldsymbol{p})\right]\gamma_5\,, \qquad E(p) \equiv \sqrt{\boldsymbol{p}^2+m^2}\,, \qquad p \equiv |\boldsymbol{p}|\,.$$

2. Upon supposing that the quark interactions in the kernel respect spherical and Fierz symmetries, our bound-state equation for $\phi(\boldsymbol{p})$ collapses to the system of *coupled radial eigenvalue equations*

$$2\,E(p)\,\varphi_2(p) + 2\int_0^\infty \frac{\mathrm{d}q\,q^2}{(2\pi)^2}\,V(p,q)\,\varphi_2(q) = \widehat{M}\,\varphi_1(p)\,, \qquad 2\,E(p)\,\varphi_1(p) = \widehat{M}\,\varphi_2(p)\,, \qquad q \equiv |\boldsymbol{q}|\,,$$

for the bound-state mass eigenvalue $\widehat{M}$ [6]. Therein, $V(r)$ enters via its Fourier–Bessel transform

$$V(p,q) \equiv \frac{8\pi}{p\,q}\int_0^\infty \mathrm{d}r\,\sin(p\,r)\,\sin(q\,r)\,V(r)\,.$$

3. In the strictly massless (Goldstone) case $\widehat{M}=0$, the system decouples: one Salpeter component, $\varphi_1(p)$, is doomed to vanish, $\varphi_1(p)\equiv 0$, whereas the surviving Salpeter component $\varphi_2(p)$ satisfies

$$E(p)\,\varphi_2(p) + \int_0^\infty \frac{\mathrm{d}q\,q^2}{(2\pi)^2}\,V(p,q)\,\varphi_2(q) = 0\,.$$

Denoting the Fourier–Bessel transform of the kinetic term $E(p)\,\varphi_2(p)$ by $T(r)$, the potential $V(r)$ may be simply read off from the configuration-space representation of this bound-state equation:

$$T(r) + V(r)\,\varphi_2(r) = 0 \qquad \Longrightarrow \qquad V(r) = -\frac{T(r)}{\varphi_2(r)}\,.$$

3 Constraints on lightest-pseudoscalar-meson Bethe–Salpeter amplitudes

Information on the input Salpeter component $\varphi_2(p)$ can be gained from the *full quark propagator* $S(p)$, which is determined by its mass function $M(p^2)$ and its wave-function renormalization function $Z(p^2)$:

$$S(p) = \frac{\mathrm{i}\,Z(p^2)}{\not p - M(p^2) + \mathrm{i}\,\varepsilon}\,, \qquad \not p \equiv p^\mu\,\gamma_\mu\,, \qquad \varepsilon \downarrow 0\,.$$

Studies of $S(p)$ within the Dyson–Schwinger framework, preferably done in Euclidean space signalled by underlined quantities, allow for pivotal insights. *In the chiral limit*, a Ward–Takahashi identity links [7] this quark propagator to the flavour-nonsinglet *pseudoscalar-meson Bethe–Salpeter amplitude* [3]:

$$\Phi(\underline{k}) \approx \frac{M(\underline{k}^2)}{\underline{k}^2 + M^2(\underline{k}^2)}\,\underline{\gamma}_5 + \text{subleading contributions}\,.$$

First, in order to devise *analytically* accessible scenarios, we exploit two crucial pieces of information:

1. *In the chiral limit*, phenomenologically sound Dyson–Schwinger studies [8] imply, for the quark mass function $M(\underline{k}^2)$, at large Euclidean momenta $\underline{k}^2$ a decrease essentially proportional to $1/\underline{k}^2$.
2. From axiomatic quantum field theory, we may deduce [9] that the presence of an *inflection point at finite space-like momenta* $\underline{k}^2 > 0$ in the quark mass function $M(\underline{k}^2)$ entails *colour confinement*.

Of course, any imposition of such kind of requirements on $M(\underline{k}^2)$ has to be *reflected by* $\Phi(\underline{k})$. An *ansatz* for $\Phi(\underline{k})$ compatible with both constraints, involving a mass parameter, μ, and a mixing parameter, η, is

$$\Phi(\underline{k}) = \left[\frac{1}{(\underline{k}^2+\mu^2)^2} + \frac{\eta\,\underline{k}^2}{(\underline{k}^2+\mu^2)^3}\right]\underline{\gamma}_5\,, \qquad \mu > 0\,, \qquad \eta \in \mathbb{R}\,.$$

An integration of this $\Phi(\underline{k})$ with respect to the time component of the Euclidean momentum $\underline{k}$ results in

$$\varphi_2(p) \propto \frac{1}{(p^2+\mu^2)^{3/2}} + \eta\,\frac{p^2+\mu^2/4}{(p^2+\mu^2)^{5/2}}\,, \qquad p \equiv |\mathbf{p}|\,,$$

in configuration space expressible in terms of modified Bessel functions of the second kind $K_\sigma(z)$ [10]:

$$\varphi_2(r) \propto 4\,(1+\eta)\,K_0(\mu\,r) - \eta\,\mu\,r\,K_1(\mu\,r)\,. \tag{1}$$

For η values satisfying $\eta < -1$ or $\eta > 0$, $\varphi_2(r)$ has one zero, which clearly induces a *singularity* in $V(r)$.

4 Analytic outcomes [3, 4] for interquark potentials exhibiting confinement

For a few particular values of the dimensionless ratio m/μ, the *analytic* expression of $V(r)$ can be found [3, 4]. (Throughout this section, any quantity has to be understood in units of the adequate power of μ.) As a consequence of our ansatz for $\Phi(\underline{k})$, giving rise to the particular form (1) of $\varphi_2(r)$, for $\eta \neq -1$ each extracted $V(r)$ will develop, at the spatial origin $r = 0$, a logarithmically softened Coulomb singularity:

$$V(r) \xrightarrow[r\to 0]{} \frac{\text{const}}{r\ln r} \xrightarrow[r\to 0]{} -\infty \qquad (\text{const} > 0) \qquad \text{for } \eta \neq -1\,.$$

4.1 Analytically manageable scenario of massless quarks, i.e., of constituent mass $m = 0$

For our choice of $\varphi_2(r)$, $V(r)$ involves both modified Bessel (I_n) and Struve ($\mathbf{L}_n$) functions [10] ($n \in \mathbb{N}$), and rises in a confinement-betraying manner to infinity either at the zero of $\varphi_2(r)$ or for $r \to \infty$ (Fig. 1):

$$V(r) = \frac{\pi\,[4+\eta\,(4+r^2)]\,[\mathbf{L}_0(r) - I_0(r)] + \pi\,(4+5\,\eta)\,r\,[\mathbf{L}_1(r) - I_1(r)] + 4\,(2+3\,\eta)\,r}{2\,r\,[4\,(1+\eta)\,K_0(r) - \eta\,r\,K_1(r)]}\,.$$

4.2 Analytically expressible observation for quarks with common constituent mass $m = \mu$

For $m = \mu$, the kinetic term $T(r)$ is a mixture of Yukawa and exponential behaviour, whence (cf. Fig. 2)

$$V(r) = -\frac{\pi\,[8+\eta\,(8-3\,r)]\exp(-r)}{4\,r\,[4\,(1+\eta)\,K_0(r) - \eta\,r\,K_1(r)]} \xrightarrow[r\to\infty]{} -\frac{\text{const}}{\sqrt{r}} \xrightarrow[r\to\infty]{} 0 \qquad (\text{const} > 0)\,.$$

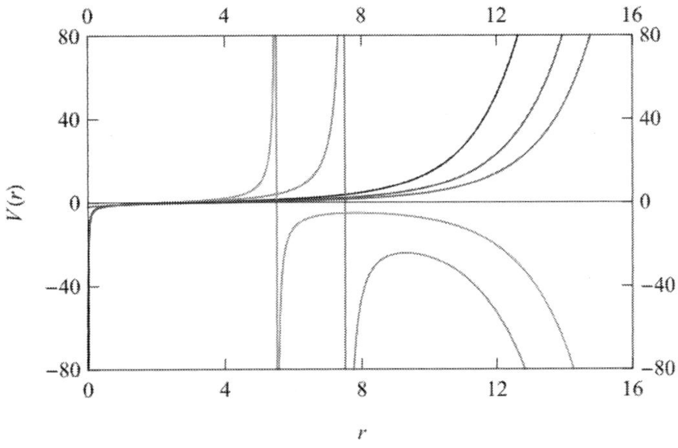

Figure 1. Configuration-space interquark potential $V(r)$ of the Fierz-symmetric kernel $K(p, q)$, for the constituent quark mass $m = 0$ and mixture $\eta = 0$ [3] (black), $\eta = 1$ (red), $\eta = 2$ (magenta), $\eta = -0.5$ (blue), or $\eta = -1$ (violet).

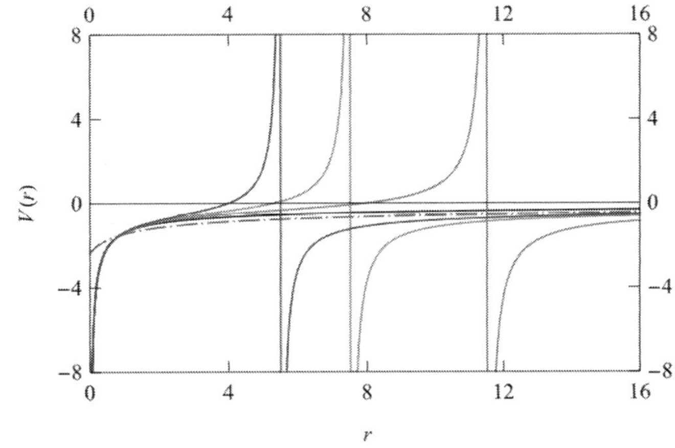

Figure 2. Configuration-space interquark potential $V(r)$ of the Fierz-symmetric kernel $K(p, q)$, for the constituent quark mass $m = 1$ and mixture $\eta = 0$ [3] (black), $\eta = 0.5$ (red), $\eta = 1$ (magenta), $\eta = 2$ (blue), and $\eta = -1$ (violet).

5 Reliability check of findings: numerical determination of the potential [5]

Our findings may be scrutinized by use of the chiral-limit quark mass function's pointwise form $M(\underline{k}^2)$, provided graphically in Ref. [8] and shown in Fig. 3 as $M(\underline{k})$ with $\underline{k} \equiv (\underline{k}^2)^{1/2}$, which we parametrize by

$$M(\underline{k}) = 0.708\ \text{GeV} \exp\left(-\frac{\underline{k}^2}{0.655\ \text{GeV}^2}\right) + \frac{0.0706\ \text{GeV}}{\left[1 + \left(\frac{\underline{k}^2}{0.487\ \text{GeV}^2}\right)^{1.48}\right]^{0.752}}\ .$$

Note that the *product* of the two exponents in the second term above yields $1.48 \times 0.752 \approx 1.1$, which is pretty close to unity, as demanded by the large-$\underline{k}$ constraint. Feeding this $M(\underline{k})$ parametrization into our inversion procedure, we obtain potentials that are finite at $r = 0$ and, for sufficiently small m, rise with r to infinity but, for large m, remain negative, as illustrated in Fig. 3 for selected constituent mass values.

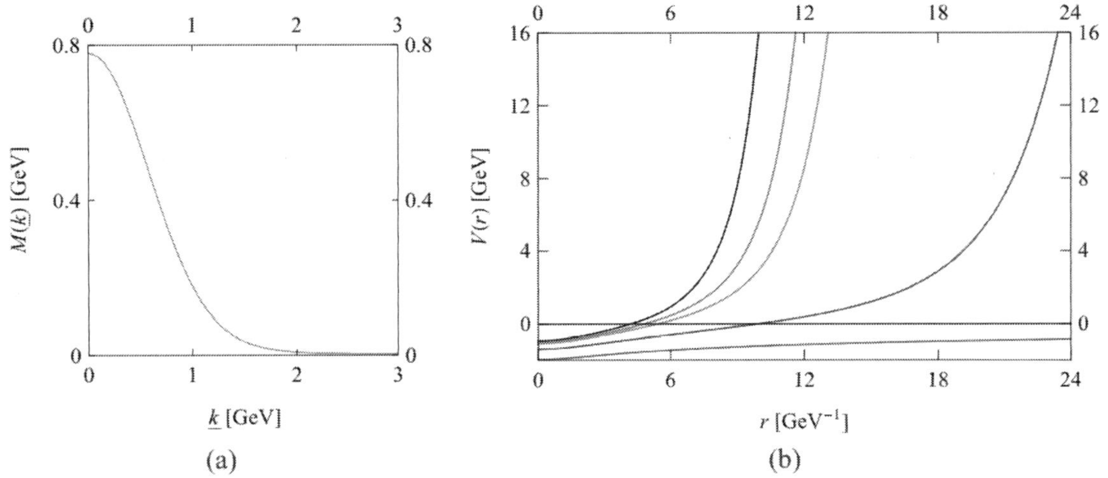

Figure 3. (a) Mass function $M(\underline{k})$ deduced from the Dyson–Schwinger model of Ref. [8] for the quark propagator. (b) Configuration-space interquark potential $V(r)$ numerically determined from $M(\underline{k}^2)$, for constituent quark mass $m = 0$ (black), $m = 0.35$ GeV (red), $m = 0.5$ GeV (magenta), $m = 1.0$ GeV (blue), and $m = 1.69$ GeV (violet) [5].

6 Summary of results, observations, discussion, conclusion, perspectives

We constructed confining potentials $V(r)$ that in cooperation with a Fierz-symmetric interaction kernel describe massless pseudoscalar quark–antiquark bound-state solutions of the Bethe–Salpeter equation. This is possible even analytically if focusing to specific aspects of the quark mass function's behaviour. Two obstacles call for a particularly careful treatment: Numerically, $M(p^2)$ is known for only a limited range of p^2. For large r, both $T(r)$ and $\varphi_2(r)$ approach zero; thus, pinning down $V(r)$ in the limit $r \to \infty$ boils down to a division of zero by zero. Dropping the free quark propagation constraint [11] allows us to thoroughly take into account the effects of $M(p^2)$ *and* the quark wave-function renormalization [12].

References

[1] W. Lucha and F. F. Schöberl, Phys. Rev. D **87** (2013) 016009, arXiv:1211.4716 [hep-ph].
[2] W. Lucha, Proc. Sci., EPS-HEP 2013 (2013) 007, arXiv:1308.3130 [hep-ph].
[3] W. Lucha and F. F. Schöberl, Phys. Rev. D **92** (2015) 076005, arXiv:1508.02951 [hep-ph].
[4] W. Lucha and F. F. Schöberl, Phys. Rev. D **93** (2016) 056006, arXiv:1602.02356 [hep-ph].
[5] W. Lucha and F. F. Schöberl, Phys. Rev. D **93** (2016) 096005, arXiv:1603.08745 [hep-ph].
[6] Z.-F. Li, W. Lucha, and F. F. Schöberl, Phys. Rev. D **76** (2007) 125028, arXiv:0707.3202 [hep-ph].
[7] P. Maris, C. D. Roberts, and P. C. Tandy, Phys. Lett. B **420** (1998) 267, arXiv:nucl-th/9707003.
[8] P. Maris and C. D. Roberts, Phys. Rev. C **56** (1997) 3369, arXiv:nucl-th/9708029.
[9] C. D. Roberts, Prog. Part. Nucl. Phys. **61** (2008) 50, arXiv:0712.0633 [nucl-th].
[10] *Handbook of Mathematical Functions*, edited by M. Abramowitz and I. A. Stegun (Dover, New York, 1964).
[11] W. Lucha and F. F. Schöberl, J. Phys. G **31** (2005) 1133, arXiv:hep-th/0507281.
[12] W. Lucha and F. F. Schöberl, preprint HEPHY-PUB 968/16 (2016), arXiv:1606.04781 [hep-ph].

EPJ Web of Conferences **129**, 00048 (2016) DOI: 10.1051/epjconf/201612900048
QCD@Work 2016

A new measurement of the leading hadronic corrections to the muon g-2*

Luca Trentadue[1,2,**]

[1] *Dipartimento di Fisica e Scienze della Terra "Macedonio Melloni"*
Universitá di Parma, Parma
[2] *INFN Sezione di Milano Bicocca, Milano, Italy*

Abstract. A novel approach to determine the leading hadronic corrections to the muon g-2 is proposed. It consists in a measurement of the effective electromagnetic coupling in the space-like region. This method may become feasible at flavor factories resulting in a determination potentially competitive with the dispersive approach via time-like data.

1 Introduction

The motivation of this work [1] is due to a long-standing discrepancy between experiment and the Standard Model (SM) prediction for a_μ, the muon anomalous magnetic moment. For this reason the hadronic corrections have been kept under close scrutiny [2–5]. The hadronic contribution represents the largest uncertainty of the SM value and is comparable with the experimental one. When the new results from the g-2 experiments at Fermilab and J-PARC will reach the unprecedented precision of 0.14 parts per million (or better) [6–8], the uncertainty of the hadronic corrections will become the main limitation of this formidable test of the SM.
Vacuum Polarization makes α_{em} running assuming a well defined effective value at any scale. Vacuum polarization and the effective charge are defined by:

$$e^2 \to e^2(q^2) = \frac{e^2}{1 + (\Pi(q^2) - \Pi(0))} \tag{1}$$

and

$$\alpha(q^2) = \frac{\alpha(0)}{1 - \Delta\alpha} \quad \Delta\alpha = -Re(\Pi(q^2) - \Pi(0)). \tag{2}$$

$\Delta\alpha$ takes contributions from leptonic and hadronic elementary states among these the non-perturbative $\Delta\alpha_{had}$

$$\Delta\alpha = \Delta\alpha_{lept} + \Delta\alpha_{had} + \Delta\alpha_{top} \tag{3}$$

$$\Delta\alpha^{(5)}_{had}(M_z^2) = -\frac{\alpha\, M_z^2}{3\pi} Re \int_{4m_\pi^2}^{\infty} ds \frac{R(s)}{s(s - M_z^2 - i\epsilon)} \tag{4}$$

*This talk is based on a work done in collaboration with C. Carloni Calame, M. Passera, and G. Venanzoni.
**e-mail: luca.trentadue@unipr.it

A direct measurement of the running of $\alpha_{em}(s/t)$ in space/time-like regions can show the running of α_{em}. It can provide a test of duality far away form resonances and ita has been done in the past by few experiments at e^+e^- colliders by comparing well known QED processes with some reference (either data or MonteCarlo one) normalization:

$$(\frac{\alpha(q^2)}{\alpha(q_0^2)})^2 \simeq \frac{N_{signal}(q^2)}{N_{norm}(q_0^2)}. \quad (5)$$

N_{signal} can be any QED process as muon pairs production or others similar ones and N_{norm} can be the Bhabha process, a pure QED channel as $\gamma\gamma$ pair production as well as theory, or any other reference process. q_0 represents a scale chosen as starting point for the running.

2 a_{mu}^{HLO} calculation

The leading-order hadronic contribution to the muon g-2 is given by the well-known formula [5, 12]

$$a_\mu^{\rm HLO} = \frac{\alpha}{\pi^2} \int_0^\infty \frac{ds}{s}\, K(s)\ {\rm Im}\Pi_{\rm had}(s + i\epsilon), \quad (6)$$

where $\Pi_{\rm had}(s)$ is the hadronic part of the photon vacuum polarization, $\epsilon > 0$,

$$K(s) = \int_0^1 dx\, \frac{x^2(1-x)}{x^2 + (1-x)(s/m_\mu^2)} \quad (7)$$

is a positive kernel function, and m_μ is the muon mass. As the total cross section for hadron production in low-energy e^+e^- annihilations is related to the imaginary part of $\Pi_{\rm had}(s)$ via the optical theorem, the dispersion integral in eq. (6) is computed integrating experimental time-like ($s > 0$) data up to a certain value of s [3, 16, 17]. The high-energy tail of the integral is calculated using perturbative QCD [18].

Alternatively, if we exchange the x and s integrations in eq. (6) we obtain [19]

$$a_\mu^{\rm HLO} = \frac{\alpha}{\pi} \int_0^1 dx\,(x-1)\,\overline{\Pi}_{\rm had}[t(x)]\,, \quad (8)$$

where $\overline{\Pi}_{\rm had}(t) = \Pi_{\rm had}(t) - \Pi_{\rm had}(0)$ and

$$t(x) = \frac{x^2 m_\mu^2}{x-1} < 0 \quad (9)$$

is a space-like squared four-momentum. If we invert eq. (9), we get $x = (1-\beta)\,(t/2m_\mu^2)$, with $\beta = (1 - 4m_\mu^2/t)^{1/2}$, and from eq. (8) we obtain

$$a_\mu^{\rm HLO} = \frac{\alpha}{\pi} \int_{-\infty}^0 \overline{\Pi}_{\rm had}(t) \left(\frac{\beta-1}{\beta+1}\right)^2 \frac{dt}{t\beta}. \quad (10)$$

Equation (10) has been used for lattice QCD calculations of $a_\mu^{\rm HLO}$ [20]; while the results are not yet competitive with those obtained with the dispersive approach via time-like data, their errors are expected to decrease significantly in the next few years [21].

DOI: 10.1051/epjconf/201612900048

The effective fine-structure constant at squared momentum transfer q^2 is defined above and is $\Delta\alpha(q^2) = -\mathrm{Re}\overline{\Pi}(q^2)$. The purely leptonic part, $\Delta\alpha_{\mathrm{lep}}(q^2)$, can be calculated order-by-order in perturbation theory – it is known up to three loops in QED [22] (and up to four loops in specific q^2 limits [23]). As $\mathrm{Im}\overline{\Pi}(q^2) = 0$ for negative q^2, eq. (8) can be rewritten in the form [24]

$$a_\mu^{\mathrm{HLO}} = \frac{\alpha}{\pi} \int_0^1 dx\,(1-x)\,\Delta\alpha_{\mathrm{had}}[t(x)]\,. \tag{11}$$

Equation (11), involving the hadronic contribution to the running of the effective fine-structure constant at space-like momenta, can be further formulated in terms of the Adler function [25], defined as the logarithmic derivative of the vacuum polarization, which, in turn, can be calculated via a dispersion relation with time-like hadroproduction data and perturbative QCD [24, 26]. We will proceed differently, proposing to calculate eq. (11) by measurements of the effective electromagnetic coupling in the space-like region. The hadronic contribution to the running of α in the space-like region, $\Delta\alpha_{\mathrm{had}}(t)$ (see eq. (3)), can be extracted comparing Bhabha scattering data to Monte Carlo (MC) predictions. The LO Bhabha cross section receives contributions from t- and s-channel photon exchange amplitudes. At NLO in QED, it is customary to distinguish corrections with an additional virtual photon or the emission of a real photon (photonic NLO) from those originated by the insertion of the vacuum polarization corrections into the LO photon propagator (VP). let us consider a few simple points. In fig. 1 (left) we plot the integrand $(1-x)\Delta\alpha_{\mathrm{had}}[t(x)]$ of eq. (11) using the output of the routine `hadr5n12` [29] (which uses time-like hadroproduction data and perturbative QCD). The range $x \in (0,1)$ corresponds to $t \in (-\infty, 0)$, with $x = 0$ for $t = 0$. The peak of the integrand occurs at $x_{\mathrm{peak}} \simeq 0.914$ where $t_{\mathrm{peak}} \simeq -0.108$ GeV^2 and $\Delta\alpha_{\mathrm{had}}(t_{\mathrm{peak}}) \simeq 7.86 \times 10^{-4}$ (see fig. 1 (right)). Such relatively low t values

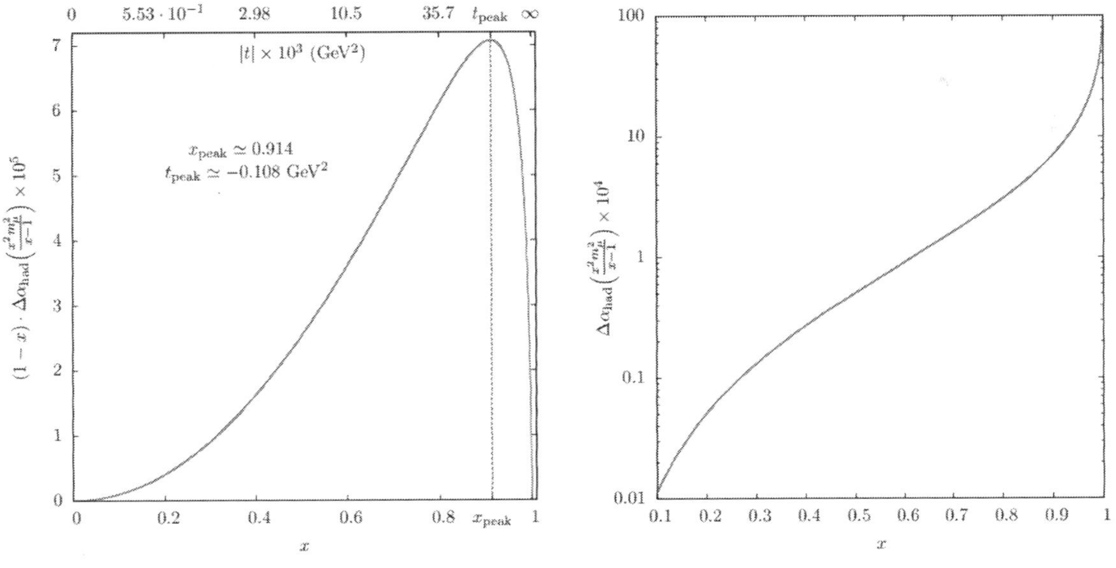

Figure 1. Left: The integrand $(1-x)\Delta\alpha_{\mathrm{had}}[t(x)] \times 10^5$ as a function of x and t. Right: $\Delta\alpha_{\mathrm{had}}[t(x)] \times 10^4$.

can be explored at e^+e^- colliders with center-of-mass energy $\sqrt{s}$ around or below 10 GeV where

$$t = -\frac{s}{2}\,(1 - cos\theta)\left(1 - \frac{4m_e^2}{s}\right), \tag{12}$$

θ is the electron scattering angle and m_e is the electron mass. Depending on s and θ, the integrand of eq. (11) can be measured in the range $x \in [x_{\min}, x_{\max}]$, as shown in fig. 2 (left). Note that to span low x intervals, larger θ ranges are needed as the collider energy decreases. In this respect, $\sqrt{s} \sim 3$ GeV appears to be very convenient, as an x interval $[0.30, 0.98]$ can be measured varying θ between $\sim 2°$ and $28°$. It is also worth remarking that data collected at flavor factories, such as DAΦNE (Frascati), VEPP-2000 (Novosibirsk), BEPC-II (Beijing), PEP-II (SLAC) and SuperKEKB (Tsukuba), and possibly at a future high-energy e^+e^- collider, like FCC-*ee* (TLEP) [30] or ILC [31], can help to cover different and complementary x regions.

Furthermore, given the smoothness of the integrand, values outside the measured x interval may be interpolated with some theoretical input. In particular, the region below $x_{\min}$ will provide a relatively small contribution to a_μ^{HLO}, while the region above $x_{\max}$ may be obtained by extrapolating the curve from $x_{\max}$ to $x = 1$, where the integrand is null, or using perturbative QCD.

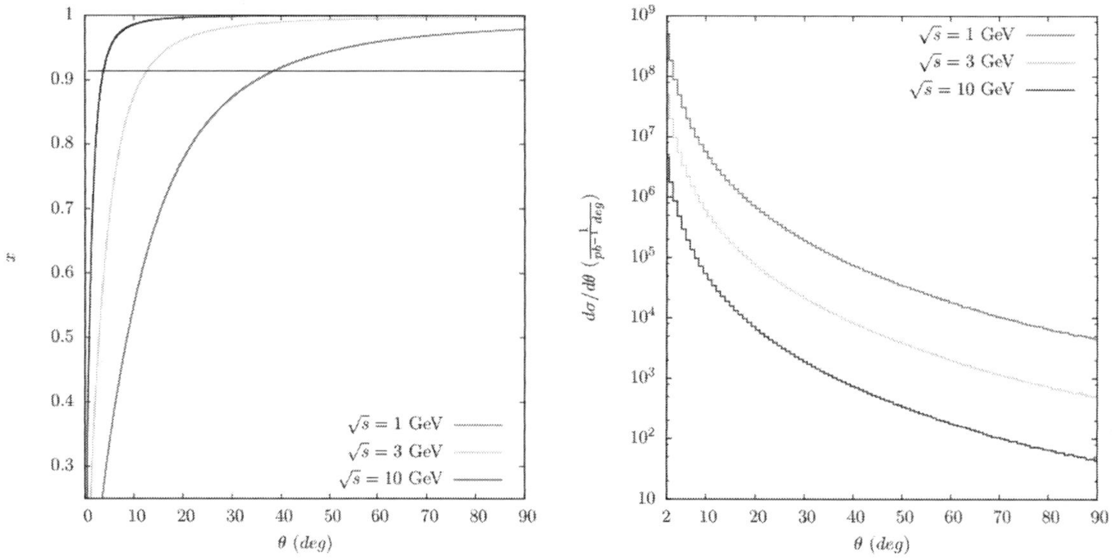

Figure 2. Left: Ranges of x values as a function of the electron scattering angle θ for three different center-of-mass energies. The horizontal line corresponds to $x = x_{\mathrm{peak}} \simeq 0.914$. Right: Bhabha differential cross section obtained with BabaYaga [27] as a function of θ for the same three values of $\sqrt{s}$ in the angular range $2° < \theta < 90°$.

The analytic dependence of the MC Bhabha predictions on $\alpha(t)$ (and, in turn, on $\Delta\alpha_{\mathrm{had}}(t)$) is not trivial, and a numerical procedure has to be devised to extract it from the data.This was not the case for example in [13, 14]: there $\alpha(t)$ was extracted from Bhabha data in the very forward region at LEP, where the t channel diagrams are by far dominant and $\alpha(t)$ factorizes (see for example [15]). In formulae, we have to find a function $\alpha(t)$ such that

$$\left.\frac{d\sigma}{dt}\right|_{\mathrm{data}} = \left.\frac{d\sigma}{dt}\big(\alpha(t), \alpha(s)\big)\right|_{\mathrm{MC}}, \tag{13}$$

where we explicitly kept apart the dependence on the time-like VP $\alpha(s)$ because we are only interested in $\alpha(t)$. We emphasise that, in our analysis, $\alpha(s)$ is an input parameter. Being the Bhabha cross section in the forward region dominated by the t-channel exchange diagrams , we checked that the present $\alpha(s)$ uncertainty induces in this region a relative error on the θ distribution of less than $\sim 10^{-4}$ (which is part of the systematic error).

We propose to perform the numerical extraction of $\Delta\alpha_{\rm had}(t)$ from the Bhabha distribution of the t Mandelstam variable. The idea is to let $\alpha(t)$ vary in the MC sample around a reference value and choose, bin by bin in the t distribution, the value that minimizes the difference with data. The procedure can be sketched as follows:

1. choose a reference function returning the value of $\Delta\alpha_{\rm had}(t)$ (and hence $\alpha(t)$) to be used in the MC sample, we call it $\bar{\alpha}(t)$;

2. for each generated event, calculate N MC weights by rescaling $\bar{\alpha}(t) \to \bar{\alpha}(t) + \frac{i}{N}\delta(t)$, where $i \in [-N, N]$ and $\delta(t)$ is for example the error induced on $\bar{\alpha}(t)$ by the error on $\Delta\alpha_{\rm had}(t)$. Being done on an event by event basis, the full dependence on $\alpha(t)$ of the MC differential cross section can be kept;

3. for each bin j of the t distribution, compare the experimental differential cross section with the MC predictions and choose the i_j which minimizes the difference;

4. $\bar{\alpha}(t_j) + \frac{i_j}{N}\delta(t_j)$ will be the extracted value of $\alpha(t_j)$ from data in the jth bin. $\Delta\alpha_{\rm had}(t_j)$ can then be obtained through the relation between $\alpha(t)$ and $\Delta\alpha_{\rm had}(t)$.

We finally find, for each bin j of the t distribution,

$$\left.\frac{d\sigma}{dt}\right|_{j,\rm data} = \left.\frac{d\sigma}{dt}\left(\bar{\alpha}(t) + \frac{i_j}{N}\delta(t), \alpha(s)\right)\right|_{j,\rm MC}. \tag{14}$$

We remark that the algorithm does not assume any simple dependence of the cross section on $\alpha(t)$, which can in fact be general, mixing s, t channels and higher order radiative corrections, relevant (or not) in different t domains.

In order to test our procedure, we perform a pseudo-experiment: we generate pseudo-data using the parameterization $\Delta\alpha^{I}_{\rm had}(t)$ of refs. [17, 32] and check if we can recover it by inserting in the MC the (independent) parameterization $\Delta\alpha^{II}_{\rm had}(t)$ (corresponding to $\bar{\alpha}(t)$ of eq. 14) of ref. [29] by means of the method described above. For this exercise, we use the generator BabaYaga in its most complete setup, generating events at $\sqrt{s} = 1.02$ GeV, requiring $10^\circ < \theta_\pm < 170^\circ$, $E_\pm > 0.4$ GeV and an acollinearity cut of 15°. We choose $\delta(t)$ to be the error induced on $\alpha(t)$ by the 1-σ error on $\Delta\alpha_{\rm had}(t)$, which is returned by the routine of ref. [29], we set $N = 150$, and we produce distributions with 200 bins. We note that in the present exercise $\alpha(s)$ and all the radiative corrections both in the pseudo-data and in the MC samples are exactly the same, because we are interested in testing the algorithm rather than assessing the achievable accuracy, at least at this stage.

In fig. 3, $\Delta\alpha^{\rm extr}_{\rm had}$ is the result extracted with our algorithm, corresponding to the minimizing set of i_j: the figure shows that our method is capable of recovering the underlying function $\Delta\alpha_{\rm had}(t)$ inserted into the "data". As the difference between $\Delta\alpha^{I}_{\rm had}$ and $\Delta\alpha^{\rm extr}_{\rm had}$ is hardly visible on an absolute scale, in fig. 3 all the functions have been divided by $\Delta\alpha^{II}_{\rm had}$ to display better the comparison between $\Delta\alpha^{I}_{\rm had}$ and $\Delta\alpha^{\rm extr}_{\rm had}$.

In order to assess the achievable accuracy on $\Delta\alpha_{\rm had}(t)$ with the proposed method, we remark that the LO contribution to the cross section is quadratic in $\alpha(t)$, thus we have

$$\frac{1}{2}\frac{\delta\sigma}{\sigma} \simeq \frac{\delta\alpha}{\alpha} \simeq \delta\Delta\alpha_{\rm had} \tag{15}$$

Equation (15) relates the *absolute* error on $\Delta\alpha_{\rm had}$ with the *relative* error on the Bhabha cross section. Let us stress here that:

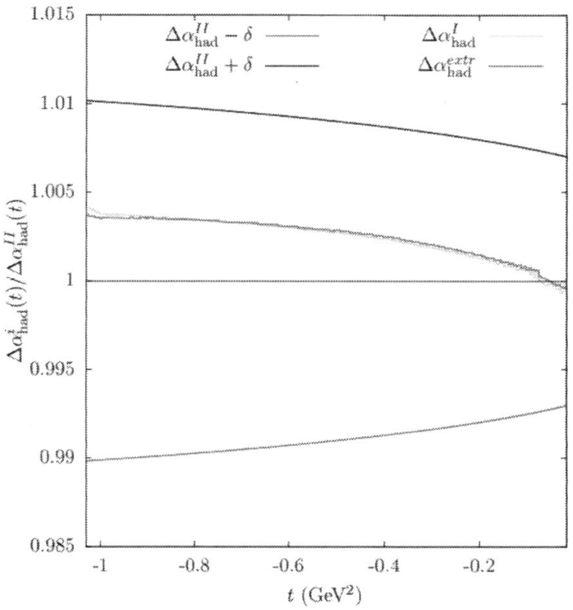

Figure 3. The extracted function $\Delta\alpha^{\rm extr}_{\rm had}(t)$ compared to the function $\Delta\alpha^{I}_{\rm had}(t)$ used in the pseudo-data (see text). The functions $\Delta\alpha^{II}_{\rm had}(t) \pm \delta(t)$ are shown to display the range spanned by the MC samples. All functions have been divided by $\Delta\alpha^{II}_{\rm had}(t)$. The tiny difference between $\Delta\alpha^{I}_{\rm had}$ and $\Delta\alpha^{\rm extr}_{\rm had}$ is due to the binning discretization.

1. From the theoretical point of view, the present accuracy of the MC predictions [28] is at the level of about 0.05%, which implies that the precision that our method can, at best, set on $\Delta\alpha_{\rm had}(t)$ is $\delta\Delta\alpha_{\rm had}(t) \simeq 2 \cdot 10^{-4}$. Any further improvement requires the inclusion of the NNLO QED corrections into the MC codes (see Ref. [28]).

2. Experimentally a measurement of $a_\mu^{\rm HLO}$ from space-like data competitive with the current time-like evaluations would require an $O(1\%)$ accuracy. Statistical considerations show that a 3% fractional accuracy on the $a_\mu^{\rm HLO}$ integral can be obtained by sampling the integrand $(1-x)\Delta\alpha_{\rm had}[t(x)]$ in $\sim$ 10 points around the x peak with a fractional accuracy of 10%. Given the value of $O(10^{-3})$ for $\Delta\alpha_{\rm had}$ at $x = x_{\rm peak}$, this implies that the cross section must be known with relative accuracy of $\sim 2\times10^{-4}$. Such a statistical accuracy, although challenging, can be obtained at flavor factories, as shown in fig. 2 (right). With an integrated luminosity of $O(1)$, $O(10)$, $O(100)$ fb^{-1} at $\sqrt{s} = 1$, 3 and 10 GeV, respectively, the angular region of interest can be covered with a 0.01% accuracy per degree. The experimental systematic error must match the same level of accuracy.

A source of experimental systematic errors comes from the machine luminosity, which can be normalized by calculating a theoretical cross section in principle not depending on $\Delta\alpha_{\rm had}$. We devise two possible options for the normalization process:

1. To use the $e^+e^- \to \gamma\gamma$ process, which has no dependence on $\Delta\alpha_{\rm had}$, at least up to NNLO order;

2. To use the Bhabha process at $t \sim 10^{-3}$ GeV^2 ($x \sim 0.3$), where the dependence on $\Delta\alpha_{\rm had}$ is of $O(10^{-5})$ and can be safely neglected.

It is worth quoting that a detailed analysis of the systematic errors involved in the measurement of the luminosity has been carried out at LEP by the OPAL collaboration reaching the final accuracy of $O(10^{-4})$ [14, 33].

3 Conclusions

We presented a novel approach to determine the leading hadronic correction to the muon g-2 based on measurements of the running of $\alpha(t)$ in the space-like region.

This approach, even if challenging, may become feasible by using data collected at present flavor factories as well as those at future high-energy e^+e^- colliders [1].

The proposed determination can become competitive with the accuracy of the present results obtained with the dispersive approach via time-like data.

An alternative formula for a_μ^{HLO} in the space-like region has been studied in detail. It emphasizes low values of $t \leq 1GeV^2$ which can be explored at low energies machines.

We have also argued that this requires a measurement of the Bhabha cross section, at relatively small angles, with an accuracy of the order 10^{-4} or better.

Reaching such an accuracy demands a dedicated experimental and theoretical work for the next few years.

Acknowledgements

I wish to thank Fulvia de Fazio and Pietro Colangelo for inviting me to this lively Workshop.

References

[1] C. Carloni Calame, M. Passera, G. Venanzoni, L. Trentadue, Phys. Lett. B **746** (2015) 325-329.
[2] G.W. Bennett *et al.* [Muon g-2 Collaboration], Phys. Rev. D **73** (2006) 072003.
[3] F. Jegerlehner, A. Nyffeler, Phys. Rept. **477** (2009) 1.
[4] T. Blum *et al.*, arXiv:1311.2198 [hep-ph]; K. Melnikov, A. Vainshtein, Springer Tracts Mod. Phys. **216** (2006) 1; M. Davier, W.J. Marciano, Ann. Rev. Nucl. Part. Sci. **54** (2004) 115; M. Passera, J. Phys. G **31** (2005) R75; M. Knecht, Lect. Notes Phys. **629** (2004) 37.
[5] F. Jegerlehner, "The anomalous magnetic moment of the muon," Springer Tracts Mod. Phys. **226**, 2008.
[6] J. Grange *et al.* [Muon g-2 Collaboration], arXiv:1501.06858 [physics.ins-det].
[7] G. Venanzoni [Muon g-2 Collaboration], arXiv:1411.2555 [physics.ins-det].
[8] N. Saito [J-PARC g-2/EDM Collaboration], AIP Conf. Proc. **1467** (2012) 45.
[9] G. Venanzoni, Nuovo Cim. C **037** (2014) 02, 165; G. Venanzoni, Frascati Phys. Ser. **54** (2012) 52.
[10] G.V. Fedotovich [CMD-2 Collaboration], Nucl. Phys. Proc. Suppl. **181-182** (2008) 146.

[1]The possibility of using a fixed target μ e scattering process is also under investigation [34].

[11] G. Colangelo, M. Hoferichter, A. Nyffeler, M. Passera, P. Stoffer, Phys. Lett. B **735** (2014) 90.

[12] C. Bouchiat, L. Michel, J. Phys. Radium 22 (1961) 121; L. Durand, Phys. Rev. **128** (1962) 441 [Erratum-ibid. **129** (1963) 2835]; M. Gourdin, E. De Rafael, Nucl. Phys. B **10** (1969) 667.

[13] A.B. Arbuzov, D. Haidt, C. Matteuzzi, M. Paganoni and L. Trentadue, Eur. Phys. J. C **34** (2004) 267.

[14] G. Abbiendi *et al.* [OPAL Collaboration], Eur. Phys. J. C **45** (2006) 1.

[15] A.B. Arbuzov, V.S. Fadin, E.A. Kuraev, L.N. Lipatov, N.P. Merenkov, L. Trentadue, Nucl. Phys. B**485** (1997) 457-502.

[16] M. Davier, A. Hoecker, B. Malaescu, Z. Zhang, Eur. Phys. J. C **71** (2011) 1515 [Erratum-ibid. C **72** (2012) 1874].

[17] K. Hagiwara, R. Liao, A.D. Martin, D. Nomura, T. Teubner, J. Phys. G **38** (2011) 085003.

[18] R.V. Harlander, M. Steinhauser, Comput. Phys. Commun. **153** (2003) 244.

[19] B.E. Lautrup, A. Peterman, E. de Rafael, Phys. Rept. **3** (1972) 193.

[20] C. Aubin, T. Blum, Phys. Rev. D **75** (2007) 114502; P. Boyle, L. Del Debbio, E. Kerrane, J. Zanotti, Phys. Rev. D **85** (2012) 074504; X. Feng, K. Jansen, M. Petschlies, D.B. Renner, Phys. Rev. Lett. **107** (2011) 081802; M. Della Morte, B. Jager, A. Juttner, H. Wittig, JHEP **1203** (2012) 055.

[21] T. Blum, M. Hayakawa, T. Izubuchi, PoS LATTICE **2012** (2012) 022.

[22] M. Steinhauser, Phys. Lett. B **429** (1998) 158.

[23] P.A. Baikov, K.G. Chetyrkin, J.H. Kuhn and C. Sturm, Nucl. Phys. B **867** (2013) 182; C. Sturm, Nucl. Phys. B **874** (2013) 698; P.A. Baikov, A. Maier, P. Marquard, Nucl. Phys. B **877** (2013) 647.

[24] F. Jegerlehner, in Proceedings of "Fifty years of electroweak physics: a symposium in honour of Professor Alberto Sirlin's 70th birthday", New York University, 27-28 October 2000, J. Phys. G **29** (2003) 101.

[25] S.L. Adler, Phys. Rev. D **10** (1974) 3714.

[26] S. Eidelman, F. Jegerlehner, A. L. Kataev, O. Veretin, Phys. Lett. B **454** (1999) 369.

[27] G. Balossini, C.M. Carloni Calame, G. Montagna, O. Nicrosini, F. Piccinini, Nucl. Phys. B **758** (2006) 227.

[28] S. Actis *et al.*, Eur. Phys. J. C **66** (2010) 585.

[29] S. Eidelman, F. Jegerlehner, Z. Phys. C **67** (1995) 585; F. Jegerlehner, Nucl. Phys. Proc. Suppl. **181-182** (2008) 135.

[30] M. Bicer *et al.* [TLEP Design Study Working Group Collaboration], JHEP **1401** (2014) 164 [arXiv:1308.6176 [hep-ex]].

[31] G. Aarons *et al.* [ILC Collaboration], Physics at the ILC, arXiv:0709.1893 [hep-ph].

[32] K. Hagiwara, A.D. Martin, D. Nomura, T. Teubner, Phys. Lett. B **649** (2007) 173; Phys. Rev. D **69** (2004) 093003.

[33] G. Abbiendi et al. [OPAL collaboration], Eur. Phys. J. C**14** (2000) 373-425.

[34] G. Abbiendi, C. M. Carloni Calame, U. Marconi, C. Matteuzzi, G. Montagna, O. Nicrosini, M. Passera, F. Piccinini, R. Tenchini, L. Trentadue, G. Venanzoni et al., hep-ex, arXiv:1609.08987.

EPJ Web of Conferences **129**, 00049 (2016) DOI: 10.1051/epjconf/201612900049
QCD@Work 2016

Flavour anomalies

Giancarlo D'Ambrosio[1,*]

[1] *INFN-Sezione di Napoli, Via Cintia, 80126 Napoli, Italia*

Abstract. The LHC observed deviations from the Standard Model (SM) in the flavour sector: LHCb found a 4.3 σ discrepancy compared to the SM in $b \to s\mu\mu$ transitions and CMS reported a non-zero measurement of $h \to \mu\tau$ with a significance of 2.4 σ. Here we discuss how these deviations from the SM can be explained, focusing on two models with gauged $L_\mu - L_\tau$ symmetry. The first model contains two scalar doublets and vector-like quarks while the second one employs three scalar doublets but does not require vector-like fermions. In both models, interesting correlations between $b \to s\mu\mu$ transitions, $h \to \mu\tau$, and $\tau \to 3\mu$ arise. We have also studied lepton flavor (universality) violation in rare kaon decays and the Bardeen Buras Gerard approach to describe the $K^\pm \to \pi^\pm l^+ l^-$ form factor.

1 Introduction

LHC is confirming fully the Standard Model. Still LHC did observe 'hints' for new physics (NP) in the flavor sector, which are sensitive to virtual effects of new particles and can be used as guidelines towards specific NP models: $h \to \mu\tau$, $B \to K^*\mu^+\mu^-$, $B_s \to \phi\mu^+\mu^-$ and $R(K) = B \to K\mu^+\mu^-/B \to Ke^+e^-$ [1] It is therefore interesting to examine if a specific NP model can explain these four anomalies simultaneously. In Refs. [5, 6], two variants of such a model were presented, which we want to review here. Also $B \to D(D^*)\tau\nu$ at Belle , BaBar and LHCb maybe shows 4-5 σ's discrepancy [1].

LHCb reported deviations from the SM predictions [2, 3] (mainly in an angular observable called P_5' [4]) in $B \to K^*\mu^+\mu^-$ [7] with a significance of 2–3 σ. In addition, the measurement of $B_s \to \phi\mu\mu$ disagrees with the SM predictions by about 3 σ [8, 9]. This discrepancy can be explained in a model independent approach by rather large contributions to the Wilson coefficient C_9 [8, 10, 11], i.e. an operator $(\bar{s}\gamma_\alpha P_L b)(\bar{\mu}\gamma^\alpha\mu)$, which can be achieved in models with an additional heavy neutral Z' gauge boson [12–16]. Furthermore, LHCb [17] recently found indications for the violation of lepton flavour universality in B meson decays, averaging to all experiments [1]

$$R(K) = \frac{B \to K\mu^+\mu^-}{B \to Ke^+e^-} = 0.785 \pm 0.080\,, \tag{1}$$

which disagrees from the theoretically rather clean SM prediction $R_K^{\rm SM} = 1.0003 \pm 0.0001$ [18] by 2.6 σ. A possible explanation is again a NP contributing to $C_9^{\mu\mu}$ involving muons, but not electrons [19–21]. Interestingly, the value for C_9 required to explain $R(K)$ is of the same order as the

*e-mail: gdambros@na.infn.it

one required by $B \to K^*\mu^+\mu^-$ [22, 23]. The global fit to the $b \to s\mu\mu$ data presented at this conference gives a 4.3 σ better fit to data for the assumption of NP in $C_9^{\mu\mu}$ only, compared to the SM fit [24].

Concerning Higgs decays, CMS measured a lepton-flavour violating (LFV) channel [25]

$$\mathrm{Br}[h \to \mu\tau] = \left(0.84^{+0.39}_{-0.37}\right)\%\,, \tag{2}$$

which disagrees from the SM (where this decay is forbidden) by about 2.4 σ. Most attempts to explain this decay rely on models with an extended Higgs sector [26–30]. One particular interesting solution employs a two-Higgs-doublet model (2HDM) with gauged $L_\mu - L_\tau$ [31].

2 The models

Our models under consideration are multi-Higgs-doublet models with a gauged $U(1)_{L_\mu - L_\tau}$ symmetry [31].[1] The $L_\mu - L_\tau$ symmetry with the gauge coupling g' is broken spontaneously by the vacuum expectation value (VEV) of a scalar Φ with $Q^{\Phi}_{L_\mu - L_\tau} = 1$, leading to the Z' mass

$$m_{Z'} = \sqrt{2}g'\langle\Phi\rangle \equiv g' v_\Phi\,, \tag{3}$$

and Majorana masses for the right-handed neutrinos.[2]

In both models at least two Higgs doublets are introduced which break the electroweak symmetry: Ψ_1 with $Q^{\Psi_1}_{L_\mu - L_\tau} = -2$ and Ψ_2 with $Q^{\Psi_2}_{L_\mu - L_\tau} = 0$. Therefore, Ψ_2 gives masses to quarks and leptons while Ψ_1 couples only off-diagonally to $\tau\mu$:

$$\mathcal{L}_Y \supset -\overline{\ell}_f Y_i^\ell \delta_{fi} \Psi_2 e_i - \xi_{\tau\mu}\overline{\ell}_3 \Psi_1 e_2 - \overline{Q}_f Y^u_{fi}\tilde{\Psi}_2 u_i - \overline{Q}_f Y^d_{fi}\Psi_2 d_i + \text{h.c.} \tag{4}$$

Here Q (ℓ) is the left-handed quark (lepton) doublet, u (e) is the right-handed up quark (charged lepton) and d the right-handed down quark while i and f label the three generations. The scalar potential is the one of a $U(1)$-invariant 2HDM [36] with additional couplings to the SM-singlet Φ, which most importantly generate the doublet-mixing term

$$V(\Psi_1, \Psi_2, \Phi) \supset 2\lambda\Phi^2\Psi_2^\dagger\Psi_1 \to \lambda v_\Phi^2 \Psi_2^\dagger\Psi_1 \equiv m_3^2\Psi_2^\dagger\Psi_1\,,$$

that induces a small vacuum expectation value for Ψ_1 [31]. We define $\tan\beta = \langle\Psi_2\rangle/\langle\Psi_1\rangle$ and α is the usual mixing angle between the neutral CP-even components of Ψ_1 and Ψ_2 (see for example [36]). We neglect the additional mixing of the CP-even scalars with Re[Φ].

Quarks and gauge bosons have standard type-I 2HDM couplings to the scalars. The only deviations are in the lepton sector: while the Yukawa couplings $Y_i^\ell\delta_{fi}$ of Ψ_2 are forced to be diagonal due to the $L_\mu - L_\tau$ symmetry, $\xi_{\tau\mu}$ gives rise to an off-diagonal entry in the lepton mass matrix:

$$m^\ell_{fi} = \frac{v}{\sqrt{2}}\begin{pmatrix} y_e \sin\beta & 0 & 0 \\ 0 & y_\mu \sin\beta & 0 \\ 0 & \xi_{\tau\mu}\cos\beta & y_\tau \sin\beta \end{pmatrix}. \tag{5}$$

[1]The abelian symmetry $U(1)_{L_\mu - L_\tau}$ is an anomaly-free global symmetry within the SM [32], and also a good zeroth-order approximation for neutrino mixing with a quasi-degenerate mass spectrum, predicting a maximal atmospheric and vanishing reactor neutrino mixing angle [33]. Breaking $L_\mu - L_\tau$ is mandatory for a realistic neutrino sector, and such a breaking can also induce charged LFV processes, such as $\tau \to 3\mu$ [34, 35] and $h \to \mu\tau$ [31].

[2]Neutrino masses arise via seesaw with close-to-maximal atmospheric mixing and quasi-degenerate masses [31].

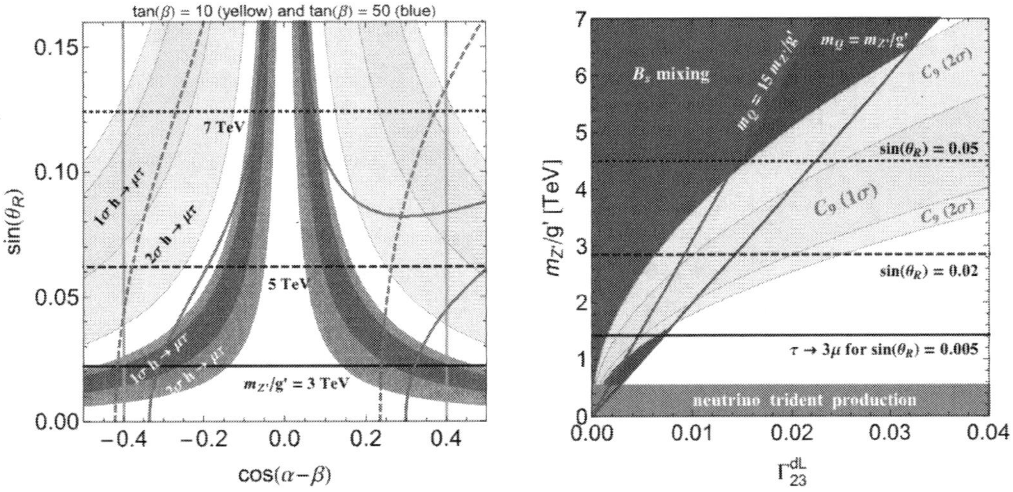

Figure 1. Left: Allowed regions in the $\cos(\alpha-\beta)$–$\sin(\theta_R)$ plane. The blue (light blue) region corresponds to the 1σ (2σ) region of the CMS measurement of $h \to \mu\tau$ for $\tan\beta = 50$; yellow stands for $\tan\beta = 10$. The (dashed) red contours mark deviations of $h \to \tau\tau$ by 10% compared to the SM for $\tan\beta = 50$ (10). The vertical green lines illustrate the naive LHC limit $|\cos(\alpha-\beta)| \leq 0.4$, horizontal lines denote the 90% C.L. limit on $\tau \to 3\mu$ via Z' exchange. Right: Allowed regions for model 1 in the Γ_{23}^{dL}-$m_{Z'}/g'$ plane from $b \to s\mu^+\mu^-$ data (yellow) and B_s–$\overline{B}_s$ mixing (blue). For B_s–$\overline{B}_s$ mixing, (light) blue corresponds to ($m_Q = 15 m_{Z'}/g'$) $m_Q = m_{Z'}/g'$. The horizontal lines denote the lower bounds on $m_{Z'}/g'$ from $\tau \to 3\mu$ for $\sin(\theta_R) = 0.05,\ 0.02,\ 0.005$. The gray region is excluded by neutrino trident production.

It is this τ–μ entry that leads to the LFV couplings of h and Z' of interest to this letter. The lepton mass basis is obtained by simple rotations of (μ_R, τ_R) and (μ_L, τ_L) with the angles θ_R and θ_L, respectively:

$$\sin\theta_R \simeq \frac{v}{\sqrt{2}m_\tau}\xi_{\tau\mu}\cos\beta\,, \qquad\qquad \frac{\tan\theta_L}{\tan\theta_R} = \frac{m_\mu}{m_\tau} \ll 1\,. \tag{6}$$

The angle θ_L is automatically small and will be neglected in the following.[3] A non-vanishing angle θ_R not only gives rise to the LFV decay $h \to \mu\tau$ due to the coupling

$$\frac{m_\tau}{v}\frac{\cos(\alpha-\beta)}{\cos(\beta)\sin(\beta)}\sin(\theta_R)\cos(\theta_R)\bar{\tau}P_R\mu h \equiv \Gamma^h_{\tau\mu}\bar{\tau}P_R\mu h\,, \tag{7}$$

in the Lagrangian, but also leads to off-diagonal Z' couplings to right-handed leptons

$$g' Z'_\nu (\bar{\mu}, \bar{\tau}) \begin{pmatrix} \cos 2\theta_R & \sin 2\theta_R \\ \sin 2\theta_R & -\cos 2\theta_R \end{pmatrix} \gamma^\nu P_R \begin{pmatrix} \mu \\ \tau \end{pmatrix}, \tag{8}$$

while the left-handed couplings are to a good approximation flavour conserving. $m_{Z'}/g'$ needs to be in the TeV range in order to suppress $\tau \to 3\mu$ if we want to explain $h \to \mu\tau$ [31] (see Fig. 1 (left)), which gives stronger bounds than neutrino trident production [14]. In order to explain the observed anomalies in the B meson decays, a coupling of the Z' to quarks is required as well, not inherently part of $L_\mu - L_\tau$ models.

[3]Choosing $Q_{L_\mu - L_\tau} = +2$ for Ψ_2 would essentially exchange $\theta_L \leftrightarrow \theta_R$ [31], with little impact on our study.

2.1 Model 1: vector-like quarks

In order to couple the Z' to quarks we follow Ref. [14] and generate effective couplings via heavy vector-like quarks [37] charged under $L_\mu - L_\tau$. As a result, the couplings of the Z' to quarks are in principle free parameters and can be parametrized as:

$$g'\left(\bar{d}_i\gamma^\mu P_L d_j Z'_\mu \Gamma^{dL}_{ij} + \bar{d}_i\gamma^\mu P_R d_j Z'_\mu \Gamma^{dR}_{ij}\right) \,. \tag{9}$$

In the limit of decoupled vector-like quarks with the quantum numbers of right-handed quarks, only C_9 is generated, giving a very good fit to data. The results are shown in the right plot of Fig. 1 depicting that for small values of Γ^L_{sb} and θ_R, $b \to s\mu^+\mu^-$ data can be explained without violating bounds from B_s–$\overline{B}_s$ mixing or $\tau \to 3\mu$. In the left plot of Fig. 2 the correlations of $b \to s\mu^+\mu^-$ and $h \to \mu\tau$ with $\tau \to 3\mu$ are shown, depicting that consistency with $\tau \to 3\mu$ requires large values of $\tan\beta$ (not being in conflict with any data as the decoupling limit is a type I model) and future searches for $\tau \to 3\mu$ are promising to yield positive results. While this model predict tiny branching ratios for lepton-flavour-violating B decays, these branching ratios can be sizable in generic Z' models in the presence of fine tuning in the B_s–$\overline{B}_s$ system [38].

2.2 Model 2: horizontal quark charges

In order to avoid the introduction of vector-like quarks, one can assign flavour-dependent charges to baryons as well [6]. Here, the first two generations should have the same charges in order to avoid very large effects in K–$\overline{K}$ or D–$\overline{D}$ mixing, generated otherwise unavoidably due to the breaking of the symmetry necessary to generate the measured Cabibbo angle of the CKM matrix. If we require in addition the absence of anomalies, we arrive at the following charge assignment for baryons $Q'(B) = (-a, -a, 2a)$, while leptons are still assigned $L_\mu - L_\tau$. Here $a \in Q$ is a free parameter of the model with important phenomenological implications. In this model, one additional Higgs doublet, which breaks the flavour symmetry in the quark sector, is required compared to the model with vector-like quarks. In case the mixing among the doublets is small, the correlations among $h \to \mu\tau$, $b \to s\mu^+\mu^-$ and $\tau \to 3\mu$ are similar as in the model with vector-like quarks discussed in the last subsection (left plot of Fig. 2).

The low-energy phenomenology is rather similar to the one of the model with vector-like quarks (model 1), but the contributions to B_s–$\overline{B}_s$ mixing are directly correlated to B_d–$\overline{B}_d$ and K–$\overline{K}$ mixing, because all flavour violation is due to CKM factors. (These constraints are evaded for $a \lesssim 1$.) However, the implications concerning direct LHC searches are very different, as the Z' boson couples to quarks of the first generation and can be directly produced on-shell as a resonance in $p\bar{p}$ collisions. The resulting strong bounds are shown in right plot of Fig. 2, where they are compared to the allowed regions from B_s–$\overline{B}_s$ mixing and $b \to s\mu^+\mu^-$ data for different values of a.

3 Lepton flavor (universality) violation in rare kaon decays

Recent anomalies in the decays of B-mesons and the Higgs boson provide hints towards lepton flavor (universality) violating physics beyond the Standard Model. We observe that 4-fermion operators which can explain the B-physics anomalies have corresponding analogs in the kaon sector, and we analyze their effect on $K \to \pi\ell\ell'$ and $K \to \ell\ell'$ decays ($\ell = \mu, e$). For these processes, we point out the corresponding physics opportunities at the NA62 and KOTO experiments. In particular, assuming minimal flavor violation, we comment on the required improvements in sensitivity necessary to test the B-physics anomalies in the kaon sector [39].

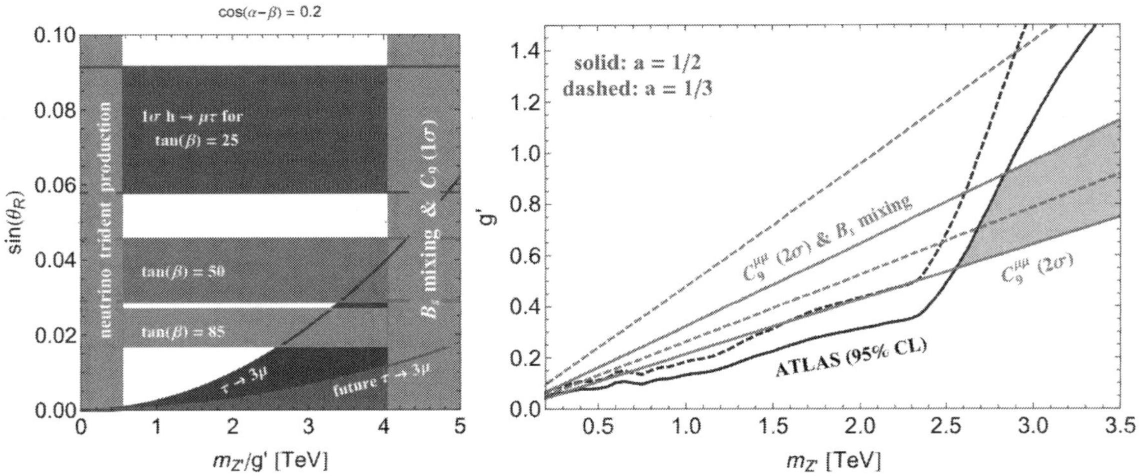

Figure 2. Left: Allowed regions for model 1 in the $m_{Z'}/g'$–$\sin(\theta_R)$ plane: the horizontal stripes correspond to $h \to \mu\tau$ (1σ) for different $\tan\beta$ and $\cos(\alpha-\beta) = 0.2$, (light) blue stands for (future) $\tau \to 3\mu$ limits at 90% C.L. The gray regions are excluded by neutrino trident production or B_s–$\overline{B}_s$ mixing in combination with the $1\,\sigma$ range for C_9. Right: Limits for model 2 on $q\overline{q} \to Z' \to \mu\overline{\mu}$ from ATLAS (black, allowed region down right) and the $2\,\sigma$ limits on $C_9^{\mu\mu}$ to accommodate $b \to s\mu\mu$ data (allowed regions inside the red cone). Solid (dashed) lines are for $a = 1/2$ ($1/3$). For $a = 1/2$, the green shaded region is allowed (similar for $a = 1/3$ using the dashed bounds).

4 Bardeen Buras Gerard approach and $K^\pm \to \pi^\pm l^+ l^-$ form factor

In Refs. [41, 42], the authors compute loop calculations employing a cut-off regulator instead of the dimensional regularization method. Consequently, their results exhibit a quadratic dependence on the physical cut-off M which is lost in the usual chiral perturbative calculations. This quadratic dependence is a crucial ingredient in the matching of the meson and quark pictures. They argue that one can obtain a parametrization of non-perturbative QCD effects by matching a low-energy Lagrangian valid up to the scale M, to the logarithmic behavior of relevant Wilson coefficients at high-energy. In this work we refer to this computational method as the *Bardeen-Buras-Gérard framework* (BBG). Recentely Buras has advocated this determination to claim New Physics to explain the experimental value of ϵ' [43]. BBG evaluate $K \to \pi\pi$ chiral loop with a dimensional cut-off (M) and match the quadratic divergences of the $Q_i(M)$ weak matrix elements with the log of the short distance Wilson coefficient as described by the equation

$$H_{\text{eff}} = \sum_i c_i(\mu)\, Q_i(\mu) \tag{10}$$

We have studied in this BBG contest the $K^\pm \to \pi^\pm l^+ l^-$ form factor described in CHPT by a loop function, $W_+^{\pi\pi}(z)$ and a polynomial in z with experimentally determined coefficients a_+ and b_+ [44]. These parameterize the intermediate region between low and high energy regimes. Our goal is to predict the values of these two coefficients using BBG framework [40]. Indeed in Fig. 3 we show: i) (left plot) the a_+ as a function of M without the vectors contributions and ii) (right plot) adding the vectors contributions particularly important are the novel weak vector couplings.

DOI: 10.1051/epjconf/201612900049

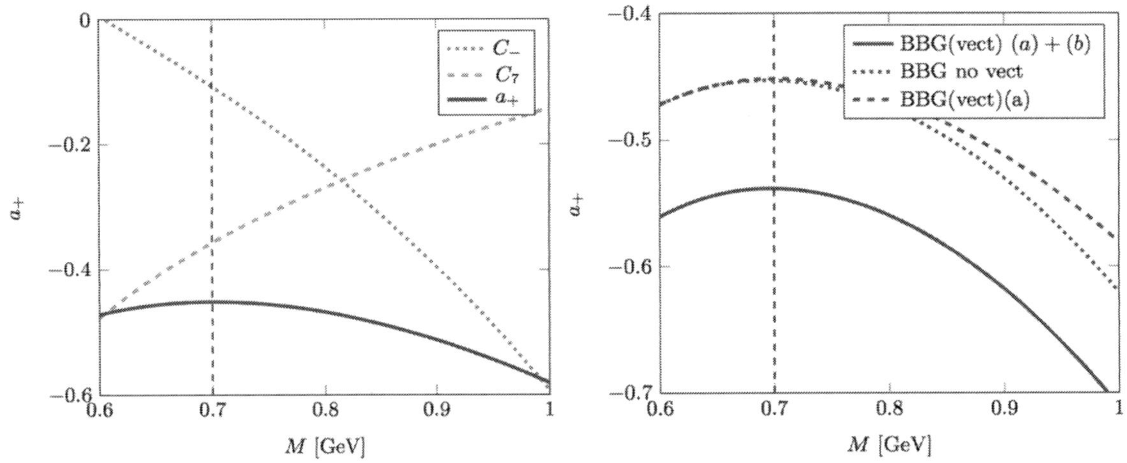

Figure 3. Left: In blue, the variation of a_+ as a function of M in GeV. The dotted green curve represents the contribution proportional to $C_-(M^2)$ and the dashed orange curve the one proportional to $C_7(M^2)$. The vertical dashed line stands for the matching scale Right: a_+ as a function of M in the three different frameworks: 'BBG no vect.' where vectors are not included, 'BBG(vect)(a)' represents the contribution coming only from diagrams with vectors coupled strongly and 'BBG(vect) (a) + (b)' is the case where vectors with weak couplings were included. The vertical line indicates the value $M = 0.7$ GeV.

5 Conclusions

In these proceedings we reviewed two variants of a model with a gauged $L_\mu - L_\tau$ symmetry which can explain all LHC anomalies in the flavour sector simultaneously: 1) a 2HDM with effective $Z'\bar{s}b$ couplings induced by heavy vector-like quarks, 2) a 3HDM with horizontal charges for baryons. The models can account for the deviations from the SM in $b \to s\mu^+\mu^-$ data and $h \to \mu\tau$ simultaneously, giving also the desired effect in $R(K)$. Due to the small values of the τ–μ mixing angle θ_R, sufficient to account for $h \to \mu\tau$, the Z' contributions to $\tau \to 3\mu$ are not in conflict with present bounds for large $\tan\beta$ in wide rages of parameter space. Interestingly, $b \to s\mu^+\mu^-$ data combined with B_s–$\overline{B}_s$ put a upper limit on $m_{Z'}/g'$ resulting in a lower limit on $\tau \to 3\mu$ if $\mathrm{Br}[h \to \mu\tau] \neq 0$: for lower values of $\tan\beta$ the current experimental bounds are reached and future sensitivities will allow for a more detailed exploration of the allowed parameter space. The possible range for the $L_\mu - L_\tau$ breaking scale further implies the masses of the Z' and the right-handed neutrinos to be at the TeV scale, potentially testable at the LHC with interesting additional consequences for LFV observables. While the low energy phenomenology of both models is rather similar, the variant with horizontal charges for baryons predicts sizable Z' production rates testable at the next LHC run. We have also studied lepton flavor (universality) violation in rare kaon decays and the Bardeen Buras Gerard approach to describe the $K^{\pm} \rightarrow \pi^{\pm} l^{+} l^{-}$ form factor.

Acknowledgements

I would like to thank the organizers of the workshop QCD@work, in particular Fulvia De Fazio and Pietro Colangelo. I also acknowledge collaboration with L- Cappiello, E. Coluccio Leskow, A. Crivellin, D. Greynat, A. Nath, M. Hoferichter and L. C. Tunstall

References

[1] https://indico.cern.ch/event/549031/timetable/ Implications of LHCb measurements and future prospects, CERN 12th-14th October.

[2] U. Egede, T. Hurth, J. Matias, M. Ramon and W. Reece, JHEP **0811** (2008) 032 [arXiv:0807.2589].

[3] W. Altmannshofer, P. Ball, A. Bharucha, A. J. Buras, D. M. Straub and M. Wick, JHEP **0901** (2009) 019 [arXiv:0811.1214].

[4] S. Descotes-Genon, T. Hurth, J. Matias and J. Virto, JHEP **1305** (2013) 137 [arXiv:1303.5794].

[5] A. Crivellin, G. D'Ambrosio and J. Heeck, Phys. Rev. Lett. **114** (2015) 151801 [arXiv:1501.00993].

[6] A. Crivellin, G. D'Ambrosio and J. Heeck, Phys. Rev. D **91** (2015) 075006 [arXiv:1503.03477].

[7] R. Aaij *et al.* [LHCb Collaboration], Phys. Rev. Lett. **111** (2013) 191801 [arXiv:1308.1707].

[8] R. R. Horgan, Z. Liu, S. Meinel and M. Wingate, Phys. Rev. Lett. **112** (2014) 212003 [arXiv:1310.3887].

[9] R. R. Horgan, Z. Liu, S. Meinel and M. Wingate, arXiv:1501.00367.

[10] S. Descotes-Genon, J. Matias and J. Virto, Phys. Rev. D **88** (2013) 074002 [arXiv:1307.5683].

[11] W. Altmannshofer and D. M. Straub, Eur. Phys. J. C **73** (2013) 2646 [arXiv:1308.1501].

[12] R. Gauld, F. Goertz and U. Haisch, Phys. Rev. D **89** (2014) 015005 [arXiv:1308.1959].

[13] A. J. Buras, F. De Fazio and J. Girrbach, JHEP **1402** (2014) 112 [arXiv:1311.6729].

[14] W. Altmannshofer, S. Gori, M. Pospelov and I. Yavin, Phys. Rev. D **89** (2014) 095033 [arXiv:1403.1269].

[15] C. Niehoff, P. Stangl and D. M. Straub, arXiv:1503.03865.

[16] S. D. Aristizabal, F. Staub and A. Vicente, arXiv:1503.06077.

[17] R. Aaij *et al.* [LHCb Collaboration], Phys. Rev. Lett. **113** (2014) 151601 [arXiv:1406.6482].

[18] C. Bobeth, G. Hiller and G. Piranishvili, JHEP **0712** (2007) 040 [arXiv:0709.4174].

[19] R. Alonso, B. Grinstein and J. Martin Camalich, Phys. Rev. Lett. **113** (2014) 241802 [arXiv:1407.7044].

[20] G. Hiller and M. Schmaltz, Phys. Rev. D **90** (2014) 054014 [arXiv:1408.1627].

[21] D. Ghosh, M. Nardecchia and S. A. Renner, JHEP **1412** (2014) 131 [arXiv:1408.4097].

[22] T. Hurth, F. Mahmoudi and S. Neshatpour, JHEP **1412** (2014) 053 [arXiv:1410.4545].

[23] W. Altmannshofer and D. M. Straub, arXiv:1411.3161.

[24] W. Altmannshofer and D. M. Straub, arXiv:1503.06199.

[25] V. Khachatryan *et al.* [CMS Collaboration], arXiv:1502.07400.

[26] A. Dery, A. Efrati, Y. Nir, Y. Soreq and V. Susič, Phys. Rev. D **90** (2014) 115022 [arXiv:1408.1371].

[27] M. D. Campos, A. E. C. Hernández, H. Päs and E. Schumacher, arXiv:1408.1652.

[28] A. Celis, V. Cirigliano and E. Passemar, arXiv:1409.4439.

[29] D. Aristizabal Sierra and A. Vicente, Phys. Rev. D **90** (2014) 11, 115004 [arXiv:1409.7690].
[30] C. J. Lee and J. Tandean, JHEP **1504** (2015) 174 [arXiv:1410.6803].
[31] J. Heeck, M. Holthausen, W. Rodejohann and Y. Shimizu, Nucl. Phys. B **896** (2015) 281 [arXiv:1412.3671].
[32] X. G. He, G. C. Joshi, H. Lew and R. R. Volkas, Phys. Rev. D **43** (1991) 22.
[33] P. Binetruy, S. Lavignac, S. T. Petcov and P. Ramond, Nucl. Phys. B **496** (1997) 3 [hep-ph/9610481].
[34] G. Dutta, A. S. Joshipura and K. B. Vijaykumar, Phys. Rev. D **50** (1994) 2109 [hep-ph/9405292].
[35] J. Heeck and W. Rodejohann, Phys. Rev. D **84** (2011) 075007 [arXiv:1107.5238].
[36] G. C. Branco, P. M. Ferreira, L. Lavoura, M. N. Rebelo, M. Sher and J. P. Silva, Phys. Rept. **516** (2012) 1 [arXiv:1106.0034].
[37] P. Langacker, Rev. Mod. Phys. **81** (2009) 1199 [arXiv:0801.1345].
[38] A. Crivellin, L. Hofer, J. Matias, U. Nierste, S. Pokorski and J. Rosiek, arXiv:1504.07928.
[39] A. Crivellin, G. D'Ambrosio, M. Hoferichter and L. C. Tunstall, Phys. Rev. D **93**, no. 7, 074038 (2016) doi:10.1103/PhysRevD.93.074038 [arXiv:1601.00970 [hep-ph]].
[40] E. Coluccio Leskow, G. D'Ambrosio, D. Greynat and A. Nath, Phys. Rev. D **93**, no. 9, 094031 (2016) doi:10.1103/PhysRevD.93.094031 [arXiv:1603.09721 [hep-ph]].
[41] W. A. Bardeen, A. J. Buras and J. M. Gerard, Nucl. Phys. B **293** (1987) 787. doi:10.1016/0550-3213(87)90091-5; A. J. Buras, J. M. Gérard and W. A. Bardeen, "Large N Approach to Kaon Decays and Mixing 28 Years Later: $\Delta I = 1/2$ Rule, $\hat{B}_K$ and ΔM_K", Eur. Phys. J. C **74**, 2871 (2014).
[42] J. M. Gerard, Acta Phys. Polon. B **21** (1990) 257.
[43] A. J. Buras, M. Gorbahn, S. Jager and M. Jamin, JHEP **1511**, 202 (2015) doi:10.1007/JHEP11(2015)202 [arXiv:1507.06345 [hep-ph]].
[44] G. D'Ambrosio, G. Ecker, G. Isidori and J. Portoles, JHEP **9808**, 004 (1998) doi:10.1088/1126-6708/1998/08/004 [hep-ph/9808289].

EPJ Web of Conferences **129**, 00050 (2016) DOI: 10.1051/epjconf/201612900050
QCD@Work 2016

The Revival of Kaon Flavour Physics

Andrzej J. Buras[1,2,*]

[1] *TUM Institute for Advanced Study, Lichtenbergstr. 2a, D-85748 Garching, Germany*
[2] *Physik Department, Technische Universität München, James-Franck-Straße, D-85748 Garching, Germany*

Abstract. After years of silence we should witness in the rest of this decade and in the next decade the revival of kaon flavour physics. This is not only because of the crucial measurements of the branching ratios for the rare decays $K^+ \to \pi^+\nu\bar{\nu}$ and $K_L \to \pi^0\nu\bar{\nu}$ by NA62 and KOTO that being theoretically clean and very sensitive to new physics (NP) could hint for new phenomena even beyond the reach of the LHC without any significant theoretical uncertainties. Indeed simultaneously the advances in the calculations of perturbative and in particular non-perturbative QCD effects in ε'/ε, ε_K, ΔM_K, $K_L \to \mu^+\mu^-$ and $K_L \to \pi^0\ell^+\ell^-$ will increase the role of these observables in searching for NP. In fact the hints for NP contributing to ε'/ε have been already signalled last year through improved estimates of hadronic matrix elements of QCD and electroweak penguin operators Q_6 and Q_8 by lattice QCD and large N dual QCD approach. This talk summarizes in addition to this new flavour anomaly the present highlights of this field including some results from concrete NP scenarios.

1 Introduction

In two recent reports [1, 2] I have stressed the importance of kaon flavour physics in the search for new physics (NP) expressing my excitement in view of the revival of this field which played such an important role in the construction of the present theory of elementary particle physics represented by the Standard Model (SM) of strong and electroweak interactions. The goal of this writing is to list the most important advances in this field which have been made in the last two years since the last workshop of this series. Many of the topics listed below have already been discussed in [1–3] but I will present them from a different perspective and will add new ones. Moreover, I think it is useful to have a list of most important findings which can be looked up faster than in the longer expositions in [1–3]. Further details, in particular numerous references, can be found there.

2 Important Messages

2.1 ε'/ε

Presently in kaon flavour physics the most exciting appears to be the anomaly in ε'/ε and we will look at it first. The present status of ε'/ε in the SM can be summarized as follows. The RBC-UKQCD

*e-mail: aburas@ph.tum.de

lattice collaboration calculating hadronic matrix elements of all operators but not including isospin breaking effects finds [4, 5]

$$(\varepsilon'/\varepsilon)_{\rm SM} = (1.38 \pm 6.90) \times 10^{-4}, \qquad ({\rm RBC-UKQCD}). \tag{1}$$

Using the hadronic matrix elements of QCD- and EW-penguin $(V-A)\otimes(V+A)$ operators from RBC-UKQCD lattice collaboration but extracting the matrix elements of penguin $(V-A)\otimes(V-A)$ operators from the CP-conserving $K \to \pi\pi$ amplitudes and including isospin breaking effects one finds [6]

$$(\varepsilon'/\varepsilon)_{\rm SM} = (1.9 \pm 4.5) \times 10^{-4}, \qquad ({\rm BGJJ}). \tag{2}$$

This result differs by 2.9σ significance from the experimental world average from NA48 [7] and KTeV [8, 9] collaborations,

$$(\varepsilon'/\varepsilon)_{\rm exp} = (16.6 \pm 2.3) \times 10^{-4}, \tag{3}$$

suggesting that models providing enhancement of ε'/ε are favoured. A new analysis in [10] confirms these findings

$$(\varepsilon'/\varepsilon)_{\rm SM} = (0.96 \pm 4.96) \times 10^{-4}, \qquad ({\rm KNT}). \tag{4}$$

All these results are based on NLO calculations of the Wilson coefficients of the relevant operators that have been completed 23 years ago [11–16]. Partial NNLO calculations have been performed in [17–19]. Complete NNLO result from Maria Cerda-Sevilla, Martin Gorbahn, Sebastian Jäger and Ahmet Kokulu should be available soon.

While these results, based on the hadronic matrix elements from RBC-UKQCD lattice collaboration, suggest some evidence for the presence of NP in hadronic K decays, the large uncertainties in the hadronic matrix elements in question do not yet preclude that eventually the SM will agree with data. In this context the upper bounds on the matrix elements of the dominant penguin operators from large N dual QCD approach [20] are important as they give presently the strongest support to the anomaly in question, certainly stronger than present lattice results. To see this in explicit terms let us look at the parameters $B_6^{(1/2)}$ and $B_8^{(3/2)}$ that represent the relevant hadronic matrix elements of the QCD penguin operator Q_6 and the electroweak penguin operator Q_8, respectively.

In the strict large N limit [21–23] one simply has

$$B_6^{(1/2)} = B_8^{(3/2)} = 1, \qquad (\text{large N Limit}). \tag{5}$$

But RBC-UKQCD results [4, 5] imply [6, 24]

$$B_6^{(1/2)} = 0.57 \pm 0.19, \qquad B_8^{(3/2)} = 0.76 \pm 0.05, \qquad (\text{RBC-UKQCD}), \tag{6}$$

and this suppression of both parameters below unity, in particular of $B_6^{(1/2)}$, is the main origin of the strong suppression of ε'/ε within the SM below the data. Yet in view of the large error in $B_6^{(1/2)}$ one could be sceptical about claims made by me and my collaborators that there is NP in ε'/ε. Future lattice results could in principle raise $B_6^{(1/2)}$ towards its large N value and above $B_8^{(3/2)}$ bringing the SM result for ε'/ε close to its experimental value.

However, the analyses of $B_6^{(1/2)}$ and $B_8^{(3/2)}$ within the dual QCD approach in [20, 25] show that such a situation is rather unlikely. Indeed, in this approach going beyond the strict large N limit one can understand the suppression of $B_6^{(1/2)}$ and $B_8^{(3/2)}$ below the unity as the effect of the meson evolution from scales $\mu = \mathcal{O}(m_\pi, m_K)$ at which (5) is valid to $\mu = \mathcal{O}(1\,\text{GeV})$ at which Wilson coefficients of

Q_6 and Q_8 are evaluated [20]. This evolution has to be matched to the usual perturbative quark evolution for scales higher than 1 GeV and in fact the supressions in question and the property that $B_6^{(1/2)}$ is more strongly suppressed than $B_8^{(3/2)}$ are consistent with the perturbative evolution of these parameters above $\mu = O(1\,\text{GeV})$. Thus we are rather confident that [20]

$$B_6^{(1/2)} < B_8^{(3/2)} < 1 \qquad \text{(dual QCD)}. \tag{7}$$

Explicit calculation in this approach gives $B_8^{(3/2)}(m_c) = 0.80 \pm 0.10$. The result for $B_6^{(1/2)}$ is less precise but in agreement with (6). For further details, see [20].

It should be recalled that in the past values $B_6^{(1/2)} = B_8^{(3/2)} = 1.0$ have been combined in phenomenological applications with the Wilson coefficients evaluated at scales $\mu = O(1\,\text{GeV})$. The discussion above shows that this is incorrect. The meson evolution from $\mu = O(m_\pi, m_K)$ to $\mu = O(1\,\text{GeV})$ has to be performed and this effect turns out to be stronger than the scale dependence of $B_6^{(1/2)}$ and $B_8^{(3/2)}$ in the perturbative regime, where it is very weak.

Additional support for the small value of ε'/ε in the SM comes from the recent reconsideration of the role of final state interactions (FSI) in ε'/ε [25]. Already long time ago the chiral perturbation theory practitioners put forward the idea that both the amplitude $\text{Re}A_0$, governed by the current-current operator $Q_2 - Q_1$ and the Q_6 contribution to the ratio ε'/ε could be enhanced significantly through FSI in a correlated manner [26–32] bringing the values of ε'/ε close to its experimental value. However, as shown recently in [25] FSI are likely to be important for the $\Delta I = 1/2$ rule, in agreement with [26–32], but much less relevant for ε'/ε. Even if the analysis in [25] is rather qualitative, it makes us belive that future more precise calculations will find ε'/ε well below its SM value. It should also be emphasized that the authors of [26–32] did not include the meson evolution of $B_6^{(1/2)}$ and $B_8^{(3/2)}$ in their analysis and already this effect would significantly lower their predictions for ε'/ε.

It should finally be noted that even without lattice results, varying all input parameters, the bound in (7) implies the upper bound on ε'/ε in the SM

$$(\varepsilon'/\varepsilon)_{\text{SM}} < (8.6 \pm 3.2) \times 10^{-4}, \qquad \text{(BG)}. \tag{8}$$

On the other hand employing the lattice value for $B_8^{(3/2)}$ in (6) and $B_6^{(1/2)} = B_8^{(3/2)} = 0.76$, one obtains $(6.0 \pm 2.4) \times 10^{-4}$ instead of (8), well below the data.

All these findings give strong motivation for searching for NP which could enhance ε'/ε above its SM value. We will summarize the present efforts in this direction below.

2.2 Tensions between ε_K and $\Delta M_{s,d}$ in the SM and CMFV Models

In [33] we have pointed out a significant tension between ε_K and $\Delta M_{s,d}$ within the SM and models with constrained MFV (CMFV) implied by new lattice QCD results from Fermilab Lattice and MILC Collaborations [34] on $B^0_{s,d} - \bar{B}^0_{s,d}$ hadronic matrix elements. Not everybody agrees on this tension as in allover fits this tension is not transparently seen. But plots in [33], in particular in Fig. 5 of that paper, show that there is a clear tension between ε_K and $\Delta M_{s,d}$ in the SM and CMFV models. Our strategy was to ignore the present tree-level values of $|V_{cb}|$ and $|V_{ub}|$ in view of some discrepancies between their exclusive and inclusive determinations and to show that this tension persists independently of the values of CKM parameters. For smaller (exclusive) values of $|V_{cb}|$ one finds $\Delta M_{s,d}$ to agree well with the data, while ε_K is roughly 25% below its experimental value. For $|V_{cb}|$ in the ballpark of inclusive determinations one finds ε_K to agree with the data, while $\Delta M_{s,d}$ are then typically by 15% larger than their experimental values. These numbers are for the SM, in all other CMFV models the situation gets worse.

But on the whole this tension is certainly not as large as is the case of the ε'/ε anomaly. We are looking forward to improved hadronic matrix elements in question from other lattice collaborations and to improved values of $|V_{cb}|$ and $|V_{ub}|$ which would tell us whether this tension persists and if this will turn out to be the case, whether there is a ε_K anomaly and/or a $\Delta M_{s,d}$ anomaly.

2.3 $K^+ \to \pi^+\nu\bar{\nu}$ and $K_L \to \pi^0\nu\bar{\nu}$ in the SM

These two rare decays allow to test the short distance scales far beyond the reach of the LHC. Even scales of $O(100)$ TeV can be probed in this manner [35]. The present status of $K^+ \to \pi^+\nu\bar{\nu}$ and $K_L \to \pi^0\nu\bar{\nu}$ within the SM has been presented in [24] with the result

$$\mathcal{B}(K^+ \to \pi^+\nu\bar{\nu}) = (8.4 \pm 1.0) \times 10^{-11}, \tag{9}$$

$$\mathcal{B}(K_L \to \pi^0\nu\bar{\nu}) = (3.4 \pm 0.6) \times 10^{-11}. \tag{10}$$

But the most important outcome of this paper are parametric expressions for the branching ratios of these two decays in terms of the CKM input and the correlations between $K^+ \to \pi^+\nu\bar{\nu}$ and $B_s \to \mu^+\mu^-$ and between $K^+ \to \pi^+\nu\bar{\nu}$ and ε_K in the SM. These formulae should be useful for monitoring the numerical values for these branching ratios within the SM when the CKM input improves. On the other hand the results for both decays obtained in simplified models with flavour violating couplings of the SM Z and of a heavy Z' can be found in [36]. A more general study of such models, performed in [37], will be discussed now.

2.4 Strategy for ε'/ε and Lessons

The present error on ε'/ε within the SM is still very large and it is rather inconvenient to carry it to NP models. Therefore in order to investigate the implications of ε'/ε anomaly on rare decays $K^+ \to \pi^+\nu\bar{\nu}$ and $K_L \to \pi^0\nu\bar{\nu}$ in a systematic fashion a strategy has been proposed in [37]. While ε'/ε plays the dominant role in this strategy it was useful to assume that there is also a modest ε_K anomaly.

Then ε'/ε and ε_K in the presence of NP contributions are given by

$$\frac{\varepsilon'}{\varepsilon} = \left(\frac{\varepsilon'}{\varepsilon}\right)^{\text{SM}} + \left(\frac{\varepsilon'}{\varepsilon}\right)^{\text{NP}}, \qquad \varepsilon_K \equiv e^{i\varphi_\varepsilon}\left[\varepsilon_K^{\text{SM}} + \varepsilon_K^{\text{NP}}\right] \tag{11}$$

with NP contributions parametrized as follows:

$$\left(\frac{\varepsilon'}{\varepsilon}\right)^{\text{NP}} = \kappa_{\varepsilon'} \cdot 10^{-3}, \qquad 0.5 \le \kappa_{\varepsilon'} \le 1.5, \tag{12}$$

and

$$\varepsilon_K^{\text{NP}} = \kappa_\varepsilon \cdot 10^{-3}, \qquad 0.1 \le \kappa_\varepsilon \le 0.4\,. \tag{13}$$

The ranges for $\kappa_{\varepsilon'}$ and κ_ε indicate the required size of this contribution but can be kept as free parameters. They will be determined one day when the theory on ε'/ε and the CKM input improve.

In the simplest NP scenarios with tree-level Z and Z' exchanges, the imaginary parts of flavour-violating Z or Z' couplings to quarks are then determined as functions of $\kappa_{\varepsilon'}$. As ε_K is governed by the product of imaginary and real parts, invoking (13) allows then to determine the corresponding real parts as functions of $\kappa_{\varepsilon'}$ and κ_ε.

Having fixed the flavour violating couplings of Z or Z' in this manner, one can express NP contributions to the branching ratios for $K^+ \to \pi^+\nu\bar{\nu}$, $K_L \to \pi^0\nu\bar{\nu}$, $K_L \to \mu^+\mu^-$ and to ΔM_K in terms of $\kappa_{\varepsilon'}$

and κ_ε. Explicit formulae can be found in [37]. In this manner one can directly study the impact of ε'/ε and ε_K anomalies in Z and Z' scenarios on these four observables. The pattern of flavour violation depends in a given NP scenario on the relative size of real and imaginary parts of the couplings involved and we will see this explicitly in the lessons below.

In [37] numerous plots for the ratios

$$R_+^{\nu\bar\nu} \equiv \frac{\mathcal{B}(K^+ \to \pi^+\nu\bar\nu)}{\mathcal{B}(K^+ \to \pi^+\nu\bar\nu)_{\rm SM}}, \qquad R_0^{\nu\bar\nu} \equiv \frac{\mathcal{B}(K_L \to \pi^0\nu\bar\nu)}{\mathcal{B}(K_L \to \pi^0\nu\bar\nu)_{\rm SM}} \tag{14}$$

as functions of $\kappa_{\varepsilon'}$ and κ_ε within the models with tree-level Z and Z' exchanges have been presented. In view of space limitations we will not repeat them here but instead we will list the most important lessons from this study defining flavour violating couplings $\Delta_{L,R}^{sd}(Z)$ by [38]

$$i\mathcal{L}(Z) = i\left[\Delta_L^{sd}(Z)(\bar s\gamma^\mu P_L d) + \Delta_R^{sd}(Z)(\bar s\gamma^\mu P_R d)\right] Z_\mu, \qquad P_{L,R} = \frac{1}{2}(1 \mp \gamma_5) \tag{15}$$

with analogous definitions for Z' couplings. Moreover, we will use the abbreviations:

$$\text{LHS} \equiv \text{left-handed scenario}, \qquad \text{RHS} \equiv \text{right-handed scenario} \tag{16}$$

for NP scenarios in which only left-handed (LH) or right-handed (RH) flavour-violating couplings are present.

Lesson 1: In the LHS, a given request for the enhancement of ε'/ε determines the coupling $\text{Im}\Delta_L^{sd}(Z)$.

Lesson 2: In LHS there is a direct unique implication of an enhanced ε'/ε on $K_L \to \pi^0\nu\bar\nu$: *suppression* of $\mathcal{B}(K_L \to \pi^0\nu\bar\nu)$. This property is known from NP scenarios in which NP to $K_L \to \pi^0\nu\bar\nu$ and ε'/ε enters dominantly through the modification of Z-penguins.

Lesson 3: The imposition of the $K_L \to \mu^+\mu^-$ constraint in LHS determines the range for $\text{Re}\Delta_L^{sd}(Z)$ which with the already fixed $\text{Im}\Delta_L^{sd}(Z)$ allows to calculate the shifts in ε_K and ΔM_K. These shifts turn out to be very small for ε_K and negligible for ΔM_K.

Lesson 4: With fixed $\text{Im}\Delta_L^{sd}(Z)$ and the allowed range for $\text{Re}\Delta_L^{sd}(Z)$, the range for $\mathcal{B}(K^+ \to \pi^+\nu\bar\nu)$ can be obtained. Both an enhancement and a suppression of $\mathcal{B}(K^+ \to \pi^+\nu\bar\nu)$ are possible. $\mathcal{B}(K^+ \to \pi^+\nu\bar\nu)$ can be enhanced by a factor of 2 at most.

Lesson 5: Analogous pattern is found in RHS, although the numerics is different. See Fig. 1 in [37]. In particular the suppression of $\mathcal{B}(K_L \to \pi^0\nu\bar\nu)$ for a given $\kappa_{\varepsilon'}$ is smaller. Moreover, an enhancement of $\mathcal{B}(K^+ \to \pi^+\nu\bar\nu)$ up to a factor of 5.7 is possible.

Lesson 6: In a general Z scenario with LH and RH flavour-violating couplings the pattern of NP effects changes because of the appearance of LR operators dominating NP contributions to ε_K and ΔM_K. The main virtue of the general scenario is the possibility of enhancing simultaneously ε'/ε, ε_K, $\mathcal{B}(K^+ \to \pi^+\nu\bar\nu)$ and $\mathcal{B}(K_L \to \pi^0\nu\bar\nu)$ which is not possible in LHS and RHS. Thus the presence of both LH and RH flavour-violating currents is essential for obtaining simultaneously the enhancements in question. The correlations between ε'/ε and $K^+ \to \pi^+\nu\bar\nu$ and $K_L \to \pi^0\nu\bar\nu$ depend sensitively on the ratio of real and imaginary parts of the flavour-violating couplings involved. But the main message from this analysis is that in the presence of both LH and RH flavour-violating couplings of Z to quarks, large departures from SM predictions for $K^+ \to \pi^+\nu\bar\nu$ and $K_L \to \pi^0\nu\bar\nu$ are possible and ε'/ε anomaly can be explained.

Z' models exhibit quite different pattern of NP effects in the K meson system than the LH and RH Z scenarios. In Z scenarios only electroweak penguins (EWP) can contribute to ε'/ε in an important manner because of flavour dependent diagonal Z coupling to quarks. But in Z' models the diagonal quark couplings can be flavour universal so that QCD penguin operators (QCDP) can dominate NP contributions to ε'/ε. Interestingly, the pattern of NP in rare K decays depends on whether NP in ε'/ε is dominated by QCDP or EWP operators. Moreover, the striking difference from Z scenarios, known already from previous studies, is the increased importance of the constraints from $\Delta F = 2$ observables.

The new finding in [37] is a large hierarchy between real and imaginary parts of the flavour violating couplings implied by anomalies in QCDP and EWP scenarios. In the case of QCDP imaginary parts dominate over the real ones, while in the case of EWP this hierarchy is opposite unless the ε_K anomaly is absent. Because of these different patterns there are striking differences in the implications of the ε'/ε anomaly for the correlation between $K^+ \to \pi^+\nu\bar{\nu}$ and $K_L \to \pi^0\nu\bar{\nu}$ in these two NP scenarios if significant NP contributions to ε'/ε are required. The plots in [37] and in particular analytic derivations presented there illustrate these differences in a spectacular manner. The main lessons are as follows.

Lesson 7: In the case of QCDP scenario the correlation between $\mathcal{B}(K_L \to \pi^0\nu\bar{\nu})$ and $\mathcal{B}(K^+ \to \pi^+\nu\bar{\nu})$ takes place along the branch parallel to the Grossman-Nir bound [39].

Lesson 8: In the EWP scenario this correlation between $\mathcal{B}(K_L \to \pi^0\nu\bar{\nu})$ and $\mathcal{B}(K^+ \to \pi^+\nu\bar{\nu})$ proceeds away from this branch for diagonal quark couplings $\mathcal{O}(1)$ if NP in ε_K is present and it is very different from the one of the QCDP case. NP effects in rare K decays turn out to be modest in this case unless the diagonal quark couplings are $\mathcal{O}(10^{-2})$ and then the requirement of shifting upwards ε'/ε implies large effects in $K^+ \to \pi^+\nu\bar{\nu}$ and $K_L \to \pi^0\nu\bar{\nu}$ also in the EWP scenario.

Lesson 9: For fixed values of the neutrino and diagonal quark couplings in ε'/ε the predicted enhancements of $\mathcal{B}(K_L \to \pi^0\nu\bar{\nu})$ and $\mathcal{B}(K^+ \to \pi^+\nu\bar{\nu})$ are much larger when NP in QCDP is required to remove the ε'/ε anomaly than it is the case of EWP.

Lesson 10: In QCDP scenario ΔM_K is *suppressed* and this effect increases with increasing $M_{Z'}$ whereas in the EWP scenario ΔM_K is *enhanced* and this effect decreases with increasing $M_{Z'}$ as long as real couplings dominate. Already on the basis of this property one could differentiate between these two scenarios when the SM prediction for ΔM_K improves.

3 Results in specific NP models

3.1 Preliminaries

While the ε'/ε anomaly identified in 2015 was shadowed until recently by the 750 GeV resonance, the death of the latter will likely increase the interest in this new flavour anomaly. In particular the recent indications from the dual QCD approach that FSI are much less relevant for ε'/ε than previously expected and the upper bounds on $B_6^{(1/2)}$ and $B_8^{(3/2)}$ from this approach diminished significantly hopes that improved lattice calculations would bring the SM prediction for ε'/ε to agree with the experimental data, opening thereby an arena for important NP contributions to this ratio. The latest analyses of such contributions in the context of models with tree-level Z and Z' exchanges like 331 models, Littlest Higgs model with T-parity can be found in [36, 37, 40–42]. The analyses in supersymmetric models can be found in [43–45]. In view of space limitations we will only briefly summarize the results in 331 models and the very recent results in models with vector-like quarks.

3.2 331 Flavour News

The 331 models are based on the gauge group $SU(3)_C \times SU(3)_L \times U(1)_X$ [46–50]. In these models new contributions to ε'/ε and other flavour observables are dominated by tree-level exchanges of a Z' with non-negligible contributions from tree-level Z exchanges generated through the $Z - Z'$ mixing.

The size of these NP effects depends not only on $M_{Z'}$ but in particular on a parameter β, which distinguishes between various 331 models, on fermion representations under the gauge group and a parameter $\tan\bar{\beta}$ present in the $Z-Z'$ mixing. Extensive recent analyses in these models can be found in [41, 42, 51–53]. References to earlier analyses of flavour physics in 331 models can be found there and in [54, 55].

A detailed analysis of 331 models with different values of β, $\tan\bar{\beta}$ for two fermion representations F_1 and F_2, with the third SM quark generation belonging respectively to an antitriplet and a triplet under the $SU(3)_L$, has been presented in [53]: 24 models in total. Requiring that these models perform at least as well as the SM, as far as electroweak tests are concerned, seven models have been selected for a more detailed study of FCNC processes. Recent updated analyses of these seven models, that address the ε'/ε anomaly, have been presented in [41, 42] and we summarize the main results of these two papers putting the emphasis on the last analysis in [42] which could take into account new lattice QCD results from Fermilab Lattice and MILC Collaborations [34] on $B^0_{s,d}-\bar{B}^0_{s,d}$ hadronic matrix elements.

The new analyses in [41, 42] show that the impact of a required enhancement of ε'/ε on other flavour observables is significant. The one in [42] also shows that the results are rather sensitive to the value of $|V_{cb}|$ which has been illustrated there by choosing two values: $|V_{cb}| = 0.040$ and $|V_{cb}| = 0.042$.

The main findings of [41, 42] for $M_{Z'} = 3\,\text{TeV}$ are as follows:

- Among seven 331 models singled out through electroweak precision study only three (M8, M9, M16) can provide for both choices of $|V_{cb}|$, significant shift of ε'/ε but not larger than 6×10^{-4}, that is $\kappa_{\varepsilon'} \le 0.6$.
- The tensions between $\Delta M_{s,d}$ and ε_K, discussed previously can be removed in these models (M8, M9, M16) for both values of $|V_{cb}|$.
- Two of them (M8 and M9) can simultaneously suppress $B_s\to\mu^+\mu^-$ by at most 10% and 20% for $|V_{cb}| = 0.042$ and $|V_{cb}| = 0.040$, respectively. This can still bring the theory within 1σ range of the combined result from CMS and LHCb and for $|V_{cb}| = 0.040$ one can even reach the present central experimental value of this rate. The most recent result from ATLAS [56], while not accurate, appears to confirm this picture. On the other hand the maximal shifts in the Wilson coefficient C_9 are $C_9^{\rm NP} = -0.1$ and $C_9^{\rm NP} = -0.2$ for these two $|V_{cb}|$ values, respectively. This is only a moderate shift and these models do not really help in the case of $B_d\to K^*\mu^+\mu^-$ anomalies that require shifts as high as $C_9^{\rm NP} = -1.0$ [57, 58].
- In M16 the situation is opposite. The rate for $B_s\to\mu^+\mu^-$ can be reduced for $M_{Z'} = 3\,\text{TeV}$ for the two $|V_{cb}|$ values by at most 3% and 10%, respectively but with the corresponding values $C_9^{\rm NP} = -0.3$ and -0.5 the anomaly in $B_d\to K^*\mu^+\mu^-$ can be significantly reduced.
- The maximal shifts in ε'/ε decrease fast with increasing $M_{Z'}$ in the case of $|V_{cb}| = 0.042$ but are practically unchanged for $M_{Z'} = 10\,\text{TeV}$ when $|V_{cb}| = 0.040$ is used.
- On the other hand for higher values of $M_{Z'}$ the effects in $B_s\to\mu^+\mu^-$ and $B_d\to K^*\mu^+\mu^-$ are much smaller. NP effects in rare K decays and $B\to K(K^*)\nu\bar{\nu}$ remain small in all 331 models even for $M_{Z'}$ of a few TeV. This could be challenged by NA62, KOTO and Belle II experiments in this decade.

All these results are valid for $|V_{ub}| = 0.0036$. For its inclusive value of $|V_{ub}| = 0.0042$, we find that for $|V_{cb}| = 0.040$ the maximal shifts in ε'/ε are increased to 7.7×10^{-4} and 8.8×10^{-4} for $M_{Z'} = 3\,\text{TeV}$ and $M_{Z'} = 10\,\text{TeV}$, respectively. Renormalization group effects are responsible for this enhancement of ε'/ε for increased $M_{Z'}$. A recent analysis in the MSSM in [44] identifies this effect as well. But as explained in [41] eventually for very high $M_{Z'}$, NP effects in ε'/ε will be suppressed.

Thus the main message from [41, 42] is that NP contributions in 331 models can simultaneously solve $\Delta F = 2$ tensions, enhance ε'/ε and suppress either the rate for $B_s\to\mu^+\mu^-$ or C_9 Wilson coefficient without any significant NP effects on $K^+ \to \pi^+ \nu\bar{\nu}$ and $K_L \to \pi^0 \nu\bar{\nu}$ and $b \to s\nu\bar{\nu}$ transitions. While sizable NP effects in $\Delta F = 2$ observables and ε'/ε can persist for $M_{Z'}$ outside the reach of the LHC, such effects in $B_s \to \mu^+\mu^-$ will only be detectable provided Z' will be discovered soon.

In this context, following our recent analysis of VLQ models (see below), let us make the following observation. Determining one day the amount of NP contributions to $\Delta F = 2$ and $\Delta F = 1$ processes necessary to explain the data, one will be able in a given 331 model to determine $M_{Z'}$ necessary to explain these contributions independently of the new mixing parameters in that model. This is clearly seen in the formula (161) in [51]. It should be stressed that this is only possible by invoking both $\Delta F = 2$ and $\Delta F = 1$ transitions which exhibit different $M_{Z'}$ dependence. Within $\Delta F = 2$ transitions or $\Delta F = 1$ transitions alone this is not possible as clearly seen in the expressions in Section 7.2 in [51].

3.3 Models with vector-like quarks(VLQs)

Very recently we have analysed flavour violation patterns in the K and $B_{s,d}$ sectors in eleven models with VLQs [59]. I will describe here mainly the results obtained in five of them in which the gauge group is the SM one and the only new particles are VLQs in a single complex representation under the SM gauge group. A general classification of such models and references to the rich literature can be found in [60]. In these models $\Delta F = 1$ FCNCs are dominated by tree-level Z exchanges, while $\Delta F = 2$ transitions by box diagrams with VLQs and scalars provided $M_{\text{VLQ}} \geq 5\,\text{TeV}$. Otherwise tree-level Z contributions cannot be neglected.

The summary of patterns of flavour violation in these models can be found in three DNA tables (Tables 5, 6, 10 in [59]) and the numerical results in Tables 8 and 9 in that paper. Our extensive numerical analysis has shown that NP effects in several of these models can be still very large and that simultaneous consideration of several flavour observables should allow to distinguish between these models. In particular models with left-handed and right-handed flavour violating currents can be distinguished from each other in this manner. Here we summarize the highlights of this paper.

- All tensions between $\Delta M_{s,d}$ and ε_K can be easily removed in these models because the usual CMFV correlations between $\Delta M_{s,d}$ and ε_K are not valid in them. The box diagrams with VLQs and Higgs scalar exchanges are dominantly responsible for it.
- Tree-level Z contributions to ε'/ε can be large so that significant upward shift in ε'/ε can easily be obtained bringing the theory to agree with data.
- Simultaneously the branching ratio for $K^+ \to \pi^+\nu\bar{\nu}$ can be significantly enhanced over its SM prediction, but only in models with flavour violating RH currents. In models with only LH currents $K^+ \to \pi^+\nu\bar{\nu}$ branching ratio can have at most its SM value because of the $K_L \to \mu\bar{\mu}$ constraint. On the other hand the positive shift in ε'/ε implies uniquely suppression of the $K_L \to \pi^0\nu\bar{\nu}$ branching ratio with the suppression being smaller in models with RH currents. The fact that in models with RH currents $K^+ \to \pi^+\nu\bar{\nu}$ can be enhanced, while $K_L \to \pi^0\nu\bar{\nu}$ suppressed is a clear signal of non-MFV sources at work. But also in models with LH currents only the correlations between the branching ratios of these two decays differ from the MFV one.
- These features distinguish VLQ-models from 331 models, discussed above, in which NP effects are dominated by Z' exchanges with the maximal shift in ε'/ε amounting to 0.8×10^{-3} and NP effects in rare K decays being very small.
- Significant suppressions of the branching ratio for $B_s \to \mu^+\mu^-$, in particular in models with LH currents are possible. While such effects are also possible in 331 models, they cannot be as large as in VLQ models.

- On the other hand while 331 models can provide solutions to some LHCb anomalies, this is not possible in the VLQs models with SM gauge group and future confirmation of these anomalies could turn out to be a problem for the latter models.

Having the latter possibility in mind we have considered also two VLQ models with a heavy Z' related to $U(1)_{L_\mu - L_\tau}$ symmetry and a single heavy scalar necessary to generate the Z' mass. These models, considered already in [61], can explain LHCb anomalies by providing sufficient suppression of the coefficient C_9 but NP effects in $B_s \to \mu^+\mu^-$ and $K_L \to \mu^+\mu^-$ are absent because of the absence of tree-level Z FCNCs and vector diagonal Z' couplings to charged leptons. NP effects in $b \to s\nu\bar{\nu}$ transitions are small and the ones in $K^+ \to \pi^+\nu\bar{\nu}$ and $K_L \to \pi^0\nu\bar{\nu}$ much smaller than in the models with the SM gauge group. Most importantly these models fail badly in explaning the ε'/ε anomaly. This is the consequence of the absence of tree-level Z contributions in this model and of the Dirac structure of the loop generated Z' couplings to quarks that do not allow for NP contributions to the Wilson coefficients of $Q_6(Q'_6)$ and $Q_8(Q'_8)$ operators.

Adding a second Higgs doublet allows to generate tree-level Z FCNCs with a different pattern of departures from the SM than in VLQ models summarized above. While NP effects in these remaining four models in $B_{s,d} \to \mu^+\mu^-$ and $K_L \to \pi^0\nu\bar{\nu}$ turn out to be small, ε'/ε anomaly can be explained and $K^+ \to \pi^+\nu\bar{\nu}$ can be enhanced both in the case of left-handed and right-handed couplings. Moreover, NP effects in ΔM_K can be larger than in the remaining seven models.

Future experimental results on $K^+ \to \pi^+\nu\bar{\nu}$, $K_L \to \pi^0\nu\bar{\nu}$, $B_s \to \mu^+\mu^-$ and LHCb anomalies and improved theoretical results on ε'/ε will tell us which of these VLQ models, if any, is selected by nature.

While the discovery of VLQs at the LHC would give a strong impetus to the models considered by us, non-observation of them at the LHC would not preclude their importance for flavour physics. In fact we have shown that large NP effects in flavour observables can be present for $M_{\text{VLQ}} = 10$ TeV and in the flavour precision era one could even be sensitive to higher masses. In this context we have pointed out that

- the combination of $\Delta F = 2$ and $\Delta F = 1$ observables in a given meson system allows to determine the masses of VLQs in a given representation independently of the size of Yukawa couplings.

4 Outlook

Our outlook is very short. The future of kaon flavour physics looks great and the coming years should be very exciting. I am looking forward to QCD@Work 2018 when the landscape of NP will be more transparent than it is now.

Acknowledgements

It is a pleasure to thank Monika Blanke, Fulvia De Fazio and Jean-Marc Gérard for very efficient studies discussed in this write-up and Christoph Bobeth, Alejandro Celis and Martin Jung for a very extensive analysis of VLQ models. I would also like to thank the organizers of this workshop for inviting me to this interesting event and for an impressive hospitality. This research was done and financed in the context of the ERC Advanced Grant project "FLAVOUR"(267104). It was also partially supported by the DFG cluster of excellence "Origin and Structure of the Universe".

References

[1] A.J. Buras, PoS **EPS-HEP2015**, 602 (2015), 1510.00128.

[2] A.J. Buras, *The Renaissance of Kaon Flavour Physics* (2016), 1606.06735.
[3] A.J. Buras, PoS **FWNP**, 003 (2015), 1505.00618.
[4] T. Blum et al., Phys. Rev. **D91**, 074502 (2015), 1502.00263.
[5] Z. Bai et al. (RBC, UKQCD), Phys. Rev. Lett. **115**, 212001 (2015), 1505.07863.
[6] A.J. Buras, M. Gorbahn, S. Jäger, M. Jamin, JHEP **11**, 202 (2015), 1507.06345.
[7] J. Batley et al. (NA48), Phys. Lett. **B544**, 97 (2002), hep-ex/0208009.
[8] A. Alavi-Harati et al. (KTeV), Phys. Rev. **D67**, 012005 (2003), hep-ex/0208007.
[9] E. Abouzaid et al. (KTeV), Phys. Rev. **D83**, 092001 (2011), 1011.0127.
[10] T. Kitahara, U. Nierste, P. Tremper (2016), 1607.06727.
[11] A.J. Buras, M. Jamin, M.E. Lautenbacher, P.H. Weisz, Nucl. Phys. **B370**, 69 (1992).
[12] A.J. Buras, M. Jamin, M.E. Lautenbacher, P.H. Weisz, Nucl. Phys. **B400**, 37 (1993), hep-ph/9211304.
[13] A.J. Buras, M. Jamin, M.E. Lautenbacher, Nucl. Phys. **B400**, 75 (1993), hep-ph/9211321.
[14] M. Ciuchini, E. Franco, G. Martinelli, L. Reina, Phys. Lett. **B301**, 263 (1993), hep-ph/9212203.
[15] A.J. Buras, M. Jamin, M.E. Lautenbacher, Nucl. Phys. **B408**, 209 (1993), hep-ph/9303284.
[16] M. Ciuchini, E. Franco, G. Martinelli, L. Reina, Nucl. Phys. **B415**, 403 (1994), hep-ph/9304257.
[17] A.J. Buras, P. Gambino, U.A. Haisch, Nucl. Phys. **B570**, 117 (2000), hep-ph/9911250.
[18] M. Gorbahn, U. Haisch, Nucl. Phys. **B713**, 291 (2005), hep-ph/0411071.
[19] J. Brod, M. Gorbahn, Phys. Rev. **D82**, 094026 (2010), 1007.0684.
[20] A.J. Buras, J.M. Gerard, JHEP **12**, 008 (2015), 1507.06326.
[21] A.J. Buras, J.M. Gérard, Nucl. Phys. **B264**, 371 (1986).
[22] W.A. Bardeen, A.J. Buras, J.M. Gérard, Phys. Lett. **B180**, 133 (1986).
[23] A.J. Buras, J.M. Gérard, Phys. Lett. **B192**, 156 (1987).
[24] A.J. Buras, D. Buttazzo, J. Girrbach-Noe, R. Knegjens, JHEP **11**, 033 (2015), 1503.02693.
[25] A.J. Buras, J.M. Gerard (2016), 1603.05686.
[26] V. Antonelli, S. Bertolini, M. Fabbrichesi, E.I. Lashin, Nucl. Phys. **B469**, 181 (1996), hep-ph/9511341.
[27] S. Bertolini, J.O. Eeg, M. Fabbrichesi, Nucl. Phys. **B476**, 225 (1996), hep-ph/9512356.
[28] E. Pallante, A. Pich, Phys. Rev. Lett. **84**, 2568 (2000), hep-ph/9911233.
[29] E. Pallante, A. Pich, Nucl. Phys. **B592**, 294 (2001), hep-ph/0007208.
[30] M. Buchler, G. Colangelo, J. Kambor, F. Orellana, Phys. Lett. **B521**, 29 (2001), hep-ph/0102289.
[31] M. Buchler, G. Colangelo, J. Kambor, F. Orellana, Phys. Lett. **B521**, 22 (2001), hep-ph/0102287.
[32] E. Pallante, A. Pich, I. Scimemi, Nucl. Phys. **B617**, 441 (2001), hep-ph/0105011.
[33] M. Blanke, A.J. Buras, Eur. Phys. J. **C76**, 197 (2016), 1602.04020.
[34] A. Bazavov et al. (Fermilab Lattice, MILC), Phys. Rev. **D93**, 113016 (2016), 1602.03560.
[35] A.J. Buras, D. Buttazzo, J. Girrbach-Noe, R. Knegjens, JHEP **1411**, 121 (2014), 1408.0728.
[36] A.J. Buras, D. Buttazzo, R. Knegjens, JHEP **11**, 166 (2015), 1507.08672.
[37] A.J. Buras, JHEP **04**, 071 (2016), 1601.00005.
[38] A.J. Buras, F. De Fazio, J. Girrbach, JHEP **1302**, 116 (2013), 1211.1896.
[39] Y. Grossman, Y. Nir, Phys. Lett. **B398**, 163 (1997), hep-ph/9701313.
[40] M. Blanke, A.J. Buras, S. Recksiegel, Eur. Phys. J. **C76**, 182 (2016), 1507.06316.

[41] A.J. Buras, F. De Fazio, JHEP **03**, 010 (2016), 1512.02869.
[42] A.J. Buras, F. De Fazio, JHEP **08**, 115 (2016), 1604.02344.
[43] M. Tanimoto, K. Yamamoto (2016), 1603.07960.
[44] T. Kitahara, U. Nierste, P. Tremper (2016), 1604.07400.
[45] M. Endo, S. Mishima, D. Ueda, K. Yamamoto (2016), 1608.01444.
[46] M. Singer, J.W.F. Valle, J. Schechter, Phys. Rev. **D22**, 738 (1980).
[47] F. Pisano, V. Pleitez, Phys. Rev. **D46**, 410 (1992), hep-ph/9206242.
[48] P.H. Frampton, Phys. Rev. Lett. **69**, 2889 (1992).
[49] R. Foot, O.F. Hernandez, F. Pisano, V. Pleitez, Phys. Rev. **D47**, 4158 (1993), hep-ph/9207264.
[50] J.C. Montero, F. Pisano, V. Pleitez, Phys. Rev. **D47**, 2918 (1993), hep-ph/9212271.
[51] A.J. Buras, F. De Fazio, J. Girrbach, M.V. Carlucci, JHEP **1302**, 023 (2013), 1211.1237.
[52] A.J. Buras, F. De Fazio, J. Girrbach, JHEP **1402**, 112 (2014), 1311.6729.
[53] A.J. Buras, F. De Fazio, J. Girrbach-Noe, JHEP **1408**, 039 (2014), 1405.3850.
[54] R.A. Diaz, R. Martinez, F. Ochoa, Phys. Rev. **D72**, 035018 (2005), hep-ph/0411263.
[55] A. Carcamo Hernandez, R. Martinez, F. Ochoa, Phys. Rev. **D73**, 035007 (2006), hep-ph/0510421.
[56] M. Aaboud et al. (ATLAS) (2016), 1604.04263.
[57] W. Altmannshofer, D.M. Straub, Eur. Phys. J. **C75**, 382 (2015), 1411.3161.
[58] S. Descotes-Genon, L. Hofer, J. Matias, J. Virto (2015), 1510.04239.
[59] C. Bobeth, A.J. Buras, A. Celis, M. Jung (2016), 1609.04783.
[60] K. Ishiwata, Z. Ligeti, M.B. Wise, JHEP **10**, 027 (2015), 1506.03484.
[61] W. Altmannshofer, S. Gori, M. Pospelov, I. Yavin, Phys. Rev. **D89**, 095033 (2014), 1403.1269.

AUTHOR INDEX